Applications of Advanced Omics Technologies: From Genes to Metabolites

Comprehensive Analytical Chemistry

Volume 64

Applications of Advanced Omics Technologies: From Genes to Metabolites

Comprehensive Analytical Chemistry

Volume 64

Edited by

Virginia García-Cañas
Laboratory of Foodomics, Institute of Food Science Research (CIAL),
CSIC. Nicolás Cabrera 9, Madrid, Spain

Alejandro Cifuentes
Laboratory of Foodomics, Institute of Food Science Research (CIAL),
CSIC. Nicolás Cabrera 9, Madrid, Spain

Carolina Simó
Laboratory of Foodomics, Institute of Food Science Research (CIAL),
CSIC. Nicolás Cabrera 9, Madrid, Spain

ELSEVIER   AMSTERDAM • BOSTON • HEIDELBERG • LONDON • NEW YORK • OXFORD
PARIS • SAN DIEGO • SAN FRANCISCO • SINGAPORE • SYDNEY • TOKYO

Elsevier
The Boulevard, Langford Lane, Kidlington, Oxford OX5 1GB, UK
Radarweg 29, PO Box 211, 1000 AE Amsterdam, The Netherlands

British Library Cataloguing in Publication Data
A catalog record for this book is available from the British Library

Library of Congress Cataloging-in-Publication Data
A catalogue record for this book is available from the Library of Congress

ISBN: 978-0-444-62650-9
ISSN: 0166-526X   Z 6466406

For information on all Elsevier publications
visit our website at store.elsevier.com

Printed and bound in Poland

14 15 16 17   10 9 8 7 6 5 4 3 2 1

V. García-Cañas: *To my beloved parents: Julia and Manuel*
"For their unconditional love"

A. Cifuentes: *To my treasured Claudia*
"Just keep swimming, just keep swimming"

C. Simó: *This book is dedicated to my parents,*
Mª Inmaculada and José Luis

Contents

Part I
Genomics and Transcriptomics

5. Emerging RNA-Seq Applications in Food Science   107

Alberto Valdés, Carolina Simó, Clara Ibáñez, and Virginia García-Cañas

Part II
Proteomics and Peptidomics

6. Making Progress in Plant Proteomics for Improved Food Safety   131

Pier Giorgio Righetti, Elisa Fasoli, Alfonsina D'Amato, and Egisto Boschetti

7. Advantages and Applications of Gel-Free Proteomic Approaches in the Study of Prokaryotes   157

John P. Bowman

Part IV
Other Omics Strategies, Data Treatment, Integration and Systems Biology

13. Profiling of Genetically Modified Organisms Using Omics Technologies

Alberto Valdés, Carolina Simó, Clara Ibáñez,
and Virginia García-Cañas

14. MS-Based Lipidomics

Päivi Pöhö, Matej Oresic, and Tuulia Hyötyläinen

Contributors to Volume 64

Gloria Alvarez-Llamas, Department of Immunology, IIS-Fundación Jiménez Díaz, Madrid, Spain

Francisco Amado, QOPNA, Mass Spectrometry Center, Department of Chemistry, University of Aveiro, Aveiro, Portugal, and QOPNA, School of Health Sciences, University of Aveiro, Aveiro, Portugal

Maria G. Barderas, Department of Vascular Physiopathology, Hospital Nacional de Parapléjicos, SESCAM, Toledo, Spain

Egisto Boschetti, EB Consulting, JAM-Conseil, Neuilly s/Seine, France

John P. Bowman, Food Safety Centre, Tasmania Institute of Agriculture, Hobart, Tasmania, Australia

Benito Cañas, Complutense University of Madrid, Madrid, Spain

Mónica Carrera, Spanish National Research Council (CSIC), Marine Research Institute (IIM), Vigo, Pontevedra, Spain

Enrique Casado, Medical Oncology, Infanta Sofia Hospital, Madrid, Spain

Armando Caseiro, QOPNA, Mass Spectrometry Center, Department of Chemistry, University of Aveiro, Aveiro, Portugal, and College of Health Technology of Coimbra, Polytechnic Institute of Coimbra, Coimbra, Portugal

Donald H. Chace, The MEDNAX Center for Research, Education, and Quality, Pediatrix Medical Group of MEDNAX, Inc., Sunrise, Florida, USA

Ilseung Cho, NYU School of Medicine, New York, USA, and Division of Gastroenterology, Department of Medicine, NYU School of Medicine, New York, USA

Philip Chuang, NYU School of Medicine, New York, USA

Alejandro Cifuentes, Laboratory of Foodomics, Institute of Food Science Research (CIAL), CSIC. Nicolás Cabrera 9, Madrid, Spain

Ana Conesa, Genomics of Gene Expression Laboratory, Centro de Investigación Príncipe Felipe, Valencia, Spain

Alfonsina D'Amato, Department of Chemistry, Materials and Chemical Engineering "Giulio Natta", Politecnico di Milano, Milano, Italy

Ana Ramírez de Molina, Molecular Oncology and Nutritional Genomics of Cancer, IMDEA Food Institute CEI UAM+CSIC, Madrid, Spain

Elisa Fasoli, Department of Chemistry, Materials and Chemical Engineering "Giulio Natta", Politecnico di Milano, Milano, Italy

Rita Ferreira, QOPNA, Mass Spectrometry Center, Department of Chemistry, University of Aveiro, Aveiro, Portugal

José M. Gallardo, Spanish National Research Council (CSIC), Marine Research Institute (IIM), Vigo, Pontevedra, Spain

Virginia García-Cañas, Laboratory of Foodomics, Institute of Food Science Research (CIAL), CSIC. Nicolás Cabrera 9, Madrid, Spain

Rafael Hernández, Genomics of Gene Expression Laboratory, Centro de Investigación Príncipe Felipe, Valencia, Spain

Tuulia Hyötyläinen, Steno Reseach Center, Niels Steensens Vej 2, 2820 Gentofte, Denmark

Clara Ibáñez, Laboratory of Foodomics, Institute of Food Science Research (CIAL), CSIC. Nicolás Cabrera 9, Madrid, Spain

Elena Ibáñez, Laboratory of Foodomics, Institute of Food Science Research (CIAL), CSIC. Nicolás Cabrera 9, Madrid, Spain

Marta Martin-Lorenzo, Department of Immunology, IIS-Fundación Jiménez Díaz, Madrid, Spain

Juan Moreno-Rubio, Medical Oncology, Infanta Sofia Hospital, Madrid, Spain

David W. Murray, Department of Physiology and Medical Physics, Royal College of Surgeons in Ireland, Dublin 2, Ireland

Matej Oresic, Steno Reseach Center, Niels Steensens Vej 2, 2820 Gentofte, Denmark

Päivi Pöhö, Steno Reseach Center, Niels Steensens Vej 2, 2820 Gentofte, Denmark

Maria Posada-Ayala, Department of Immunology, IIS-Fundación Jiménez Díaz, Madrid, Spain

Ricardo Ramos, Genomic Unit, Scientific Park of Madrid, Madrid, Spain

Guillermo Reglero, Molecular Oncology and Nutritional Genomics of Cancer, IMDEA Food Institute CEI UAM+CSIC, and Instituto de Investigación en Ciencias de la Alimentación (CIAL), CEI UAM+CSIC, Madrid, Spain

Pier Giorgio Righetti, Department of Chemistry, Materials and Chemical Engineering "Giulio Natta", Politecnico di Milano, Milano, Italy

Carolina Simó, Laboratory of Foodomics, Institute of Food Science Research (CIAL), CSIC. Nicolás Cabrera 9, Madrid, Spain

Sinead M. Smith, Department of Clinical Medicine and School of Pharmacy and Pharmaceutical Sciences, Trinity College Dublin, Dublin 2, Ireland

Alan R. Spitzer, The MEDNAX Center for Research, Education, and Quality, Pediatrix Medical Group of MEDNAX, Inc., Sunrise, Florida, USA

Alberto Valdés, Laboratory of Foodomics, Institute of Food Science Research (CIAL), CSIC. Nicolás Cabrera 9, Madrid, Spain

Joy D. Van Nostrand, Department of Microbiology and Plant Biology, Institute for Environmental Genomics, University of Oklahoma, Norman, Oklahoma, USA

Rui Vitorino, QOPNA, Mass Spectrometry Center, Department of Chemistry, University of Aveiro, Aveiro, Portugal

Fernando Vivanco, Department of Immunology, IIS-Fundación Jiménez Díaz, Madrid, Spain, and Department of Biochemistry and Molecular Biology I, UCM, Madrid, Spain

Emily Walsh, NYU School of Medicine, New York, USA

Timothy Yen, NYU School of Medicine, New York, USA

Jizhong Zhou, Department of Microbiology and Plant Biology, Institute for Environmental Genomics, University of Oklahoma, Norman, Oklahoma, USA

This volume, edited by C. Simó, A. Cifuentes and V. García-Cañas, is a direct consequence of the leading-edge anaytical chemistry research carried out in Spain during the last 20 years. When my colleague at CSIC in Madrid and old friend Alejandro Cifuentes told me about his project, I immediately proposed that we include a book on it in the CAC series. Why? We did not have many titles on Omics in our series. In this respect we should consider this volume a follow up to volumes 46 and 52 on Proteomics and Peptidomics, and on Protein Mass Spectrometry, respectively. This new CAC volume on the applications of advanced omics technologies is an excellent addition to these two previous books and complements the volume on Fundamentals published earlier this year.

As mentioned by the editors in their introduction to Part A, this field was developed in tandem with the instrumental and methodological developments achieved at the end of the 20th Century. Powerful bioinformatic tools were also integrated in the systems, allowing us to understand the relevant mechanisms of protein synthesis. Omics tools are used nowadays for high throughput assessment of changes at genome (genomics), epigenome (epigenomics), transcript (transcriptomics), protein (proteomics) and metabolite (metabolomics) levels.

The book contains 16 chapters covering a broad range of topics. It contains a balanced cocktail of chapters on applications of new microarray technologies, proteomics, peptidomics and metabolomics. Other omics strategies, data treatment and systems biology are presented in the last chapters, showing the importance of these tools for omics technologies.

Although most of the applications are in the food science and nutrition area, cancer research and other clinical examples are also reported. In this respect the book offers a unique and comprehensive vision on the applications of omic technologies into two fields extremely important for humanity: food safety and clinical research. By incorporating these new tools better diagnostic solutions can be achieved for an improvement of human health. This is an excellent example of how investment in research, development and innovation on state-of-the art technologies over the last few years has lead to a better quality of life.

Finally, I would like to thank the editors and all the authors for their contributions to this needed state of the art book in the CAC series. I am sure it will enjoy success among the scientific community and I expect to see many citations in the literature in the next coming years.

Prof. Dr. D. Barceló
Barcelona, Spain, March 24, 2014

As mentioned in the Preface to our previous book in this series, *Fundamentals of Advanced Omics Technologies: From Genes to Metabolites* (ISBN: 9780444626516), the impressive advances observed in omics tools in recent years have brought about new analytical instruments and methodologies for genomics, epigenomics, transcriptomics, proteomics, and metabolomics studies. These advances also include powerful bioinformatics tools to integrate and interrogate the multiple omic datasets usually generated in these studies, making it possible to investigate changes at the genome, epigenome, transcriptome, proteome, and metabolome level that were unthinkable a few years ago.

The interest of the scientific community in the development and application of these new omics technologies in different and hot areas of research is well documented in the 16 chapters of this book, *Applications of Advanced Omics Technologies: From Genes to Metabolites*. Namely, this volume covers recent applications of genomics, epigenomics, and transcriptomics (Chapters 1–5), proteomics and peptidomics (Chapters 6–9), and metabolomics (Chapters 10–12), and it includes updated information on other omics strategies, data treatment, data integration, and Systems Biology approaches (Chapters 13–16). More concretely, the application of functional gene microarrays for profiling microbial communities is presented in Chapter 1, while Chapter 2 presents an overview of current technologies and applications of microRNA profiling. The use of novel genotyping microarray technologies in cancer research is discussed in Chapter 3, while Chapters 4 and 5 present next-generation sequencing technologies to study the human microbiome and the use of emerging RNA-Seq applications in food science, respectively. The following three chapters review different proteomics approaches and their applications in plant proteomics for improved food safety (Chapter 6), in the study of prokaryotes (Chapter 7), or for food fingerprints with special emphasis on food quality and authenticity (Chapter 8). Salivary peptidomics and its clinical applications are discussed in Chapter 9. Metabolomics is shown as an impressive tool to discover new biomarkers related to Alzheimer's disease (Chapter 10), cardiovascular and renal diseases (Chapter 11), or together with proteomics to address new challenges in neonatal medicine (Chapter 12). The last four chapters are devoted to reviewing the use of omics technologies for profiling genetically modified organisms (Chapter 13), the latest advances and applications in MS-based lipidomics (Chapter 14), the fundamentals and recent uses of Foodomics (Chapter 15), and the methods and applications for omics data integration in Systems Biology (Chapter 16).

Together the two books we have edited for this series should provide readers with the necessary background as well as relevant and updated information on the fundamentals and applications of modern omics technologies. We expect that these two volumes will be useful to a broad audience of experts working on (or studying) different topics related to omics technologies. This audience includes, among others, scientists, Ph.D. students, and technicians from control laboratories, universities, hospitals, research laboratories, and regulatory agencies.

We, the editors, would like to thank all the authors for their inspiring contributions, Damiá Barceló for inviting us to prepare this work; Derek Coleman, Susan Dennis, and Mohanapriyan Rajendran for their help and support; and to all the others at Elsevier whose efforts have contributed to the publication of this volume. *Gracias de nuevo!*

Virginia García-Cañas, Alejandro Cifuentes, and Carolina Simó
Laboratory of Foodomics, CIAL, Madrid
National Research Council of Spain (CSIC)

Part I

Genomics and Transcriptomics

Chapter 1

Applications of Functional Gene Microarrays for Profiling Microbial Communities

Joy D. Van Nostrand and Jizhong Zhou

Department of Microbiology and Plant Biology, Institute for Environmental Genomics, University of Oklahoma, Norman, Oklahoma, USA

1 INTRODUCTION

Microorganisms play important roles in global biogeochemical cycling of carbon, nitrogen, sulfur, phosphorous, and metals as well as being involved in the stabilization or degradation of anthropogenic contaminants. With an estimated 2000–50,000 microbial species per gram of soil *(1–4)*, microorganisms are also the most phylogenetically and functionally diverse group of organisms on the planet. However, our understanding of many of these functions is not well understood due to the difficulty in examining and monitoring microbial activity in the environment. This is due, in part, to the uncultured status of many (>99%) microorganisms *(5–7)*. As most microorganisms cannot be examined directly, culture-independent approaches are needed to determine what microorganisms are present and what functions they are performing. There are many culture-independent methods available, such as 16S rRNA gene-based cloning, quantitative PCR, denaturing (or temperature) gradient gel electrophoresis, terminal-restriction fragment length polymorphism

Comprehensive Analytical Chemistry, Vol. 64. http://dx.doi.org/10.1016/B978-0-444-62650-9.00001-4

(T-RFLP), quantitative PCR, and *in situ* hybridization. However, the resolution and coverage these methods provide are limited. Clone libraries, for example, are often too small by a factor of 10 or more to capture the true diversity of microbial communities *(8,9)*. In addition, many of these methods require a PCR amplification step, which introduces well-known biases *(10–12)*. The use of PCR also limits the number of genes that can be examined since there is so much variance in gene sequences or too few sequences are available, making the use of conserved PCR primers impractical or impossible. Lastly, often phylogenetic markers are used, such as 16S rRNA or DNA gyrase (GyrB) genes *(5,13–16)*, which provide information on the identity of microorganisms within the community but provide limited information on the community's functional abilities and activity. Because of their design, microarrays can overcome many of these limitations and have been shown to be valuable for the study of microbial communities *(17)*.

Microarrays are comprised of probes for specific genes, sequences, or genomes on a solid surface. They are similar to traditional Northern or Southern blots, except that instead of a labeled probe hybridizing to target nucleic acid attached to a membrane, labeled nucleic acids are hybridized to the probes that are attached to the array surface. Microarrays can be produced on glass slides *(18–20)* or nylon membranes *(21)*; although most arrays use glass slides because they produce less background fluorescence *(22,23)* and allow higher probe density *(24)*. Microarrays allow for the examination of thousands of genes at one time without the need for PCR amplification of individual genes and can provide detection for a wide range of target microorganisms. With the use of oligonucleotide probes, any gene can be added to the array as long as sequences are available without the need for primers. Since microarrays have a defined probe set that all samples are tested against, they are ideal for comparison of samples across different conditions, sites, or times.

Since they were first reported, several new types of microarrays have been developed for use on microbial communities. These include phylogenetic oligonucleotide arrays (POA), which are designed to examine phylogenetic relatedness or community composition using 16S rRNA or other conserved genes *(16,25–27)*. The most comprehensive POA to date is the PhyloChip *(16,28–30)*. The current version of PhyloChip, G3, has 1.1 million 25-mer probes covering 59,959 taxa *(30)*. Other POAs, designed to study specific environments such as compost *(31)* or specific groups such as the "*Rhodocyclales*" order *(26)*, have also been developed. Community genome arrays (CGA) use whole genomic DNA as probes *(32,33)* and can be used to determine the relatedness of microbial species or strains or to identify community members. CGAs have been used to compare microbial communities from different environments *(32)*, to examine communities from acid mine drainage and bioleaching communities *(34)*, and to determine the relatedness of different *Escherichia coli* strains *(33)* and 55 different *Azoarcus, Pseudomonas,* and *Shewanella* strains *(35)*. Metagenomic arrays (MGAs) use environmental clone library inserts as probes and can be used as a high-throughput screening method *(36–38)*. MGAs have been used to examine communities from marine environments *(39)*. Whole-genome ORF arrays (WGAs) have probes for all ORFs in one or more genomes and are used for gene expression analysis *(40)*. However, WGAs have also been used for comparative genomics *(41)*. Several *Shewanella* strains were compared using a WGA containing 192 ORFs from *Shewanella oneidensis* MR-1 *(41)*, an *E. coli* K-12 WGA was used to examine similar genes in *Klebsiella pneumoniae (42)*, *Pseudomonas syringae* strains were compared using a WGA of 353 virulence factors *(43)*, and a *Pyrococcus furiosus* WGA was used to examine 7 *Pyrococcus* isolates *(44)*. Functional gene arrays (FGAs) contain probes for specific functional genes involved in various processes of interest, such as carbon cycling or metal resistance, and can provide information on the genes and populations within a community and provide direct linkages between gene functions and ecosystem processes *(45)*. FGAs will be the primary focus of this chapter.

2 COMPARISON OF MICROARRAYS WITH HIGH-THROUGHPUT SEQUENCING

Although this chapter deals with microarrays, high-throughput sequencing, which has become an increasingly popular choice for studying microbial communities even for metatranscriptome analysis *(46)*, will be discussed here very briefly. High-throughput sequencing has many advantages; however, there are a few distinct disadvantages that bear mention. These disadvantages can be overcome with microarray technology, while high-throughput sequencing can overcome one of the greatest disadvantages of microarrays, making microarrays and high-throughput sequencing ideal as complementary approaches to the study of microbial communities.

All high-throughput techniques can be divided into two main types based on the sampling strategy: open format or closed format. Open format methods are those such as sequencing or proteomics in which you collect a sample and then detect as many sequences, proteins, metabolites, etc., as the method allows. In closed format methods, such as microarrays, the samples are interrogated against a known list of genes or sequences to determine if they are present. Pitfalls of either method include (1) random sampling errors. With true random sampling, the probability of sampling the same fraction of the community multiple times is theoretically low *(47)*; however, in general, only a small portion of the microbial community is actually sampled in sequencing studies *(48)*. As such, dominant populations within the community would be expected to have a greater chance of being sampled multiple times. Sampling errors can lead to low reproducibility of sequencing results between technical replicates ($17.2 \pm 2.3\%$ for two replicates; $8.2 \pm 2.3\%$ for three) *(49)*. In contrast, microarrays use the same set of sequences (probes) to interrogate all samples so that the same population is sampled each time. (2) Relative abundance. Individual species within a community will vary in abundance, and the most abundant species would be expected to have a higher representation while lesser abundant species/sequences may not be represented at all. Microarrays tend to have less abundance bias since lesser abundant sequences will still hybridize to their corresponding probe, and as long as it is above the detection limit, it will be detected. (3) New sequence detection. One of the biggest advantages of sequencing is the detection of new sequences since any sequence in the sample can be sequenced (open system). Microarrays, on the other hand, can only detect sequences covered by the array probe set (closed system). However, some novel approaches have been used to overcome this limitation such as capture microarrays *(50,51)* or use of hybridization patterns to classify novel sequences *(52,53)*.

3   FUNCTIONAL GENE ARRAYS

FGAs were first reported over a decade ago. These arrays are used to monitor genes that are protein coding genes whose protein(s) are responsible for specific processes of interest. This prototype FGA used PCR-amplicon probes to target *nirS*, *nirK*, *amoA*, and *pmoA* *(45)*. Because the probes were PCR-amplicons, only a limited number of genes or sequences can be used because of primer limitation. Later, FGAs typically used oligonucleotide probes since they provide more specificity and primers are not needed *(54–56)*. Many FGAs have since been reported including those targeting microbial processes such as N cycling *(18,21,57,58)*, methane cycling *(59–61)*, or hydrocarbon degradation *(62,63)* and for monitoring or detecting specific microbial groups such as pathogens *(64–68)*, *nodC* variants *(69)*, microorganisms containing benzene monooxygenase genes *(70)*, hydrogenases *(71)*, and for specific environments such as acid mine drainage sites *(72)*. A recent report described an FGA designed to examine soil health which contained almost 3000 probes covering N and P cycling, C degradation, antibiotic production, and organic toxin degradation *(73)*. Currently, the most comprehensive FGAs are the GeoChip arrays *(19,20,74–76)*. The GeoChip has been updated several times, and current versions have greatly increased the number and type of genes and gene variants covered *(75,76)*.

3.1   GeoChip Microarrays

GeoChips were first designed to provide a comprehensive probe set for many functional gene categories and to provide increased specificity for highly homologous gene variants *(74)*. The first step in the design process is to select processes of interest (C fixation, Ni resistance, denitrification, sulfite reduction, etc.), determine what genes are involved in these processes, and which are critical for this function. GeoChip was designed to examine functional potential, so only genes for those proteins containing catalytic subunits or active sites are selected. Keywords are then chosen to search public sequence databases. The sequences are downloaded and then confirmed by HMMER alignment (http:/hmmer.wustl.edu/) with sequences whose protein function has been experimentally confirmed (seed sequences). Probes are then designed from the confirmed sequences using experimentally determined criteria based on sequence homology ($\leq$90% identity for gene-specific probes and $\geq$96% for group-specific probes), continuous stretch length ($\leq$20 bases for gene-specific probes and $\geq$35 for group-specific probes), and free energy ($\geq-35\,\text{kJ mol}^{-1}$ for gene-specific probes and $\leq-60\,\text{kJ mol}^{-1}$ for group-specific probes) (77,78) with new versions of the CommOligo software (79). Probe specificity is confirmed by BLASTing against the GenBank database. The final probe set is then spotted or printed onto the array surface. Earlier GeoChip versions used premade oligonucleotides spotted onto glass slides (74,75), while later versions were manufactured using probes printed directly onto the array slide by Nimblegen (76) or Agilent. Current GeoChips have probes for genes involved in the biogeochemical cycling of carbon (C), nitrogen (N), phosphorus (P), sulfur (S), and metals, antibiotic resistance, biodegradation of environmental contaminants, energy processing, and stress response among others.

GeoChip arrays have been used in numerous studies to examine microbial communities from a variety of sites. These include studies examining metals-impacted sites such as uranium contaminated aquifers (32,80–85), rhizosphere communities associated with metal accumulating plants (86,87), zinc-contaminated lake sediments (88), arsenic contaminated soils (89,90), and to probe bacteria for metal resistance genes (91). Numerous studies examining oil- or hydrocarbon-contaminated environments have used GeoChips (92–95) including sites in the high Arctic (96), phenanthrene-spiked soil microcosms (97), atrazine-contaminated aquifers (98), polycyclic aromatic hydrocarbon-contaminated sites (99), industrial pollutant, and pesticide-contaminated river sediments (100) and to elucidate microbial bioconversion of coal to methane (101). Several of these studies evaluated the effects of the 2010 Gulf of Mexico oil spill on microbial communities from open ocean areas near the spill (30,76) and at on shore locations that received oil influxes such as salt marshes (102). Marine environments examined include those studying communities from deep-sea basalt (103), hydrothermal vents (104), sediments from Puget Sound (105), the Gulf of Mexico sediments (106), soils surrounding mangroves (107), those involved in N cycling near the Florida shelf (108), and microbial communities from diseased and healthy corals (109) as well as Cr resistant bacteria and sulfate reducing bacteria in the Great Salt Lake (110). GeoChip has been used to study the effects of global climate change on microbial communities and how those changes affect C and N cycling (111–115), microbial responses to different land management strategies, uses (116–120) or locations such as Antarctica (62,121) or the rhizosphere *(122)*, as well as to examine gene-area relationships (GARs) in forest soils *(47)*. Other studies have examined communities from acid mine drainage sites *(123)*, a leachate-contaminated landfill *(124)*, microbial electrolysis cells *(125)*, and wastewater treatment systems *(99)*. A few of these are described in more detail below.

Several studies utilizing GeoChip have examined marine environments. Two interesting studies in particular examined two different deep-sea environments. Deep-sea basalt microbial communities were characterized in one study, and GeoChip detected carbon fixation, methane oxidation, methanogenesis, and nitrogen fixation genes *(103)*. These processes had not been previously associated with this environment, but these results indicated they may be important. Another study examined deep-sea hydrothermal vent communities and analyzed samples collected from a mature chimney and inside and outside sections of a 5-day-old chimney *(104)*. GeoChip results indicated that communities from the inner chimney were less diverse than those from the outer portion of the young chimney or the mature chimney. The microbial communities associated with the deep-sea vents appear to have the potential to fix carbon or nitrogen and oxidize and produce methane, none of which have been observed in these environments. Overall, the array analyses indicated that microbial communities from deep-sea hydrothermal vents and basalts have unexpected metabolic and physiological potential.

The impact of the 2010 Gulf of Mexico oil spill on the associated microbial communities was monitored using GeoChip 4.0 *(30,76)*. Samples were collected from the hydrocarbon plume (1100 m depth) approximately 40 days after the spill occurred and from uncontaminated water at the same depth. Results showed that there had been a significant shift in the microbial community functional structure and composition. Genes associated with aerobic and anaerobic hydrocarbon degradation as well as microbial strains similar to known petroleum-degraders and were stimulated within the hydrocarbon plume *(30,76)*. Enriched genes included those for naphthalene 1,2-dioxygenase, β-oxidation of benzylsuccinate, cyclohexanone 1,2-monooxygenase, and alkene monooxygenase *(76)*. These findings demonstrate that microbial communities within the Gulf of Mexico are able to bioremediate the contamination and that the added oil stimulated the oil degrading microbes important in determining the fate of the deep-sea oil spill.

Microbial communities associated with the surface mucopolysaccharide layer and tissue of healthy and yellow band diseased coral, *Montastraea faveolata*, were examined with GeoChip to determine the microbial functional structures and understand how changes in the microbial community may impact disease status *(109)*. Diseased corals had increased numbers of cellulose degradation and nitrification genes, suggesting that these processes may provide a competitive advantage to coral pathogens.

Antarctic microbial communities associated with soil and rock surface niches were examined with GeoChip to better understand their potential to perform geobiological processes *(121)*. Samples were collected from open soil, hypolithic, chasmoendolithic, and cryptoendolithic communities, and results showed significant differences among these communities. The detected genes indicated that the microbial communities utilized autotrophic and heterotrophic strategies for carbon utilization. Results also showed that communities that were more exposed to the environment (hypolithic and chasmoendolithic communities) had greater numbers of stress response genes. There was also evidence that microbial communities could utilize denitrification and annamox pathways during more favorable conditions.

The GeoChip has also been used to answer important ecological questions. GeoChip 2.0 was used to determine the GAR of microbial communities in forest soils. The GAR is similar to the taxa–area relationship which is used in ecology to examine changes in diversity at different spatial scales. For macroorganisms, spatial scaling has been fairly well studied and it is known that species richness tends to increase as the area increases. However, this area of study is still poorly understood for microorganisms. To examine spatial scaling at the micro level, samples were collected from forest soils in a nested design with samples taken over a scale of less than 1 m to 1 km. Functional genes were used to measure richness rather than phylogenetic genes, and results suggested that the forest soil microbial community demonstrated a relatively flat GAR with less turnover than observed for plants and animals although the GAR varied across different functional and phylogenetic groups *(47)*.

The impact of elevated atmospheric CO_2 on biological systems is a central issue in ecology. How belowground microbial communities respond to elevated CO_2 is critical in determining whether carbon loss or sequestration will occur in terrestrial ecosystems, but these processes are poorly understood, and controversial. GeoChip was used to examine the response of soil microbial communities to elevated CO_2 in a grassland system after a 10-year field exposure of grass species to elevated CO_2 *(74)*. Results showed that the elevated CO_2 significantly altered the belowground microbial community functional structure. The community changes significantly correlated with soil carbon and nitrogen content, and plant productivity. Nitrogen and carbon fixation genes were significantly increased under elevated CO_2, while genes for recalcitrant soil carbon degradation were unchanged. These results have important implications for developing accurate global climate change modeling systems.

Two recent extensions of the GeoChip include the StressChip *(126)* and the PathoChip *(127)*. The StressChip is comprised of 22,855 probes for 46 genes related to stress responses for osmotic, oxidative, and protein stress, oxygen, glucose, nitrogen, and phosphate limitation, temperature shock, stringent response, general stress, as well as sigma factors associated with stress response *(126)*. Evaluation of seven metagenome databases from relatively nonstressful (marine samples) to highly stressful (desert soil) environments indicated an increase in the number of stress response genes in the more stressful environments, suggesting the selected genes would be a good indicator of stress response in microbial communities. Further examination of Gulf of Mexico samples from within oil plumes and those in nonplume sites indicated an increase in stress-response genes in the oil plume samples, most likely from the sudden carbon influx *(126)*.

The PathoChip contains 3715 probes covering 7417 sequences for 13 bacterial virulence factors from known pathogens, including those for toxins, capsule proteins, adhesins, pilins, fimbrea, colonization factors, siderophores, type III secretion proteins, invasins, sortases, and general virulence proteins *(127)*. The pathogenic potential of several environments was examined with the PathoChip, and it was found that there were an increased number of virulence factor genes detected under conditions of warmed (+2 °C from ambient) and that warmed and control communities showed statistically significant differences in structure and composition. Oil-contaminated sea samples also showed higher numbers of virulence genes than noncontaminated samples, and the communities were significantly different. Examination of oral samples from individuals that were caries active versus caries free showed similar numbers of positive probes (individual sequences) in both groups but a greater variety of virulence genes in the caries active group.

While, at present, most studies using GeoChip have focused on changes in DNA to infer activity, GeoChips can also be used to directly measure activity. For example, stable isotope probing (SIP) can be used in combination with Geo-Chip to detect active populations *(128)*. In this study, ^{14}C-labeled and unlabeled biphenyl were added to microcosms set up using soil collected from the root zone of a tree growing in a hydrocarbon-contaminated soil and incubated at 25 °C. Samples were collected at 14 days, and extracted DNA was hybridized to the GeoChip to detect active PCB-degrading microbial populations. After genes detected in the background sample (fed unlabeled biphenyl) were subtracted out, 30 genes involved in organic-contaminant degradation were detected in the ^{13}C-labeled sample. These genes were involved in biphenyl degradation as well as degradation of benzoate, catechol, naphthalene, and phenol.

mRNA levels are often very low in environmental samples and are easily degraded, making these samples difficult to analyze. To overcome these problems, a whole community RNA amplification protocol was developed *(80)*. This method was tested using samples from a denitrifying fluidized bed reactor at a uranium contaminated site *(80)*. Genes that would be expected to be active at this site were detected including nitrate reduction genes and several organic contaminant degradation genes, *dsr*, and polyphosphate kinase.

3.2 Other FGAs

3.2.1 Hydrogenase Chip

A Hydrogenase Chip was developed for evaluating the diversity of hydrogenase genes *(71)*. To select sequences for the initial test array, known sequences of [NiFe]- and [FeFe] hydrogenases were used for BLAST queries of the NCBI nr database. The resulting sequences were then grouped at 97% similarity with CD-HIT *(129)*, and then the longest sequence within each cluster was used for probe design. Probes (60-mer) with 30 nt overlap between neighboring probes and 1.67–2× coverage were manufactured by Agilent. Subsequent arrays used sequences obtained from the Integrated Microbial Genomes and Microbiomes database *(130)*. The specificity of the array was tested using a mix of genomic DNA from *E. coli, S. oneidensis, Shewanella sediminis, Pseudomonas aeruginosa*, and *Bacillus subtilis*. Of the 845 genes covered on the array, 49 were expected to be positive based on the genes present in each of these strains. Signal intensities were ranked based on the bright probe fraction (BPF) with "bright spots" having intensities greater than [3 × (median intensity of all spots)] and "dim spots" having intensities below that threshold. BPF was calculated by summing the number of bright spots probes for a given gene and dividing by the total number of probes for that gene. All 49 of the expected probes had BPFs >90%. Two false positives had BPFs above this level but were found to have 89% sequence identity to one of the expected genes. Sensitivity testing indicated a detection limit of 1–10 ng genomic DNA or 0.9–9% of the added DNA.

Community DNA from a reductive dechlorinating soil column was used to test the applicability of the Hydrogenase Chip *(71)*. Samples were collected from the column during three different amendments (lactate, formate, and propionate). Fourteen genes were positive above the 90% BPF threshold in at least one sample, and there were significant differences observed in signal intensity between the different amendments. All of the genes detected by the Hydrogenase Chip were from genera detected using 16S rRNA sequencing. Next, the array was tested with RNA from reactor experiments and 36 genes had BPFs >90%. Expression levels of a *Dehalococcoides hupL* gene was compared between the Hydrogenase Chip and reverse-transcriptase quantitative PCR, and there was a linear correlation between both. In addition, transcription levels correlated with tetrachloroethene degradation rates. There was a larger signal intensity standard deviation in one sample, which was likely due to cross hybridization of closely related strains. For highly diverse communities (phototrophic mat community), there was a loss of sensitivity. The authors compensated for this by decreasing the cutoff for bright spots from 3 × to 2 × the median intensity of all spots.

3.2.2   Food Safety and Health Applications

Use of microarrays in the diagnosis and prevention of disease and illness is an intense area of interest *(131)*. Several experimental arrays, many FGAs, have been developed to detect pathogens in either patient samples or food and water supplies. A *Salmonella*-specific array was reported for detecting virulence genes (*invA* and *sopB*) *(64)*. Using 70-mer oligonucleotide probes and hybridization of whole genome amplified DNA at 42 °C and 40% formamide, the array was able to detect *Salmonella* populations comprising as little as 1% of the original community. Cleven *et al. (67)* developed an array for detecting common bloodstream pathogens. PCR-amplicon probes (200–800 bp) were developed from 120 virulence, antibiotic resistance, and metabolic or structural genes from *Staphylococcus aureus*, *E. coli*, and *P. aeruginosa*. The presence of antibiotic resistance genes detected in *S. aureus* strains were confirmed using traditional culture-based methods, confirming the array's specificity. An array designed to detect virulence markers from 29 waterborne pathogens was designed. Samples were PCR amplified using primers specific for genes covered on the array and was shown to have a sensitivity of 0.01% of the original community for pathogens in tap water, wastewater treatment plant effluent, and river water *(65)*.

A pathogen detection array for the detection of 10 common pathogens (*Staphylococcus* spp., *Streptococcus* spp., *Enterococcus* spp., *Proteus* spp., *Klebsiella* spp., *Stenotrophomonas* spp., *Enterobacter* spp., *Acinetobacter* spp., *E. coli*, *P. aeruginosa*, and *Candida albicans*) was developed by Palka-Santini *et al. (68)*. The array was comprised of 930 genes specific to the covered pathogens, additional genes for antimicrobial resistance and control probes, including bacterial 16S rRNA gene probes as positive controls and human, mouse, amoeba, and grass genes as negative controls. A multiplex PCR amplification strategy using 800 primer pairs covering all genes on the array was used and increased sensitivity to 1 ng of DNA at 42 °C and 25% formamide. In concurrence with culturing results, samples from an infected wound hybridized to *Enterococcus faecium* and *Staphylococcus epidermidis*.

Other types of arrays have been used for the detection of pathogens. To determine which *P. syringae* virulence factors were associated with host specificity, a WGA array containing probes for 353 virulence factors from 91 *P. syringae* strains was used *(43)*. There was a statistically significant association between several genes and hosts. For example, the *hopAA1-1* gene was significantly associated with *P. syringae* strains from rice, while *shcA* was strongly associated with strains from soybean. A POA using 23S rRNA gene-based probes was able to detect *K. pneumoniae*, *P. aeruginosa*, and *Clostridium perfringens* in municipal wastewater *(132)*. The PhyloChip has been used to monitor aerosols for microorganisms as part of a biosurveillance program to detect potential bioterrorism threats and was able to detect sequences similar to several potential pathogens including *Campylobacteraceae*, *Helicobacteraceae*, *Francesella*-like, and bacteria related to *Bacillus anthracis*, *Rickettsia*, and *Clostridium (29)*. A CGA of fragmented whole genomic DNA from an *E. coli* reference collection was used to test *E. coli* strains for *hly*, a hemolysin gene, and was able to determine copy number based on signal intensity increase *(33)*. A POA to identify fungal pathogens using the internal transcribed spacer 1 and 2 (ITS1, ITS2) has also been developed *(133)*. This array allowed the identification of 24 fungal species.

A microfluidic device has been developed, which allows hybridization in 15 min and was able to discriminate between four *Staphylococcus* strains *(134)*. A similar device has also been developed for cell lysis *(135)*. The goal of these devices is to allow a clinical lab to take a patient specimen; dispense it into a cassette; and have DNA extraction, labeling, hybridization, and scanning occur automatically *(136)*.

3.2.3   Detection of Novel Sequences

One of the major perceived drawbacks of using microarrays is the inability to detect sequences that are not covered on the array. However, over the past few years, several groups have sought to overcome this limitation by developing new techniques and designs for using arrays to identify new sequences. One example of this is a viral array developed to identify novel viral pathogens *(52,53)*. The array probes were designed by taking full viral genomes from individual families, selecting highly conserved sequences, dividing them into overlapping 70-nt fragments, and performing a pair-wise BLASTN alignment for each fragment and genome within a given family *(52)*. In general, the five highest ranking 70-mers (based on the number of genomes each 70-mer had >20-nt homology) and the corresponding complimentary sequence were selected. The resulting array covered ~140 viral genomes. Even viruses that were not used for probe design could be detected because each virus serotype gave a distinct hybridization pattern. This array has been used in the identification of the SARS virus *(53)*, for the diagnosis of viral infections in patients with acute exacerbation of idiopathic pulmonary fibrosis *(137)*, to detect viral pathogens in patients with respiratory infections *(138)*, and to identify a possible etiologic agent for proventricular dilation disease *(139)*.

Another option for detecting novel sequences is the use of capture microarrays. These arrays are designed to cover entire loci using oligonucleotide probes (>60-mer) spaced approximately 1–10 bases apart across the entire sequence as probes *(50)*, similar to tiling arrays. DNA is hybridized under low stringency conditions to hybridize or "capture" sequence variants, the unhybridized DNA is then washed off, and the captured DNA is eluted using high-temperature (95 °C) water. The eluted DNA is then sequenced to identify the new sequences *(50,51)*. While designed for use in human and mouse samples, this method has been used for studying *E. coli (140)*. A capture array covering 68 genes involved in fatty acid biosynthesis, β-oxidation, and glycolysis was designed and used on 750 ng of *E. coli* K12 DNA. Sequencing results indicated an enrichment of ~50-fold and an error rate of <0.2%. This method has since evolved from the use of solid-capture devices (microarrays) to a liquid-capture medium.

3.2.4   Other Microarrays: Unique Applications

Some interesting developments in the application of microarrays for the identification of microorganisms using "bar code" arrays have been presented.

One such array was a DNA fingerprinting array developed by Kingsley *et al.* *(141)*. To design this array, 2000 9-mer probes were randomly selected from the *E. coli* K-12 genome. This initial list was reduced to a final number of 47 probes by removing hairpins, repeats greater than 3 bases, G+C content within 44–55% or any GGGCCC sequences, and palindrome sequences. DNA from *Xanthanmonas* spp. was amplified using repetitive extragenic palindromic (REP) PCR primers and then hybridized to the array overnight at 4 °C. Based on the hybridization pattern, a "bar code" is generated that can then be used to separate closely related strains using multivariate statistics. This method was able to separate *Xanthomonas oryzae* strains that could be distinguished using gel electrophoresis of REP–PCR products, indicating a higher resolution for differentiating closely related microbial strains.

An amplification array was developed to allow for amplification, labeling, and hybridization to occur in a single reaction vessel *(142)*. This array was developed as a way to perform microarray hybridization in the field with minimal equipment and rapid turnaround time. The 16S rRNA probes were designed for bacteria that were fermenters; dechloronators; and nitrate, sulfur, and metal reducers. The reaction buffer combined PCR amplification reagent with labeling reagents in an asymmetric ratio to favor the production of single-stranded DNA that would be able to hybridize to the probes. The arrays were incubated in a thermal cycler for 25 and 35 cycles with annealing temperatures of 59 and 64 °C, respectively, and then hybridized for 6–8 h at 37 °C. Samples were collected from the Super 8 experiment at the DOE's Integrated Field Research Center site in Rifle, CO after the transition to iron reducing conditions. Microarray results showed an increase in iron-reducing bacteria and were consistent with qPCR results.

4 KEY ISSUES IN MICROARRAY APPLICATION

While many technical challenges have been overcome for the use of microarrays in microbial community studies, there are still some important issues that can be problematic including probe coverage, nucleic acid quality, specificity, sensitivity, quantitative capability, and detection of microbial activity. The number of sequences for a given gene continues to increase as new sequences are submitted. This is especially true since sequencing has become such a common place technique in microbial studies. To remain relevant, continuous updates are necessary, especially for FGAs such as the GeoChip that expect to provide comprehensive coverage of genes and gene groups. Nucleic acid quality is another concern as contaminants in the DNA may inhibit amplification, labeling, and hybridization steps. The $A_{260}:A_{230}$ ratio has been shown to be most critical for hybridization success *(143)*, so care must be taken to process the DNA/RNA in such a manner that minimal contaminants remain.

Sensitivity and specificity are critical to microarray success. Environmental communities can be extremely diverse, yet many environmental sequences are unknown and many gene variants are highly homologous. Studies have provided several ways to improve sensitivity and specificity although most of these require a balance as often increasing specificity results in a decrease in sensitivity and vice versa. An obvious first step is designing specific probes using experimentally determined design criteria *(78,144)* as mentioned previously. The current detection limit for FGAs is 5% or 25 ng of the microbial community *(19,20,45,59)*. Hybridization conditions (temperature and formamide concentration) affect specificity by varying the stringency. Mixing during hybridization will also increase sensitivity *(145)*. Increasing probe length *(54,77)* or probe concentration per spot *(146–148)* would increase sensitivity although specificity would decrease. Amplification of community DNA can also increase sensitivity by increasing the concentration of low-abundance sequences. Whole community genome amplification (WCGA) is commonly used for GeoChip arrays. This method of amplification can representatively amplify 1–250 ng of community DNA *(149)*. For some FGAs containing relatively few genes, a multiplex PCR-based amplification using primers for all genes contained on the array can be used *(68)*. However, this would not work if large numbers of genes were used or if the covered genes were not easily amplified. Labeling techniques could be used to improve sensitivity. Cyanine dye-doped nanoparticles or tyramide signal amplification labeling can increase sensitivity 10-fold *(54,150)*. Other strategies include development of more sensitive signal detection systems *(146,148)*, decreasing background fluorescence by the use of unmodified array slides *(151,152)* or improved washing protocols, or decreasing ozone levels, which can cause degradation of Cy-dye signal *(153)*.

Other issues of importance are the ability to provide quantitative information and to provide information about the activity of the bacterial communities. If no amplification is performed or if the amplification provides minimal bias, then quantitative information can be provided. GeoChips, however, have been shown to provide quantitative information with or without WCGA for both DNA and RNA *(19,20,45,80,149)*. However, signal intensity can also be affected by sequence divergence. Nonspecific hybridization is usually caused by highly homologous sequences, such as those from the same gene families or multispliced variants *(154,155)*. While changes in DNA concentration between conditions can be used to infer microbial activity, it does not provide direct proof of functional gene activity. Some very active microbes may not change in number. RNA can be used for FGA analysis *(61,80,156)*, although it can be difficult to work with. Hybridization with SIP *(128)* can be used as well.

5   FUTURE ADVANCEMENTS

The technological ability to produce a "SuperChip" is expected within the next 5 years *(157)*. This array would provide both phylogenetic information using 16S rRNA gene sequences for Bacteria and Archaea and the small subunit rRNA gene sequence, the intergenic transcribed spacer region, and the D1/D2 region of the large subunit rRNA gene sequence for Fungi and functional ability using the GeoChip microarray *(157)*. Advances in bioengineering to reduce the size and increase the probe density of microarrays, microfluidics advancements to allow miniaturization of sample hybridization and in bioinformatics to utilize the exponential expansion of bioinformatics data *(157)*.

6 CONCLUSIONS

Microarrays are powerful tools to examine microbial communities and gain insight into ecosystem processes. While there are still technical challenges to overcome, there have been great improvements in microarray technology, design, and application over the past decade. Numerous studies reported over the past several years have shown the applicability and power of microarrays in the field of microbial ecology, pathogen detection, and environmental monitoring. Exciting developments in the use of microarrays and array processing have been reported including rapid processing, in field applications, and novel sequence detection.

ACKNOWLEDGMENTS

The effort for preparing this review was supported by the Office of Science, Office of Biological and Environmental Research, of the U.S. Department of Energy under Contract No. DE-AC02-05CH11231 through ENIGMA- Ecosystems and Networks Integrated with Genes and Molecular Assemblies (http:/enigma.lbl.gov), a Scientific Focus Area Program at Lawrence Berkeley National Laboratory, the National Science Foundation's Macrosystems biology program (EF-1065844), and the U.S. Department of Energy, Biological Systems Research on the Role of Microbial Communities in Carbon Cycling Program (DE-SC0004601).

REFERENCES

1. Torsvik, V.; Goksoyr, J.; Daae, F. L. *Appl. Environ. Microbiol.* **1990**, *56*, 782–787.
2. Hong, S.-H.; Bunge, J.; Jeon, S.-O.; Epstein, S. S. *Proc. Natl. Acad. Sci. U.S.A.* **2006**, *103*, 117–122.
3. Schloss, P. D.; Handelsman, J. *PLoS Comput. Biol.* **2006**, *2*, 786–793.
4. Roesch, L. F. W.; Fulthorpe, R. R.; Riva, A.; Casella, G.; Hadwin, A. K. M.; Kent, A. D.; Daroub, S. H.; Camargo, F. A.; Farmerie, W. G.; Triplett, E. W. *ISME J.* **2007**, *1*, 283–290.
5. Amann, R. I.; Ludwig, W.; Schleifer, K. H. *Microbiol. Rev.* **1995**, *59*, 143–169.
6. Fuhrman, J. A.; Steele, J. A. *Aquat. Microb. Ecol.* **2008**, *53*, 69–81.
7. Whitman, W. B.; Coleman, D. C.; Wiebe, W. J. *Proc. Natl. Acad. Sci. U.S.A.* **1998**, *95*, 6578–6583.
8. Dunbar, J.; Barns, S. M.; Ticknor, L. O.; Kuske, C. R. *Appl. Environ. Microbiol.* **2002**, *68*, 3035–3045.

9. Jenkins, B. D.; Steward, G. F.; Short, S. M.; Ward, B. B.; Zehr, J. P. *Appl. Environ. Microbiol.* **2004**, *70*, 767–1776.

10. Suzuki, M. T.; Giovannoni, S. J. *Appl. Environ. Microbiol.* **1996**, *62*, 625–630.

11. Warnecke, P. M.; Stirzaker, C.; Melki, J. R.; Millar, D. S.; Paul, C. L.; Clark, S. J. *Nucleic Acids Res.* **1997**, *25*, 4422–4426.

12. Lueders, T.; Friedrich, M. W. *Appl. Environ. Microbiol.* **2003**, *69*, 320–326.

13. Wilson, K. H.; Blitchington, R. B.; Greene, R. C. *J. Clin. Microbiol.* **1990**, *28*, 1942–1946.

14. Yamamoto, S.; Harayama, S. *Appl. Environ. Microbiol.* **1995**, *61*, 1104–1109.

15. Hugenholtz, P.; Goebel, B. M.; Pace, N. R. *J. Bacteriol.* **1998**, *180*, 4765–4774.

16. Brodie, E. L.; DeSantis, T. Z.; Joyner, D. C.; Baek, S. M.; Larsen, J. T.; Andersen, G. L.; Hazen, T. C.; Richardson, P. M.; Herman, D. J.; Tokunaga, T. K.; Wan, J. M.; Firestone, M. K. *Appl. Environ. Microbiol.* **2006**, *72*, 6288–6298.

17. Guschin, D. Y.; Mobarry, B. K.; Proudnikov, D.; Stahl, D. A.; Rittmann, B. E.; Mirzabekov, A. D. *Appl. Environ. Microbiol.* **1997**, *63*, 2397–2402.

18. Taroncher-Oldenburg, G.; Griner, E. M.; Francis, C. A.; Ward, B. B. *Appl. Environ. Microbiol.* **2003**, *69*, 1159–1171.

19. Rhee, S. K.; Liu, X.; Wu, L.; Chong, S. C.; Wan, X.; Zhou, J. *Appl. Environ. Microbiol.* **2004**, *70*, 4303–4317.

20. Tiquia, S. M.; Wu, L.; Chong, S. C.; Passovets, S.; Xu, D.; Xu, Y.; Zhou, J. *Biotechniques* **2004**, *36*, 1–8.

21. Steward, G. F.; Jenkins, B. D.; Ward, B. B.; Zehr, J. P. *Appl. Environ. Microbiol.* **2004**, *70*, 1455–1465.

22. Schena, M.; Shalon, D.; Davis, R. W.; Brown, P. O. *Science* **1995**, *270*, 467–470.

23. Schena, M.; Shalon, D.; Heller, R.; Chai, A.; Brown, P. O. *Proc. Natl. Acad. Sci. U.S.A.* **1996**, *93*, 10614–10619.

24. Ehrenreich, A. *Appl. Microbiol. Biotechnol.* **2006**, *73*, 255–273.

25. Small, J.; Call, D. R.; Brockman, F. J.; Straub, T. M.; Chandler, D. P. *Appl. Environ. Microbiol.* **2001**, *67*, 4708–4716.

26. Loy, A.; Lehner, A.; Lee, N.; Adamcsyk, J.; Meier, H.; Ernst, J.; Schleifer, K. H.; Wagner, M. *Appl. Environ. Microbiol.* **2002**, *68*, 5064–6081.

27. Wilson, K. H.; Wilson, W. J.; Radosevich, J. L.; DeSantis, T. Z.; Viswanathan, V. S.; Kuczmarski, T. A.; Anderson, G. L. *Appl. Environ. Microbiol.* **2002**, *68*, 2535–2541.

28. DeSantis, T. Z.; Brodie, E. L.; Moberg, J. P.; Zubieta, I. X.; Piceno, Y. M.; Andersen, G. L. *Microb. Ecol.* **2007**, *53*, 371–383.

29. Brodie, E. L.; DeSantis, T. Z.; Moberg Parker, J. P.; Zubietta, I. X.; Piceno, Y. M.; Anderson, G. L. *Proc. Natl. Acad. Sci. U.S.A.* **2007**, *104*, 299–304.

30. Hazen, T. C.; Dubinsky, E.; DeSantis, T.; Andersen, G.; Piceno, Y.; Singh, N.; Jansson, J.; Probst, A.; Borglin, S.; Fortney, J.; Stringfellow, W.; Bill, M.; Conrad, M.; Tom, L.; Chavarria, K.; Alusi, T.; Lamendella, R.; Joyner, D.; Spier, C.; Baelum, J.; Auer, M.; Zemla, M.; Chakraborty, R.; Sonnenthal, E.; D'haeseleer, P.; Holman, H.-Y.; Osman, S.; Lu, Z. M.; Van Nostrand, J.; Deng, Y.; Zhou, J. Z.; Mason, O. U. *Science* **2010**, *330*, 204–208.

31. Franke-Whittle, I. H.; Klammer, S. H.; Insam, H.; Microbiol, J. *Methods* **2005**, *62*, 37–56.

32. Wu, L.; Thompson, D. K.; Liu, X.; Fields, M. W.; Bagwell, C. E.; Tiedje, J. M.; Zhou, J. *Environ. Sci. Technol.* **2004**, *38*, 6775–6782.

33. Zhang, L.; Srinivasan, U.; Marrs, C. F.; Ghosh, D.; Gilsdorf, J. R.; Foxman, B. *BMC Microbiol.* **2004**, *4*, 12–18.

34. Chen, Q.; Huaqun, Y.; Luo, H.; Xie, M.; Qiu, G.; Liu, X. *Hydrometallurgy* **2009**, *95*, 96–103.

35. Wu, L.; Liu, X.; Fields, M. W.; Thompson, D. K.; Bagwell, C. E.; Tiedje, J. M.; Hazen, T. C.; Zhou, J.-Z. *ISME J.* **2008**, *2*, 642–655.

36. Sebat, J. L.; Colwell, F. S.; Crawford, R. L. *Appl. Environ. Microbiol.* **2003**, *69*, 4927–4934.

37. Mockler, T. C.; Ecker, J. R. *Genomics* **2005**, *85*, 1–15.

38. Gresham, D.; Dunham, M. J.; Botstein, D. *Nat. Rev. Genet.* **2008**, *9*, 291–302.

39. Rich, V. I.; Konstantinidis, K.; DeLong, E. F. *Environ. Microbiol.* **2008**, *10*, 506–521.

40. Wilson, M.; DeRisi, J.; Kristensen, H. H.; Imboden, P.; Rane, S.; Brown, P. O.; Schoolnik, G. K. *Proc. Natl. Acad. Sci. U.S.A.* **1999**, *96*, 12833–12838.

41. Murray, A. E.; Lies, D.; Li, G.; Nealson, K.; Zhou, J.; Tiedje, J. M. *Proc. Natl. Acad. Sci. U.S.A.* **2001**, *98*, 9853–9858.

42. Dong, Y.; Glasner, J. D.; Blattner, F. R.; Triplett, E. W. *Appl. Environ. Microbiol.* **2001**, *67*, 1911–1921.

43. Sarkar, S. F.; Gordon, J. S.; Martin, G. B.; Guttman, D. S. *Genetics* **2006**, *147*, 1041–1056.

44. White, J. R.; Escobar-Paramo, P.; Mongodin, E. F.; Nelson, K. E.; DiRuggiero, J. *Appl. Environ. Microbiol.* **2008**, *74*, 6447–6451.

45. Wu, L.; Thompson, D. K.; Li, G.; Hurt, R. A.; Tiedje, J. M.; Zhou, J. *Appl. Environ. Microbiol.* **2001**, *67*, 5780–5790.

46. van Vliet, A. H. M. *FEMS Microbiol. Lett.* **2010**, *302*, 1–7.

47. Zhou, J.; Kang, S.; Schadt, C. W.; Garten, C. T., Jr. *Proc. Natl. Acad. Sci. U.S.A.* **2008**, *105*, 7768–7773.

48. McKenna, P.; Hoffmann, C.; Minkah, N.; Aye, P. P.; Lackner, A.; Liu, Z. *PLoS Pathog.* **2008**, *4*, e20.

49. Zhou, J.; Wu, L.; Deng, Y.; Zhi, X.; Jiang, Y.; Tu, Q.; Xie, J.; Van Nostrand, J. D.; He, Z.; Zhou, J. *ISME J.* **2011**, *5*, 1303–1313.

50. Albert, T. J.; Molla, M. N.; Muzny, D. M.; Nazareth, L.; Wheeler, D.; Song, X.; Richmond, T. A.; Middle, C. M.; Rodesch, M. J.; Packard, C. J.; Weinstock, G. M.; Gibbs, R. A. *Nat. Methods* **2007**, *4*, 903–905.

51. Okou, D. T.; Steinberg, K. M.; Middle, C.; Cutler, D. J.; Albert, T. J.; Zwick, M. E. *Nat. Methods* **2007**, *4*, 907–909.

52. Wang, D.; Coscoy, L.; Zylberberg, M.; Avila, P. C.; Boushey, H. A.; Ganem, D.; DeRisi, J. L. *Proc. Natl. Acad. Sci. U.S.A.* **2002**, *99*, 15687–15692.

53. Ksiazek, T. G.; Erdman, D.; Goldsmith, C. S.; Zaki, S. R.; Peret, T.; Emery, S.; Tong, S.; Urbani, C.; Comer, J. A.; Lim, W.; Rollin, P. E.; Dowell, S. F.; Ling, A. E.; Humphrey, C. D.; Shieh, W. J.; Guarner, J.; Paddock, C. D.; Rota, P.; Fields, B.; DeRisi, J.; Yang, J. Y.; Cox, N.; Hughes, J. M.; LeDuc, J. W.; Bellini, W. J.; Anderson, L. J.; The SARS Working Group. *N. Engl. J. Med.* **2003**, *348*, 1953–1966.

54. Denef, V. J.; Park, J.; Rodrigues, J. L. M.; Tsoi, T. V.; Hashsham, S. A.; Tiedje, J. M. *Environ. Microbiol.* **2003**, *5*, 933–943.

55. Zhou, J. *Curr. Opin. Microbiol.* **2003**, *6*, 288–294.

56. Gentry, T. J.; Wickham, G. S.; Schadt, C. W.; He, Z.; Zhou, J. *Microb. Ecol.* **2006**, *52*, 159–175.

57. Zhang, Y. G.; Zhang, X. Q.; Liu, X. D.; Xiao, Y.; Qu, L. J.; Wu, L. Y.; Zhou, J. Z. *FEMS Microbiol. Lett.* **2007**, *266*, 144–151.

58. Abell, G. C.; Robert, S. S.; Frampton, D. M. F.; Volkman, J. K.; Rizwi, F.; Csontos, J.; Bodrossy, L. *PLoS One* **2012**, *7*, e51542.

59. Bodrossy, L.; Stralis-Pavese, N.; Murrell, J. C.; Radajewski, S.; Weilharter, A.; Sessitsch, A. *Environ. Microbiol.* **2003**, *5*, 566–582.

60. Stralis-Pavese, N.; Sessitsch, A.; Weilharter, A.; Reichenauer, T.; Riesing, J.; Csonton, J.; Murrell, J. C.; Bodrossy, L. *Environ. Microbiol.* **2004**, *6*, 347–363.

61. Bodrossy, L.; Stralis-Pavese, N.; Konrad-Köszler, M.; Weilharter, A.; Reichenauer, T. G.; Schöfer, D.; Sessitsch, A. *Appl. Environ. Microbiol.* **2006**, *72*, 1672–1676.
62. Yergeau, E.; Arbour, M.; Brousseau, R.; Juck, D.; Lawrence, J. R.; Masson, L.; Whyte, L. G.; Greer, C. W. *Appl. Environ. Microbiol.* **2009**, *75*, 6258–6267.
63. Vilchez-Vargas, R.; Geffers, R.; Suárez-Diez, M.; Conte, I.; Waliczekm, A.; Kaser, V. S.; Kralova, M.; Junca, H.; Pieper, D. H. *Environ. Microbiol.* **2013**, *15*, 1016–1039.
64. Kostić, T.; Weilharter, A.; Sessitsch, A.; Bodrossy, L. *Anal. Biochem.* **2005**, *346*, 333–335.
65. Miller, S. M.; Tourlousse, D. M.; Stedtfeld, R. D.; Baushke, S. W.; Herzog, A. B.; Wick, L. M.; Rouillard, J. M.; Gulari, E.; Tiedje, J. M.; Hashsham, S. A. *Appl. Environ. Microbiol.* **2008**, *74*, 2200–2209.
66. Call, D. R.; Bakko, M. K.; Krug, M. J.; Roberts, M. C. *Antimicrob. Agents Chemother.* **2003**, *47*, 3290–3295.
67. Cleven, B. E. E.; Palka-Santini, M.; Gielen, J.; Meembor, S.; Krönke, M.; Krut, O. *J. Clin. Microbiol.* **2006**, *44*, 2389–2397.
68. Palka-Santini, M.; Cleven, B. E.; Eichinger, L.; Krönke, M.; Krut, O. *BMC Microbiol.* **2009**, *9*, 1.
69. Bontemps, C.; Goldier, G.; Gris-Liebe, C.; Carere, S.; Talini, L.; Boivin-Masson, C. *Appl. Environ. Microbiol.* **2005**, *71*, 8042–8048.
70. Iwai, S.; Kurisu, F.; Urakawa, H.; Yagi, O.; Furumai, H. *Appl. Microbiol. Biotechnol.* **2007**, *75*, 929–939.
71. Marshall, I. P. G.; Berggren, D. R. V.; Azizian, M. F.; Burow, L. C.; Semprini, L.; Spormann, A. M. *ISME J.* **2012**, *6*, 814–826.
72. Yin, H.; Cao, L.; Qiu, G.; Wang, D.; Kellogg, L.; Zhou, J.; Dai, Z.; Liu, X. *J. Microbiol. Methods* **2007**, *70*, 165–178.
73. Chapman, R.; Hayden, H. L.; Webster, T.; Crowley, D. E.; Mele, P. M. *Pedobiologia* **2012**, *55*, 41–49.
74. He, Z.; Gentry, T. J.; Schadt, C. W.; Wu, L.; Liebich, J.; Chong, S. C.; Wu, W. M.; Gu, B.; Jardine, P.; Criddle, C.; Zhou, J. Z. *ISME J.* **2007**, *1*, 67–77.
75. He, Z.; Deng, Y.; Van Nostrand, J. D.; Tu, Q.; Xu, M.; Hemme, C. L.; Li, X.; Wu, L.; Gentry, T. J.; Yin, Y.; Leibich, J.; Hazen, T. C.; Zhou, J. *ISME J.* **2010**, *4*, 1167–1179.
76. Lu, Z.-M.; Deng, Y.; Van Nostrand, J. D.; He, Z.; Voordeckers, J.; Zhou, A.; Lee, Y. J.; Mason, O. U.; Dubinsky, E. A.; Chavarria, K. L.; Tom, L. M.; Fortney, J. L.; Lamendella, R.; Jansson, J. K.; D'haeseleer, P.; Hazen, T. C.; Zhou, J. *ISME J.* **2012**, *6*, 451–460.
77. He, Z.; Wu, L. Y.; Li, X. Y.; Fields, M. W.; Zhou, J. Z. *Appl. Environ. Microbiol.* **2005**, *71*, 3753–3760.
78. Liebich, J.; Schadt, C. W.; Chong, S. C.; He, Z.; Rhee, S. K.; Zhou, J. *Appl. Environ. Microbiol.* **2006**, *72*, 1688–1691.
79. Li, X.; He, Z.; Zhou, J. *Nucleic Acids Res.* **2005**, *33*, 6114–6123.
80. Gao, H.; Yang, Z. K.; Gentry, T. J.; Wu, L.; Schadt, C. W.; Zhou, J. *Appl. Environ. Microbiol.* **2007**, *73*, 563–571.
81. Waldron, P. J.; Wu, L.; Van Nostrand, J. D.; Schadt, C.; Watson, D.; Jardine, P.; Palumbo, T.; Hazen, T. C.; Zhou, J. *Environ. Sci. Technol.* **2009**, *43*, 3529–3534.
82. Van Nostrand, J. D.; Wu, W. M.; Wu, L.; Deng, Y.; Carley, J.; Carroll, S.; He, Z.; Gu, B.; Luo, J.; Criddle, C. S.; Watson, D. B.; Jardine, P. M.; Marsh, T. L.; Tiedje, J. M.; Hazen, T. C.; Zhou, J. *Environ. Microbiol.* **2009**, *11*, 2611–2626.
83. Van Nostrand, J. D.; Wu, L.; Wu, W. M.; Gentry, T. J.; Huang, Z.; Deng, Y.; Carley, J.; Carroll, S.; He, Z.; Gu, B.; Luo, J.; Criddle, C. S.; Watson, D. B.; Jardine, P. M.; Marsh, T. L.; Tiedje, J. M.; Hazen, T. C.; Zhou, J. *Appl. Environ. Microbiol.* **2011**, *77*, 3860–3869.

84. Xu, M.; Wu, W. M.; Wu, L.; He, Z.; Van Nostrand, J. D.; Deng, Y.; Luo, J.; Carley, J.; Ginder-Vogel, M.; Gentry, T. J.; Gu, B.; Watson, D.; Jardine, P. M.; Marsh, T. L.; Tiedje, J. M.; Hazen, T.; Criddle, C. S.; Zhou, J. *ISME J.* **2010**, *4*, 1060–1070.

85. Liang, Y.; Van Nostrand, J. D.; N'Guessan, L. A.; Peacock, A. D.; Deng, Y.; Long, P. E.; Resch, C. T.; Wu, L.; He, Z.; Li, G.; Hazen, T. C.; Lovley, D. R.; Zhou, J. *Appl. Environ. Microbiol.* **2012**, *78*, 2966–2972. http://dx.doi.org/10.1128/AEM.06528-11.

86. Xiong, J. B.; Wu, L. Y.; Tu, S. X.; Van Nostrand, J. D.; He, Z. H.; Zhou, J.-Z.; Wang, G. J. *Appl. Environ. Microbiol.* **2010**, *76*, 7277–7284.

87. Epelde, L.; Becerril, J. M.; Kowalchuk, G. A.; Deng, Y.; Zhou, J.-Z.; Garbisu, C. *Appl. Environ. Microbiol.* **2010**, *76*, 7843–7853.

88. Kang, S.; Van Nostrand, J. D.; Gough, H. L.; He, Z.; Hazen, T. C.; Stahl, D. A.; Zhou, J. *FEMS Microbiol. Ecol.* **2013**, *86*, 200–214. http://dx.doi.org/10.1111/1574-6941.12152.

89. Xiong, J.; He, Z.; Van Nostrand, J. D.; Luo, G.; Tu, S.; Zhou, J.; Wang, G. *PLoS One* **2012**, *7*, e50507.

90. Zhao, F.-J.; Harris, E.; Jia, Y.; Ma, J.; Wu, L.; Liu, W.; McGrath, S. P.; Zhou, J.; Zhu, Y.-G. *Environ. Sci. Technol.* **2013**, *47*, 7147–7154.

91. Van Nostrand, J. D.; Khijniak, T. V.; Gentry, T. J.; Novak, M. T.; Sowder, A. G.; Zhou, J. Z.; Bertsch, P. M.; Morris, P. J. *Microb. Ecol.* **2007**, *53*, 670–682.

92. Rodríguez-Martínez, E. M.; Pérez, E. X.; Schadt, C. W.; Zhou, J.; Massol-Deyá, A. A. *Int. J. Environ. Res. Public Health* **2006**, *3*, 292–300.

93. Liang, Y.; Li, G.; Van Nostrand, J. D.; He, Z.; Wu, L.; Deng, Y.; Zhang, X.; Zhou, J. *FEMS Microbiol. Ecol.* **2009**, *70*, 324–333.

94. Liang, Y.; Wang, J.; Van Nostrand, J. D.; Zhou, J.; Zhang, X.; Li, G. *Chemosphere* **2009**, *75*, 193–199.

95. Liang, Y. T.; Van Nostrand, J. D.; Deng, Y.; He, Z. H.; Wu, L. Y.; Zhang, X.; Li, G.; Zhou, J. *ISME J.* **2011**, *5*, 403–413.

96. Yergeau, E.; Bokhorst, S.; Kang, S.; Zhou, J.-Z.; Greer, C. W.; Aerts, R.; Kowalchuk, G. A. *ISME J.* **2012**, *6*, 692–702.

97. Ding, G. C.; Heuer, H.; He, Z.; Xie, J.; Zhou, J.; Smalla, K. *FEMS Microbiol. Ecol.* **2012**, *82*, 148–156. http://dx.doi.org/10.1111/j.1574-6941.2012.01413.x.

98. Liebich, J.; Wachtmeister, T.; Zhou, J.; Burauel, P. *Vadose Zone J.* **2009**, *8*, 703–710.

99. Zhang, Z.; Zhao, X.; Liang, Y.; Li, G.; Zhou, J. *Environ. Chem. Lett.* **2013**, *11*, 11–17.

100. Taş, N.; van Eekert, M. H. A.; Schraa, G.; Zhou, J.; de Vos, W. M.; Smidt, H. *Appl. Environ. Microbiol.* **2009**, *75*, 4696–4704.

101. Wawrik, B.; Mendivelso, M.; Parisi, V. A.; Suflita, J. M.; Davidova, I. A.; Marks, C. R.; Van Nostrand, J. D.; Liang, Y.; Zhou, J.-Z.; Huizinga, B. J.; Stropoc, D.; Callaghan, A. V. *FEMS Microbiol. Ecol.* **2012**, *81*, 26–42.

102. Beazley, M. J.; Martinez, R. J.; Rajan, S.; Powell, J.; Piceno, Y. M.; Tom, L. M.; Andersen, G. L.; Hazen, T. C.; Van Nostrand, J. D.; Zhou, J.; Mortazavi, B.; Sobecky, P. A. *PLoS One* **2012**, *7*, e41305.

103. Mason, O. U.; DiMeo-Savoie, C. A.; Van Nostrand, J. D.; Zhou, J.-Z.; Fisk, M. R.; Giovannoni, S. J. *ISME J.* **2009**, *3*, 231–242.

104. Wang, F.; Zhou, H.; Meng, J.; Peng, X.; Jiang, L.; Sun, P.; Zhang, C.; Van Nostrand, J. D.; Deng, Y.; He, Z.; Wu, L.; Zhou, J.; Xiao, X. *Proc. Natl. Acad. Sci. U.S.A.* **2009**, *106*, 4840–4845.

105. Tiquia, S. M.; Gurczynski, S.; Zholi, A.; Devol, A. *Environ. Technol.* **2006**, *27*, 1377–1389.

106. Wu, L.; Kellogg, L.; Devol, A. H.; Palumbo, A. V.; Tiedje, J. M.; Zhou, J. *Appl. Environ. Microbiol.* **2008**, *74*, 4516–4529.

107. Bai, S.; Li, J.; He, Z.; Van Nostrand, J. D.; Tian, Y.; Lin, G.; Zhou, J.; Zheng, T. *Appl. Microbiol. Biotechnol.* **2013**, *97*, 7035–7048.

108. Wawrik, B.; Boling, W.; Van Nostrand, J.; Xie, J. P.; Zhou, J.-Z.; Bronk, D. A. *FEMS Microbiol. Ecol.* **2012**, *79*, 400–411.

109. Kimes, N. E.; Van Nostrand, J. D.; Weil, E.; Zhou, J.-Z.; Morris, P. J. *Environ. Microbiol.* **2010**, *12*, 541–556.

110. Parnell, J. J.; Rompato, G.; Latta, L. C., IV; Pfrender, M. E.; Van Nostrand, J. D.; He, Z.; Zhou, J.; Andersen, G.; Champine, P.; Ganesan, B.; Weimer, B. C. *PLoS One* **2010**, *5*, e12919.

111. He, Z. H.; Xu, M.; Ye, D.; Kang, S.; Kellogg, L.; Wu, L. Y.; Van Nostrand, J. D.; Hobbie, S. E.; Reich, P. B.; Zhou, J.-Z. *Ecol. Lett.* **2010**, *13*, 564–575.

112. Zhou, J.-Z.; Deng, Y.; Luo, F.; He, Z.-H.; Tu, Q.; Zhi, X. Y. *mBio* **2010**, *1*, e00169-10.

113. Li, X.; Deng, Y.; Li, Q.; Lu, C.; Wang, J.; Zhang, H.; Zhu, J.; Zhou, J.; He, Z. *ISME J.* **2012**, *7*, 660–667.

114. He, Z.; Xiong, J.; Kent, A. D.; Deng, Y.; Xue, K.; Wang, G.; Wu, L.; Van Nostrand, J. D.; Zhou, J. *ISME J.* **2013**, *8*, 714–726. http://dx.doi.org/10.1038/ismej.2013.177.

115. Xu, M.; He, Z.; Deng, Y.; Wu, L.; Van Nostrand, J. D.; Hobbie, S. E.; Reich, P. B.; Zhou, J. *BMC Microbiol.* **2013**, *13*, 124.

116. Zhang, L.; Hurek, T.; Reihold-Hurek, B. *Microb. Ecol.* **2007**, *53*, 456–470.

117. Zhang, Y.; Xie, J.; Liu, M.; Tian, Z.; He, Z.; Van Nostrand, J. D.; Ren, L.; Zhou, J.; Yang, M. *Water Res.* **2013**, *47*, 6298–6308.

118. Xue, K.; Wu, L.; Deng, Y.; He, Z.; Van Nostrand, J.; Robertson, P. G.; Schmidt, T. M.; Zhou, J. *Appl. Environ. Microbiol.* **2013**, *79*, 1284–1294.

119. Yang, Y.; Wu, L.; Lin, Q.; Yuan, M.; Xu, D.; Yu, H.; Hu, Y.; Duan, J.; Li, X.; He, Z.; Xue, K.; Van Nostrand, J.; Wang, S.; Zhou, J. *Glob. Chang. Biol.* **2013**, *19*, 637–648.

120. Wakelin, S. A.; Barratt, B. I. P.; Gerard, E.; Gregg, A. L.; Brodie, E. L.; Andersen, G. L.; DeSantis, T. Z.; DeSantis, T. Z.; Zhou, J.; He, Z.; Kpwalchuk, G. A.; O'Callaghan, M. Shifts in the phylogenetic structure and functional capacity of soil microbial communities follow alteration of native tussock grassland ecosystems. *Soil Biol. Biochem.* **2013**, *57*, 675–682.

121. Chan, Y.; Van Nostrand, J. D.; Zhou, J.; Pointing, S. B.; Farrell, R. L. *Proc. Natl. Acad. Sci. U.S.A.* **2013**, *110*, 8990–8995.

122. Trivedi, P.; He, Z.; Van Nostrand, J. D.; Albrigo, G.; Zhou, J.-Z.; Wang, N. *ISME J.* **2012**, *6*, 363–383.

123. Xie, J. P.; He, Z. H.; Liu, X. X.; Liu, X. D.; Van Nostrand, J. D.; Deng, Y.; Wu, L.; Qiu, G.; Zhou, J. *Appl. Environ. Microbiol.* **2011**, *77*, 991–999.

124. Lu, Z.; He, Z.; Parisi, V. A.; Kang, S.; Deng, Y.; Van Nostrand, J. D.; Masoner, J. R.; Cozzarelli, I. M.; Suflita, J. M.; Zhou, J. *Environ. Sci. Technol.* **2012**, *46*, 5824–5833.

125. Liu, W.; Wang, A.; Sun, D.; Ren, N.; Zhang, Y.; Zhou, J.-Z. *J. Biotechnol.* **2012**, *157*, 628–632.

126. Zhou, A.; He, Z.; Qin, Y.; Lu, Z.; Deng, Y.; Tu, Q.; Hemme, C. L.; Van Nostrand, J. D.; Wu, L.; Hazen, T. C.; Arkin, A. P.; Zhou, J. *Environ. Sci. Technol.* **2013**, *47*, 9841–9849.

127. Lee, Y.-J.; Van Nostrand, J. D.; Tu, Q.; Lu, Z.; Cheng, L.; Yuan, T.; Deng, Y.; Carter, M. Q.; He, Z.; Wu, L.; Yang, F.; Xu, J.; Zhou, J. *ISME J.* **2013**, *7*, 1974–1984.

128. Leigh, M. B.; Pellizari, V. H.; Uhlík, O.; Sutka, R.; Rodrigues, J.; Ostrom, N. E.; Zhou, J.; Tiedje, J. M. *ISME J.* **2007**, *1*, 134–148.

129. Li, X.; Godzik, A. *Bioinformatics* **2006**, *22*, 1658–1659.

130. Markowitz, M.; Ivanova, N. N.; Szeto, E.; Palaniappan, K.; Chu, K.; Dalevi, D.; Chen, I.-M. A.; Grechkin, Y.; Dubchak, I.; Anderson, I.; Lykidis, A.; Mavromatis, K.; Hugenholtz, P.; Kyrpides, N. C. *Nucleic Acids Res.* **2008**, *32*, D534–D538.

131. Van Nostrand, J. D.; Gentry, T. J.; Zhou, J.-Z. In: *Advanced Techniques in Diagnostic Microbiology;* Tang, Y.-W.; Stratton, C. W., Eds, 2nd ed.; Springer: New York, 2013.

132. Lee, D. Y.; Shannon, K.; Beaudette, L. A. *J. Microbiol. Methods* **2006**, *65*, 453–467.

133. Tavanti, S. Landi; Senesi, S. In: *Fungal Diagnostics: Methods and Protocols;* O'Connor, L.; Glynn, B., Eds.; Methods in Molecular Biology; Vol. 968; Springer Sci: New York, 2013.

134. Peytavi, R.; Raymond, F. R.; Gagné, D.; Picard, F. J.; Jia, G.; Zoval, J.; Madou, M.; Boissinot, K.; Boissinot, M.; Bissonnette, L.; Ouellette, M.; Bergeron, M. G. *Clin. Chem.* **2005**, *51*, 1836–1844.

135. Kido, H.; Micic, M.; Smith, D.; Zoval, J.; Norton, J.; Madou, M. *Colloids Surf. B: Biointerfaces* **2007**, *58*, 44–51.

136. Gorkin, R.; Park, J.; Siegris, J.; Amasia, M.; Lee, B. S.; Park, J.-M.; Kim, J.; Kim, H.; Madou, M.; Cho, Y.-K. *R. Soc. Chem.* **2010**, *10*, 1758–1773.

137. Wootton, S. C.; Kim, D. S.; Kondoh, Y.; Chen, E.; Lee, J. S.; Song, J. W.; Huh, J. W.; Taniquchi, H.; Chui, C.; Boushey, H.; Lancaster, L. H.; Wolters, P. J.; DeRisis, J.; Ganem, D.; Collard, H. R. *Am. J. Respir. Crit. Care Med.* **2011**, *183*, 1698–1702.

138. Kistler, A.; Avila, P. C.; Rouskin, S.; Wang, D.; Yaqi, S.; Schnurr, D.; DeRisi, J. L.; Boushey, H. A. *J. Infect. Dis.* **2007**, *196*, 817–825.

139. Kistler, A. L.; Gancz, A.; Clubb, S.; Skewes-Cox, P.; Fischer, K.; Sorber, K.; Chiu, C. Y.; Lublin, A.; Mechani, S.; Farnoushi, Y.; Greninger, A.; Wen, C. C.; Karlene, S. B.; Ganem, D.; DeRisi, J. L. *Virol. J.* **2008**, *5*, 88.

140. Schracke, N.; Kornmeyer, T.; Kränzle, M.; Stähler, P. F.; Summerer, D.; Beier, M. *New Biotechnol.* **2009**, *26*, 229–233.

141. Kingsley, T.; Straub, T. M.; Call, D. R.; Daly, D. S.; Wunschel, S. C.; Chandler, D. P. *Appl. Environ. Microbiol.* **2002**, *68*, 6361–6370.

142. Chandler, D. P.; Knickerbocker, C.; Bryant, L.; Golova, J.; Wiles, C.; Willians, K. H.; Peacock, A. D.; Long, P. E. *Appl. Environ. Microbiol.* **2013**, *79*, 799–807.

143. Ning, J.; Liebich, J.; Kästner, M.; Zhou, J.; Schäffer, A.; Burauel, P. *Appl. Microbiol. Biotechnol.* **2009**, *82*, 983–993.

144. He, Z.; Wu, L.; Fields, M. W.; Zhou, J. *Appl. Environ. Microbiol.* **2005**, *71*, 5154–5162.

145. Adey, N. B.; Lei, M.; Howard, M. T.; Jenson, J. B.; Mayo, D. A.; Butel, D. L.; Coffin, S. C.; Moyer, T. C.; Slade, D. E.; Spute, M. K.; Hancock, A. M.; Eisenhoffer, G. T.; Dalley, B. K.; McNeely, M. R. *Anal. Chem.* **2002**, *74*, 6413–6417.

146. Cho, J. C.; Tiedje, J. M. *Appl. Environ. Microbiol.* **2002**, *58*, 1425–1430.

147. Relógio, A.; Schwager, C.; Richter, A.; Ansorge, W.; Valcárcel, J. *Nucleic Acids Res.* **2002**, *30*, e51.

148. Zhou, J.; Thompson, D. K. *Curr. Opin. Biotechnol.* **2002**, *13*, 204–207.

149. Wu, L.; Liu, X.; Schadt, C. W.; Zhou, J. *Appl. Environ. Microbiol.* **2006**, *72*, 4931–4941.

150. Zhou, X.; Zhou, J. *Anal. Chem.* **2004**, *76*, 5302–5312.

151. Kumar, M. A.; Larsson, O.; Parodi, D.; Liang, Z. *Nucleic Acids Res.* **2000**, *28*, e71.

152. Gudnason, H.; Dufva, M.; Bang, D. D.; Wolff, A. *Biotechniques* **2008**, *45*, 261–271.

153. Branham, W. S.; Melvin, C. D.; Han, T.; Gesai, V. G.; Moland, C. L.; Scully, A. T.; Fuscoe, J. C. *BMC Biotechnol.* **2007**, *7*, 8.

154. Evertsz, E. M.; Au-Young, J.; Ruvolo, M. V.; Lim, A. C.; Reynolds, M. A. *Biotechniques* **2001**, *31*, 1182, 1184, 1186 passim.

155. Modrek, B.; Lee, C. *Nat. Genet.* **2002**, *30*, 13–19.

156. Dennis, P.; Edwards, E. A.; Liss, S. N.; Fulthorpe, R. *Appl. Environ. Microbiol.* **2003**, *69*, 769–778.

157. Hazen, T. C. *Microb. Biotechnol.* **2013**, *6*, 450–452. http://dx.doi.org/10.1111/1751-7915.12045.

microRNA Profiling: An Overview of Current Technologies and Applications

Sinead M. Smith* and David W. Murray†

*Department of Clinical Medicine and School of Pharmacy and Pharmaceutical Sciences, Trinity College Dublin, Dublin 2, Ireland
†Department of Physiology and Medical Physics, Royal College of Surgeons in Ireland, Dublin 2, Ireland

1 INTRODUCTION

Recent insight into the transcriptional landscape of mammalian genomes gained from high-throughput next-generation sequencing (NGS) technologies has revealed that although 80% of the genome is transcribed, less than 2% is subsequently translated into protein, resulting in the generation of a large number of ncRNA transcripts (1–5). Although originally considered to be nonfunctional, recent data have highlighted that noncoding RNAs (ncRNAs) play key roles in gene regulation by influencing transcription and translation of target genes, thus regulating a diverse range of biological processes under normal physiological conditions and in disease settings (6). Broadly speaking,

Comprehensive Analytical Chemistry, Vol. 64. http://dx.doi.org/10.1016/B978-0-444-62650-9.00002-6

ncRNAs may be subdivided according to their size. First, long ncRNAs (lncRNAs) ranging in size from a few hundred nucleotides (nt) to multiple kilobases in length represent the largest class of ncRNAs and account for much of the transcribed genome (6). In contrast to lncRNAs, small ncRNAs are less than 300 nt in length and include microRNAs (miRNAs), piwi-interacting RNAs, small-interfering RNAs, and small nuclear RNAs (snRNAs) among others (7). The miRNAs (18–25 nt) are the most widely studied family of the small ncRNAs. Since their discovery in 1993 (8,9), the diversity and significance of this class of regulatory molecule has become increasingly appreciated. miRNAs posttranscriptionally decrease the expression of thousands of target genes by binding to specific messenger RNA (mRNA) targets and promoting their degradation and/or inhibiting their translation (10–12). Although a relatively limited number of miRNAs (approximately 1000) have been identified in humans compared with the number of mRNAs and proteins (approximately 30,000), a single miRNA may regulate hundreds of mRNAs and thus has the potential to greatly impact gene-expression networks (10).

The importance of miRNAs in biological processes is demonstrated by their high levels of evolutionary conservation (13). Accumulating evidence supports a role for miRNAs in normal cellular physiology, where they act as key regulators of development (14), differentiation (15,16), cell proliferation, and apoptosis (17,18). For example, both specific miRNAs and proteins involved in miRNA processing play fundamental roles in the development and function of B- and T-cells within the immune system (19–23). Additionally, miRNAs coordinate cellular responses in innate immune cells (24) and play key roles in the regulation of host–pathogen interactions during infection (25). miRNA expression is also regulated during inflammation (25–27) and as a consequence perturbed miRNA expression is associated with a number of autoimmune diseases, including multiple sclerosis (28), rheumatoid arthritis (29,30), and systemic lupus erythematosus (31,32). Furthermore, expression profiling has implicated miRNAs in numerous cancers, including B-cell chronic lymphocytic leukemia (33) and cancer of the breast (34–37), colon (35,38,39), liver (40,41), and lung (35,42–47), among others. miRNAs function as both tumor suppressors and oncogenes (48,49). Given the key roles of miRNAs in normal cellular homeostasis and their dysregulation under disease conditions, investigation of miRNA expression profiles and regulation of their mRNA targets is essential for a complete understanding of the cell signaling mechanisms that mediate cell function in health and disease, with the potential to identify new therapeutic agents or drug targets. Moreover, as miRNAs are stable in a variety of clinical specimens, including formalin-fixed paraffin-embedded (FFPE) tissue, blood, and urine, there is substantial interest in their development as biomarkers for diagnostic applications. Indeed, profiling of circulating miRNAs has already shown promise for detection of cancers (36,44,45,50–53).

2   miRNA BIOGENESIS AND NOMENCLATURE

An understanding of the multistep processes involved in miRNA processing and biogenesis is required for the design and application of analytical techniques for miRNA detection and quantitation. Primary miRNA (pri-miRNA) transcripts are either transcribed by RNA polymerase II from independent genes or represent introns of protein-coding genes *(54–56)*. A pri-miRNA transcript can contain a single miRNA or multiple miRNAs that are processed from the same transcript *(25)*. The pri-miRNA folds into a hairpin structure, which acts as a substrate for cleavage by the endonuclease Drosha resulting in an approximately 70–100 nt long precursor miRNA (pre-miRNA) (Figure 1). Following the Drosha cleavage, the pre-miRNA is exported to the cytoplasm by Exportin, where it is further processed by the endonuclease Dicer to produce an miRNA duplex comprised of miRNA strands derived from the 5′ and 3′ regions of the precursor duplex *(7,10,55)*. One strand of this duplex, representing a mature miRNA, is incorporated into the RNA-induced silencing complex (RISC), while the other passenger strand is usually degraded. As part of the RISC, miRNAs base pair with complete or partial

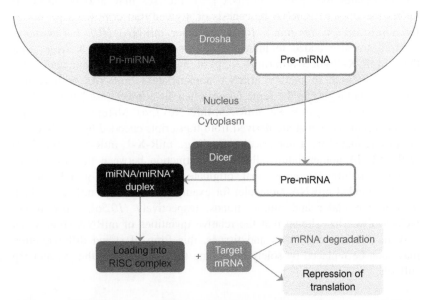

FIGURE 1   miRNA Biogenesis. In the nucleus, pri-miRNA transcripts are either transcribed by RNA polymerase II from independent genes or represent introns of protein-coding genes. Drosha mediates the processing of a pri-miRNA to a pre-miRNA. Following Drosha cleavage, the pre-miRNA is transported to the cytoplasm. Once in the cytoplasm, Dicer processes the pre-miRNA to produce a miRNA duplex derived from the 5′ and 3′ regions of the precursor. Alternatively, the two strands of the miRNA duplex may be distinguished by the nomenclature miRNA and miRNA*, which indicate the major and the minor strands respectively. Mature miRNAs are loaded into the RISC complex to target specific mRNA molecules for degradation or inhibition of translation.

complementarity to sequences in the 3′ untranslated region (3′-UTR) of target mRNAs and induce mRNA translational repression or instability by deadenylation and degradation *(55)*.

Different mature miRNA species can be produced from a single pre-miRNA molecule, as distinct miRNAs are generated from the 3′ and 5′ arms of the pre-miRNA duplex. In addition, a given mature miRNA may comprise a distribution of sizes centered around 22 nt rather than a discrete length. The variation in mature miRNA length is due to 3′ or 5′ end posttranscriptional modifications that include addition or deletion of nucleotides *(10,57,58)*, which have been shown to affect miRNA stability and function. Addition or deletion of nucleotides to the 5′ end of the mature miRNA can have significant effects on miRNA function by shifting the sequence of the seed region, which is the +2 to +8 nucleotide position from the 5′ end of the miRNA that determines the mRNA target within the RISC *(59)*. Profiling approaches therefore need to distinguish among pri-miRNAs, pre-miRNAs, and mature miRNAs and take into account mature miRNA sequence variations.

miRNAs are usually designated with a three letter species prefix (e.g., hsa-miR-X for *Homo sapiens*; mmu-miR-X for *Mus musculus*) and a number that designates the specific miRNA *(56)*. Prefixes may also be added to reflect the stage of miRNA biogenesis of a given transcript (e.g., pri-mir-X for a pri-miRNA; pre-mir-X for a precursor miRNA) *(10)*. Furthermore, suffixes are also added to indicate whether a mature miRNA arose from the 3′ (miR-X-3p) or 5′ arm (miR-X-5p) of the pre-miRNA hairpin, and a capital "R" is used in mature miRNA nomenclature. miRNAs with related sequences belonging to the same family often have lower case letters following the name (e.g., miR-Xa; miR-Xb, miR-Xc) *(56)*. Mature miRNAs with identical sequences that are derived from transcripts encoded by multiple loci are differentiated by a numerical suffix (e.g., miR-X-1, miR-X-2, miR-X-3) *(10)*. miRNA nomenclature allows for distinction between the two single-stranded mature sequences that originate from the two strands of the double-stranded miRNA molecule, for example, miR-X and miR-X*, which indicate the major and minor strands, respectively *(10,56)*. Since recent NGS analysis has shown that the relative quantities of miRNA strands can vary in different cell types and tissues, this form of strand differentiation may soon be replaced by solely describing the strands using the -5p and -3p suffixes.

3 CONSIDERATIONS FOR miRNA PROFILING

Properties specific to miRNA molecules must be taken into account during miRNA profiling. The short length of mature miRNAs (approximately 22 nt) is not sufficient for annealing of traditional primers commonly used to generate complementary DNA (cDNA) during reverse transcription (RT) for subsequent downstream profiling. Mature miRNAs also lack a poly(A) tail for use as a universal primer binding site or in selective enrichment processes. This has implications, given that miRNAs must be selectively detected in a background of diverse RNA molecules, including pri-miRNAs and pre-miRNAs that contain the same sequence as the mature miRNA *(10,58)*. miRNAs are heterogeneous in their GC content, resulting in variations in their melting temperatures (Tm). miRNAs within a family can differ by a single nucleotide. Additionally, as mentioned in Section 2, posttranscriptional modifications can result in mature miRNAs of different lengths centered around 22 nt *(10,57,58)*.

There has been a rapid advancement in the analysis platforms available for miRNA profiling and analysis in recent years. The three major approaches to detect miRNA expression are RT quantitative PCR (qPCR), hybridization-based microarrays, and RNA-sequencing (RNA-seq) (Table 1). The method of choice depends on the specific aims of a given experiment and local expertise, as well as the budget and access to specific equipment. qPCR methods allow for detection and quantitation of miRNAs through methods that are sensitive and specific, and allow for small quantities of starting material, which is particularly important in situations where samples are limited. Hybridization-based methods involve incubating labeled miRNAs with custom-made microarrays or commercial high-density oligonucleotide microarrays *(60)*. Microarray technology allows for the detection and quantitation of thousands of miRNAs simultaneously but is limited by high background levels due to cross-hybridization, a limited dynamic range due to both background and saturation of signals, and a reliance upon existing knowledge about genome sequences *(7,60,61)*. Moreover, miRNAs that display sequence length variations or base substitution differences go undetected by these approaches *(7,57)*. Recently developed RNA-seq technologies have revolutionized our ability to characterize and quantify transcriptomes by enabling RNA analysis through cDNA sequencing on a massive scale *(61)*. RNA-seq approaches provide a more precise measurement of transcript levels and their isoforms than microarray technology *(60,61)*. RNA-seq technologies provide single-base resolution and have limited background signal compared with microarrays, as the cDNA sequence reads are unambiguously mapped to unique regions of the reference genome. RNA-seq does not have an upper quantification limit, resulting in a large dynamic range of expression levels *(60)*. RNA-seq results correlate well with qPCR assays and demonstrate high levels of reproducibility for both technical and biological replicates *(60)*. Perhaps, the most exciting aspect of next-generation RNA-seq technology is the detection of both known and novel transcripts, which is particularly important in an era when novel genes and their splice variants are continually being described. Additionally, RNA-seq is particularly useful when working with poorly characterized genomes from different species. The extensive volume of data generated by RNA-seq, however, requires substantial data storage systems and computational expertise and support.

TABLE 1 Examples of miRNA Profiling Platforms

Method	Comments	Source	Platform
qPCR	Highly specific and sensitive	Qiagen	miScript miRNA PCR system
	Low quantities of starting material required	Exiqon	miRCURY LNA Universal RT microRNA PCR system
	Broad dynamic range of detection	Life Technologies	Taqman small RNA assays
	Customizable		
	Cost effective		
	Low throughput		
qPCR arrays	Highly specific	Qiagen	miScript miRNA PCR arrays
	Low quantities of starting material required	Exiqon	miRNA qPCR panels
	Broad dynamic range of detection	Life Technologies	Taqman low-density miRNA arrays
	Customizable		
	Cost effective		
	Medium throughput		
miRNA microarray	Less specific than qPCR or RNA-seq	Agilent	SurePrint microRNA microarray
	Limited dynamic range of detection	Life Technologies	NCode miRNA microarrays
	High background	Affymetrix	GeneChip microRNA array
	Cost effective	Exiqon	miRCURY LNA microRNA arrays
	High throughput		
	Relies on existing knowledge on genome sequence		
RNA-seq	Highly specific and sensitive	Illumina	TruSeq small RNA sequencing
	Broad dynamic range	Roche	454 Sequencing
	Low background	Life Technologies	SOLID next-generation sequencing
	Expensive		
	Extremely high throughput		
	Identifies known and novel miRNAs		
	Bioinformatically challenging		

4   miRNA EXTRACTION

High-quality miRNA isolation is a fundamental step in the analytical process. miRNAs can be successfully extracted and purified from cell lines and a variety of tissue specimens and biofluids, including blood products and urine

(62,63). Cell lines and tissue samples yield far greater amounts of miRNA than plasma or urine, which contain high levels of endogenous ribonuclease (RNase) enzymes that degrade RNA molecules even in very small quantities *(10,64)*. Generally speaking, the principles for miRNA isolation are similar to those for total RNA isolation, except that miRNA purification protocols are modified to retain or enrich the fraction of small RNA molecules less than 200 nt. As with extracting total RNA, extreme care must be taken to avoid degradation of the RNA sample by exposure to RNase enzymes *(65)*. Samples for RNA processing should be harvested as rapidly as possible to protect against RNase activity and to prevent unwanted changes in gene expression. Samples should be frozen immediately at $-80\,°C$, placed in a suitable RNA stabilization solution such as RNAlater for tissue samples (Qiagen, Life Technologies, and Sigma Aldrich) and RNAprotect Cell Reagent for cultured cells (Qiagen), or lysed and homogenized immediately upon harvesting in the presence of RNase-denaturing buffers. RNA work should be performed in a designated area of the laboratory using dedicated equipment and pipettors with nuclease-free aerosol-resistant tips. RNase decontamination solutions, such as RNaseZap (Life Technologies) or RNaseKiller (Qiagen) should be used to remove RNase contamination from bench surfaces and laboratory equipment. Aseptic technique is recommended when working with RNA samples, and powder-free latex or vinyl gloves should be worn and changed frequently to prevent the introduction of RNase contamination. Samples and reagents should be prepared on ice to inhibit RNase activity and sterile, disposable, certified RNase-free plasticware should be used.

A variety of isolation kits are commercially available that involve solvent-based miRNA extraction, followed by solid-phase purification on columns *(62)*. The miRVana miRNA Isolation kit (Life Technologies) utilizes an organic extraction procedure followed by purification on a glass fiber filter to isolate total RNA ranging from 10 nt to multiple kilobases from cells and fresh or frozen tissue samples. The miRNeasy range of miRNA isolation kits (Qiagen) employ a combined phenol/guanidine-based lysis procedure, followed by a silica-membrane-based purification of total RNA from 18 nt upward. The miRNeasy mini Kit isolates total RNA including the small RNA fraction from cultured cells, laser capture microdissected specimens, and fresh or frozen tissue samples, while the miRNeasy Serum/Plasma Kit copurifies total RNA and small RNAs from serum/plasma or urine samples. The lysis buffer provided with these kits enables sample lysis, RNase activity inhibition, and cellular DNA and protein removal by organic extraction.

Histopathology archives of FFPE tissue samples are valuable resources for retrospective studies of disease. The miRNeasy FFPE kit is available from Qiagen to purify total RNA together with the small RNA fraction from FFPE tissue specimens. As nucleic acids in FFPE samples are usually heavily fragmented and chemically modified by formaldehyde, they are often a lower molecular weight than those obtained from fresh or frozen tissue samples. Prior to miRNA isolation using the miRNeasy FFPE kit, the FFPE tissue specimens are treated with a solution provided with the kit that reverses formaldehyde modification as much as possible without further RNA degradation. Proteinase K is contained within the lysis buffer to release RNA from the tissue sections. A short high-temperature incubation step partially reverses formalin cross-linking of the released nucleic acids and is followed by a DNase treatment to eliminate genomic DNA, and subsequent ethanol precipitation and purification on columns. Using either the miRVana miRNA Isolation kit or the miRNeasy range, further enrichment of the small RNA species (<200 nt) may be performed to increase the sensitivity of downstream analyses.

The concentration of the RNA sample is determined by measuring the absorbance at 260 nm in a spectrophotometer. Because larger RNA species will dominate the absorbance measurement, it is not recommended to quantify purified small RNA fractions by spectrophotometry. RNA concentration and quality should be assessed using the total RNA preparation before enriching for the small RNA fraction. Based on the fact that a 40-μg/mL sample of pure RNA has an absorbance of 1 at 260 nm, the concentration of the sample can be calculated (65). The ratio of the absorbance at 260 and 280 nm is used to determine the quality of the preparation. A ratio of between 1.8 and 2.1 is indicative of high-quality RNA (65). Recently developed NanoDrop spectro-photometers (Thermo Scientific) are useful for the measurement of limited sample preparations as they accurately determine RNA concentration and quality in samples volumes as small as 0.5 μL. The integrity and size distribution of total RNA can be determined using the Bioanalyzer 2100 (Agilent Technologies) or by agarose gel electrophoresis and ethidium bromide staining. In a high-quality RNA sample, the 28s ribosomal RNA (rRNA) and 18s rRNA bands should appear as sharp peaks on the Bioanalyzer or sharp bands on a gel. The desirable ratio of 28s rRNA to 18s rRNA is approximately 2:1. The Bioanalyser 2100 also provides an RNA Integrity Number that should be close to 10, especially if the RNA samples are to be considered for use in downstream applications such as microarrays or RNA-seq.

5   qPCR ANALYSIS OF miRNAs

Quantitation of the mature miRNA in its functionally active form is often per-formed rather than that of the pri-miRNA or pre-miRNA. However, it is sometimes of interest to investigate pri-miRNA expression in order to eluci-date transcriptional control of miRNA expression. In addition, combined dif-ferential expression analysis of pri-miRNAs, pre-miRNAs, and mature miRNAs may determine whether altered mature miRNA expression is a result of regulation in miRNA biogenesis. Many commercially available assays for qPCR analysis of pri- and pre-miRNAs are available. The first step in qPCR analysis of miRNAs involves the RT of RNA to generate a cDNA sequence. Because of the small size of mature miRNAs and their lack of a poly(A) tail, certain considerations need to be taken into account for primer design for mature miRNA analysis by qPCR. One method involves tailing of miRNAs with a common sequence and performing RT using a universal primer. One such approach involves the addition of a poly(A) tail to the mature miRNA molecule, for example, Qiagen's miScript PCR System *(66)*. Total RNA containing the small RNA fraction is used as a starting material. Mature miRNAs are polyadenylated by poly(A) polymerase and reverse transcribed into cDNA in a first-strand cDNA synthesis reaction using oligo-dT primers containing a universal tag on the 5′ end. miRNA abundance is then detected by SYBR green qPCR using a miRNA-specific forward primer and a universal reverse primer. As the oligo-dT primer detects all poly(A)-tailed miRNA molecules during the RT reaction, this assay enables detection of multiple miRNAs within a single cDNA preparation. The miRCURY LNA Universal RT micro-RNA PCR system (Exiqon) also employs a universal RT approach involving the addition of a 5′ universal tagged poly(A) tail, followed by qPCR with miRNA-specific locked nucleic acid (LNA)-enhanced forward and reverse primers. LNA is a synthetic RNA/DNA analog characterized by increased thermostability of nucleic acid duplexes. As each incorporated LNA monomer increases the Tm of a given primer *(58)*, this technology overcomes PCR specificity and sensitivity issues due to Tm variations resulting from miRNAs that are heterogeneous in their GC content.

Another approach for detecting miRNA levels by qPCR is through the use of stem-loop RT primers that are specific for the 3′ end of miRNAs *(67)*. Taqman Small RNA assays (Life Technologies) are based on this approach. Following first-strand cDNA synthesis using a stem-loop primer, the RT cDNA product is quantified by Taqman PCR using an miRNA-sequence-specific forward primer, a reverse primer, and a Taqman probe. Because this approach utilizes a sequence-specific stem-loop RT primer, a separate RT reaction is required for quantitation of each miRNA of interest. However, there are advantages to using the stem-loop RT primer. By annealing a short RT priming sequence to the 3′ end of the miRNA, better specificity for discriminating similar mature miRNA sequences is achieved *(58,67,68)*. In addition, the double-stranded stem-loop structure inhibits hybridization of the RT primer to pre-miRNAs and other long RNAs. Further, the base stacking of the stem-loop enhances the stability of miRNA and DNA duplexes, improving the efficiency of the RT reaction. Lastly, the stem-loop structure when unfolded adds sequence downstream of the miRNA after RT, resulting in a longer RT template more suitable for TaqMan assay design *(68)*.

qPCR-based platforms that allow simultaneous profiling of hundreds of miRNAs from the one sample using customizable or predesigned microfluidic cards/plates are also available (Table 1). Examples of these include miScript miRNA PCR Arrays (Qiagen), miRNA PCR panels (Exiqon), and TaqMan Low Density miRNA Arrays (Life Technologies). These PCR-based arrays enable quantification of up to 752 human miRNAs from small quantities of starting total RNA. Following RT, the resulting cDNA templates are mixed with a PCR master mix and added to the microfluidic array cards or multiwell plates prior to standard qPCR amplification and analysis.

In order to control for variations in RNA input or RT efficiency when performing qPCR using either individual miRNA assays or PCR-based arrays, normalization to endogenous control genes is recommended. Inclusion of controls enables normalization of qPCR results for relative quantitation analysis. An ideal control demonstrates gene expression that is highly abundant and constant across cell types and tissues. Endogenous control assays may be custom designed or are available predesigned commercially and include members of the snRNA (e.g., U6 snRNA (26,69)) and snoRNA (e.g., snoRNA202 (70)) families. In addition to low-medium throughput miRNA profiling, qPCR is usually the method of choice for validating results from microarray analysis (Section 6) or RNA-seq profiling (Section 7) of miRNAs.

6   MICROARRAY PROFILING OF miRNAs

miRNA microarray technology is based on complementary hybridization of labeled miRNAs to an array of immobilized synthetic oligonucleotide probes on a gene chip. This technology allows for high-throughput analysis of miRNA expression and is cheaper to perform than RNA-seq. However, it is not as sensitive as qPCR or RNA-seq approaches. As it is based on current knowledge on genome sequence, miRNA microarray profiling does not permit the identification of novel previously uncharacterized miRNAs. Various commercial platforms are available for miRNA microarray profiling (Table 1). These platforms employ different approaches to label miRNAs for subsequent hybridization to the DNA probes on the microarray chip. One approach involves the enzymatically catalyzed ligation of a fluorophore-conjugated oligonucleotide to the 3′ end of the miRNA molecule using T4 RNA ligase prior to hybridization (SurePrint microRNA microarray; Agilent). Because Dicer processing during miRNA biogenesis results in the exposure of a 5′ phosphate on the mature miRNA that could lead to circularization during the ligation reaction, procedures involving T4-mediated ligation are preceded with a dephosphorylation step to remove the 5′ phosphate.

Alternative microarray approaches are available that involve the 3′ tailing of the miRNA molecule with poly(A), followed by labeling and hybridization. The NCode miRNA microarray platform (Life Technologies) involves the addition of the poly(A) tail to the miRNA prior to ligation of fluorescent dye molecules and subsequent hybridization to the antisense miRNA probes on the microarray chip. In the case of Affymetrix GeneChip microRNA arrays, the poly(A)-tailed miRNAs are labeled with a biotinylated signal molecule. A limitation of hybridization-based microarrays is that because of the small size of miRNAs, the Tm of the probes has a wide range that may diminish specificity and/or sensitivity for miRNAs with a low GC content (10,56).

As a result, it may be difficult to use a temperature that suits the annealing of all the probes to their target miRNAs during hybridization. To overcome the issue of diminished specificity, LNA probes may be used to increase the Tm value (miRCURY LNA microRNA arrays, Exiqon).

7   SMALL RNA-SEQ

Next-generation RNA-seq provides a more precise measurement of levels of transcripts and their isoforms than other methods and is not limited by variability in Tm, small variations in miRNA sequences among families or post-transcriptional modifications. NGS technologies rely on a combination of procedures involving template preparation, sequencing, and imaging, followed by sequence alignment to a reference genome and data analysis (71). In the case of miRNAs, adapter sequences are firstly ligated to the 5' and 3' ends of the mature miRNA molecule, taking advantage of the natural structure common to most mature miRNA molecules of a 3'-hydroxyl group and 5'-phosphate that result from cleavage by the enzyme Dicer (56). Adapter ligation is followed by an RT reaction, a PCR amplification, and a purification step. Sample quality should be assessed on the Agilent Bioanalyzer 2100 at each step of the cDNA library preparation to ensure that the integrity of the samples is sufficient for sequencing. The high-quality cDNA template preparation is attached to a solid support surface and this immobilization of spatially separated template sites allows for thousands of sequencing reactions to be performed simultaneously in order to obtain millions of short-sequence reads (60,71). Both Illumina and Roche platforms (Table 1) sequence by synthesis. The Illumina platform supports massively parallel sequencing using a reversible terminator-based cyclic method that comprises nucleotide incorporation, fluorescence imaging, and cleavage. A fluorescently labeled terminator is imaged as each dNTP is added and then removed to allow incorporation of the next base. Since all four reversible terminator-bound dNTPs are present during each sequencing cycle, natural competition minimizes incorporation bias. By contrast, Roche sequencing technology involves pyrosequencing, where each nucleotide addition results in the activation of luciferase activity that produces light that is detected during the imaging process. As this technique does not require terminator removal at each step, increased sequencing speed can be achieved (56). The SoLID platform sequences by ligation. Primers hybridize to the adapter sequences on template DNA and a set of four fluorescently labeled probes compete for ligation to the sequencing primer.

NGS machines usually produce raw data files in FASTA or FASTQ format, which contain millions of short-sequence reads. A significant level of computational expertise is thus required for the analysis of NGS small RNA-seq datasets, with many command line and web-based analysis platforms available (Table 2). Following sequencing, quality is assessed in order to determine whether the sequencing run was successful. Resources for

TABLE 2 Analysis Software Available for Small RNA-seq Data Analysis

Resource	URL	Tools
Galaxy	https://usegalaxy.org/	Sequencing data quality assessment
FASTX-Toolkit	http://hannonlab.cshl.edu/ fastx_toolkit/index.html	Sequencing data quality assessment and preprocessing, for example, adapter sequence removal, sequence trimming, and conversion of sequence reads to reverse complement
FastQC	http://www.bioinformatics. babraham.ac.uk/projects/fastqc/	Sequence data quality assessment
BWA	http://bio-bwa.sourceforge.net/	Alignment of sequence reads to a reference genome
Bowtie	http://bowtie-bio.sourceforge.net/ index.shtml	Alignment of sequence reads to a reference genome
TopHat	http://tophat.cbcb.umd.edu/	Alignment of sequence reads to a reference genome
UCSC genome browser	http://genome.ucsc.edu/cgi-bin/ hgGateway	Visualization of aligned sequence reads in the genome of interest
Integrative Genomics Viewer	http://www.broadinstitute.org/igv/	Visualization of aligned sequence reads in the genome of interest Integration with clinical and phenotypic data
Gene Expression Omnibus	http://www.ncbi.nlm.nih.gov/geo/	Publicly available functional genomics repository
Cufflinks	http://cufflinks.cbcb.umd.edu/ index.html	Differential expression analysis
Gobyweb	http://campagnelab.org/software/ gobyweb/	Differential expression analysis

quality control assessment are available from web sites including Galaxy *(72)*, FASTX-Toolkit, and FastQC. These web sites provide useful tools that determine whether there are any problems with the datasets before commencing with further downstream analysis. In addition, FASTX-Toolkit enables file preprocessing, including shortening of reads to remove barcodes or noise (FASTQ/A Trimmer), conversion of sequence reads to the reverse complement (FASTQ/A Reverse-Complement), or removal of adapter sequences from the sequence reads (FASTQ/A Clipper). Following quality control, sequence reads are then aligned to a reference genome to produce a genome-wide transcription map using bioinformatics tools such as BWA *(73)*, Bowtie *(74)*, or TopHat *(75)*. Aligned files may be visualized using the UCSC genome browser *(76,77)* or the Integrative Genomics Viewer (IGV) *(78)*. The UCSC genome browser provides a web-based tool for the display of any requested portion of the genome at the desired scale. Multiple annotation tracks of sequence data are publicly available on the web site. Alternatively, users may add their own NGS files to the browser as custom tracks for visualization, or custom tracks of interest that have been published in the literature and made available through resources such as the NCBI Gene Expression Omnibus database *(79)*. The IGV is freely available for download to personal computers as a desktop application, and also enables visualization of genomic data from public sources as well as visualization of users own datasets. In addition, the IGV allows for integration of clinical and phenotypic data. Finally, relative quantitation of differential miRNA expression in terms of the number of sequence reads for a particular miRNA relative to the total reads in a sample can be determined using differential expression platforms such as Cufflinks *(80)* and Gobyweb *(81)*. Following differential expression analysis, differentially regulated miRNAs identified by RNA-seq analysis are usually validated by qPCR approaches.

Bioinformatics analysis of the sequence reads identifies both known and novel miRNAs and provides precise sequence information, which is important to distinguish between miRNA molecules that differ by as little as a single nucleotide. Not all novel miRNAs identified by small RNA-seq are genuine functional mature miRNAs. A number of criteria have been suggested before a novel miRNA may be considered genuine and entered into the miRBase miRNA repository, including a sequence length of approximately 22 nt, conservation across species, a genomic origin that predicts a hairpin pre-miRNA sequence and identification in the data reads that correspond to both the -5p -3p arms of the pre-miRNA hairpin *(82,83)*.

8   *IN SILICO* miRNA ANALYSIS RESOURCES

A complete understanding of the function of miRNAs and their role in biological processes requires elucidation of the mRNA targets that they regulate. Having identified differentially expressed miRNAs through qPCR, microarray or RNA-seq approaches, a wide variety of computational resources exist for *in silico* data mining in relation to miRNA expression patterns, miRNA target prediction, and the identification of miRNA–disease linkages *(59)* (Table 3). miRBase is the key repository of miRNA sequences and nomenclature. miRBase *(84)* provides information about predicted precursor hairpin sequences and experimentally identified mature miRNA sequences. It also provides links to miRNA target prediction databases and

TABLE 3 A Selection of *In Silico* miRNA Analysis Resources

Resource	URL	Tools
miRBase	http://www.mirbase.org	A comprehensive database of miRNA nomenclature and sequences Links to target prediction web sites and RNA-seq datasets.
TargetScan	http://www.targetscan.org/	A tool for prediction miRNA targets in multiple species
PicTar	http://pictar.mdc-berlin.de/	A tool for prediction of miRNA targets in multiple species
Tarbase	http://www.diana.pcbi. upenn.edu/tarbase.html	A database of experimentally verified miRNA targets
microRNA.org	http://www.microrna.org/ microrna/home.do	A database of miRNA targets and experimentally validated tissue expression levels
miRNAMap	http://mirnamap.mbc.nctu. edu.tw/	A database of miRNA targets and experimentally validated tissue expression levels
mir2Disease	http://www.mir2disease.org/	Comprehensive resource of miRNA dysregulation in human disease
mirWalk	http://www.umm.uni- heidelberg.de/apps/zmf/ mirwalk/	miRNA target prediction and associations with pathways and disease
Magia	http://gencomp.bio.unipd.it/ magia/start/	Integrative analysis of miRNA and gene-expression networks
mirConnex	http://mirconnx.csb.pitt.edu/ job_config	Integrative analysis of miRNA and gene-expression networks
Cytoscape	http://www.cytoscape.org/	Network data integration and analysis

RNA-seq datasets deposited in the NCBI Gene Expression Omnibus repository *(10,54,79)*. Target prediction tools involve mRNA target prediction based on an inputted miRNA sequence or accession number *(85)*. Numerous target prediction resources are available. TargetScan identifies mRNA targets in many species by detecting the presence of conserved 7mer and 8mer sites that match the seed region of each miRNA *(86)*. Similar to TargetScan, PicTar uses genome-wide alignment to predict target mRNAs in different species *(87)*. The false-positive discovery rates for PicTar and TargetScan have been reported to be between 20% and 30% with between 80% and 90% overlap in target predictions between the two for human targets *(88)*.

Many online resources collect experimentally confirmed miRNA–mRNA interactions. One of these is Tarbase *(89)*, which collects interactions reported in the literature from a variety of experimental techniques. Numerous other online resources exist that contain information on experimentally validated interactions and expression patterns including microRNA.org *(90)* and miRNAMap *(91)*. Online tools can be used to study a disease state of interest including miR2disease *(92)*, which is a curated database for miRNA dysregulation in human disease. miR2disease is updated frequently and contains information on miRNA–disease relationships, including expression patterns of miRNAs in specific disease states as well as the method used for detection of miRNA expression. miR2disease also contains information on experimentally verified miRNA target genes as well as links to other miRNA databases. miRwalk *(93)* is another comprehensive database of experimentally verified miRNA interactions associated with pathways and diseases. Information is provided on the chromosomal location of the miRNA as well as links to the original published study. Integrative analysis of miRNA expression profiles with large-scale mRNA expression datasets may also be performed using *in silico* tools such as MAGIA *(94)* and mirConnX *(95)*, which enable target prediction, integrated analysis of expression profiles, posttranscriptional regulatory network browsing, functional annotation, and enrichment analysis in combination with the network visualization tool Cytoscape *(96)*.

9   miRNA FUNCTIONAL ANALYSIS

In order to determine the function of miRNAs identified by the miRNA profiling techniques outlined above, various analytical tools are available to perform loss- and gain-of-function studies. The anti-miR (Life Technologies) and the miScript miRNA inhibitor (Qiagen) ranges consist of a panel of single-stranded molecules for specific inhibition of miRNAs currently listed in miRBase. These RNA inhibitors can be transfected into cells in order to assess the effect of decreased miRNA expression on cell phenotype and function. Furthermore, the effect of miRNA inhibition on global gene expression can be studied using microarray or NGS approaches in order to identify target mRNAs regulated by an miRNA of interest. Gain-of-function studies can also be performed by transfecting miRNA mimic molecules directly into cells. miRNA mimics in the form of pre-miR miRNA precursors (Life Technologies) or miScript miRNA mimics (Qiagen) are small, chemically modified double-stranded RNAs that mimic endogenous miRNAs and enable functional analysis by upregulation of miRNA activity. Loss- or gain-of-function studies may also be performed to investigate the effect of miRNA expression on putative mRNA targets identified by bioinformatics analysis. To directly assess miRNA binding to their target mRNAs, the ability of miRNA mimics or inhibitors to regulate expression from a luciferase reporter construct (e.g., pMIR-REPORT miRNA reporter system, Life Technologies), in which the 3′-UTR of the target mRNA has been subcloned, can be tested. Changes in target protein expression may be validated by Western blot analysis. Such experiments provide insight into the functional consequences of altered miRNA expression.

10 RECENT DEVELOPMENTS IN THE APPLICATION OF miRNA PROFILING TO CANCER RESEARCH

The role of miRNAs in cancer development and progression has been widely reported (11,97–99). Generally speaking, miRNA profiling and functional studies in cancer research can be divided into (i) those that focus on deciphering the mechanistic role of miRNA dysregulation in the tumor phenotype with the ultimate aim of developing therapeutic strategies to target these mechanisms and (ii) those that involve biomarker discovery to identify miRNA profiles associated with disease type or predicted response to therapy. Various forms of the miRNA profiling methods outlined above have been used to describe miRNAs whose expression is altered in cancer cell lines, tumor tissue, and plasma or serum samples from cancer patients. Using a combination of qPCR and miRNA microarrays, numerous studies have delineated miRNAs that are differentially expressed in cancer of the breast (34,37), lung (42,43,46,47), pancreas (100), liver (40,41), and in B-cell lymphoma (101). Additionally, dysregulated miRNAs have been associated with treatment outcome. By combining miRNA microarray analysis with stem-loop qPCR validation, Yang et al. described significantly decreased let-7i expression in chemotherapy-resistant epithelial ovarian cancer (102). Tomimura et al. showed that increased miR-21 expression in hepatocellular carcinoma leads to increased resistance to the antitumor effect of combination therapy involving interferon-α and 5-fluorouracil (40). Increased miR-21 expression has also been reported in colon adenocarcinoma and is associated with poor survival and therapeutic outcome (39).

The significant body of work involving miRNA profiling of numerous tumor types has led to the generation of a substantial body of evidence on dysregulated miRNAs in human cancers. Subsequently, there has been a dedicated effort focused on understanding the functional role of such altered miRNA expression in disease progression and outcome. miRNAs with a demonstrated functional importance in cancer can be subdivided into oncogenic miRNAs (oncomiRs) and tumor suppressor miRNA (TS-miR) (48,49). Much effort has been made to elucidate the functional consequences of oncomiR and TS-miR expression in order to unveil potential therapeutic targets for disease management. Significant progress has been reported in our understanding of the functional role of specific miRNAs and regulation of their targets. By applying miRNA profiling, target identification, and functional analysis, Liu et al. have identified miR-31 as an oncomiR in lung cancer (43). First, miRNA microarrays and qPCR were performed to identify differences in miRNA expression in lung cancer tissue compared with adjacent normal tissue in mouse models of lung carcinoma. miR-31 was shown to be significantly increased in the murine lung cancer tissue and the finding was confirmed using human tissue samples. Functional analysis indicated that miR-31 inhibition decreased lung cancer cell growth and tumorigenicity. Using the bioinformatics resources TargetScan, Pictar, and miRanda, Liu *et al.* also identified the tumor suppressor genes, large tumor suppressor 2 (LATS2) and the PP2A regulatory subunit B alpha isoform (PPP2R2A) as targets of miR-31, and validated these targets using 3'-UTR luciferase-binding assays and by demonstrating inverse expression of miR-31 and LATS2 or PPP2R2A in mouse and human lung cancers *(43)*. In another study, Volinia *et al.* used microarrays to identify dysregulated miRNAs in samples from six solid tumor types, namely lung, breast, stomach, prostate, colon, and pancreatic *(35)*. They defined a solid tumor signature of overexpression of miRNAs, including miR-17-5p, miR-20a, miR-21, miR-92, miR106a, and miR-155. Using TargetScan, the retinoblastoma 1 tumor suppressor gene and transforming growth factor, beta receptor II were identified as predicted targets. The mRNA targets were confirmed experimentally using 3'-UTR luciferase assays and by monitoring target protein expression to support a functional role for the role of miRNAs in solid cancer pathogenesis. Using a similar approach involving combined miRNA profiling, target identification, and functional analysis, decreased expression of the miRNA let-7a has been suggested to contribute to the development of prostate cancer *(103)*. More recently, RNA-seq analysis has shed further light on how miRNA expression contributes to cancer progression, in particular, in the lung. Using a mouse model of lung adenocarcinoma, Valdmanis *et al.* utilized small RNA-seq to show increased expression of a cluster of miRNAs at the Dlk1-Dio3 locus, some of which are known to be involved in key cancer-associated pathways *(46)*. Perdomo *et al.* identified an association of miR-4423 with lung cancer in primates by utilizing a small RNA-seq approach *(47)*.

Based on their disease association and functional contribution to carcinogenesis, miRNAs are attractive therapeutic targets for cancer treatment. The rationale for targeting oncomiRs that are upregulated in cancer involves the use of oligonucleotides that block their activity. Likewise, the principle behind targeting TS-miRs is to replenish their expression *(104)*. Encouraging results from preclinical models have resulted in miRNAs entering clinical trials for therapies in some disease settings. For example, an effective and safe approach for sequence-specific antagonism of miR-122, which is implicated in Hepatitis C virus (HCV) infection, in the livers of nonhuman primates has been demonstrated *in vivo (105,106)*. Based on these studies, a miR-122 inhibitor, Miravirsen, is currently in Phase 2A clinical trials in humans for chronic HCV infection *(107)*. Miravirsen treatment demonstrated good tolerability in 36 patients and a dose-dependent reduction in HCV RNA levels that endured beyond the end of active therapy was reported *(107)*. Although not yet in clinical trials, select miRNAs have emerged as potential clinical targets for cancer treatment. miR-34 has a well-defined role as a tumor suppressor and miRNA replacement therapy by means of miRNA mimics has shown promise in animal studies *(108)*. Like all therapeutics, the major barriers to targeting miRNAs in humans are ensuring high bioavailability, achieving target specificity, and reducing toxicity. The use of nanoparticles to target miRNA is a potential option for the treatment of cancers. Recently, a miR-155-targeting nanoparticle containing miR-155 antisense nucleotides, was shown to reduce the growth of B-cell lymphoma in a murine model of disease *(109)*, suggesting a promising therapeutic option for lymphoma and leukemia.

In relation to biomarker discovery, many studies have focused on identifying cancer-associated biomarkers in plasma or serum as a noninvasive means for diagnosis or predictive of response to therapy. A good example of this is miR-22, the plasma levels of which were used to distinguish lung cancer patients from healthy individuals by qPCR, thereby supporting its potential role as a diagnostic biomarker that could be further evaluated in clinical trials *(44)*. In prostate cancer, circulating levels of miR-375 and miR-141 have been reported to correlate with disease progression and lymph node metastasis *(53)*. miR-1, mir-92a, miR-133a, and miR-133b were identified as the most important diagnostic markers for breast cancer detection in a recent study by Chan *et al. (51)*. In terms of utilizing miRNAs as predictive biomarkers for response to therapy, qPCR on circulating levels of miR-125b are predictive of the chemotherapy response in breast cancer *(36)*. The translation of these findings into either noninvasive rapid diagnostic tools or as viable therapeutic targets depends on additional carefully designed and executed profiling studies.

11 CONCLUSIONS

Herein, we have described numerous approaches that may be employed to detect miRNAs, investigate their differential expression, and assess their functional properties in a wide variety of sample types. These include molecular profiling approaches such as qPCR, microarray, and next-generation RNA-seq analysis, as well as computational approaches to predict miRNA targets and investigate miRNA association with disease states. We have also described loss- and gain-of-function methods used to elucidate the functional importance of a given miRNA, together with tools that may be used to confirm direct miRNA:mRNA targeting. The methods of choice for an investigator depend on multiple factors, including the specific aims of a given experiment, the availability of local expertise, budget considerations, and access to specific platforms. The recent advances in RNA-seq technology have provided researchers with an exciting opportunity for the specific and sensitive high-throughput detection and quantification of known and previously unidentified miRNAs. As the cost of these techniques is falling, RNA-seq will become a more accessible platform for rapid and large-scale miRNA profiling, offering the potential to provide invaluable insight into the role of these key regulatory molecules a variety of settings. It is clear that the elucidation of the biological, pathological, and clinical roles of miRNA regulation, expression, and functional properties has greatly contributed to our understanding of pathogenesis. Continued investigation involving both miRNA discovery as well as miRNA targeting will be of great importance to the translation of miRNA biology into clinical applications for diagnosis of disease and novel therapeutics.

REFERENCES

1. Derrien, T.; Johnson, R.; Bussotti, G.; Tanzer, A.; Djebali, S.; Tilgner, H.; Guernec, G.; Martin, D.; Merkel, A.; Knowles, D. G.; Lagarde, J.; Veeravalli, L.; Ruan, X.; Ruan, Y.; Lassmann, T.; Carninci, P.; Brown, J. B.; Lipovich, L.; Gonzalez, J. M.; Thomas, M.; Davis, C. A.; Shiekhattar, R.; Gingeras, T. R.; Hubbard, T. J.; Notredame, C.; Harrow, J.; Guigo, R. *Genome Res.* **2012**, *22*, 1775–1789.

2. Djebali, S.; Davis, C. A.; Merkel, A.; Dobin, A.; Lassmann, T.; Mortazavi, A.; Tanzer, A.; Lagarde, J.; Lin, W.; Schlesinger, F.; Xue, C.; Marinov, G. K.; Khatun, J.; Williams, B. A.; Zaleski, C.; Rozowsky, J.; Roder, M.; Kokocinski, F.; Abdelhamid, R. F.; Alioto, T.; Antoshechkin, I.; Baer, M. T.; Bar, N. S.; Batut, P.; Bell, K.; Bell, I.; Chakrabortty, S.; Chen, X.; Chrast, J.; Curado, J.; Derrien, T.; Drenkow, J.; Dumais, E.; Dumais, J.; Duttagupta, R.; Falconnet, E.; Fastuca, M.; Fejes-Toth, K.; Ferreira, P.; Foissac, S.; Fullwood, M. J.; Gao, H.; Gonzalez, D.; Gordon, A.; Gunawardena, H.; Howald, C.; Jha, S.; Johnson, R.; Kapranov, P.; King, B.; Kingswood, C.; Luo, O. J.; Park, E.; Persaud, K.; Preall, J. B.; Ribeca, P.; Risk, B.; Robyr, D.; Sammeth, M.; Schaffer, L.; See, L. H.; Shahab, A.; Skancke, J.; Suzuki, A. M.; Takahashi, H.; Tilgner, H.; Trout, D.; Walters, N.; Wang, H.; Wrobel, J.; Yu, Y.; Ruan, X.; Hayashizaki, Y.; Harrow, J.; Gerstein, M.; Hubbard, T.; Reymond, A.; Antonarakis, S. E.; Hannon, G.; Giddings, M. C.; Ruan, Y.; Wold, B.; Carninci, P.; Guigo, R.; Gingeras, T. R. *Nature* **2012**, *489*, 101–108.

3. Cheng, J.; Kapranov, P.; Drenkow, J.; Dike, S.; Brubaker, S.; Patel, S.; Long, J.; Stern, D.; Tammana, H.; Helt, G.; Sementchenko, V.; Piccolboni, A.; Bekiranov, S.; Bailey, D. K.; Ganesh, M.; Ghosh, S.; Bell, I.; Gerhard, D. S.; Gingeras, T. R. *Science* **2005**, *308*, 1149–1154.

4. Carninci, P.; Kasukawa, T.; Katayama, S.; Gough, J.; Frith, M. C.; Maeda, N.; Oyama, R.; Ravasi, T.; Lenhard, B.; Wells, C.; Kodzius, R.; Shimokawa, K.; Bajic, V. B.; Brenner, S. E.; Batalov, S.; Forrest, A. R.; Zavolan, M.; Davis, M. J.; Wilming, L. G.; Aidinis, V.; Allen, J. E.; Ambesi-Impiombato, A.; Apweiler, R.; Aturaliya, R. N.; Bailey, T. L.; Bansal, M.; Baxter, L.; Beisel, K. W.; Bersano, T.; Bono, H.; Chalk, A. M.; Chiu, K. P.; Choudhary, V.; Christoffels, A.; Clutterbuck, D. R.; Crowe, M. L.; Dalla, E.; Dalrymple, B. P.; de Bono, B.; Della Gatta, G.; di Bernardo, D.; Down, T.; Engstrom, P.; Fagiolini, M.; Faulkner, G.; Fletcher, C. F.; Fukushima, T.; Furuno, M.; Futaki, S.; Gariboldi, M.; Georgii-Hemming, P.; Gingeras, T. R.; Gojobori, T.; Green, R. E.; Gustincich, S.; Harbers, M.; Hayashi, Y.; Hensch, T. K.; Hirokawa, N.; Hill, D.; Huminiecki, L.; Iacono, M.; Ikeo, K.; Iwama, A.; Ishikawa, T.; Jakt, M.; Kanapin, A.; Katoh, M.; Kawasawa, Y.; Kelso, J.; Kitamura, H.; Kitano, H.; Kollias, G.; Krishnan, S. P.; Kruger, A.; Kummerfeld, S. K.; Kurochkin, I. V.; Lareau, L. F.; Lazarevic, D.; Lipovich, L.;

Liu, J.; Liuni, S.; McWilliam, S.; Madan Babu, M.; Madera, M.; Marchionni, L.; Matsuda, H.; Matsuzawa, S.; Miki, H.; Mignone, F.; Miyake, S.; Morris, K.; Mottagui-Tabar, S.; Mulder, N.; Nakano, N.; Nakauchi, H.; Ng, P.; Nilsson, R.; Nishiguchi, S.; Nishikawa, S.; Nori, F.; Ohara, O.; Okazaki, Y.; Orlando, V.; Pang, K. C.; Pavan, W. J.; Pavesi, G.; Pesole, G.; Petrovsky, N.; Piazza, S.; Reed, J.; Reid, J. F.; Ring, B. Z.; Ringwald, M.; Rost, B.; Ruan, Y.; Salzberg, S. L.; Sandelin, A.; Schneider, C.; Schonbach, C.; Sekiguchi, K.; Semple, C. A.; Seno, S.; Sessa, L.; Sheng, Y.; Shibata, Y.; Shimada, H.; Shimada, K.; Silva, D.; Sinclair, B.; Sperling, S.; Stupka, E.; Sugiura, K.; Sultana, R.; Takenaka, Y.; Taki, K.; Tammoja, K.; Tan, S. L.; Tang, S.; Taylor, M. S.; Tegner, J.; Teichmann, S. A.; Ueda, H. R.; van Nimwegen, E.; Verardo, R.; Wei, C. L.; Yagi, K.; Yamanishi, H.; Zabarovsky, E.; Zhu, S.; Zimmer, A.; Hide, W.; Bult, C.; Grimmond, S. M.; Teasdale, R. D.; Liu, E. T.; Brusic, V.; Quackenbush, J.; Wahlestedt, C.; Mattick, J. S.; Hume, D. A.; Kai, C.; Sasaki, D.; Tomaru, Y.; Fukuda, S.; Kanamori-Katayama, M.; Suzuki, M.; Aoki, J.; Arakawa, T.; Iida, J.; Imamura, K.; Itoh, M.; Kato, T.; Kawaji, H.; Kawagashira, N.; Kawashima, T.; Kojima, M.; Kondo, S.; Konno, H.; Nakano, K.; Ninomiya, N.; Nishio, T.; Okada, M.; Plessy, C.; Shibata, K.; Shiraki, T.; Suzuki, S.; Tagami, M.; Waki, K.; Watahiki, A.; Okamura-Oho, Y.; Suzuki, H.; Kawai, J.; Hayashizaki, Y.; The FANTOM Consortium, RIKEN Genome Exploration Research Group and Genome Science Group. *Science* **2005**, *309*, 1559–1563.

5. Consortium, E. P.; Bernstein, B. E.; Birney, E.; Dunham, I.; Green, E. D.; Gunter, C.; Snyder, M. *Nature* **2012**, *489*, 57–74.

6. Wahlestedt, C. *Nat. Rev. Drug Discov.* **2013**, *12*, 433–446.

7. Pagani, M.; Rossetti, G.; Panzeri, I.; de Candia, P.; Bonnal, R. J.; Rossi, R. L.; Geginat, J.; Abrignani, S. *Immunol. Rev.* **2013**, *253*, 82–96.

8. Lee, R. C.; Feinbaum, R. L.; Ambros, V. *Cell* **1993**, *75*, 843–854.

9. Wightman, B.; Ha, I.; Ruvkun, G. *Cell* **1993**, *75*, 855–862.

10. Pritchard, C. C.; Cheng, H. H.; Tewari, M. *Nat. Rev. Genet.* **2012**, *13*, 358–369.

11. Lu, J.; Getz, G.; Miska, E. A.; Alvarez-Saavedra, E.; Lamb, J.; Peck, D.; Sweet-Cordero, A.; Ebert, B. L.; Mak, R. H.; Ferrando, A. A.; Downing, J. R.; Jacks, T.; Horvitz, H. R.; Golub, T. R. *Nature* **2005**, *435*, 834–838.

12. Djuranovic, S.; Nahvi, A.; Green, R. *Science* **2012**, *336*, 237–240.

13. Lim, L. P.; Glasner, M. E.; Yekta, S.; Burge, C. B.; Bartel, D. P. *Science* **2003**, *299*, 1540.

14. Reinhart, B. J.; Slack, F. J.; Basson, M.; Pasquinelli, A. E.; Bettinger, J. C.; Rougvie, A. E.; Horvitz, H. R.; Ruvkun, G. *Nature* **2000**, *403*, 901–906.

15. Chen, X. *Science* **2004**, *303*, 2022–2025.

16. Houbaviy, H. B.; Murray, M. F.; Sharp, P. A. *Dev. Cell* **2003**, *5*, 351–358.

17. Brennecke, J.; Hipfner, D. R.; Stark, A.; Russell, R. B.; Cohen, S. M. *Cell* **2003**, *113*, 25–36.

18. Xu, P.; Vernooy, S. Y.; Guo, M.; Hay, B. A. *Curr. Biol.* **2003**, *13*, 790–795.

19. O'Carroll, D.; Mecklenbrauker, I.; Das, P. P.; Santana, A.; Koenig, U.; Enright, A. J.; Miska, E. A.; Tarakhovsky, A. *Genes Dev.* **2007**, *21*, 1999–2004.

20. Cobb, B. S.; Nesterova, T. B.; Thompson, E.; Hertweck, A.; O'Connor, E.; Godwin, J.; Wilson, C. B.; Brockdorff, N.; Fisher, A. G.; Smale, S. T.; Merkenschlager, M. *J. Exp. Med.* **2005**, *201*, 1367–1373.

21. Li, Q. J.; Chau, J.; Ebert, P. J.; Sylvester, G.; Min, H.; Liu, G.; Braich, R.; Manoharan, M.; Soutschek, J.; Skare, P.; Klein, L. O.; Davis, M. M.; Chen, C. Z. *Cell* **2007**, *129*, 147–161.

22. Rodriguez, A.; Vigorito, E.; Clare, S.; Warren, M. V.; Couttet, P.; Soond, D. R.; van Dongen, S.; Grocock, R. J.; Das, P. P.; Miska, E. A.; Vetrie, D.; Okkenhaug, K.; Enright, A. J.; Dougan, G.; Turner, M.; Bradley, A. *Science* **2007**, *316*, 608–611.

23. Thai, T. H.; Calado, D. P.; Casola, S.; Ansel, K. M.; Xiao, C.; Xue, Y.; Murphy, A.; Frendewey, D.; Valenzuela, D.; Kutok, J. L.; Schmidt-Supprian, M.; Rajewsky, N.; Yancopoulos, G.; Rao, A.; Rajewsky, K. *Science* **2007**, *316*, 604–608.

24. O'Neill, L. A.; Sheedy, F. J.; McCoy, C. E. *Nat. Rev. Immunol.* **2011**, *11*, 163–175.

25. O'Connell, R. M.; Rao, D. S.; Baltimore, D. *Annu. Rev. Immunol.* **2012**, *30*, 295–312.

26. O'Connell, R. M.; Taganov, K. D.; Boldin, M. P.; Cheng, G.; Baltimore, D. *Proc. Natl. Acad. Sci. U. S. A.* **2007**, *104*, 1604–1609.

27. Taganov, K. D.; Boldin, M. P.; Chang, K. J.; Baltimore, D. *Proc. Natl. Acad. Sci. U. S. A.* **2006**, *103*, 12481–12486.

28. Junker, A. *FEBS Lett.* **2011**, *585*, 3738–3746.

29. Kurowska-Stolarska, M.; Alivernini, S.; Ballantine, L. E.; Asquith, D. L.; Millar, N. L.; Gilchrist, D. S.; Reilly, J.; Ierna, M.; Fraser, A. R.; Stolarski, B.; McSharry, C.; Hueber, A. J.; Baxter, D.; Hunter, J.; Gay, S.; Liew, F. Y.; McInnes, I. B. *Proc. Natl. Acad. Sci. U. S. A.* **2011**, *108*, 11193–11198.

30. Nakasa, T.; Miyaki, S.; Okubo, A.; Hashimoto, M.; Nishida, K.; Ochi, M.; Asahara, H. *Arthritis Rheum.* **2008**, *58*, 1284–1292.

31. Luo, X.; Yang, W.; Ye, D. Q.; Cui, H.; Zhang, Y.; Hirankarn, N.; Qian, X.; Tang, Y.; Lau, Y. L.; de Vries, N.; Tak, P. P.; Tsao, B. P.; Shen, N. *PLoS Genet.* **2011**, *7*, e1002128.

32. Dai, R.; Zhang, Y.; Khan, D.; Heid, B.; Caudell, D.; Crasta, O.; Ahmed, S. A. *PLoS One* **2010**, *5*, e14302.

33. Calin, G. A.; Dumitru, C. D.; Shimizu, M.; Bichi, R.; Zupo, S.; Noch, E.; Aldler, H.; Rattan, S.; Keating, M.; Rai, K.; Rassenti, L.; Kipps, T.; Negrini, M.; Bullrich, F.; Croce, C. M. *Proc. Natl. Acad. Sci. U. S. A.* **2002**, *99*, 15524–15529.

34. Iorio, M. V.; Ferracin, M.; Liu, C. G.; Veronese, A.; Spizzo, R.; Sabbioni, S.; Magri, E.; Pedriali, M.; Fabbri, M.; Campiglio, M.; Menard, S.; Palazzo, J. P.; Rosenberg, A.; Musiani, P.; Volinia, S.; Nenci, I.; Calin, G. A.; Querzoli, P.; Negrini, M.; Croce, C. M. *Cancer Res.* **2005**, *65*, 7065–7070.

35. Volinia, S.; Calin, G. A.; Liu, C. G.; Ambs, S.; Cimmino, A.; Petrocca, F.; Visone, R.; Iorio, M.; Roldo, C.; Ferracin, M.; Prueitt, R. L.; Yanaihara, N.; Lanza, G.; Scarpa, A.; Vecchione, A.; Negrini, M.; Harris, C. C.; Croce, C. M. *Proc. Natl. Acad. Sci. U. S. A.* **2006**, *103*, 2257–2261.

36. Wang, H.; Tan, G.; Dong, L.; Cheng, L.; Li, K.; Wang, Z.; Luo, H. *PLoS One* **2012**, *7*, e34210.

37. Tavazoie, S. F.; Alarcon, C.; Oskarsson, T.; Padua, D.; Wang, Q.; Bos, P. D.; Gerald, W. L.; Massague, J. *Nature* **2008**, *451*, 147–152.

38. Michael, M. Z.; O'Connor, S. M.; van Holst Pellekaan, N. G.; Young, G. P.; James, R. J. *Mol. Cancer Res.* **2003**, *1*, 882–891.

39. Schetter, A. J.; Leung, S. Y.; Sohn, J. J.; Zanetti, K. A.; Bowman, E. D.; Yanaihara, N.; Yuen, S. T.; Chan, T. L.; Kwong, D. L.; Au, G. K.; Liu, C. G.; Calin, G. A.; Croce, C. M.; Harris, C. C. *JAMA* **2008**, *299*, 425–436.

40. Tomimaru, Y.; Eguchi, H.; Nagano, H.; Wada, H.; Tomokuni, A.; Kobayashi, S.; Marubashi, S.; Takeda, Y.; Tanemura, M.; Umeshita, K.; Doki, Y.; Mori, M. *Br. J. Cancer* **2010**, *103*, 1617–1626.

41. Ji, J.; Yamashita, T.; Budhu, A.; Forgues, M.; Jia, H. L.; Li, C.; Deng, C.; Wauthier, E.; Reid, L. M.; Ye, Q. H.; Qin, L. X.; Yang, W.; Wang, H. Y.; Tang, Z. Y.; Croce, C. M.; Wang, X. W. *Hepatology* **2009**, *50*, 472–480.

42. Takamizawa, J.; Konishi, H.; Yanagisawa, K.; Tomida, S.; Osada, H.; Endoh, H.; Harano, T.; Yatabe, Y.; Nagino, M.; Nimura, Y.; Mitsudomi, T.; Takahashi, T. *Cancer Res.* **2004**, *64*, 3753–3756.

43. Liu, X.; Sempere, L. F.; Ouyang, H.; Memoli, V. A.; Andrew, A. S.; Luo, Y.; Demidenko, E.; Korc, M.; Shi, W.; Preis, M.; Dragnev, K. H.; Li, H.; Direnzo, J.; Bak, M.; Freemantle, S. J.; Kauppinen, S.; Dmitrovsky, E. *J. Clin. Invest.* **2010**, *120*, 1298–1309.
44. Shen, J.; Liu, Z.; Todd, N. W.; Zhang, H.; Liao, J.; Yu, L.; Guarnera, M. A.; Li, R.; Cai, L.; Zhan, M.; Jiang, F. *BMC Cancer* **2011**, *11*, 374.
45. Rani, S.; Gately, K.; Crown, J.; O'Byrne, K.; O'Driscoll, L. *Cancer Biol. Ther.* **2013**, *14*, 1104–1112.
46. Valdmanis, P. N.; Roy-Chaudhuri, B.; Kim, H. K.; Sayles, L. C.; Zheng, Y.; Chuang, C. H.; Caswell, D. R.; Chu, K.; Zhang, Y.; Winslow, M. M.; Sweet-Cordero, E. A.; Kay, M. A. *Oncogene* **2013**.
47. Perdomo, C.; Campbell, J. D.; Gerrein, J.; Tellez, C. S.; Garrison, C. B.; Walser, T. C.; Drizik, E.; Si, H.; Gower, A. C.; Vick, J.; Anderlind, C.; Jackson, G. R.; Mankus, C.; Schembri, F.; O'Hara, C.; Gomperts, B. N.; Dubinett, S. M.; Hayden, P.; Belinsky, S. A.; Lenburg, M. E.; Spira, A. *Proc. Natl. Acad. Sci. U. S. A.* **2013**, *110*, 18946–18951.
48. Nicoloso, M. S.; Spizzo, R.; Shimizu, M.; Rossi, S.; Calin, G. A. *Nat. Rev. Cancer* **2009**, *9*, 293–302.
49. Ryan, B. M.; Robles, A. I.; Harris, C. C. *Nat. Rev. Cancer* **2010**, *10*, 389–402.
50. Xie, Y.; Todd, N. W.; Liu, Z.; Zhan, M.; Fang, H.; Peng, H.; Alattar, M.; Deepak, J.; Stass, S. A.; Jiang, F. *Lung Cancer* **2010**, *67*, 170–176.
51. Chan, M.; Liaw, C. S.; Ji, S. M.; Tan, H. H.; Wong, C. Y.; Thike, A. A.; Tan, P. H.; Ho, G. H.; Lee, A. S. *Clin. Cancer Res.* **2013**, *19*, 4477–4487.
52. Corcoran, C.; Friel, A. M.; Duffy, M. J.; Crown, J.; O'Driscoll, L. *Clin. Chem.* **2011**, *57*, 18–32.
53. Mahn, R.; Heukamp, L. C.; Rogenhofer, S.; von Ruecker, A.; Muller, S. C.; Ellinger, J. *Urology* **2011**, *77*, 1265.e9–1265.e16.
54. Ghildiyal, M.; Zamore, P. D. *Nat. Rev. Genet.* **2009**, *10*, 94–108.
55. Krol, J.; Loedige, I.; Filipowicz, W. *Nat. Rev. Genet.* **2010**, *11*, 597–610.
56. Guerau-de-Arellano, M.; Alder, H.; Ozer, H. G.; Lovett-Racke, A.; Racke, M. K. *J. Neuroimmunol.* **2012**, *248*, 32–39.
57. Git, A.; Dvinge, H.; Salmon-Divon, M.; Osborne, M.; Kutter, C.; Hadfield, J.; Bertone, P.; Caldas, C. *RNA* **2010**, *16*, 991–1006.
58. Benes, V.; Castoldi, M. *Methods* **2010**, *50*, 244–249.
59. Tan Gana, N. H.; Victoriano, A. F.; Okamoto, T. *Genes Cells* **2012**, *17*, 11–27.
60. Wang, Z.; Gerstein, M.; Snyder, M. *Nat. Rev. Genet.* **2009**, *10*, 57–63.
61. Ozsolak, F.; Milos, P. M. *Nat. Rev. Genet.* **2011**, *12*, 87–98.
62. Accerbi, M.; Schmidt, S. A.; De Paoli, E.; Park, S.; Jeong, D. H.; Green, P. J. *Methods Mol. Biol.* **2010**, *592*, 31–50.
63. Doleshal, M.; Magotra, A. A.; Choudhury, B.; Cannon, B. D.; Labourier, E.; Szafranska, A. E. *J. Mol. Diagn.* **2008**, *10*, 203–211.
64. Jensen, S. G.; Lamy, P.; Rasmussen, M. H.; Ostenfeld, M. S.; Dyrskjot, L.; Orntoft, T. F.; Andersen, C. L. *BMC Genomics* **2011**, *12*, 435.
65. Smith, S. M.; Murray, D. W. *Methods Mol. Biol.* **2012**, *823*, 119–138.
66. Ach, R. A.; Wang, H.; Curry, B. *BMC Biotechnol.* **2008**, *8*, 69.
67. Chen, C.; Ridzon, D. A.; Broomer, A. J.; Zhou, Z.; Lee, D. H.; Nguyen, J. T.; Barbisin, M.; Xu, N. L.; Mahuvakar, V. R.; Andersen, M. R.; Lao, K. Q.; Livak, K. J.; Guegler, K. J. *Nucleic Acids Res.* **2005**, *33*, e179.
68. Schmittgen, T. D.; Lee, E. J.; Jiang, J.; Sarkar, A.; Yang, L.; Elton, T. S.; Chen, C. *Methods* **2008**, *44*, 31–38.

69. Cheng, Y.; Kuang, W.; Hao, Y.; Zhang, D.; Lei, M.; Du, L.; Jiao, H.; Zhang, X.; Wang, F. *Inflammation* **2012**, *35*, 1308–1313.
70. Yang, L.; Boldin, M. P.; Yu, Y.; Liu, C. S.; Ea, C. K.; Ramakrishnan, P.; Taganov, K. D.; Zhao, J. L.; Baltimore, D. *J. Exp. Med.* **2012**, *209*, 1655–1670.
71. Metzker, M. L. *Nat. Rev. Genet.* **2010**, *11*, 31–46.
72. Goecks, J.; Nekrutenko, A.; Taylor, J.; Galaxy, T. *Genome Biol.* **2010**, *11*, R86.
73. Li, H.; Durbin, R. *Bioinformatics* **2009**, *25*, 1754–1760.
74. Langmead, B.; Trapnell, C.; Pop, M.; Salzberg, S. L. *Genome Biol.* **2009**, *10*, R25.
75. Kim, D.; Salzberg, S. L. *Genome Biol.* **2011**, *12*, R72.
76. Kent, W. J.; Sugnet, C. W.; Furey, T. S.; Roskin, K. M.; Pringle, T. H.; Zahler, A. M.; Haussler, D. *Genome Res.* **2002**, *12*, 996–1006.
77. Karolchik, D.; Barber, G. P.; Casper, J.; Clawson, H.; Cline, M. S.; Diekhans, M.; Dreszer, T. R.; Fujita, P. A.; Guruvadoo, L.; Haeussler, M.; Harte, R. A.; Heitner, S.; Hinrichs, A. S.; Learned, K.; Lee, B. T.; Li, C. H.; Raney, B. J.; Rhead, B.; Rosenbloom, K. R.; Sloan, C. A.; Speir, M. L.; Zweig, A. S.; Haussler, D.; Kuhn, R. M.; Kent, W. J. *Nucleic Acids Res.* **2013**, *42*, D764–D770.
78. Thorvaldsdottir, H.; Robinson, J. T.; Mesirov, J. P. *Brief. Bioinform.* **2013**, *14*, 178–192.
79. Barrett, T.; Wilhite, S. E.; Ledoux, P.; Evangelista, C.; Kim, I. F.; Tomashevsky, M.; Marshall, K. A.; Phillippy, K. H.; Sherman, P. M.; Holko, M.; Yefanov, A.; Lee, H.; Zhang, N.; Robertson, C. L.; Serova, N.; Davis, S.; Soboleva, A. *Nucleic Acids Res.* **2013**, *41*, D991–D995.
80. Trapnell, C.; Williams, B. A.; Pertea, G.; Mortazavi, A.; Kwan, G.; van Baren, M. J.; Salzberg, S. L.; Wold, B. J.; Pachter, L. *Nat. Biotechnol.* **2010**, *28*, 511–515.
81. Dorff, K. C.; Chambwe, N.; Zeno, Z.; Simi, M.; Shaknovich, R.; Campagne, F. *PLoS One* **2013**, *8*, e69666.
82. Ambros, V.; Bartel, B.; Bartel, D. P.; Burge, C. B.; Carrington, J. C.; Chen, X.; Dreyfuss, G.; Eddy, S. R.; Griffiths-Jones, S.; Marshall, M.; Matzke, M.; Ruvkun, G.; Tuschl, T. *RNA* **2003**, *9*, 277–279.
83. Chiang, H. R.; Schoenfeld, L. W.; Ruby, J. G.; Auyeung, V. C.; Spies, N.; Baek, D.; Johnston, W. K.; Russ, C.; Luo, S.; Babiarz, J. E.; Blelloch, R.; Schroth, G. P.; Nusbaum, C.; Bartel, D. P. *Genes Dev.* **2010**, *24*, 992–1009.
84. Kozomara, A.; Griffiths-Jones, S. *Nucleic Acids Res.* **2011**, *39*, D152–D157.
85. Watanabe, Y.; Tomita, M.; Kanai, A. *Methods Enzymol.* **2007**, *427*, 65–86.
86. Lewis, B. P.; Burge, C. B.; Bartel, D. P. *Cell* **2005**, *120*, 15–20.
87. Krek, A.; Grun, D.; Poy, M. N.; Wolf, R.; Rosenberg, L.; Epstein, E. J.; MacMenamin, P.; da Piedade, I.; Gunsalus, K. C.; Stoffel, M.; Rajewsky, N. *Nat. Genet.* **2005**, *37*, 495–500.
88. Rajewsky, N. *Nat. Genet.* **2006**, *38 Suppl*, S8–S13.
89. Vergoulis, T.; Vlachos, I. S.; Alexiou, P.; Georgakilas, G.; Maragkakis, M.; Reczko, M.; Gerangelos, S.; Koziris, N.; Dalamagas, T.; Hatzigeorgiou, A. G. *Nucleic Acids Res.* **2012**, *40*, D222–D229.
90. Betel, D.; Wilson, M.; Gabow, A.; Marks, D. S.; Sander, C. *Nucleic Acids Res.* **2008**, *36*, D149–D153.
91. Hsu, S. D.; Chu, C. H.; Tsou, A. P.; Chen, S. J.; Chen, H. C.; Hsu, P. W.; Wong, Y. H.; Chen, Y. H.; Chen, G. H.; Huang, H. D. *Nucleic Acids Res.* **2008**, *36*, D165–D169.
92. Jiang, Q.; Wang, Y.; Hao, Y.; Juan, L.; Teng, M.; Zhang, X.; Li, M.; Wang, G.; Liu, Y. *Nucleic Acids Res.* **2009**, *37*, D98–D104.
93. Dweep, H.; Sticht, C.; Pandey, P.; Gretz, N. *J. Biomed. Inform.* **2011**, *44*, 839–847.

94. Sales, G.; Coppe, A.; Bisognin, A.; Biasiolo, M.; Bortoluzzi, S.; Romualdi, C. *Nucleic Acids Res.* **2010**, *38*, W352–W359.
95. Huang, G. T.; Athanassiou, C.; Benos, P. V. *Nucleic Acids Res.* **2011**, *39*, W416–W423.
96. Smoot, M. E.; Ono, K.; Ruscheinski, J.; Wang, P. L.; Ideker, T. *Bioinformatics* **2011**, *27*, 431–432.
97. Garzon, R.; Fabbri, M.; Cimmino, A.; Calin, G. A.; Croce, C. M. *Trends Mol. Med.* **2006**, *12*, 580–587.
98. Calin, G. A.; Croce, C. M. *Nat. Rev. Cancer* **2006**, *6*, 857–866.
99. Croce, C. M. *Nat. Rev. Genet.* **2009**, *10*, 704–714.
100. Lee, E. J.; Gusev, Y.; Jiang, J.; Nuovo, G. J.; Lerner, M. R.; Frankel, W. L.; Morgan, D. L.; Postier, R. G.; Brackett, D. J.; Schmittgen, T. D. *Int. J. Cancer* **2007**, *120*, 1046–1054.
101. He, L.; Thomson, J. M.; Hemann, M. T.; Hernando-Monge, E.; Mu, D.; Goodson, S.; Powers, S.; Cordon-Cardo, C.; Lowe, S. W.; Hannon, G. J.; Hammond, S. M. *Nature* **2005**, *435*, 828–833.
102. Yang, N.; Kaur, S.; Volinia, S.; Greshock, J.; Lassus, H.; Hasegawa, K.; Liang, S.; Leminen, A.; Deng, S.; Smith, L.; Johnstone, C. N.; Chen, X. M.; Liu, C. G.; Huang, Q.; Katsaros, D.; Calin, G. A.; Weber, B. L.; Butzow, R.; Croce, C. M.; Coukos, G.; Zhang, L. *Cancer Res.* **2008**, *68*, 10307–10314.
103. Dong, Q.; Meng, P.; Wang, T.; Qin, W.; Qin, W.; Wang, F.; Yuan, J.; Chen, Z.; Yang, A.; Wang, H. *PLoS One* **2010**, *5*, e10147.
104. Nana-Sinkam, S. P.; Croce, C. M. *Clin. Pharmacol. Ther.* **2013**, *93*, 98–104.
105. Elmen, J.; Lindow, M.; Schutz, S.; Lawrence, M.; Petri, A.; Obad, S.; Lindholm, M.; Hedtjarn, M.; Hansen, H. F.; Berger, U.; Gullans, S.; Kearney, P.; Sarnow, P.; Straarup, E. M.; Kauppinen, S. *Nature* **2008**, *452*, 896–899.
106. Elmen, J.; Lindow, M.; Silahtaroglu, A.; Bak, M.; Christensen, M.; Lind-Thomsen, A.; Hedtjarn, M.; Hansen, J. B.; Hansen, H. F.; Straarup, E. M.; McCullagh, K.; Kearney, P.; Kauppinen, S. *Nucleic Acids Res.* **2008**, *36*, 1153–1162.
107. Janssen, H. L.; Reesink, H. W.; Lawitz, E. J.; Zeuzem, S.; Rodriguez-Torres, M.; Patel, K.; van der Meer, A. J.; Patick, A. K.; Chen, A.; Zhou, Y.; Persson, R.; King, B. D.; Kauppinen, S.; Levin, A. A.; Hodges, M. R. *N. Engl. J. Med.* **2013**, *368*, 1685–1694.
108. Bader, A. G. *Front. Genet.* **2012**, *3*, 120.
109. Babar, I. A.; Cheng, C. J.; Booth, C. J.; Liang, X.; Weidhaas, J. B.; Saltzman, W. M.; Slack, F. J. *Proc. Natl. Acad. Sci. U. S. A.* **2012**, *109*, E1695–E1704.

Chapter 3

Application of Novel Genotyping Microarray Technologies in Cancer Research

Ricardo Ramos[*,1], Juan Moreno-Rubio[†,1], Enrique Casado[†], Guillermo Reglero[‡,§] and Ana Ramírez de Molina[‡]

[*]*Genomic Unit, Scientific Park of Madrid, Madrid, Spain*
[†]*Medical Oncology, Infanta Sofia Hospital, Madrid, Spain*
[‡]*Molecular Oncology and Nutritional Genomics of Cancer, IMDEA Food Institute CEI UAM + CSIC, Madrid, Spain*
[§]*Instituto de Investigación en Ciencias de la Alimentación (CIAL), CEI UAM + CSIC, Madrid, Spain*

Chapter Outline

1. Equal contribution.

Comprehensive Analytical Chemistry, Vol. 64. http://dx.doi.org/10.1016/B978-0-444-62650-9.00003-8

1   INTRODUCTION

Cancer is the second leading cause of death in industrialized countries, exceeded only by heart disease. Characterized by out-of-control abnormal cell growth, cancer might affect men and women of all ages and races, and actually the term "cancer" refers to a class of diseases, comprising over 100 different types of cancer classified by the organ or type of cell that is initially affected.

Despite the variability found in the different types of cancer including causes, symptoms, prognosis and treatments, a general statement can be made: cancer is a genetic disease. Cancer cells are characterized by the presence of damage/mutations on one or more genes leading to unnatural function.

In addition to cancer, most relevant chronic diseases considered almost as pandemic diseases nowadays are also genetic diseases. Thus, the search of genetic variation associated to human disease is one of the major goals of genetics worldwide. The fine-tuning that genes may display due to a variety of alleles makes relevant to study human DNA variation, to understand, and eventually find treatments for a number of diseases. Continuous development of new technologies for genome analysis is resulting in increasing amount of evidence of the relevance of genetic alterations in chronic diseases and in cancer in particular. By the hand of the discoveries of genetic variations in cancer patients, personalized therapies are being improved. This chapter will review some of the most common techniques used to analyze human variation at the level of DNA with special focus on the introduction and clinical application of the analysis of genomic alterations in cancer.

2   CANCER GENETICS: FROM BIOLOGY TO CLINICAL PRACTICE

2.1   The Discovery of Cancer Genetics

Cancer is caused by genetic alterations that lead to cellular malfunction. This evident statement was unthinkable 200 years ago when physicians could only observe and study macroscopically the tumors after postmortem extraction. Extensive and varied discoveries for decades have conducted to the concept of cancer that we have nowadays. The first assumptions trying to unveil the tumorogenic process were made by Steven Paget in 1889 proposing "The seed and soil hypothesis" (1,2). Paget analyzed more than 1000 autopsies of women with breast cancer and found out nonrandomly defined metastatic patterns. He suggested that tumor cells (seeds) have a specific affinity for certain organs (soil), and metastasis only occurs if the "seed" and "soil" are supported (1,2). In 1910, Peyton Rous discovered that virus can lead to carcinogenesis, after observing a type of sarcoma in chickens that was termed Rous Sarcoma (3,4). Four years later, the German cytologist Theodor Bovery reported that cancer was caused by abnormal chromosomes and proposed the Somatic Mutation

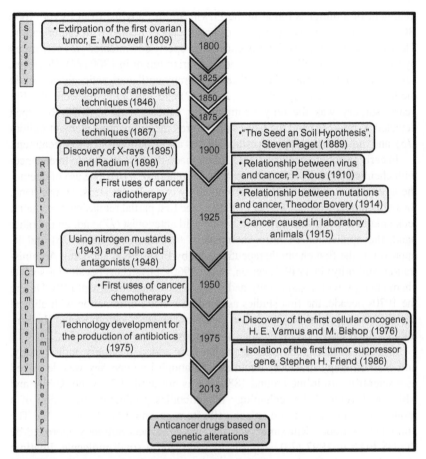

FIGURE 1   Cancer genetics: from biology to clinical practice.

Theory of Cancer *(5,6)*. In 1915, Katsusaburo Yamagiwa and Koichi Ichikawa at Tokyo University induced cancer in laboratory animals by applying coal tar to rabbit skin *(7)*, though it was not till more than 150 years later when the first carcinogen, tobacco, was recognized. In 1976, Harold E. Varmus and J. Michael Bishop described the first cellular oncogene *(8)*, and in 1986, Stephen H. Friend *et al.* isolated the first tumor suppressor gene, *Rb* (for retinoblastoma) *(9)*, one of the first genes related to inherited (familial) form of cancer *(10)*. From this date on, development of new technologies and discoveries increased demonstrating that cancer is a genetic disease (Figure 1).

2.2   The Introduction of Cancer Genetics on Cancer Therapy

Regarding therapy, as it is today, surgery has been for decades the main efficient therapeutic option. Historically, one of the first indications for surgical treatment of a tumor is found in the Edwin Smith Papyrus, from Egypt (1600 BC). Papyrus describes eight cases of breast cancer treated with cauterization. The first-known successful tumor excision properly documented was performed by Ephraim McDowell, who resected an ovarian tumor in 1809 *(11)*. However, surgery was not established in anticancer therapy until anesthetic and antiseptic methods were efficiently settled in 1846 *(12)* and 1867 *(13)*, respectively. For years, surgery was the only treatment for cancer, until in 1895 Roentgen detected X-rays *(14)* and the Curies discovered the radioactive element radium *(15)*, and radiation therapy was established as a new option for cancer treatment.

In early twentieth century, Paul Ehrlich made the first attempt to treat cancer with chemical agents coining the term "chemotherapy" *(16)*. The use of chemotherapy for cancer treatment was promoted by two relevant steps: on one hand, the publication by Sidney Farber in 1948 of the first partial remission of pediatric leukemia in a 4-year-old girl using a folic acid antagonist *(17)* and, on the other hand, the application of nitrogen mustards to treat lymphomas *(18)* and its approval as the first chemotherapeutic agent by FDA (Food and Drug Administration Authority) in 1949. Later on, in the 1960s, physicians began to associate chemotherapy with radiotherapy and surgery for cancer treatment *(19)*. During the 1970s decade, the first studies associating genetic information with clinical decisions were published *(20)*, and in the following years, chromosomal abnormalities gained increased importance as prognostic cancer markers *(21,22)*.

In recent decades, immunotherapy has been added to surgery, radiotherapy, and chemotherapy in cancer treatment. Although immunology was described as a scientific discipline around 1880, it was not until 1975 when Kohler and Milstein developed the technology of antibodies production by fusion of a mouse myeloma and mouse spleen cells from an immunized donor *(23)*. The anti-CD20 antibody Rituximab was the first anticancer antibody approved by the US FDA in 1997 *(24)*. Almost a decade later a small molecule, Imatinib showed a potent effectiveness against a chromosomal abnormality in a chronic myeloid leukemia type *(25)*, demonstrating the relevance of the application of new targeted therapeutic agents. Following this approach, the current trend is focused on the development of new drugs with specificity of action on selected molecular targets of the tumorogenic process in each concrete disease type and stage (Figure 1; Table 1). Furthermore, genetic fingerprints patterns can be recognized across different cancer types, increasingly moving therapeutics from a morphology-based therapy toward a genetic-directed approach.

In this sense, following the discovery of the essential role of genetic alterations in tumor development *(26)*, a new generation of targeted antitumoral drugs is being developed as a complementary approach to conventional treatments, focused on the specific blockage of the molecular effects mediated by these alterations. With this "intelligent drug design" of new antitumoral agents, clinical responsiveness in some cases such as specific non-small-cell lung cancer or melanoma subtypes has resulted impressive, providing evidence of the necessity of identifying subgroups of patients carrying specific

TABLE 1 Main Genetic Determinations with Clinical Relevance in Cancer Therapy

Disease	Genetic Alteration	Molecular Marker	Target	Drug Therapy
Colorectal cancer	Mutation	KRAS	EGFR	Cetuximab, Panitumumab
Melanoma	Mutation	BRAF	V600E mutated BRAF	Vemurafenib
Basal cell carcinoma	Mutation	PTCH, SMO	SMO (hedgehog signaling pathway)	Vismodegib
Breast cancer, gastric cancer	Gene amplification	ERBB2	HER2	Trastuzumab, Lapatinib
Myelofibrosis	Mutation	JAK2	JAK	Ruxolitinib
Chronic myelogenous leukemia	Translocation	BCR-ABL	BCR-ABL (tyrosine-kinase activity)	Imatinib, Dasatinib, Nilotinib, Bosutinib
Gastrointestinal stromal tumor (GIST)	Mutation	c-KIT	c-KIT(tyrosine-kinase activity)	Imatinib, Dasatinib, Sunitinib
Medullary thyroid carcinoma	Mutation	RET	RET, VEGFR2, VEGFR3, EGFR, etc. (tyrosine-kinase activity)	Vandetanib, Cabozantinib
Acute promyelocytic leukemia	Translocation	PML/RARA	PML/RARA	All-*trans* retinoic acid, arsenic trioxide
Non-small cell lung cancer	Mutation	EGFR	EGFR (tyrosine-kinase activity)	Erlotinib, Gefitinib
Non-small cell lung cancer	Translocation	EML4–ALK	ALK (tyrosine-kinase activity)	Crizotinib

genetic abnormalities that might be efficiently targeted with these drugs *(27)*, and illuminating the promise of a genetic/molecular framework for the application of successful strategies in cancer therapy.

2.3   Tumor Genome Analysis Applied to Improve Cancer Therapy

A great amount of evidence has demonstrated the relevance of the determination of genetic alterations of tumor cells and its impact on patient diagnosis, prognosis, and management. Thus, cancer genetic information is positioning itself as an indispensable tool for properly stratifying cancer patients with the aim of administrating the optimal treatment for each patient.

Though a significant amount of genetic alterations have been described in cancer patients in the past years (26,28–30), only a reduced group of them has been extensively studied and currently considered for patient management in the clinical setting (Table 1).

In colorectal cancer, the employment of anti-EGFR (epidermal growth factor receptor) antibodies is part of the backbone of therapeutic options for metastatic colorectal cancer (mCRC), with relevant efficacy in first-, second-, and third-line therapies (31). Antibody binding to the receptor inhibits ligand binding (31). The downstream RAS/RAF/MAPK and PI3K/PTEN/AKT pathways are thus inhibited, resulting in inhibition of cell growth and proliferation, cell-cycle arrest, and/or apoptosis (31). However, this therapy is poorly effective or ineffective in unselected patients (31). Mutations in KRAS (Kirsten rat sarcoma viral oncogene homolog), NRAS (Neuroblastoma), BRAF (v-raf murine sarcoma viral oncogene homolog B), and PIK3CA (phosphatidylinositol-4,5-bisphosphate 3-kinase, catalytic subunit alpha) genes have emerged as predictive factors of low/absent response to EGFR-targeted therapy (31). KRAS is a molecular marker required to select patients eligible for the monoclonal antibodies Cetuximab and Panitumumab. Very recently, it has been shown that NRAS wild-type genotype at exons 2, 3, and 4 is also required for Panitumumab clinical activity (32).

In the case of melanoma, after its identification as an oncogene in 2002, mutant BRAF has become the target of a number of drug discovery programs, primarily aimed at the treatment of late stage or unresectable melanoma (33). Actually, the RAS/RAF/MEK/ERK pathway has been reported to be activated in over 80% of all cutaneous melanomas, making it the focus of many scientific studies in this field. Discoveries of mutations and aberrant expression of components in this cascade, BRAF in particular, provides a deeper understanding of the mechanisms responsible for oncogenesis and permits new therapeutic strategies for this deadly disease (34). The most common mutation from among all reported BRAF mutations is the resultant of the substitution of glutamic acid (E) for valine (V) in codon 600, the BRAF V600E mutation. Other common BRAF mutations in melanoma, found in the same codon, are V600K (about 16% of mutations in melanoma) and V600D/R (3% of all mutations) (35). These less common variants are found at slightly higher rates in melanomas arising in older patients (35). Thus, some of the targeted developed drugs, including the BRAF inhibitors Vemurafenib and Dabrafenib, and the MEK inhibitor Trametinib have shown to improve survival in melanoma patients harboring a BRAF mutation (33). However, while first-generation BRAF inhibitors (like Vemurafenib) initially mediate excellent responses in late stage or unresectable melanoma patients bearing the V600 mutation, resistance usually occurs and patients eventually relapse (33). The current literature for new BRAF inhibitors and therapies reflect the efforts to develop second-generation drugs with the capacity to overcome this resistance and combination treatments that increase the efficiency of current mutant BRAF inhibitors *(33)*.

Besides melanoma, basal cell carcinoma is the most common human malignancy, accounting for 90% of skin cancers *(35)*. Although multiple factors contribute to the risk of developing basal cell carcinoma, the underlying mechanism is based on genetic alterations *(36)*. Mutations in the sonic hedgehog signaling pathway play a key role in the development of basal cell carcinomas *(36)*. Specifically, mutations in the patched 1 gene (*PTCH1*, also known as PTCH or PTC1) and SMO genes (smoothened, frizzled family receptor) cause tumor formation through constitutive activation of the pathway *(36)*. Misregulation of the pathway has also been implicated in the nevoid basal cell carcinoma syndrome and other tumors *(36)*. Regarding the treatment of basal cell carcinoma, in cases in which surgery and radiation therapy are not useful, a new drug targeting the hedgehog pathways, Vismodegib *(37)*, has been recently approved by the FDA.

Amplification of the ERBB2 (v-erb-b2 avian erythroblastic leukemia viral oncogene homolog 2) gene, which encodes human epidermal growth factor receptor 2 (HER2), causes the overexpression of a major proliferative driver for a subset of breast and gastric cancers *(38)*. Treatments for patients with HER2-positive cancer includes the monoclonal antibody Trastuzumab (Herceptin, a humanized monoclonal antibody that binds to HER2, blocking its function and possibly triggering the immune system to attack tumor cells), and in the case of metastatic breast cancer, the tyrosine-kinase inhibitor Lapatinib (Tykerb/Tyverb, a small-molecule inhibitor of HER2 tyrosine-kinase activity) *(38)*. HER2 is a receptor tyrosine kinase that regulates cell growth and differentiation signaling pathways and is significantly overexpressed (up to 100-fold) in around 20% of breast and gastric cancers when compared to normal tissues *(38)*. Despite significant improvement in patient outcome as a result of these therapies (Trastuzumab and Lapatinib), novel strategies are still needed, and resistance to these drugs remains a challenge, particularly in the metastatic setting *(38)*. It is likely that therapeutic resistance will not arise via the same mechanisms in all patients and this heterogeneity amplifies the challenge *(38)*. Myelofibrosis is a hematopoietic stem cell malignancy classified as a myeloproliferative neoplasm *(39)*. Historically, patients with this debilitating disease have had limited treatment options and disease-modifying agents were not available *(39)*. Hematopoietic stem cell transplantation is the only potentially curative therapy, but it is only an option for selected patients *(39)*. The discovery of an activating point mutation in the Janus kinase 2 gene (JAK2V617F) in a significant proportion of patients with Myelofibrosis led to improved understanding of the pathobiology of these disorders and promoted the rapid development of JAK inhibitors *(39)*. The JAK–STAT pathway plays a pivotal role in the differentiation and development of hematopoietic cells and the functioning of the immune system. The JAK family comprises JAK1, JAK2, JAK3, and TYK2 (Tyrosine Kinase 2) *(40)*. Cytokines and growth factors activate the extracellular portion of their cognate receptors *(40)*, which in turn promote the recruitment to the intracellular portion and activation via phosphorylation of JAK proteins. Phosphorylation of JAK proteins, in turn, leads to the phosphorylation and activation of several intracellular downstream signaling molecules including STAT (signal transducer and activator of transcription) proteins *(41)*. The dysregulation of the JAK–STAT signaling pathway in hematopoietic progenitor cells has been implicated in the pathogenesis of Myelofibrosis. JAK2 V617F is the most common mutation occurring in a large proportion (50–60%) of patients with Myelofibrosis. With JAK2-activating mutations, the pathway becomes cytokine- and growth-factor-independent; therefore, even in the absence of these ligands, the intracellular signaling proteins are constitutively active *(42)*. Ruxolitinib is the first-in-class and only JAK inhibitor currently approved by the US FDA for the treatment of patients with Myelofibrosis. Despite not constituting a curative option, Ruxolitinib offers great palliative potential, resulting in a significant reduction of splenomegaly and improving constitutional symptoms in the majority of treated patients, with a significant impact in their quality of life and performance status *(39)*. Additionally, Ruxolitinib is the only agent that has demonstrated a survival benefit in patients with Myelofibrosis *(39)*.

Chronic myelogenous leukemia (CML) is a clonal disorder caused by the malignant transformation of a pluripotent stem cell. It is characterized by the Philadelphia chromosome (Ph), a genetic abnormality which arises from the reciprocal translocation t(9;22) (q34;q11) *(43)*. This translocation fuses the genes encoding BCR (Breakpoint Cluster Region) and ABL1 (c-abl oncogene 1, nonreceptor tyrosine kinase), and results in expression of the constitutively active protein tyrosine kinase BCR–ABL1 *(43)*. The resulting BCR–ABL fusion gene leads to an ABL protein that forms a constitutively active tyrosine-kinase protein, with cell-cycle dysregulation *(44)*. The survival of patients in the chronic and advanced phase was greatly enhanced with the development of targeted therapy, such as the BCR–ABL tyrosine-kinase inhibitor Imatinib mesylate, that was approved by the US FDA in 2001 for advanced stage CML and as first-line treatment in 2002 *(44)*. Unfortunately, patients can become resistant or intolerant to Imatinib and for this reason, a second generation of tyrosine-kinase inhibitors has been recently developed: Two are currently available for first-line therapies, Dasatinib and Nilotinib, which have been associated with a higher response rate, reduced disease progression, and improved tolerability *(44)*. A third tyrosine-kinase inhibitor, Bosutinib, has recently been approved by the FDA only for second (or later) line therapy *(44)*.

Some of the mentioned tyrosine-kinase inhibitors, such as Imatinib, Sasatinib, Nilotinib, and Sunitinib, are currently used to treat other pathologies, such as gastrointestinal stromal tumors (GIST) where mutated forms in the tyrosine-kinase receptor c-KIT have been found *(45)*. The relationship between c-KIT (CD117) and GISTs was first described by Hirota *et al.* in 1998, who sequenced c-KIT complementary DNA from GISTs and found an exon 11 mutation in the region between the transmembrane and tyrosine-kinase domains *(46)*. The tyrosine-kinase receptor KIT functions as a regulator of cell growth *(46)*. The KIT ligand binds and activates two KIT receptors which results in phosphory-lation and activation of signaling pathways, leading to cell proliferation *(46)*. Mutations in KIT result in the activation of KIT in the absence of a ligand, unstoppable signal of cell growth, and consequent tumor formation *(46)*. These mutations are found in 85–90% of GIST tumors *(46)*. Survival outcomes have increased threefold since the introduction of KIT inhibitors in advanced disease and have also largely improved survival in the postoperative setting. Importantly, the mutational profile determines response to KIT inhibitors, with better responses in KIT exon 11 mutations and in the *PGFRA* gene as compared to exon 9 mutations. Specifically, the PDGFRA D842V mutation exhibits a complete resistance to KIT inhibitors *(47)*.

In the case of medullary thyroid carcinoma (MTC), germline mutations of the rearranged during transfection (RET) proto-oncogene have been described *(48)*. In hereditary MTC, these mutations are the most important precipitating events, while in sporadic MTC, the genetic or molecular biomarkers are yet to be established *(48)*. Recent findings of H-RAS mutations in 56% of RET-negative sporadic MTC and the activation of the mammalian target of rapamy-cin intracellular signaling pathway in hereditary MTC suggest that additional or alternative genetic events are important for MTC pathogenesis *(48)*. These findings show that the assessment of RAS mutational status in patients with sporadic RET-negative MTC might be useful to select targeted therapies. Cur-rently, targeted therapies focused on RET inhibition and other tyrosine-kinase receptors involved in angiogenesis have been studied for treatment of MTC *(48)*. In this sense, Andetinib (ZD6474), an inhibitor of vascular endothelial growth factor receptor (VEGFR) 2 and 3, RET, and EGFR, has been recently approved for the treatment of adults with symptomatic or progressive MTC *(48)*. Furthermore, Cabozantinib (XL184), an inhibitor of VEGFR 1 and VEGFR 2, hepatocyte growth factor receptor (MET), and RET, was associated with partial response and stable disease in 29% and 41%, respectively *(48)*, and very recently has shown an outcome benefit in a phase III trial in advanced disease *(49)*. Acute promyelocytic leukemia (APL) is driven by a chromosomal translocation that always involves the retinoic acid (RA) receptor α (RARA) gene on chromosome 17 to create a variety of X-RARA fusions *(50)*. The most common one is the t(15,17) translocation encoding the PML (promyelocytic leukemia)/RARA fusion *(50)*. The product of this translocation, the PML/RARA fusion protein, affects both nuclear receptor signaling and PML body assembly *(51)*. The case of APL must be remarked due to the fortu-itous identification of two clinically effective therapies, all-*trans*-retinoic acid and arsenic trioxide, both of which degrade PML/RARA oncoprotein and, together, have been described as a cure for the disease *(51)*.

Finally, targeted drugs have also emerged in non-small cell lung cancer (NSCLC) therapy due to the inclusion of genetic information in the clinical practice. The EGFR is a transmembrane protein with cytoplasmic kinase activity that transduces important growth factor signals from the extracellular milieu to the cell. EGFR is a key regulator of epithelial cell proliferation and excessive EGFR signaling disbalances cell growth and apoptosis contributing to tumorogenesis in a wide variety of solid tumors including NSCLC *(52)*. Somatic mutations in the tyrosine-kinase domain (exons 18–21) of *EGFR* gene can bring about constitutive activation of EGFR tyrosine-kinase activity *(52)*. Most of these mutations are associated with sensitivity to the small-molecule tyrosine-kinase inhibitors Erlotinib, Efitinib, and Afatinib, which have shown to extend survival in first- and second-line therapies *(52)*. Cancer patients harboring activating EGFR mutations benefit from this tyrosine-kinase inhibitors therapy *(53)*.

Cancer research focused on the development of new and effective targets other than EGFR in NSCLC treatment has not been successful until year 2007, when Soda *et al.* identified the transforming EML4–ALK fusion gene *(54)*. Anaplastic lymphoma kinase (ALK) is a transmembrane receptor tyrosine kinase, member of the insulin receptor superfamily. Echinoderm microtubule-associated protein-like 4 (EML4) is a protein that may modify the assembly dynamics of microtubules. A recurrent chromosome translocation, inv(2)(p21p23), found in NSCLC generates fused mRNA encoding the amino-terminal half of EML4 ligated to the intracellular region of the receptor-type protein tyrosine kinase ALK *(55)*. EML4–ALK protein oligomerizes constitutively in cells through the coiled coil domain within the EML4 region and becomes activated to exert a marked oncogenicity *(55)*. While EML4–ALK fusion was identified in NSCLC, a new ALK tyrosine-kinase inhibitor, Crizotinib, was developed at similar time *(56)*. Prospective clinical trials were designed and started quickly on NSCLC patients with EML4–ALK fusion genes *(56)*. Crizotinib was finally approved by FDA of the United States in 2011 for treatment of NSCLC patients *(56)*.

In summary, several molecular markers based on genetic alterations are currently analyzed in cancer research and treatment, with essential implications for patient management and clinical outcome (Table 1). Focused on determined genetic alterations, targeted therapies are increasingly introduced in the clinical setting. Thus, cancer treatment is moving toward a personalized medicine based on the determination of the specific individual genetic characteristics of patients.

The benefits of these new drugs in oncology are undeniable; the improvement in clinical outcome of specific cancer patients with targeted therapies has been impressive, specially to treat those tumors in which conventional chemotherapy did provide almost no benefit. As an additional advantage, toxicity of targeted drugs is generally lower than that of conventional cancer therapies, avoiding in some cases important side-effects of conventional chemotherapy. Moreover, many of these new therapies are orally administered, which constitute an additional benefit to the quality of life of the patients.

However, despite obtaining sometimes significant increases in response rates, only few complete remissions are observed, usually reaching partial remissions or stabilizations. In addition, precise conditions of tumor samples are required for genetic determinations, which often might result in added difficulties. Finally, extensive research is still required to optimize dosages and combined therapies, not only between conventional and targeted drugs but also among targeted therapies themselves.

Current advances in gene technology constitute the required tool for rapid progress toward the knowledge of cancer genetics, what could lead to truly personalized medicine capable to impact on patients prognosis and quality of life.

3   GENOTYPING TECHNOLOGIES

There are two main perspectives when facing genomic studies on cancer cells. One is the so-called discovery, based on searching for new variants or changes of any kind in the genome or transcriptome of cancer cells, without a previous definition of expected results, at least not with a higher degree of detail. The second approach has much to do with diagnostics and consists on measuring specific changes that are known to be related to disease or response; we usually refer to this approach as resequencing (57,58). Main approaches and related techniques for genome analysis are summarized in Table 2.

3.1   The Study of Defined Variants at the Level of DNA

3.1.1   Variants Discovery by Sequencing

Sequencing can be considered as the milestone from which all our genomics knowledge has been built (59). Sequencing unambiguously identifies any nucleotide by directly copying and reading the bases of a given gene or DNA fragment. From a technical point of view, modern sequencing has remarkably evolved from the first chemical and enzymatic systems developed in the 1970s and which drove the Nobel Prize to their inventors (60). But conceptually, DNA sequencing still relies on the very same basis than the enzymatic method. Thus, a DNA polymerase is used to extend a target template, occasionally stopping the nascent strand by incorporating dideoxy nucleotides. Inclusion of fluorescently labeled dideoxy nucleotides will result in a collection of molecules of different lengths, each labeled with a dye that identifies the last nucleotide incorporated and, hence, its position in the DNA chain. Using a single-base resolution matrix (initially a gel, currently a synthetic matrix mounted onto a capillary), molecules are discriminated by electrophoresis and optically detected, resulting in a series of signals which are

TABLE 2 Main Genotyping Technologies Currently Used in Cancer Research

Sequencing		CE	Discovery + Resequencing	Cornerstone Technique
PCR-based techniques	SnapShot	CE	Resequencing	One base sequencing
	Real-time PCR	MWP	Resequencing	Different platforms amenable for SNP, CNV, gene fusions, etc.
	FRET probes	MWP	Resequencing	
	Molecular beacons	MWP	Resequencing	
	FAS-PCR	MWP	Resequencing	
	castPCR		Resequencing	Minor allele determination
	Digital PCR		Resequencing	Minor allele determination
	HRM	MWP	Discovery or scanning	Heteroduplex analysis, no variant identification
Fragment analysis		CE	Resequencing	Identification purposes
MLPA		CE	Resequencing	Intragenic indels
Microarrays	SNP arrays	ARRAY	Discovery	Pharmacogenomics, LOH
	CGH arrays	ARRAY	Discovery	CNV
Pyrosequencing		MWP	Resequencing	Genotyping and methylation
NGS	DNAseq	PTP FC	Discovery + resequencing	Whole-genome sequencing
				Exome sequencing
				Target resequencing
				CNV analysis
				Methyl-seq
				ChIP-seq

interpreted in terms of order of nucleotides. In addition to reading, every specific nucleotide is assigned to a coefficient that measures the accuracy with which each base has been assigned as compared to the other three bases. This factor termed Q value and defined as $Q = -\log 10P$ (where P is the basecall error probability) gives the confidence or trueness with which each base has been determined within the sequence and is a factor of major importance to assess confidence and ensure reproducibility *(61)*.

As a basic discovery technique, DNA sequencing has been the way by which most genomic variants have been first discovered, up to the point that Human Genome Project could be accomplished by a technical revolution in sequencing technologies *(62,63)*. Development of high-throughput capillary sequencers was the instrument which allowed Celera to enter into the sequencing arena and overpass the international consortium of Human Genome in its effort to sequence the first Human Genome more than a decade ago.

Resequencing of discrete regions of DNA in groups of patients and controls worldwide has provided an enormous amount of information regarding human variation and linkage to disease. At the level of Sanger sequencing, its limits are the length of the sequencing reads and the number of sequencing runs that can be run in parallel. Core laboratories extensively using several Applied Biosystems 3730XL apparatus (Life Technologies) can afford the simultaneous sequencing of hundreds of reactions up to 1 kb long, rendering information in the order of megabases of information, but at a cost of roughly 0.2 cent per base, a cost that has dramatically dropped by means of the recent next-gene sequencing technologies that will be discussed afterwards.

A big deal of the success rate of sequencing reactions relies on an efficient binding of the sequencing oligonucleotide which primes the labeling reaction to its cognate annealing region, before the extension and synthesis of the new strand. A clean and informative reaction is not easily achieved using whole-genome DNA as a template as such, most likely due to difficulties in primer binding, inhibition of polymerases, and the possible presence of by-products in the reaction, which may impair the quality of the sequences. Instead, either cloned or amplified products are used as sequencing templates, making mandatory a previous cloning DNA work or fine adjustment of PCR reactions. This preparative step constitutes a bottleneck for Sanger sequencing, which compromises its high performance. Although sequencing has demonstrated to be a most valuable technology to discover new variants or to scan long DNA fragments in search for changes of any kind, may not be ideal for the analysis of individual polymorphisms, which may be better addressed by other methods.

3.1.2   Alternative Techniques for the Analysis of Defined Variants

Once defined a putative variant to be tested in a population of patients, different techniques may be preferred, directly focused on the base(s) of interest. They are usually performed on genomic DNA without the need of a previous amplification or cloning of specific regions, which make them a very straight and convenient way of analyzing human DNA variation. These alternative techniques would compensate lack-of-discovery for ease of use, reduction of costs, and the use of genomic DNA as the template molecule. A rational combination of sequencing and additional techniques has allowed the launching of international collaborative genotyping efforts such as the HapMap project, which has brought major advances in the knowledge of human variability at the level of DNA focused in haplotypes.

The same principle of sequencing itself may be restricted to single-DNA positions by a technique of single-nucleotide sequencing called SnapShot, a strategy applied for resequencing specific cancer-related mutations such as KRAS in mCRC *(64)*. In this case, only the four dideoxy nucleotides in the absence of standard deoxy forms are used, so amplification cannot progress further than the very first position. Variants are detected by optical measurement in a sequencer. Since a single band is expected, SnapShot may be easily multiplexed accommodating different reactions for individual polymorphisms, under the single condition that product lengths can be distinguished in the capillary electrophoresis matrix.

3.1.3   Real-Time PCR-Based Techniques

The most common methods to detect predefined variants are based on PCR amplification *(65)*. Real-time PCR reactions (rtPCR) incorporate the measurement of DNA amplification in a standard PCR reaction by incorporating fluorescent molecules, allowing the follow-up of amplification process or end-point detection. The simplest rtPCR reaction dedicated to variant analysis would be addressed by the use of a set of three primers, two of them distinguishing between two alleles, the third just serving to maintain PCR progression. The only requisite is that primers must be designed so that a single-base mismatch has a marked influence on annealing. A further refinement of this technique is based upon FRET probes, where a pair of probes labeled in 3' and 5' ends will generate a fluorescent signal only if they are close enough (in a range of some 50 nm), a situation that only happens if they are fully hybridized upon the template. These probes will be designed so that the polymorphic position lies in the region where those probes would connect, so a single mismatch would impair complete annealing and therefore FRET energy transfer, indicating the presence of a genetic alteration at that point, as it has been used for detecting EGFR mutation status in lung cancer patients *(66)*.

However, the most extended way of analyzing individual SNPs in cancer research is by far the use of TaqMan® probes *(67,68)*. This system relies in the use of a set of four oligonucleotides for each polymorphism that is to be detected. Two of the oligonucleotides are standard PCR primers flanking the target position and therefore should be independent of the allele present. The two other oligonucleotides are hydrolysis probes, each labeled with a different dye, which will specifically hybridize with one of the allelic versions of the gene but not the other, providing a fluorescent signal according to the presence of one specific variant. In addition to fluorescent signals generated during PCR progression by each of the TaqMan® probes, allelic discrimination is achieved by plotting end-point fluorescence levels, which should result in the discrimination of a population of patients in a three-group fashion of single positives (homozygous) or double positives (heterozygous), providing the minor allele is sufficiently represented and the population of patients enough covered. Including appropriate controls, the assignation of each individual to their corresponding group is easily performed and a confidence value can be assigned according to how well a DNA sample fits within each of the groups. Combining simplicity of rtPCR with the use of high-throughput equipment such as 384-or 1.536-position thermocyclers, the quick and easy assignment of hundreds to thousands of patients has been a common task in many cancer genomics laboratories (69).

Larger genotyping experiments would need still quicker and cheaper analytical tools. Some of the systems which have been developed in the last few years impact on the reduction of reaction volumes, to minimize costs, and increment the number of reactions that can be run in parallel The Fluidigm® BioMark platform is a clear example of this, reducing volume reactions down to several nL and implementing a platform able to individually amplify 48/96 DNA samples with 48/96 TaqMan®-based assays mentioned above, therefore increasing throughput for a factor of >10 and correspondingly reducing costs and the needs of sample availability, as it has been recently described when studying new prognostic biomarkers in gastric tumors (70,71). BioTrove (now incorporated into Life Technologies) has also developed a miniaturized system to accelerate SNP measurement, known as the Open Array chip. In both cases, the availability of a broad collection of predefined dual color TaqMan® probe human assays is an irreplaceable tool which helps in the setup of most genotyping projects, avoiding the tedious, difficult, and time-wasting activity of optimizing a discriminatory PCR reaction.

Despite simplicity, rtPCR-based genotyping also presents some issues. On one hand, TaqMan® probes must be very precisely designed since they must discriminate between single changes which distinguish the two alleles, a task that is not always easy to achieve. On the other hand, the unlabeled PCR primers of the set must be independent from the searched variant and lay in an absolutely nonvariant DNA region. Allelic plots with a classical arrowhead or the division of allelic groups into two populations are the reflection of such issues which clearly deserve further refinement in PCR conditions and reagent design. PCR reactions have been also used for detecting a different type of DNA alteration consisting in the abnormal fusion of pieces of DNA found in some types of cancer such as prostate or melanoma as well as circulating tumor cells (72,73). Again, a rational design of PCR primers is a key step which supports specific detection of those chimeric genes.

High-Resolution Melting (HRM) is a different technique also based on PCR but excluding the use of TaqMan® probes and the requirement to define a single-target base for analysis. Instead, HRM fixes a test region and scans for possible alterations which may result in heteroduplex formation *(74,75)*. HRM is a quick and inexpensive scanning system very useful to search mutations in defined gene regions where a number of alterations might be found, as it happens in the diagnostics of hereditary cancers *(76,77)*. In addition, a number of alternate techniques such as molecular beacons, dHPLC, mass spectrometric, and others have also been used to analyze DNA variants associated to disease.

A common theme in SNP analysis is the need to associate genotype to phenotype, a task that usually deserves the concurrent analysis of several SNPs at the same time. Thus, some higher throughput systems were developed for genotyping an increased number of polymorphic positions in large cohorts. These are, for instance, based in oligonucleotide ligation followed by capillary electrophoresis (SNPlex) or flow cytometry *(78,79)*, associated to a higher genotyping productivity.

3.1.4   Whole-Genome SNP Analysis

A major breakthrough in DNA research was reached when wide-genome systems were developed. Small-size devices known as microarrays were designed to study DNA variation at a global scale *(80,81)*. These devices consist of a big collection of probes covalently linked to a solid surface that is exposed and hybridized to the test DNA, which has been previously labeled. Probes have been spotted at defined coordinates within the solid surface, so that the specific binding of DNA to each of the probes can be monitored. Controls as well as postmeasurement corrections are needed to calibrate the amount of test DNA bound to each specific probe as well as specificity of hybridization *(82)*. Thus, the end result is a map of variation at a global genome scale. The most remarkable microarrays devices in terms of DNA variation are SNP arrays and CGH arrays. SNP arrays interrogate a high number of thousands of SNPs in every chip, while CGH arrays distribute a collection of probes all throughout the genome. Both serve to construct an elaborate picture of variation of a given DNA sample at different levels. In particular, SNP arrays scan the complete genome in search for single-base variations by the use of probes able to distinguish between alleles and then serve to construct a genotyping map throughout the complete genome. SNP arrays can be used to detect variants in multiple positions linked to response against drug administration or markers that are associated to effectiveness of treatments, a fruitful discipline termed Pharmacogenomics *(83,84)*.

3.2   Analysis of DNA Variation Independent from Single Nucleotide Polymorphisms

In addition to single-nucleotide substitutions or small insertions or deletions detected by DNA sequencing, rtPCR or SNP arrays, different sources of

DNA variation can be equally important. Capillary electrophoresis is able to detect to a high precision and resolution a sort variant that is found in mammalian genomes, based on the presence of short DNA elements that are repeated throughout the genome and present in form of tandems (STRs; short tandem repeats). Out of the different repetitive elements found in genomes, STRs have been mostly associated with individual identification, in forensic medicine, or criminal investigation *(85)*. Capillary electrophoresis is a powerful technique very well issued to analyze changes in this sort of variation, where the number of repeated elements can be taken as a different allele. In this particular case, the range of possible "alleles" is increased so that a relatively reduced set of STRs becomes highly informative marker. Choosing a combination of labeled PCR primers with different dyes and groups of PCR products of different sizes allows the simultaneous detection of more than a dozen markers in a single capillary, making .it a powerful identification technique.

Other important DNA variation that has been associated to cancer is related to larger gain or losses of DNA material, referred to as copy-number variation *(86,87)*. Short scale or intragenic loss of genetic material is commonly addressed by a technique known as multiplex ligation-dependent probe amplification (MLPA) *(88)*. MLPA uses a combination of ligation and amplification to scan for possible deletions within one gene that may impact on gene expression, produce aberrant transcripts, or unravel genetic variations that now would appear like a homozygous phenotype in a context where heterozygosis may have a null impact. Microarray assays are also very well suited to detect these large rearrangements in the DNA. It is the case for CGH arrays, that will result in an accumulation of hybridization signals in regions with extra DNA copies and a reduction in deficient areas, pointing to big deletions or overrepresentation of genomic regions *(89,90)*. Resolution of CGH arrays is estimated to be within 1 Mb for the whole-genome approach, but tuning the number of probes or selecting specific chromosomal regions would result in a higher resolution if needed *(91)*.

SNP arrays result useful to detect DNA rearrangements by an indirect approach. In this case, the result will be measured as a loss of heterozygosity, a statistical value which measures DNA deletions due to statistical deviations from the expected rate in the different alleles. Loss of heterozygosity can also be measured at a shorter scale by capillary electrophoresis, in this case restricted to individual gene analysis *(92)*.

3.3   Variants Found in Small Numbers

With the notable exception of quimerism, germline cells have an obvious restricted outcome for only three possible genotypes of SNPs (AA, Aa, aa). However, cancer samples have an inherent analysis difficulty, due to the cellular nature of the own samples, very often mixed with an undefined number of nontumor cells. Due to the concomitant presence of nonmutated DNA coming from normal tissue or even by the clonal diversity of tumor cells themselves, tumor samples with a specific point mutation will separate from their cognate genotyping group in allelic discrimination plots, will show chromatographic peaks of lower intensity in Sanger sequencing, difficult to be correctly detected against the noise signal, or may have a doubtful (or even incorrect) genotyping result when analyzed by SNP arrays or HRM. Therefore, some other techniques are clearly needed to accurately type the presence of minority variants in tumor-derived DNAs.

Out of them, a powerful strategy is pyrosequencing *(93,94)*, as it has been reported from a comparison of several techniques for detecting KRAS and EGFR mutations. This technique provides a different method for base to base sequencing. Instead of giving the chance to a polymerase to use any of the four nucleotides, each labeled to a different color, pyrosequencing introduces free nucleotides one by one, measuring the signal of base polymerization due to the pyrophosphate release that occurs after the covalent addition of any base. Therefore, a nucleotide that binds in a low percentage due to the presence of a low proportion of mutated molecules would not result concealed by the major base. Moreover, pyrosequencing can become quantitative, a situation very well suited to estimate the clonal proportion of cells harboring a mutation. Pyrosequencing has also been used for epigenetic studies, measuring the degree of DNA methylation following bisulfite-mediated transformation of methyl-Cyt into Ura and detecting the proportion of C or U as a measure of cytosine methylation and hence promoter activity *(95)*. One drawback of pyrosequencing is that only small stretches of DNA can be analyzed every time, so it is a powerful technique for resequencing minor alleles but not very well suited to discover new variants. A careful design of amplification and sequencing primers, as well as the order of dispensation of nucleotides during the sequencing run, is an important parameter that must be finely adjusted to obtain accurate genotyping or methylation results.

Allele-specific PCR is another powerful methodology designed to measure minor variants in DNA. PCR primers are designed to specifically amplify one of the two alleles, taking advantage of the greatest dynamic range of qPCR to detect even very minor populations *(96)*.

Finally, digital PCR is a novel technology also directed to detect minor populations. Using nanoliter-size wells by means of nanofluidic valves, droplet formation, or simple capillarity disposal, digital PCR works by dividing a standard PCR reaction into hundreds to thousands of nanoscale reactions that are followed and measured individually. Target DNAs are severely diluted so that only a limited subset of those microreactors will render a specific product that will be detected. Providing a minor representation of positive wells to ensure statistical evenness, counting the specific positive positions against the total number of possibilities gives an accurate number used to deduce the initial concentration.

3.4 Gene Expression Studies

All the techniques described so far are used to study a relatively static situation, searching for variants present in the germline of patients in familiarly transmitted cancers or changes in tumors which accumulate during malignant evolution and dissemination. However, there is also an area to apply genomic technologies to study gene expression as well. rtPCR is universally used as the gold standard method to measure gene expression, accomplishing ease of use, reduced cost, reproducibility, dynamic range, and simple conceptual analysis. A correct experimental design must consider a detailed selection of reference genes and an accurate optimization of every parameter, a situation that is not always performed *(97)*, but results can be usually sound and robust. The use of rtPCR to measure gene expression has been extensively reviewed in terms of optimization and correct performance and interpretation *(97)*, quite similar to a previous effort to homogenize data coming from microarray technologies. Current knowledge of Gene Ontology and metabolic routes allow the use of phenotype-directed gene panels, a situation that has been recently also adapted to the study of gene profiles at the level of small noncoding regulatory miRNAs *(98)*, one main current focus of cancer research nowadays. Microarrays have been used to discover gene-expression profiles of mRNA or miRNAs that should have been eventually validated by rtPCR methods. OncoDx, Oncochip, or Mamaprint are good examples of such a workflow related to cancer where gene-expression arrays have been used to find a gene signature that has been refined by rtPCR in order to be used as diagnostic or prognostic tool *(99,100)*.

Noteworthy, validation studies made by rtPCR on selected genes found by microarrays technologies have not always been fruitful. A number of issues must be taken into account, such as the differences in dynamic range, differences in target areas within each specific gene, or differences in transcription reaction rates. The higher complexity of human transcriptome, as unraveled by the Encode project *(101)*, also makes difficult to interpret gene-expression plasticity in terms of RNA levels. In any case, gene-expression analysis is one of the most fruitful approaches for the discovery of cancer biomarkers *(102,103)*.

4 NEXT-GENERATION GENOMIC TECHNIQUES

The post-Human Genome era has brought a complete revolution in genomics. The combination of a whole-genome approach with the 1-base resolution of sequencing, together to amazing improvements in optical, computational, and electronics has brought the termed Second Generation of Genomics (or next-generation sequencing, NGS) which is called to overcome previous high-throughput technologies *(104,105)*. Conceptually, NGS can be envisioned quite simply: DNA is sequenced (by the same methods of either pyrosequencing or dideoxy-sequencing already described), but in a parallel fashion, so that a number from millions (Ion Torrent-Life technologies; 454-Roche) pyrosequences to billions of Sanger-type sequences (Illumina) can be achieved within a single run. The cost of sequencing has dropped to less than 0.1 USD per megabase in the most productive sequencers *(106)*, a more than 10,000-fold reduction that has made available highly ambitious projects such as the sequencing of human diversity worldwide *(107)* or the deep mapping of cancer in the International Cancer Atlas Consortium. In addition, NGS skips one of the bottleneck steps of large-scale sequencing, making *in vitro* cloning of molecules instead of physical cloning, avoiding cost and labor-consuming construction of physical libraries.

5 APPLICATIONS OF NOVEL SNP TECHNOLOGIES IN CANCER RESEARCH

Massive parallel sequencing provides different fields for application within cancer research *(108)*. NGS can address genomic analysis of cancer tissues at the genomic level, a level of resolution which other high-throughput technologies such as microarrays cannot afford. On one hand, it serves as a discovery platform. Using the lack-of-hypothesis strategy, NGS will accelerate the discovery of biomarkers by the sequencing of cancer genomes at a higher depth and coverage. On the other hand, genome sequencing can discover new pharmacogenomic profiles which serve to predict the individuals' behavior in terms of expected effectiveness of anticancer treatments or disease outcome *(109–111)*.

Moreover, NGS might be tuned to get a rapid and accurate answer to a genetic question with potential application in the clinical setting. The power of high throughput and cost-effectiveness can be focused to specific regions, coupled to efficient high-throughput amplification or specific hybridization-based capture. A number of companies (Illumina, Rain Dance, Multiplicon, Fluidigm, Life Technologies) offer predefined panels for the study of a defined set of cancer-related genes. SAB Biosciences offers predesigned amplicons to choose from for any set of genes of interest. At this moment, custom-selected panels seem to be the preferred option for most research projects. For instance, Fluidigm, TSCA-Illumina, Agilent-Haloplex, NimbleGen, or iCommunity (Life) offer platforms to develop custom panels which cover the selected target regions defined by the specific needs *(112,113)*. In a higher level, Agilent, Illumina, Life, or NimbleGen have developed their own reagents for whole-exome analysis, as a reasonable compromise between sequencing depth and lack-of-premises for discovery of variants when no suspected genes are assumed *(114)*.

Combination of exome sequencing and noncoding SNP microarray analysis can provide a good picture of clinical relevant variation *(115,116)*, which is already used as a tool to find genetic alterations associated to different disorders (for instance, tumoral, neurological, or genetic). However, whole-genome sequencing is predicted to be in a close future the method of choice to access the genetic information in a more comprehensive level. With the current prices of sequencing, the bottleneck seems to be in the pipelines for data analysis which have to gain further access to the clinical. For the moment, our functional knowledge of the genome does not make always evident the possible relationships between new genetic variants that might be observed and the specific phenotype that can be derived, a situation that is going to evolve rapidly as soon as more information will become available. Efforts such as the GoNL program, which combines biobanking activities with deep genetic analysis is much on the way to fill this gap, resolving "harmless" variants that can be intrinsic of a certain population and real alterations that may have impact on health as driver mutations.

Cancer, as well as rare diseases, has been the area where NGS has brought most benefits. Independent of the potential use of NGS in the clinical setting, cancer is now better known, prominent, and diagnosed; discoveries have resulted in better molecular classifications. As mentioned above, of major importance is the discovery of biomarkers based on NGS findings and the identification of therapeutic targets associated to specific genetic alterations (117).

The field of RNA profiling has been also relevant to cancer research (main applied techniques for RNA analysis are summarized in Table 3). RNA-seq adds a number of improvements over hybridization technologies. In addition to higher (and configurable) sequencing depth, RNA-seq provides information about coding SNPs, splicing junctions, start and stop ends, and even strand information. RNA profiling is of higher confidence and opens the path for new discoveries providing a more detailed map of transcription, splicing, allele-associated expression, and identification of transcribed strand, which is helping to understand the big complexity of RNA expression in human cancer (118). Due to the nature of RNA-seq, also small RNAs or noncoding

TABLE 3 Main Techniques for RNA Analysis Applied to Cancer Research

	Minor/ Medium Scale	Up to Few Hundred Genes	Defined Set of Genes	mRNA miRNA ncRNA	Tools: Custom or Predefined Specific Primer Pairs, Panels of Genes
Real-Time PCR					
Gene expression arrays	Large scale	Known genes and isoforms	All mRNAs	mRNA small RNA	Tools: Predefined hybridization probes
RNA-seq	Very large scale	No limits	Complete transcriptome	mRNA smallRNA lncRNA	Tools: Open, universal adaptors

RNAs are equally amenable for analysis *(119)* up to the point that RNA sequencing has dramatically changed our vision of transcriptome status (ENCode). As another example, RNA sequencing has unrevealed the existence of gene fusions up to an unexpected level, an achievement that has emerged due to the high read depth that can be reached *(120)*. This enormous throughput and simplicity might even be improved by combining RNA analysis and single-cell capture, obtaining the maximal resolution in genomic or transcriptomic analysis.

Finally, NGS can also be applied to epigenome analysis, currently one of the main areas of cancer research. By one of three different approaches, immunoprecipitation of methyl-cytosine after DNA fragmentation, capture of CpG islands and associated sequences, or complete genome sequencing after bisulfite treatment, all of them are directed to envision what promoter regions may be modified by methylation-based silencing *(121,122)*.

In summary, from the simplest nucleotide substitution to the complex whole-genome analysis, Genomics has developed tools which discover and measure to the highest precision any possible alteration of genomes, transcriptomes, or epigenomes, making available an invaluable tool to explain and eventually find a way to treat cancer.

6   CONCLUSIONS AND FUTURE DIRECTIONS

Genetics has emerged as an essential discipline to gain knowledge conducting to further understand and properly manage some of the most prevalent chronic diseases of our society such as cancer.

Impressive scientific and technological breakthroughs have been performed in genome analysis in the past years going from the analysis of single-nucleotide substitution to the complex whole genome. These advances have resulted in outstanding discoveries of the tumor genome, with relevant implications in the management of cancer patients. However, further work is still necessary to understand the basic mechanisms underlying the carcinogenic process. In addition, new easy-to-use and cost-effectiveness genotyping techniques are still necessary to introduce tumor genome analysis as a routine test for cancer patients.

In this sense, the following generation of sequencing is in the shortcoming future. The so-called third-generation equipments will avoid one of the drawbacks of second generation that need an amplification step in order to get a light, fluorescent, or electronic signal that can be detected. Pacific Bio has recently released to the market one equipment with those characteristics. Among their properties, we should remark the possibility of directly measuring modified nucleotides, which would be translated into the native bases during the PCR steps performed in library preparation. Future advanced technologies like Oxford Nanopores or Genya *(123–125)* are attempts to bring sequencing to rates and costs that would make sequencing absolutely affordable for clinical purposes. Next technological advances are expected to be so relevant as to be announced to the entire community through their Web pages *(126,127)*. Both systems rely in nanometer-size channels sensitive to molecule transit, either that of specific dyes released by each nucleotide during DNA polymerization or the DNA molecule itself. This enormous throughput and simplicity might even be improved by adding the option of single-cell analysis, obtaining the maximal resolution in genomic or transcriptomic analysis.

The continuous development of genotyping technologies provide a unlimited value tool to generate scientific knowledge leading to new and more effective diagnostic and therapeutic approaches for cancer patients.

REFERENCES

1. Paget, S. *Lancet* **1889**, *1*, 571–573.
2. Paget, S. *Cancer Metastasis Rev.* **1989**, *8* (2), 98–101.
3. Rous, P. *J. Exp. Med.* **1910**, *12*, 696–705.
4. Rous, P. *Clin. Orthop. Relat. Res.* **1993**, *289*, 3–8.
5. Boveri, T. Gustav-Verlag: Jena, Germany, 1914.
6. Boveri, T. *J. Cell Sci.* **2008**, *121* (Suppl 1), 1–84.
7. Huff, J. *Ann. N.Y. Acad. Sci.* **1999**, *895*, 56–79.
8. Marx, J. L. *Science* **1989**, *246* (4928), 326–327.
9. Friend, S. H.; Bernards, R.; Rogelj, S.; Weinberg, R. A.; Rapaport, J. M.; Albert, D. M.; Dryja, T. P. *Nature* **1986**, *323*, 643–646.
10. Knudson, A. G., Jr. *Proc. Natl. Acad. Sci. U.S.A.* **1971**, *68* (4), 820–823.
11. Othersen, H. B., Jr. *Ann. Surg.* **2004**, *239* (5), 648–650.
12. Moore, F. D. *Ann. Surg.* **1999**, *229* (2), 187–196.
13. Jessney, B. *J. Med. Biogr.* **2012**, *20* (3), 107–110.
14. Wong, V. S.; Tan, S. Y. *Singapore Med. J.* **2009**, *50* (9), 851–852.
15. Diamantis, A.; Magiorkinis, E.; Papadimitriou, A.; Androutsos, G. *Hell. J. Nucl. Med.* **2008**, *11* (1), 33–38.
16. Maruta, H. *Drug Discov. Ther.* **2009**, *3* (2), 37–40.
17. Farber, S.; Diamond, L. K. *N. Engl. J. Med.* **1948**, *238* (23), 787–793.
18. Goodman, L. S.; Wintrobe, M. M.; Dameshek, W.; Goodman, M. J.; Gilman, A.; McLennan, M. T. *JAMA* **1984**, *251* (17), 2255–2261.
19. Moxley, J. H., 3rd.; De Vita, V. T.; Brace, K.; Frei, E., 3rd. *Cancer Res.* **1967**, *27* (7), 1258–1263.
20. Sakurai, M.; Sandberg, A. A. *Cancer* **1976**, *37* (1), 285–299.
21. Bloomfield, C. D.; Lindquist, L. L.; Arthur, D.; McKenna, R. W.; LeBien, T. W.; Nesbit, M. E.; Peterson, B. A. *Cancer Res.* **1981**, *41* (11 Pt. 2), 4838–4843.
22. Secker-Walker, L. M.; Swansbury, G. J.; Hardisty, R. M.; Sallan, S. E.; Garson, O. M.; Sakurai, M.; Lawler, S. D. *Br. J. Haematol.* **1982**, *52* (3), 389–399.
23. Köhler, G.; Milstein, C. *Nature* **1975**, *256* (5517), 495–497.
24. Golay, J.; Semenzato, G.; Rambaldi, A.; Foà, R.; Gaidano, G.; Gamba, E.; Pane, F.; Pinto, A.; Specchia, G.; Zaja, F.; Regazz, M. *MAbs* **2013**, *5* (6), 826–837.
25. Druker, B. J.; et al. *N. Engl. J. Med.* **2006**, *355* (23), 2408–2417.
26. Greenman, C.; et al. *Nature* **2007**, *446*, 153–158.

27. Sos, M. L.; Michel, K.; Zander, T.; et al. *J. Clin. Invest.* **2009**, *119*, 1727–1740.

28. Ding, L.; et al. *Nature* **2008**, *455*, 1069–1075.

29. Sjoblom, T.; et al. *Science* **2006**, *314*, 268–274.

30. Thomas, R. K.; et al. *Nat. Genet.* **2007**, *39*, 347–351.

31. Berg, M.; Soreide, K. *Discov. Med.* **2012**, *14* (76), 207–214.

32. Douillard, J. Y.; Oliner, K. S.; Siena, S.; Tabernero, J.; Burkes, R.; Barugel, M.; Humblet, Y.; Bodoky, G.; Cunningham, D.; Jassem, J.; Rivera, F.; Kocákova, I.; Ruff, P.; Błasińska-Morawiec, M.; Šmakal, M.; Canon, J. L.; Rother, M.; Williams, R.; Rong, A.; Wiezorek, J.; Sidhu, R.; Patterson, S. D. *N. Engl. J. Med.* **2013**, *369* (11), 1023–1034.

33. Zambon, A.; Niculescu-Duvaz, D.; Niculescu-Duvaz, I.; Marais, R.; Springer, C. J. *Expert Opin. Ther. Pat.* **2013**, *23* (2), 155–164.

34. Wang, A. X.; Qi, X. Y. *IUBMB Life* **2013**, *65* (9), 748–758.

35. Kudchadkar, R. R.; Gonzalez, R.; Lewis, K. *Cancer Control* **2013**, *20* (4), 282–288.

36. Iwasaki, J. K.; Srivastava, D.; Moy, R. L.; Lin, H. J.; Kouba, D. J. *J. Am. Acad. Dermatol.* **2012**, *66* (5), e167–e178.

37. Dubey, A. K.; Dubey, S.; Handu, S. S.; Qazi, M. A. *J. Postgrad. Med.* **2013**, *59* (1), 48–50.

38. Stern, H. M. *Sci. Transl. Med.* **2012**, *4* (127), 127rv2.

39. Kremyanskaya, M.; Atallah, E. L.; Hoffman, R.; Mascarenhas, J. O. *Oncology (Williston Park)* **2013**, *27* (7), 706–714.

40. Quintás-Cardama, A.; Kantarjian, H.; Cortes, J.; Verstovsek, S. *Nat. Rev. Drug Discov.* **2011**, *10* (2), 127–140.

41. Agrawal, M.; Garg, R. J.; Cortes, J.; Kantarjian, H.; Verstovsek, S.; Quintas-Cardama, A. *Cancer* **2011**, *117* (4), 662–676.

42. Levine, R. L.; et al. *Cancer Cell* **2005**, *7* (4), 387–397.

43. Goldman, J. M.; Melo, J. V. *N. Engl. J. Med.* **2003**, *349* (15), 1451–1464.

44. Abruzzese, E.; Breccia, M.; Latagliata, R. *BioDrugs* **2013**, *28* (1), 17–26.

45. Ashman, L. K.; Griffith, R. *Expert Opin. Investig. Drugs* **2013**, *22* (1), 103–115.

46. Adekola, K.; Agulnik, M. *Curr. Oncol. Rep.* **2012**, *14* (4), 327–332.

47. Corless, C. L.; Ballman, K. V.; Antonescu, C.; et al. *J. Clin. Oncol.* **2010**, *28*, 10006.

48. Almeida, M. Q.; Hoff, A. O. *Curr. Opin. Oncol.* **2012**, *24* (3), 229–234.

49. Elisei, R.; Schlumberger, M. J.; Müller, S. P.; Schöffski, P.; Brose, M. S.; Shah, M. H.; Licitra, L.; Jarzab, B.; Medvedev, V.; Kreissl, M. C.; Niederle, B.; Cohen, E. E.; Wirth, L. J.; Ali, H.; Hessel, C.; Yaron, Y.; Ball, D.; Nelkin, B.; Sherman, S. I. *J. Clin. Oncol.* **2013**, *31* (29), 3639–3646.

50. Piazza, F.; Gurrieri, C.; Pandolfi, P. P. *Oncogene* **2001**, *20* (49), 7216–7222.

51. deThé, H.; Le Bras, M.; Lallemand-Breitenbach, V. *J. Cell Biol.* **2012**, *198* (1), 11–21.

52. Liu, Y.; Wu, B. Q.; Zhong, H. H.; Hui, P.; Fang, W. G. *Int. J. Clin. Exp. Pathol.* **2013**, *6* (9), 1880–1889.

53. Cadranel, J.; Ruppert, A. M.; Beau-Faller, M.; Wislez, M. *Crit. Rev. Oncol. Hematol.* **2013**.

54. Soda, M.; Choi, Y. L.; Enomoto, M.; Takada, S.; Yamashita, Y.; Ishikawa, S.; Fujiwara, S.; Watanabe, H.; Kurashina, K.; Hatanaka, H.; Bando, M.; Ohno, S.; Ishikawa, Y.; Aburatani, H.; Niki, T.; Sohara, Y.; Sugiyama, Y.; Mano, H. *Nature* **2007**, *448* (7153), 561–566.

55. Mano, H. *Cancer Sci.* **2008**, *99* (12), 2349–2355.

56. Wu, Y. C.; Chang, I. C.; Wang, C. L.; Chen, T. D.; Chen, Y. T.; Liu, H. P.; Chu, Y.; Chiu, Y. T.; Wu, T. H.; Chou, L. H.; Chen, Y. R.; Huang, S. F. *PLoS One* **2013**, *8* (8), e70839.

57. Mardis, E. R.; Wilson, R. K. *Hum. Mol. Genet.* **2009**, *18* (2), 163–168.

58. Vogelstein, B.; Papadopoulos, N.; Velculescu, V. E.; Zhou, S.; Diaz, L. A.; Kinzler, K. W. *Science* **2013**, *339*, 1546–1558.
59. Mardis, E. R. *Nature* **2011**, *470*, 198–203.
60. Sanger, F.; Nicklen, S.; Coulson, A. R. *Proc. Natl. Acad. Sci. U.S.A.* **1977**, *74* (12), 5436–5437.
61. Ewing, B.; Hillier, L. D.; Wendl, M. C.; Green, P. *Genome Res.* **1998**, *8*, 175–185.
62. Venter, J. C.; Adams, M. D.; Myers, E. W.; et al. *Science* **2001**, *291*, 1304–1351.
63. International Human Genome Sequencing Consortium. *Nature* **2001**, *409*, 860–921.
64. DiFiore, F.; Blanchard, F.; Charbonnier, F.; Le Pessot, F.; Lamy, A.; Galais, M. P.; Bastit, L.; Killian, A.; Sesboüe, R.; Tuech, J. J.; Queuniet, A. M.; Paillot, B.; Sabourin, J. C.; Michot, F.; Michel, P.; Frebourg, T. *Br. J. Cancer* **2007**, *96*, 1166–1169.
65. Mocellin, S.; Rossi, C. R.; Pilati, P.; Nitti, D.; Marincola, F. M. *Trends Mol. Med.* **2003**, *9* (5), 189–195.
66. Sasaki, H.; Endo, K.; Konishi, A.; Takada, M.; Kawahara, M.; Iuchi, K.; Matsumura, A.; Okumura, M.; Tanaka, H.; Kawaguchi, T.; Shimizu, T.; Takeuchi, H.; Yano, M.; Fukai, I.; Fujii, Y. *Clin. Cancer Res.* **2005**, *11* (8), 2924–2929.
67. Endo, K.; Konishi, A.; Sasaki, H.; Takada, M.; Tanaka, H.; Okumura, M.; Kawahara, M.; Sugiura, H.; Kuwabara, Y.; Fukai, I.; Matsumura, A.; Kobayashi, Y.; Mizuno, K.; Haneda, H.; Suzuki, E.; Iuchi, K.; Fujii, Y. *Lung Cancer* **2005**, *50* (3), 375–384.
68. Ranade, K.; Chang, M. S.; Ting, C. T.; Pei, D.; Hsiao, C. F.; Olivier, M.; Pesich, R.; Hebert, J.; Chen, Y. D. I.; Dzau, V. J.; Curb, D.; Olshen, R.; Risch, N.; Cox, D. R.; Botstein, D. *Genome Res.* **2001**, *11*, 1262–1268.
69. James, E.; Seeb, J. E.; Pascal, C. E.; Ramakrishnan, R.; Seeb, L. W. In: *Single Nucleotide Polymorphisms;* Komar, A. A., Ed, 2nd ed.; Humana Press: New York, 2011; pp 277–292, Ch. 18.
70. Shitara, K.; Ito, S.; Misawa, K.; Ito, Y.; Ito, H.; Hosono, S.; Watanabe, M.; Tajima, K.; Tanaka, H.; Muro, K.; Matsuo, K. *Ann. Oncol.* **2012**, *23*, 659–664.
71. Kocarnik, J. D.; Hutter, C. M.; Slattery, M. L.; Berndt, S. I.; Hsu, L.; Duggan, D. J.; Muehling, J.; Caan, B. J.; Beresford, S. A. A.; Rajkovic, A.; Sarto, G. A.; Marshall, J. R.; Hammad, N.; Wallace, R. B.; Makar, K. W.; Prentice, R. L.; Potter, J. D.; Hayes, R. B.; Peters, U. *Cancer Epidemiol. Biomarkers Prev.* **2010**, *19* (12), 3131–3139.
72. Demichelis, F.; Fall, K.; Perner, S.; Andrén, O.; Schmidt, F.; Setlur, S. R.; Hoshida, Y.; Mosquera, J. M.; Pawitan, Y.; Lee, C.; Adami, H. O.; Mucci, L. A.; Kantoff, P. W.; Andersson, S. O.; Chinnaiyan, A. M.; Johansson, J. E.; Rubin, M. A. *Oncogene* **2007**, *26*, 4596–4599.
73. Tan, R.; Brzoska, P. *J. Biomol. Tech.* **2010**, *21* (Suppl 3), S30.
74. Wittwer, C. T. *Hum. Mutat.* **2009**, *30*, 857–859.
75. Erali, M.; Wittwer, C. T. *Methods* **2010**, *50*, 250–261.
76. de Juan, I.; Esteban, E.; Palanca, S.; Barragán, E.; Bolufer, P. *Breast Cancer Res. Treat.* **2009**, *115*, 405–414.
77. Vossen, R. H. A. M.; Aten, E.; Roos, A.; den Dunnen, J. T. *Hum. Mutat.* **2009**, *30* (6), 860–866.
78. Tobler, A. R.; Short, S.; Andersen, M. R.; Paner, T. M.; Briggs, J. C.; Lambert, S. M.; Wu, P. P.; Wang, Y.; Spoonde, A. Y.; Koehler, R. T.; Peyret, N.; Chen, C.; Broomer, A. J.; Ridzon, D. A.; Zhou, H.; Hoo, B. S.; Hayashibara, K. C.; Leong, L. N.; Ma, C. N.; Rosenblum, B. B.; Day, J. P.; Ziegle, J. S.; De La Vega, F. M.; Rhodes, M. D.; Hennessy, K. M.; Wenz, H. M. *J. Biomol. Tech.* **2005**, *16* (4), 396–404.
79. Taylor, J. D.; Briley, D.; Nguyen, Q.; Long, K.; Iannone, M. A.; Li, M. S.; Ye, F.; Afshari, A.; Lai, E.; Wagner, M.; Chen, J.; Weiner, M. P. *Biotechniques* **2001**, *30* (3), 661–669.

80. Syvänen, A. C. *Nat. Genet.* **2005**, *37*, S5–S10.
81. Perkel, J. *Nat. Methods* **2008**, *5* (5), 447–453.
82. Hoheisel, J. D. *Nat. Rev. Genet.* **2006**, *7* (3), 200–210.
83. Walther, A.; Johnstone, E.; Swanton, C.; Midgley, R.; Tomlinson, I.; Kerr, D. *Nat. Rev. Cancer* **2009**, *9*, 489–499.
84. McLeod, H. L.; Yu, J. *Pharmacogenomics* **2003**, *21* (4), 630–640.
85. Butler, J. M.; Buel, E.; Crivellente, F.; McCord, B. R. *Electrophoresis* **2004**, *25*, 1397–1412.
86. Zhang, F.; Gu, W.; Hurles, M. E.; Lupski, J. R. *Annu. Rev. Genomics Hum. Genet.* **2009**, *10*, 451–481.
87. Hastings, P. J.; Lupski, J. R.; Rosenberg, S. M.; Ira, G. *Nat. Rev. Genet.* **2009**, *10* (8), 551–564.
88. Sellner, L. N.; Taylor, G. R. *Hum. Mutat.* **2004**, *23*, 413–419.
89. Iafrate, A. J.; Feuk, L.; Rivera, M. N.; Listewnik, M. L.; Donahoe, P. K.; Qi, Y.; Scherer, S. W.; Lee, C. *Nat. Genet.* **2004**, *36* (9), 949–951.
90. Davies, J. J.; Wilson, I. M.; Lam, W. L. *Chromosome Res.* **2005**, *13*, 237–248.
91. Berg, M.; Ågesen, T. H.; Thiis-Evensen, E.; Merok, M. A.; Texeira, M. R.; Vatn, M. H.; Nesbakken, A.; Skotheim, R. I.; Lothe, R. A. *Mol. Cancer* **2010**, *9*, 100–113.
92. Medintz, I. L.; Lee, C. C. R.; Wong, W. W.; Pirkola, K.; Sidransky, D.; Mathies, R. A. *Genome Res.* **2000**, *10*, 1211–1218.
93. Tsiatis, A. C.; Norris-Kirby, A.; Rich, R. G.; Hafez, M. J.; Gocke, C. D.; Eshleman, J. R.; Murphy, K. M. *J. Mol. Diagn.* **2010**, *12* (4), 425–432.
94. Dufort, S.; Richard, M. J.; Lantuejoul, S.; de Fraipont, F. *J. Exp. Clin. Cancer Res.* **2011**, *30*, 57–63.
95. England, R.; Pettersson, M. *Nat. Methods* **2005**, *2*, i–ii.
96. Myakishev, M. V.; Khripin, Y.; Hu, S.; Hamer, D. H. *Genome Res.* **2001**, *11*, 163–169.
97. Bustin, S. A.; Benes, V.; Garson, J. A.; Hellemans, J.; Huggett, J.; Kubista, M.; Mueller, R.; Nolan, T.; Pfaffl, M. W.; Shipley, G. L.; Vandesompele, J.; Wittwer, C. T. *Clin. Chem.* **2009**, *55* (4), 611–622.
98. van Schooneveld, E.; Wouters, M. C. A.; Van der Auwera, I.; Peeters, D. J.; Wildiers, H.; Van Dam, P. A.; Vergote, I.; Vermeulen, P. B.; Dirix, L. Y.; Van Laere, S. *J. Breast Cancer Res.* **2012**, *14* (1), R34.
99. Glas, A. M.; Floore, A.; Delahay, L. J. M. J.; Witteveen, A. T.; Pover, R. C. F.; Bakx, N.; Lahti-Domenici, J. S. T.; Bruinsma, T. J.; Warmoes, M. O.; Bernards, R.; Wessels, L. F. A.; Van't Veer, L. J. *BMC Genomics* **2006**, *7*, 278–287.
100. Cuadros, M.; Dave, S. S.; Jaffe, E. S.; Honrado, E.; Milne, R.; Alves, J.; Rodríguez, J.; Zajac, M.; Benitez, J.; Staudt, L. M.; Martinez-Delgado, B. *J. Clin. Oncol.* **2007**, *25* (22), 3321–3329.
101. Ecker, J. R.; Bickmore, W. A.; Barroso, I.; Pritchard, J. K.; Gilad, Y.; Segal, E. *Nature* **2012**, *489*, 52–55.
102. Sawyers, C. L. *Nature* **2008**, *452*, 548–552.
103. Dalton, W. S.; Friend, S. H. *Science* **2006**, *312* (5777), 1165–1168.
104. Voelkerding, K. V.; Dames, S. A.; Durtschi, J. D. *Clin. Chem.* **2009**, *55* (4), 641–658.
105. Metzker, M. L. *Nat. Rev. Genet.* **2010**, *11*, 31–46.
106. Wetterstrand, K. A. Available at www.genome.gov/sequencingcosts.
107. The 1000 Genomes Project Consortium. *Nature* **2010**, *467*, 1061–1073.
108. Stratton, M. R.; Campbell, P. J.; Futreal, P. A. *Nature* **2009**, *458*, 719–724.
109. Chin, L.; Andersen, J. N.; Futreal, P. A. *Nat. Med.* **2011**, *17* (3), 297–303.

110. Wheeler, H. E.; Maitland, M. L.; Dolan, M. E.; Cox, N. J.; Ratain, M. J. *Nat. Rev. Genet.* **2013**, *14*, 23–34.

111. Bottillo, I.; Morrone, A.; Grammatico, P. J. *Pharmacovigilance* **2013**, *1* (2), 109. http://dx. doi.org/10.4172/2329-6887.1000109.

112. Ross, J. S.; Ali, S. M.; Wang, K.; Palmer, G.; Yelensky, R.; Lipson, D.; Miller, V. A.; Zajchowski, D.; Shawver, L. K.; Stephens, P. J. *Gynecol. Oncol.* **2013**, *130* (3), 554–559.

113. Chang, F.; Li, M. M. *Cancer Genet.* **2013**, *206* (12), 413–419. http://dx.doi.org/10.1016/j. cancergen.2013.10.003.

114. Chang, H.; Jackson, D. G.; Kayne, P. S.; Ross-Macdonald, P. B.; Ryseck, R. P.; Siemers, N. O. *PLoS One* **2011**, *6* (6), e21097.

115. Ku, C. S.; Cooper, D. N.; Polychronakos, C.; Naidoo, N.; Wu, M.; Soong, R. *Ann. Neurol.* **2012**, *71*, 5–14.

116. Majewski, J.; Schwartzentruber, J.; Lalonde, E.; Montpetit, A.; Jabado, N. *J. Med. Genet.* **2011**, *48*, 580–589.

117. Ballard-Barbash, R.; Friedenreich, C. M.; Courneya, K. S.; Siddiqi, S. M.; McTiernan, A.; Alfano, C. M. *J. Natl. Cancer Inst.* **2012**, *104*, 815–840.

118. Burgess, D. J. *Nat. Rev. Genet.* **2013**, *14*, 154–155.

119. Volinia, S.; Galasso, M.; Sana, M. E.; Wise, T. F.; Palatini, J.; Huebner, K.; Crocea, C. M. *Proc. Natl. Acad. Sci.* **2012**, *109* (8), 3024–3029.

120. Kangaspeska, S.; Hultsch, S.; Edgren, H.; Nicorici, D.; Murumägi, A.; Kallioniemi, O. *PLoS One* **2012**, *7* (10), e48745.

121. Ruike, Y.; Imanaka, Y.; Sato, F.; Shimizu, K.; Tsujimoto, G. *BMC Genomics* **2010**, *11*, 137–147.

122. Gu, H.; Bock, C.; Mikkelsen, T. S.; Jäger, N.; Smith, Z. D.; Tomazou, E.; Gnirke, A.; Lander, E. S.; Meissner, A. *Nat. Methods* **2010**, *7*, 133–136.

123. Venkatesan, B. M.; Bashir, R. *Nat. Nanotechnol.* **2011**, *6*, 615–624. http://dx.doi.org/ 10.1038/nnano.2011.129.

124. Maitra, R. D.; Kim, J.; Dunbar, W. B. *Electrophoresis* **2012**, *33*, 1–11.

125. Kumar, S.; Tao, C.; Chien, M.; Hellner, B.; Balijepalli, A.; Robertson, J. W. F.; Li, Z.; Russo, J. J.; Reiner, J. E.; Kasianowicz, J. J.; Ju, J. *Sci. Rep.* **2012**, *2*, 648. http://dx.doi. org/10.1038/srep00684.

126. www.geniachip.com.

127. https://www.nanoporetech.com/.

Chapter 4

Applications of Next-Generation Sequencing Technologies to the Study of the Human Microbiome

Philip Chuang*, Emily Walsh*, Timothy Yen* and Ilseung Cho*,†

*NYU School of Medicine, New York, USA
†Division of Gastroenterology, Department of Medicine, NYU School of Medicine, New York, USA

Chapter Outline

Comprehensive Analytical Chemistry, Vol. 64. http://dx.doi.org/10.1016/B978-0-444-62650-9.00004-X

1 INTRODUCTION

In clinical practice, the presence and identification of bacterial species use culturing and phenotypic testing as the gold standard. For many decades, however, there has been a large incongruity regarding the number and distribution of bacteria observed microscopically and the number of bacteria able to be cultured on a plate. This "great plate count anomaly," a term coined by Staley and Konopka in the 1980s, revealed the limitations of culture-dependent studies for identifying the vast majority of microbial species in complex communities (1). Although some species-specific parameters are known, such as the sensitivity of *Vibrio cholerae* and *Salmonella enteritidis* to salt content and temperature, there is a plethora of incompletely characterized species and strains due to the restrictions inherent with bacterial culture (2).

1.1 16S Sequencing Technologies

Based on the work of Woese *et al.* in the 1970s and 1980s, it was discovered that phylogenetic relationships amongst bacteria could be measured based on the sequence of the well-conserved 16S rRNA gene (3,4). The 16S rRNA gene is approximately 1500 base pairs long and contains nine conserved regions (C1–C9) and nine hypervariable regions (V1–V9) interposed amongst one another. Initially, the rRNA genes were examined via fingerprinting technologies based on gradient gel electrophoresis and terminal restriction fragment length polymorphism (T-RFLP) analysis (5,6). Although relatively inexpensive, these techniques had low specificity and were soon passed over in favor of "long-read" Sanger sequencing with the capability of sequencing all 1500 bases. Currently, the favored method of "next-generation high-throughput sequencing" is pyrosequencing due to its speed, lower unit cost, and greater depth of information (7). These various sequencing methods target the bacterial 16S rRNA gene for selective amplification, cloning, and sequencing. The data would then be analyzed, usually focusing on the hypervariable regions to differentiate amongst taxa (8).With such analyses, phylogenetic trees could be generated and taxonomic relationships revealed.

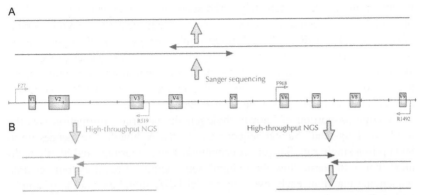

FIGURE 1   Map of 16S rRNA gene. (A) Sanger methods allow for sequencing of the complete 16S rRNA but has limited depth compared to high-throughput sequencing methods. (B) Current high-throughput sequencing methods sequence shorter reads and focus on specific hypervariable regions within the 16S rRNA. Studies suggest that sufficient taxonomic discrimination can be achieved without sequencing the entire gene.

To date, the Ribosomal Database Project (RDP) release contains over 2.7 million sequences and counting *(9)* (Figure 1).

Although most 16S rRNA sequencing is now done via pyrosequencing, traditional sequencing technologies such as sequence-based fingerprinting remain important. Fingerprinting relies on the separation of amplified gene sequences, such as 16S rRNA, via gel electrophoresis based on fragment size and sequence. For example, denaturing gradient gel electrophoresis and temperature gradient gel electrophoresis separate amplified DNA fragments based on melting temperature along a linear gradient of denaturant concentration or temperature, respectively. Another common fingerprinting technique is T-RFLP, in which the gene fragments are amplified using fluorescently-tagged 5′ primers, digested by specific restriction enzymes, and separated by gel (or capillary) electrophoresis before determination of fragment lengths by a DNA sequencer. Fingerprinting has the capability to test a large number of samples simultaneously and is much cheaper and faster than sequencing *(6)*. However, not only is the resolution and depth limited to the most dominant bacterial lines of the sample, it is often not informative regarding changes at lower taxonomic levels, and inherently utilizes an output format that makes the assimilation of results from different experiments difficult *(10)*. As will be discussed, it is precisely these characteristics that have made high-throughput sequencing the preferred method for microbiome analyses *(11)*.

Other 16S rRNA analytic techniques such as short-sequence tags and DNA microarrays provide bridge technologies between conventional fingerprinting and sequencing. One short-sequence tagging technique is automated ribosomal intergenic spacer analysis, in which a specific noncoding highly variable internal transcribed spacer region between the 16S and 23S rRNA gene is subjected to fingerprinting *(12)*. Alternately, short-sequence tagging can lean toward the other end of the spectrum, such as the selective sequencing of the V6 hypervariable region of 16S rRNA *(13)*. 16S DNA microarrays utilize the hybridization of amplified 16S gene fragments to a large library of 16S sequence probes *(14)*. While these techniques leverage the strength of fingerprinting for simultaneous analysis of many samples, they are limited to taxa with known 16S sequences and fail to identify previously uncharacterized microbes.

A large proportion of known phylogenetic bacterial information was the result of "long-read" Sanger-sequencing efforts. Although it sequences up to 800 bp in a single run, Sanger sequencing is slow, laborious, and highly technical (hence the term "low-throughput" sequencing). It requires four separate sequencing reactions each involving partial DNA strand elongation of amplified 16S gene fragments prematurely terminated by the incorporation of dNTPs at variable steps of the process, subsequent separation on a capillary-based polymer gel, and then sequence determination *(15)*. Therefore, when it was first suggested that as few as 500 bp may be adequate for studies at the genus or possibly species level, avid interest was redirected toward the idea of "next-generation" or "high-throughput" sequencing that would target specific regions of the 16S rRNA gene *(16)*.

The "high-throughput" next-generation sequencing technologies most commonly include pyrosequencing (ex. Roche 454 Genome Sequencer GS, FLX, and FLX Titanium), Illumina clonal arrays (ex. Illumina GAIIx and HiSeq2000), and others that vary in parameters such as read length, insert size, read length/depth, cost, error rate, etc. *(17)*. Although the next-generation high-throughput technologies tend to have shorter read lengths, 100–200 bp reads turn out to be just as accurate as the longer 800-bp reads of Sanger sequencing *(18)*, particularly with the most common method, pyrosequencing *(19)*. Furthermore, the new technologies are often much cheaper and do not rely on bacterial cloning techniques. One of the few situations in which Sanger sequencing may be advantageous, however, is when entirely novel bacterial lines completely unrelated to known species in databases is discovered. While there are numerous sequencing technologies available to study the microbiome, this discussion will primarily focus on pyrosequencing, while still discussing other "second generation" high-throughput next-generation technologies such as sequencing by synthesis (SBS) and sequencing by oligo ligation detection (SOLiD) and "third generation" technologies like single molecule real time (SMRT) and Ion Torrent.

2 NEXT-GENERATION SEQUENCING

2.1 Roche 454 Pyrosequencing

Pyrosequencing utilizes the release of inorganic pyrophosphate from DNA polymerase during the addition of dNTPs onto a growing sequence. Each released inorganic pyrophosphate is used to generate oxyluciferin, which luminesces with an amplitude specific to the dNTP incorporated. A camera detects this spectrum of light, which can then be analyzed and translated into a sequence *(20)*. Since the development of high-throughput pyrosequencing for metagenomics, performance has improved due to the advances both in the biochemistry of pyrosequencing (e.g., the addition *(21)* and modification *(22)* of single-stranded-binding protein into the reaction) and the instrumentation itself. It started with 100-bp reads at 30–60 Mb/run (first Generation, GS), then to 250-bp reads at 150 Mb/run (second generation, FLX Standard), and currently 350–500-bp reads at 400 Mb/run (third generation, FLX Titanium) *(17,20)*. Even though each run can easily span multiple hypervariable and conserved regions of the 16S rRNA gene, the theoretical maximum read length is 600 bases per run, well below the 1500 bp span of the entire 16S gene *(17)*. While this is shorter than the 800-bp read capability of Sanger sequencing, it is longer than the other next-generation sequencing techniques and allows for precise mapping of highly complex samples without sacrificing cost or speed.

2.1.1   Barcoding

A particularly useful characteristic of "high-throughput" pyrosequencing is multiplexing, which involves running multiple samples in a single run. This can be achieved through physical allocation of samples on a plate, barcoded pyrosequencing, or both. Some studies may choose to pool extracted DNA from multiple samples prior to sequencing in an effort to reduce sample variability as well as overall cost. However, multiplexing (with the option of post-analytic pooling via bioinformatics techniques) is strongly preferred in human microbiome studies. Although initially developed before the advent of pyrosequencing, barcoding uses a diverse set of hybrid primers containing two sequences—one portion is universal to all the primers necessary for DNA polymerase activity, while the other portion is a short predetermined sequence assigned to a specific sample of interest. While there are a wide variety of barcode types *(23–26)*, one type of particular interest is the error-detecting/correcting barcodes. By applying mathematical approaches (such as the Hamming code) used commonly in electronics toward barcode design, several hundred samples (potentially thousands) can be run at once, and errors in barcodes can be detected and corrected *(24)*. While studies on bias *(23)* and confounding errors *(19,24)* are limited, the use of barcoding is promising for both 16S sequencing as well as other techniques of metagenomics such as shotgun sequencing *(27)*.

2.1.2   Sequencing Parameters

Beyond the speed and cost considerations of each sequencing technology, parameters including read length, sampling depth, and primer selection must be optimized. As mentioned previously, 16S rRNA sequencing can be done with as few as a couple hundred bases while maintaining the ability to perform subsequent analysis for community and taxonomic analysis, given an appropriate primer selection (to be discussed below) (28). However, if the data are to be used in metagenomic studies of function or species, longer reads such as the FLX Titanium's 350–500 bp should be utilized (29). When considering sampling depth, the number of sequences required also depends on the goal at hand. In studies taking place at the phyla level, many fewer sequences are required while metagenomic studies require more scrutiny. At a depth of approximately 1000 sequences/sample, only species comprising at least 1% of the microbiome population can be seen (10).

Because next-generation sequencing technologies are fairly limited in read length, primer selection that determines what region of the 16S rRNA gene will be sequenced is crucial. Unfortunately, there is no consensus regarding the optimal region to sequence, which significantly hinders comparisons of data among various studies except in broad perspectives (11). As mentioned previously, there are nine hypervariable regions V1–V9 interspersed throughout the 16S rRNA gene, with each approximately 50–100 bp in length. However, usually only 1–4 sequential bases in any given region of the 16S rRNA gene are conserved, making primer selection crucial in the accuracy of sequencing results (30). Among the hypervariable regions, the more commonly used regions are V2, V4, and V6 (10), but several others are also utilized as well (13,30–34). In order to maximize the accuracy of sequencing studies, combinations of hypervariable regions often give lower error rates, such as V2 and V4 (28,35). Equally important in the accuracy of sequencing is the primer design itself, a fact well known to almost all the applications of sequencing and PCR, and particularly important in the microbiome containing rare sequences of interest. Commonly used primers known to be effective for the major bacterial phyla include 8F (31,36,37), 337F (32), 338R (24,38), 515F (37,39), 915F (39), 930R (39), 1046R (13), and 1061R (34). However, there are primers that miss Actinobacteria/Proteobacteria (1492R) (37), Bacteroidetes (967F) (13), or Verrucomicrobia (784F) (34). A promising approach to primer design involves the use of miniprimers, approximately 10-bp long primers (compared to the 20-bp minimum of conventional primers), in conjunction with a modified DNA polymerase. This new technique is still in development, but initial studies have shown greater accuracy and depth without increased nonspecific amplification (40).

Other than the limitations in sequencing parameters, pyrosequencing has particular problems with homopolymer repeats, which are strings of the same base pair repeated multiple times in a row. Not surprisingly, these errors made in sequencing increase with longer homopolymers. When this occurs in pyrosequencing, the barrage of pyrophosphates released from multiple incorporations of the same base decreases the accuracy of the detected signal amplitude due to asynchronous molecules and photon shot noise (41).

As a result, 16S sequences with long homopolymers are still better handled by, with the exception of Ion Torrent, other technologies, such as Sanger sequencing.

2.2   Illumina SBS Platform (Genome Analyzer IIx, HiSeq, MiSeq)

SBS circumvents the use of PCR in favor of a bridge amplification approach. First, DNA fragments ligated to single-stranded oligo adaptors are added to the surface of a glass flow cell containing eight lanes via a microfluidic station. Each lane is coated with covalently bonded oligos complementary to the aforementioned oligo adaptors, with hybridization occurring with temperature change. An isothermal polymerase then amplifies each fragment into a discrete cluster on the surface of the flow cell in preparation for sequencing. To sequence the flow cell, each cluster is denatured into single strands and provided with polymerase and four base-specific fluorescently labeled nucleotides with inactivated 3'-OH ends that allow only one base to be added each cycle. The sequencer reads out the labeled base, and then the fluorescent group is chemically removed and the 3' end deblocked. The process repeats again and again to yield a read length of 32–40 bases after quality filtering *(42–44)*. Although the use of SBS provides an extremely high number of reads per sample and samples per run (more than 454 pyrosequencing), the short read length is a significant disadvantage. Even though the addition of paired-end sequencing (using a set of primers that are directed toward each other, also available in Sanger sequencing) has improved this shortcoming (particularly when reads overlap in sequence), the maximum read length is still under 200 bases *(17)* (Table 1).

2.3   AB (Applied Biosystems) SOLiD Platform (AB 5500, 5500xl)

SOLiD sequencer utilizes DNA ligase to identify 2 bases per cycle. In preparation for sequencing, 16S DNA fragments are first linked to oligo adaptors that facilitate coupling with 1-μm magnetic beads containing complementary oligos. The fragment-bead complexes then undergo emulsion PCR and are then covalently bonded to a glass slide for sequencing. A universal primer anneals to the oligo adaptors, a set of semidegenerate 8-mer oligonucleotides is added to hybridize adjacent to the 3' end of the universal primer, and DNA ligase seals the phosphate backbone between the primer and 8-mer. A fluorescent readout will be generated for the fifth base of the 8-mer. The 8-mer is then cleaved between bases 5 and 6, thus removing bases 6–8 including the fluorescent group, and the process repeats again. This identifies the sequences at intervals of 5 bases each cycle until the end of the fragment. A second round is then performed with the primer position at $n-1$ (one position upstream of first round), and so on until the entire sequence is read out. Due to the use of known fixed bases in the 8-mers, SOLiD "2 base encoding"

TABLE 1 Comparison of Current Sequencing Technologies Used for Studies in the Human Microbiome

	Sanger/ABI 3730	Roche 454	Illumina SBS	PacBio SMRT	Ion Torrent Semiconductor
Mechanism	Dideoxy chain termination	Pyrosequencing	Sequencing by synthesis (SBS)	Single molecule real-time SBS	Semiconductor-based SBS
Amplification	PCR	Emulsion PCR	Bridge amplification	n/a	n/a
Read length	800 b	300–600 b	36–151 b	1100 b	200 b
Maximum insert size	>1 Kb	800–1200 b	500 b	>1 Kb	400
Run time	2 h	9–23 h	4 h–15 d	1.5 h	2–3 h
Reads per run	96	10^6	10^7–3×10^9	3.5×10^7	1.5–3×10^6
Relative cost factor per Mb	100	0.7–1	0.002–0.1	1.5	0.4
Scale of reads per sample	10^2	10^3	10^4–10^6	10^3	10^3
Scale of samples per run	10^1	10^2	10^2–10^4	10^1	10^2
Error rate	0.001%	1%	<1%	15%	2%

Adapted from Kuczynski et al. (17) and Chandra et al. (45).

is quite adept at identifying mistakes in sequencing with an accuracy that surpasses the 454 pyrosequencing. However, the reliance on 8-mers limits the read capabilities to short read assemblies and slows the sequencing process down such that it takes several days per run (compared to hours in 454 pyrosequencing) (42–44). When studying novel 16S sequences, this same reliance on 8-mers is also much less feasible as compared to the other next-generation sequencing technologies.

2.4   Pacific Bioscience SMRT Platform (PacBio RS II)

SMRT sequencer is a variation on SBS using a sequencing chip containing thousands of zero-mode waveguide (ZMW) detectors each with a single DNA polymerase attached at the bottom of its detection zone. First, fluorescent markers specific to each type of dNTP are attached to terminal phosphates. Next the sequencing reaction is initiated, and each labeled dNTP to be incorporated diffuses into the vicinity of the DNA polymerase within the

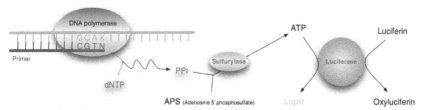

FIGURE 2   Pyrosequencing diagram. Pyrosequencing utilize the release of inorganic pyrophosphate from DNA polymerase in a PCR-based reaction to generate oxyluceferin, which luminesces and is detected using charge-coupled device imagers.

ZMW detection zone. This fluorescence is detected in real-time by a charge-coupled device array even before the dNTP is chemically incorporated. Once the dNTP is incorporated, the labeled component (the inorganic phosphate) diffuses out of the detection zone and the fluorescence signal drops back to baseline *(45–48)*. SMRT has the capability of producing long reads over 1000 bases and is quite fast, but currently suffers from poor accuracy and a modest scale of reads per sample and samples per run *(17)* (Figure 2).

2.5   Ion Torrent (PostLight™) Semiconductor Platform (Ion PGM, Ion Proton)

Using Moore's Law of semiconductor technology in conjunction with biochemical techniques, Ion Torrent semiconductor sequencing is another variation on SBS focusing on the release of hydrogen ions upon dNTP incorporation. 16S fragments are placed into a high-density array of microwells (one DNA template per well) that reside over an ion-sensitive layer covering a proprietary Ion sensor (ion-sensitive field-effect transistor, or ISFET) *(45,49)*. In each cycle, one of the four dNTPs is introduced into the well. The resulting change in pH is directly detected by the ion-sensing component and recorded, and the unincorporated dNTPs washed out before the next cycle involving another one of the remaining three dNTPs *(45)*. Ion Torrent sequencing is very fast and low cost due to its direct detection capabilities and can virtually sequence in real-time. It is assumed that this method will improve with developments in semiconductor technology. However, it is currently limited by short read lengths (compared to pyrosequencing) and has considerable problems with homopolymers *(50)* (thought to be due to larger changes in pH within a single cycle that impedes interpretation).

Although the cost of sequencing has dropped dramatically from initial 16S studies, there is still potential for improvement in sequencing parameters such as read lengths and sequencing accuracy *(51)*. Furthermore, improvements must be made in the quality and yield during the initial DNA extraction process. There are several studies comparing various extraction methods, but a consensus method has yet to emerge *(52–56)*. For example, one study found protocols utilizing bead beating or mutanolysin to provide the highest yields for bacterial community representation *(53)*, while another supported the use of lysostaphin and lysozyme *(54)*. Few investigators would disagree that the process of DNA extraction remains time-consuming, laborious, and suffers from lack of precision. Although improvement remains minimal, one study suggested the use of a direct PCR approach that is 6–8 times faster than conventional extraction protocols without detriment to the quality of bacterial community representation; albeit with caveats regarding application and storage *(57)*.

While 16S rRNA sequencing is the primary technique for microbiome studies, it remains unable to detect rare members of a community despite improvements in technology. Although there is still no effective method of detecting these rare taxa, whole-genome metagenomic shotgun sequencing offers a possible approach toward the functional characteristics of these bacteria (and others) because it sequences the entire genome and therefore is not readily restricted by specific sequences and primers. In the process of whole-genome sequencing, the entire length of extracted DNA is amplified, randomly sheared into fragments, and subjected to high-throughput parallel sequencing for subsequent bioinformatics analysis *(58)*. Just as 16S PCR is required for several 16S-sequencing methods, whole-genome amplification is often required due to the low-yield of microbiome samples after extraction *(20)*. Although this introduces amplification bias, current techniques have attempted to minimize bias and increase yield *(59)*. Similarly, advances in these technologies have also attempted to address contamination by host human DNA *(60–62)* while simultaneously improving sequencing parameters and efficiency *(63,64)*.

3　BIOINFORMATICS

High-throughput sequencing, like 454 pyrosequencing, Illumina and SoliD have the potential to generate enormous depth of coverage, sequencing hundreds to thousands of amplicons per sample in a single run. However, the short reads that make this possible also have numerous drawbacks. They make phylogenetic analyses less specific, particularly in context of using conserved 16S regions, which increases the difficulty in identifying bacteria at lower taxonomic levels and potentially confounds the actual population distribution in a given sample *(28,17)*. In addition, assessing microbial diversity through short-chain sequencing increases the risk of error. Unlike conventional sequencing, there are no consensus assemblies to minimize misidentification of base pairs *(32)*. Furthermore, any errors in base pair matching could lead to overestimation of rare phylotypes *(65)*. Finally one last issue commonly associated with 16S high-throughput sequencing is chimera formation. During the amplification of the 16S sequence, hybrid products between multiple sequences can form and may inflate microbial diversity of a given sample *(66)*. Thus, it is important to filter out sequences with any errors or chimeras prior to analysis.

3.1   Denoising

Depending on the modality of choice, high-throughput sequencing can have error rates that can be as high as 4% prior to processing *(32)*. In order to reduce the number of problematic sequences, algorithms are used to discard likely inaccurate results. A common source of error in techniques that use DNA capture beads is whether it has incorporated more than one template. This can be detected when a significant number of nucleotide flow falls into a range of 0.5–0.7. Using this criteria, one could filter out this type of error by using only samples where less than 5% of nucleotide flows lie in this over-lap region *(67)*. In addition, higher quality reads typically have their signal correspond to integer values for the number of incorporated nucleotides *(67)*. Further along the process, reads can be filtered by looking for other pat-terns of problematic sequences. It was found that error rates increased as read lengths diverged from predicted values. In addition, the presence of Ns in the sequence also correlates well with sequencing errors. By filtering out these two types of problem sequences, one could additionally reduce error rate from 0.49% to approximately 0.16% *(66)*. Although one can expect an additional 10% of sequences to be filtered out using this method, the volume of data generated from high-throughput sequencing compensates for this loss.

One of the first programs introduced to the microbial community to denoise sequencing information was PyroNoise. Rather than focusing on sequence analysis, it analyzed flowgrams *(68)*. Using an iterative process of sequence analysis, PyroNoise identified the most probable sequence based on each flowgram. Unfortunately, this is also a computationally intensive pro-cess that makes it difficult to implement. Addressing this issue, the creators built AmpliconNoise. This program couples two algorithms: fast flowgram clustering without alignment (PyroNoise) and sequence-based clustering with alignments (SeqNoise). This achieves more computational efficiency because the fast flowgram clustering reduces the computational burden on the sequence clustering algorithm *(69)*.

An alternative tool, Denoiser, focuses on analyzing the aligned sequence data, often after using tools like NAST *(70)*, but in a rank-abundance method. The theory it implements is that microbial communities tend to consist mostly of a few taxa. Using this, one can then assume that sequences similar to, but not exactly matching highly represented taxa, are likely from those taxa but with sequence errors. This allows for easier assignment of sequences to clus-ters while decreasing the number of erroneous taxa. This is performed by an initial sequence analysis that generates a list of prefix sequences and sorts them in descending order of abundance. It then clusters similar reads by

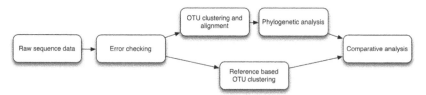

FIGURE 3   Sequence analysis flow diagram. Raw sequencing data are first processed to account for erroneous reads and to filter out chimeras. Sequences are then clustered into *de novo* OTUs and aligned prior to inferring phylogenetic relationships. Alternatively, error checked sequences are clustered into taxa using reference-based OTU clustering with a curated database. The phylogenetic data, or even sequence data, are then compared to other libraries or communities.

comparing them to the most common clusters. The remaining sequences then denote the taxa that are less commonly represented and are clustered accordingly *(71)*.

A similar method, termed sequence link preclustering (SLP) also analyzes sequenced data and clusters sequences based on abundance and similarity. As it sorts through the sequences, it clusters those with a pairwise distance less than 0.02 (which equates to a single difference in a 16S variable region) together. After this first round of clustering, sequences in low-abundance clusters are tested against larger clusters by a pairwise distance of less than 0.02 *(72)*. After all of the clusters have been assigned, SLP then assigns the sequence with the highest frequency to represent that cluster in downstream operational taxonomic unit (OTU) clustering (Figure 3).

3.2   Detecting Chimeras

Addressing the issues of chimeras are three popular programs designed for high-throughput sequencing, ChimeraSlayer *(66)*, Perseus *(69)*, and UCHIME *(73)*. ChimeraSlayer works by comparing the ends of query sequences to a database of reference 16S sequences to identify potential parents of a chimera sequence based on alignment score. It then simulates chimeric 16s sequence construction with a less than 10% global sequence divergence and compares them to the sequenced data. Sequences that match more closely to a simulated chimera than to reference sequences were then identified as potential chimeras *(66)*.

Another algorithm, Perseus, approaches chimera detection without the use of databases. It instead identified chimeras based on the observation that chimeric sequences should undergo one less PCR cycle than its parent strands, and thus the parent strands should be at least equal in frequency to the chimeric strands. Perseus then compares each sequence to every sequence with equal or greater frequency in order to identify a set of potential parents and break points. Based on the three-way alignment data, an optimum chimera is calculated. If the distance from the sequence to the optimum chimera is less than 0.15, then PCR error could possibly account for the differences. Then another calculation is performed to determine the probability that this chimera could have evolved. If this probability is greater than 50% then it is classified as a chimeric sequence *(69)*.

A mixture of both approaches, UCHIME, also uses chimera-free sequence data to detect chimeras, but can additionally detect them *de novo* using abundance data. The algorithm involves splitting a query into four segments, each of which is then used to search a reference database. If a pair of segments match more closely to separate database sequences than to either candidate sequence alone, it is reported as a chimera based on preset thresholds. In *de novo* mode, it uses a strategy similar to Perseus, which involves first constructing a database in order of decreasing abundance. It then analyzes for chimeras using the same method, but with the theory that parent candidate sequences will have at least $2 \times$ the quantity of the query sequence. Sequences that are not classified as a chimera are then added to the dynamic database *(73)*.

3.3   Operational Taxonomic Units

Only after denoising and chimera detection have been completed is the data ready to be analyzed. Most programs will first do this by assigning amplicon sequences to OTUs. Assigning sequences to OTUs acts as a proxy to species assignments as it accounts for those taxa who may be underrepresented in culture data *(17,74)*. This allows for better diversity estimates and allows for further exploration into unknown microbiota. These sequences are assigned to OTUs based on sequence similarity, but the significance of the groupings and cutoff thresholds are less definite when analyzed *de novo (75)*. Reference-based OTU analysis addresses this issue by analyzing only sequences represented in or similar to known taxa. By referencing curated databases such as Greengenes *(76)*, SILVA *(77)*, or RDP *(9)*, sequences can be labeled based on data generated from previously cultured bacteria. In addition, it serves as another check against chimeras and other sources of errors. While reference-based OTU databases provide their own web-based tools to align and match sequences, *de novo* OTU analysis requires its own alignment prior to phylogenetic analysis.

3.4   Sequence Alignment

One of the most useful tools used early in the analysis of sequence data is NAST, a multiple sequence alignment server specialized for 16S rRNA genes stored on the GreenGenes database. It takes unaligned sequences and matches them by comparing seven nucleotide units via BLAST to a core set of 10,000 nonchimeric, near full-length 16S rRNA sequences, favoring a long match. It then trims sequences of flanking data such as tRNA, spacer regions, vectors, and any sequence outside of the high-scoring pair range. The alignments are then returned via e-mail and further phylogenetic analysis can then be pursued *(70)*.

3.5 Phylogenetic Analysis

Phylogenetic analysis has become the mainstay for inferring relationships amongst species and is accomplished using various methods. The distance matrix method involves calculations of distance between every pair of sequences, followed by an algorithm, such as the neighbor joining *(78,79)*, minimum evolution, or least squares *(79,80)* methods to generate phylogenetic clusters. The least squares method minimizes the difference between the calculated distance in the matrix, and the expected distance on the tree such that the tree with the smallest deviation of the sum of branch lengths represents the best estimate of the true tree *(79,80)*. The minimum evolution method is similar to the least squares method in that it minimizes a measure of difference, but focuses on the tree length itself. A shorter tree in this method is more likely to be correct compared to a longer tree *(80)*. Finally the most popular method is the neighbor-joining method. It operates by linking each sequence outward from a single point and using a cluster algorithm to successively pair taxas together until a tree is obtained. In addition the taxa are chosen to minimize tree length *(78,81)*. Although not necessarily the most accurate, the main advantage of nearest neighbor analysis is the speed at which the phylogenetic tree can be generated. FastTree2 *(82)*, a nearest neighbor maximum-likelihood algorithm which also incorporates minimum evolution subtree analysis, is one of the best known examples. It is several magnitudes faster than the more accurate, by approximately 4% *(82)*, maximum-likelihood program RAXmL *(83)* depending on the number of sequences analyzed. Whereas FastTree2 can create a tree for 15,000 distinct 16S rRNA sequences in roughly an hour, it would take RAXmL roughly 64 h. For further comparison, it takes FastTree2 22 h to create a 237,000 16S rRNA sequence tree with a desktop computer whereas it is estimated to take RAXmL half a year *(82)*.

3.6 Estimating Diversity

After establishing a phylogenetic tree, further analysis can be pursued to determine patterns in diversity within and across communities. Phylogenetic diversity (PD) can be determined with the aid of programs like DOTUR *(84)*, and DIVERSITY *(85)*. DOTUR calculates the Shannon-Weiner index, a calculation of diversity based on how predictable the community is, and the Simpson diversity index, a calculation based on both the number of different samples in a community and their proportions in addition to several richness estimators. DIVERSITY calculates PD a diversity index that summates the lengths of all phylogenetic branches required to span the taxa represented by the tree. To visualize diversity, one can also plot a rarefaction curve, which is an option in DOTUR, and can be used to visualize sufficiency in sample collection. Although these indices can help compare between communities, direct comparisons can be performed by programs like UniFrac *(86)* and LIBSHUFF *(87)*.

3.7   Comparing Between Communities

One of the earlier and popular tools used to compare between communities is LIBSHUFF. It performs its analysis by first determining the coverage of a community (or library), comparing it to the coverage of sequences shared between two different libraries and graphing it over evolutionary distance. Similarities between libraries are reflected in similarities between the curves representing the shared sequences and the original sequences. It then calculates a difference in coverage (ΔC). Significance is determined by a Monte Carlo resampling method. LIBSHUFF then shuffles the two libraries together and randomly divides them into new libraries with the same number of sequences as the original with random distributions and calculates a ΔC. The program repeats this process for a total of 1000 values. If ΔC of the original library is greater than 95% of shuffled ΔC's, then significance is achieved at a *p*-value of 0.05 *(83,87)*.

UniFrac compares communities by integrating the phylogenetic (P) test and the F_{ST} test *(88)*. The F_{ST} test assesses genetic diversity and indicates whether less genetic diversity is present within each community when compared to both communities combined. The (P) test assesses whether the two compared communities covary. Significance indicates that the phylogenetic lineages within both communities are distinct from each other. By interpreting both tests, the investigator is better able to compare the communities. For example, significance in F_{ST} and insignificance in (P) indicates that there is less genetic diversity in the two communities individually compared to when combined, but that both communities have statistically indistinguishable PD. The main limitation of these tests is that it cannot provide information regarding the relationships of several communities. UniFrac further addresses this by introducing its own metric which measures the distance between communities as a percentage of branch length of lineages that are unique to each community *(86)*. Visualization of these patterns can be performed through principal coordinate analysis of the distance matrix data. This produces a scatterplot, either 2D or 3D, that is capable of visualizing relatedness amongst communities *(17,86)*. Further analyses can be pursued by comparing results to various databases and incorporating and comparing metadata, which may include information on how DNA extraction was performed, the sequencing pipeline used, or the subjects from which the samples were obtained *(17,89)*.

3.8 Unifying Platforms

To summarize, the analysis of high-throughput sequence data is a multistage process. The first step will typically consist of denoising and chimera checking. After this, OTU assignment can be performed, whether it is *de novo* or through reference-based methods. Taxonomic and phylogenetic analysis can then be performed, after which comparisons can be made to other libraries and databases. Although it can be fairly overwhelming to perform all of these analyses, there are several programs that have been developed to unify all the steps necessary. Two of the most popular programs are QIIME *(89)* and mothur *(90)*. Whereas QIIME integrates many of the popular tools mentioned above among others, mothur reimplements many the algorithms used for faster computing and ease of use. It is important to recognize that results from mothur's reimplementation of algorithms may not replicate results from the original tool *(91)*. Regardless, both are robust, comprehensive platforms that produce publication quality graphics and scale well to desktop and laptop use. Furthermore, one of their greatest strengths is the highly developed support and constant updating based on user feedback and progress in the field of bioinformatics.

4 THE MICROBIOME IN HEALTH AND DISEASE

Along with the dramatic shift from culture-dependent to sequencing techniques used in analyzing, the human microbiome came a wealth of information about its role in human health and disease. It is important to note, however, that while many associations have been found between the microbiome and certain disease states, a causal direction has yet to be established for many of these relationships.

4.1 The Gastric Microbiome

The belief that the acidic gastric environment is sterile was overturned with the discovery that *Helicobacter pylori* is adapted to survive in this specialized niche. In fact, in individuals not inhabited by *H. pylori*, there is a high diversity of gastric microbiota represented by several major phylotypes including *Streptococcus, Actinomyces, Prevotella,* and *Gemella (34)*. However, in those colonized by *H. pylori,* this organism usually accounts for >90% of sequence reads of the gastric microbiota *(34)*, demonstrating an evolved fitness for this specialized niche *(92)*. As described by Atherton and Blaser, *H. pylori* is a classical amphibiont; that is, colonization of the stomach by this organism is strongly associated with several human diseases *(92)*. *H. pylori*-positive individuals have an increased risk for developing peptic ulcer disease, gastric adenocarcinomas, and gastric mucosa-associated lymphoid tissue tumors *(93)*, but interestingly this organism is associated with a decreased risk of reflux esophagitis *(94)* and childhood-onset asthma *(95)*. Additionally, some evidence has pointed to an association between *H. pylori*-positive individuals and levels of the gastric metabolic hormone, ghrelin, though evidence on this topic is mixed and not clear as of yet *(96)*.

4.2   The Colonic Microbiota and Colorectal Cancer

The colonic microbiome has been extensively investigated for its potential role in the development of colorectal cancers (CRCs) *(97,98)*, and has been postulated to act through a variety of mechanisms that may alter tumor formation and growth. Microbiome studies have demonstrated that antibiotic administration affects the expression of host genes involved in cell cycle regulation, thus reducing epithelial proliferation *(99)*. Specifically, colonic bacteria are known to produce short-chain fatty acids (SCFAs) from dietary complex carbohydrates *(100)*, and SCFAs such as butyrate may induce cell cycle arrest, differentiation, and apoptosis through Wnt activity *(101)*. Wang *et al.* recently demonstrated that in patients with CRC, there appears to be a decrease in butyrate-producing gut bacteria when compared to normal controls *(102)*. This same study also demonstrated an increase in several opportunistic pathogens in patients with CRC, including *Enterococcus, Streptococcus,* and *Escherichia/Shigella (102)*. *Streptococcus bovis* has long been linked to CRC, particularly in early, culture-dependent techniques for identifying gut microbiota, and certain wall-extracted antigens from this bacterium have shown to promote carcinogenesis in human epithelial colorectal adenocarcinoma cells *(103)*. Additionally, studies with *Enterococcus faecalis* demonstrated that free radical production by this organism leads to DNA damage in rat colonic epithelial cells, which may play a role in the chromosomal instability associated with sporadic adenomatous polyps and CRC *(104)*. Finally, colonic microbes may also promote CRC by stimulating exaggerated immune responses through helper type 17 T-cells (Th17) *(105,106)*, and it has been postulated that microbial products may penetrate through mucosal defects in adenomatous lesions, promoting host immune responses and inflammation and ultimately driving tumor growth *(106)*.

Members of the *Fusobacterium* genus have recently been implicated in CRC *(102,107,108)*. *Fusobacterium nucleatum* is a mucosally adherent and proinflammatory microbe indigenous to the oral cavity *(109)*, and has been associated with periodontitis *(110)* and a number of other diseases. Two recent studies used qPCR analysis and high-throughput sequencing to compare whole genome sequences of *Fusobacterium* species between CRC tissues and matched normal colonic tissue. Castellarin *et al.* demonstrated that both *F. nucleatum* as well as other members of the *Fusobacterium* genus were enriched in the mucosal tissue of CRC compared to adjacent normal tissues *(107)* while Kostic *et al.* demonstrated an enrichment of *Fusobacterium* sequences in colonic carcinomas compared with control tissue, and a concomitant decrease in both the Bacteroidetes and Firmicutes phyla in tumor tissue as well *(108)*. It is important to note, however, that causality in these associations has not yet been determined.

4.3   The Colonic Microbiota and Inflammatory Bowel Disease

The inflammatory bowel diseases (IBDs), namely Crohn's disease and ulcerative colitis (UC), are characterized by relapsing periods of gastrointestinal inflammation. Interactions between the gut microbiome and host immune response have long been suspected in their pathogenesis. Exposure to antibiotics in childhood has been associated with an increased risk for Crohn's disease *(111)* while antibiotics such as metronidazole *(112)* and ciprofloxacin *(113)* are currently among the treatment regimens available for treating active Crohn's disease *(114)*, suggesting that gut microbial perturbations are important for IBD risk. Indeed, IBD is thought to be due to a community microbial dysbiosis, that is, a perturbation from normal gut microbiota, rather than a single causative organism *(115)*. While studies in both UC and Crohn's disease patients have shown significantly reduced gut microbial diversity when compared to healthy controls *(115,116)*, they have also demonstrated that microbial populations in these patients cluster together according to their disease and separately from healthy individuals *(117)*.

In addition to decreased bacterial diversity, patients with IBD have decreased proportions of members of the Firmicutes phylum (notably the *C. leptum* and *C. coccoides* groups) *(118)* and increased members of the Proteobacteria phylum *(115,119)*, notably of the *Enterobacteriaceae* family *(120,121)*. Additionally, members of the *Enterobacteriaceae* family may act in concert with a disordered microbiome to increase the risk of UC *(122)*. Notably, a recent metagenomic analysis identified altered microbial metabolism during oxidative stress as well as perturbed nutrient availability during tissue damage *(120)* in IBD patients. Additionally, this study identified reductions in *Roseburia*, a butyrate producer, and *Phascolarctobacterium*, a succinate consumer and proprionate producer, in patients with IBD *(120)*. These microbial metabolic patterns, however, are still being investigated and a causal direction has yet to be established.

The pathogenesis of IBDs has long been thought to be related to the interplay between a dysbiotic gut microbial population and the human host. It has been well established that the gut microbiome is required for activation of host immune responses *(123)*. For example, segmented filamentous bacteria (SFB) are sufficient to induce differentiation of Th17 cells in mouse small bowel lamina propria *(124)*, and Th17 cells have been postulated to play a role in IBD pathogenesis through their release of regulatory cytokines *(125)*. Additionally, differentiation of CD4+ T cells into regulatory T cells can be induced by *Bacteroides fragilis* through the production of polysaccharide A *(126)*. Individual susceptibility to IBD has also been associated with polymorphisms in host genes such as nucleotide-binding oligomerization domain-containing protein 2 *(127,128)* as well as toll-like receptor 4 *(129)*, both bacterial sensor genes.

4.4   Irritable Bowel Syndrome

It has been shown that patients with irritable bowel syndrome (IBS) or visceral hypersensitivity have an increased ratio of Firmicutes to Bacteroidetes when compared to normal controls, resulting from increased numbers of *Ruminococcus*, *Clostridium*, and *Dorea (130)*. Additionally, different subgroups of IBS patients have been identified with distinct gut microbial populations, some of which overlapped with those of normal controls *(131)*, and investigators have suggested that microbial "fingerprinting" of IBS subgroups may be useful in delineating prognostic or therapeutic options for patients *(131)*. Interestingly, certain Firmicutes found to be increased in patients with IBS, such as those of the *Ruminococcaceae* and *Clostridium* cluster XIVa groups, are notable SCFA producers *(132)* which may play a role in disease pathogenesis. Studies in rats have shown that rectal infusion of butyrate can produce concentration-dependent colonic hypersensitivity *(133)*. Additionally, Jeffery *et al.* note that because some SCFAs act on receptors involved in modulating certain inflammatory processes, the microbial production of SCFAs may act through unknown mechanisms to invoke a proinflammatory gut environment contributing to symptoms of IBS *(134)*. However, additional studies are needed to describe the mechanisms by which the gut microbiota or their products influence IBS pathogenesis.

4.5   The Gut Microbiota and Diseases of the Liver

The gut microbiota may also be implicated in diseases of the liver such as hepatocellular carcinoma, cirrhosis, alcoholic steatosis, and nonalcoholic fatty liver disease (NAFLD) *(135)*. *Helicobacter hepaticus*, an organism known to colonize the murine intestine without clinical signs of disease, appears to promote hepatocellular carcinoma *(136)*. Changes in bacterial taxonomic groups can be observed in association with shifts in colonic bile acids, where the ratio of Firmicutes to Bacteroidetes is altered *(137)*, as well as in cirrhosis, where enrichment of the Proteobacteria and Fusobacteria phyla, and of the *Enterobacteriaceae*, *Veillonellaceae*, and *Streptococcaceae* families is seen *(138)*. In rats, chronic alcohol consumption leads to a dysbiotic gut microbiome, though the role of the gut microbiota in alcoholic liver disease in still under investigation *(139,140)*. NAFLD is characterized by hepatic fat accumulation in the absence of chronic alcohol consumption and is associated with diseases such as obesity and type II diabetes (mechanisms by which the gut microbiome contributes to obesity and metabolic disease are further discussed in the following section). Compared with germ-free mice, presence of gut microbiota leads to hepatic triglyceride accumulation due to suppression of intestinal epithelium angiopoietin-related protein 4, an inhibitor of lipoprotein lipase *(141)*. Although many observations suggest links between microbiome composition and liver disease, definitive causal relationships in humans have yet to be defined.

4.6   Host Diet, Metabolism, and the Gut Microbiota

4.6.1   The Gut Microbiome and Obesity

Numerous studies have shown associations of the gut microbiome with diseases of metabolism such as obesity and type II diabetes. Ley and colleagues showed that genetically obese (ob/ob) mice had significant differences in the relative abundances of bacterial taxa when compared with their lean (ob/+ and +/+ wild type) siblings, with a 50% decrease in Bacteroidetes and a greater proportion of Firmicutes *(142)*. Additionally, studies of the obese gut microbiome in mice have showed an increased capacity for harvesting energy from the diet when compared to controls, as well as a transmissible obese phenotype when germ-free mice are colonized with microbiota from obese donors *(143)*. More recent murine studies have shown that subtherapeutic antibiotic administration in young mice increased adiposity and hormone levels related to metabolism, further strengthening a role for the gut microbiome in metabolic homeostasis *(144)*. Alterations in metabolic homeostasis and concomitant shifts in bacterial taxa can also be induced by dietary changes, as in supplementation with the nonfermentable fiber hydroxypropyl methylcellulose (HPMC). Cox *et al.* showed that mice given a high-fat diet (HFD) supplemented with 10% HPMC gained less weight, had lower plasma cholesterol and liver triglycerides, as well as decreased and altered microbial diversity when compared to mice on HFD alone *(145)*.

In humans studies, it has also been shown that the relative proportion of Bacteroidetes was decreased in obese compared with lean subjects *(146)*. Similarly, studies of mono- and dizygotic twins have shown that obesity was associated with phylum-level changes in microbiota, specifically lower proportions of Bacteroidetes, as well as decreased bacterial diversity and altered representation of genes for lipid and carbohydrate metabolism *(19)*. However, the association between obesity and the relative proportions of the major bacterial taxa, namely Bacteroidetes and Firmicutes, is still unclear, as some studies have shown conflicting evidence demonstrating higher proportions of Bacteroidetes in overweight and obese populations *(147)*. Additionally, different interventions in obesity have been associated with changes in microbial profiles, such as an increase in the relative proportion of Bacteroidetes with weight loss *(146)*, as well as significantly increased levels of Proteobacteria following Roux-en-Y bariatric surgery *(148)*. As in mouse-model studies, the development of obesity has also been shown to be associated with early antibiotic administration in humans, specifically in infants before the age of 6 months *(149)*. Conversely, perinatal probiotic administration with *Lactobacillus rhamnosus* decreased excessive weight gain in childhood *(150)*. Together, these studies support the concept that early, modifiable perturbations to the gut microbiota may contribute to childhood-onset diabetes.

4.6.2 The Gut Microbiome and Type II Diabetes

The worldwide burden of type II diabetes has increased dramatically in the past few decades. While the etiology of this disease has largely been considered a combination of environmental and genetic components *(151)*, there is increasing interest in the role of the human microbiome with this disease. Studies by Larsen *et al.* demonstrated reduced proportions of the Firmicutes phylum and Clostridia class in patients with type II diabetes when compared to controls, as well as a positive correlation of plasma glucose levels with the Betaproteobacteria class and ratio of Bacteroidetes to Firmicutes *(152)*. A recent metagenome-wide association study by Qin *et al.* analyzed over 300 Chinese patients to determine genetic and functional components of the gut metagenome and its association with type II diabetes *(153)*. These researchers found only a moderate degree of gut microbial dysbiosis, in diabetic patients but also a significant decrease in butyrate-producing bacteria *(153)*. The authors proposed that butyrate producers may be protective in diseases such as type II diabetes, a fact that was supported in a study using gut microbiota transplantation from lean donors to patients with metabolic syndrome that demonstrated improved insulin sensitivity in conjunction with a rise in known butyrate producing bacteria, such as those related to *Roseburia* *(154)*. Additionally, similar changes are seen in other diseases (such as CRC *(102)*), suggesting that butyrate producers may be protective for multiple disease states *(153)*. Qin *et al.* also noted an increase in multiple opportunistic pathogens (most notably those of the *Clostridium* genus) as well as an enrichment of microbial oxidative stress resistance genes in patients with type II diabetes, and suggested that gut metagenomic markers may be more useful than human genomic differences in differentiating type II diabetes cases from controls *(153)*. This hypothesis was examined by Karlsson *et al.* who employed shotgun sequencing to analyze compositional and functional variations in the fecal metagenome of 145 European women with normal, impaired, or diabetic glucose control, and developed a mathematical model to identify both type II diabetics as well as those with diabetes-like metabolism (i.e., impaired glucose tolerance) based on microbial metagenomic profiles *(155)*. Their findings demonstrated an increase in the *Clostridium clostridioforme* metagenomic cluster as well as a decrease in the butyrate-producing *Roseburia* group of type II diabetic metagenomes which confirmed that parallel changes were identified in the European and Chinese cohorts used in these two studies *(155)*.

However, the most discriminatory metagenomic clusters differed between cohorts, emphasizing the need for development of population-specific metagenomic predictive methods.

The gut microbiome is also being investigated for its role in severe malnutrition states such as kwashiorkor *(156,157)*. Indeed, a complex interplay exists between the gut microbiota and host metabolism, and likely has far-ranging implications on normal metabolic homeostasis as well as several disease states, the mechanisms of which we are just beginning to elucidate.

4.6.3   The Gut Microbiome and Cardiovascular Disease

Given the known risk that high-fat and meat-rich diets pose to cardiovascular health, as well the evolving information being uncovered regarding the role of the gut microbiome in human metabolism, researchers have looked to the potential implications of the gut microbiome and its metabolic functions in cardiovascular disease (CVD). In a metabolomic study by Wang *et al.*, three metabolites of dietary phophatidylcholine, namely choline, trimethylamine N-oxide (TMAO), and betaine, were shown to predict risk for CVD in human subjects. Supplementation of these metabolites in mice produced host changes consistent with CVD pathogenesis, such as the upregulation of macrophage scavenger receptors and atherosclerosis itself *(158)*. Furthermore, the role of the gut microbiome in this process was pivotal, as studies using germ-free mice as well as mice given antibiotic suppression demonstrated that the gut microbiota play a key role in generating TMAO from dietary phosphatidylcholine and augmenting macrophage foam cell formation *(158)*. It has also been shown that the gut microbiota play a key role in the metabolism of L-carnitine, abundant in red meat, to TMAO, also accelerating atherosclerosis in mice *(159)*. While these investigations are still early, the interaction between the gut microbiota and pathways by which dietary substances are converted to proatherogenic metabolites suggest novel possibilities for interventions to prevent or reduce CVD.

4.7   The Gut Microbiome and Renal Physiology

Hyperoxaluria is a known risk factor for renal calcium oxalate stones and attempts to identify methods for reducing urinary oxalate levels brought attention to the gut bacterium, *Oxalobacter formigenes*, which inhabits the human gut and is dependent on the metabolism of oxalate for energy *(160)*. While there is some evidence that recurrent kidney stone-formers are less likely to be inhabited by *O. formigenes*, not all individuals who test negative for *O. formigenes* have hyperoxaluria *(161)*, and some studies suggest that other oxalate metabolizers such as *Lactobaccilus* and *Bifidobacterium* species may be more protective in urolithiasis *(160,162)*. There is increasing interest in the contribution of metabolites created by colonic bacteria, including p-cresol sulfate and indoxyl sulfate as well as numerous other unidentified substances *(163)* in the development of uremic illness, as these solutes are normally secreted by the kidney and accumulate in the plasma with renal failure *(163,164)*. Thus, in patients with chronic kidney disease, the colon and its gut microbiota are being looked at as potential targets for therapy in patients with uremia through the proposed use of pro- and prebiotics, antibiotics, intestinal absorptive strategies, and other drug therapies *(165)*.

4.8 The Gut Microbiota and Rheumatoid Arthritis

Though autoimmune diseases likely result from a complex interaction between genetic and environmental factors, dysbiosis of the gut microbiome has been postulated to play a role in the development of rheumatoid arthritis and other autoimmune diseases *(166)*. In mice, it has been shown that SFB colonization in the small intestine is sufficient to produce Th17 cell expansion in the lamina propria *(167)*. Autoreactive Th17 cells can migrate to immune tissues to stimulate production of autoantibody producing plasma cells, leading to immune-mediated destruction of joints seen in RA *(166)*. In fact, it has been shown that in K/BxN germ-free mice, autoimmune arthritis was strongly attenuated with decreases in serum autoantibody titers, splenic auto-antibody secreting cells, and Th17 cell populations of both the spleen and lamina propria. Reintroduction of SFB was enough to induce rapid arthritis and population of lamina propria with Th17 cells *(168)*. Additionally, *IL1rn−/−* mice, which develop spontaneous arthritis due to excessive IL-1 signaling, do not develop arthritis in germ-free models but rapidly develop arthritis upon monocontamination with *Lactobacillus bifidus*, an indigenous microbe *(169)*. Thus, it is possible that a microbial gut dysbiosis leads to a dysregulated host immune response that may act at distant host sites to produce the autoimmune inflammation seen in diseases such as rheumatoid arthritis.

4.9 The Cutaneous Microbiome

Not surprisingly, the cutaneous microbiome has been studied for its potential role in dermatologic conditions such as psoriasis, atopic dermatitis, and acne. Psoriasis, a chronic idiopathic inflammatory disease of the skin *(170)*, has been associated with an overrepresentation of Firmicutes and under-representation of Actinobacteria when compared to both the unaffected skin of psoriatic patients and the skin of normal controls *(171)*. Additionally, Proteobacteria were detected less frequently in psoriatic lesions compared to skin of healthy controls *(171)*. Subsequent studies using pyrosequencing techniques confirmed that Actinobacteria were more abundant in controls compared to patients with psoriasis; however, Proteobacteria were significantly higher in trunk skin samples from psoriatic patients compared to controls *(172)*.

While these studies used different sampling techniques and tools for analysis that may account for their conflicting findings, a clear microbial association with psoriasis remains unidentified.

A significant increase in the prevalence of allergic diseases in developed countries has largely been attributed to a modern Western lifestyle, characterized by diminished exposure to soil and animals, nutritional bias, obesity, and increased exposure to antibiotics *(173)*. Such environmental influences have been postulated to impact intestinal microbiota *(173)*, the in turn diseases such as atopic dermatitis, a chronic allergic disease of the skin that has increased threefold in industrialized countries over the last three decades *(174)*. In fact, a recent study using 16S rRNA analysis and HTS showed that disease activity in children with atopic dermatitis correlated with shifts in the cutaneous microbiome *(15)*. During disease flares, the proportions of *Staphylococcus aureus* and *Staphylococcus epidermidis* sequences were significantly higher than at baseline or after treatment, and increases in *Streptococcus*, *Propionibacterium*, and *Corynebacterium* species occurred after therapy *(15)*.

Shifts in the cutaneous microbiome have also been noted in chronic ulcer disease, such as those caused by venous stasis or diabetes *(175)*. Patients with chronic ulcers treated with antibiotics have been shown to have an increased abundance of *Pseudomonadaceae* while an increase in *Streptococcaceae* has been noted in diabetic ulcers *(175)*. A longitudinal shift in wound microbiota has also been shown to coincide with impaired healing in diabetic mice, and may interact with aberrantly expressed host cutaneous defense response genes leading to ulcerogenesis *(176)*. When compared to controls, the feet of diabetic men has also been noted to have decreased populations of *Staphylococcus* species, a relative increase in the population of S. *aureus* and increased bacterial diversity, which may increase the risk for wound infections in diabetic patients *(177)*.

Additionally, *Propionibacterium acnes*, which thrives in the pilosebaceous glands of the skin, secretes enzymes that cause local skin irritation and inflammation, and is generally accepted to play a role in the dermatologic condition, acne *(178)*. The role of other microbiota in the development of acne, however, is still under investigation.

4.10   Emerging Research Involving the Lung Microbiome

In contrast to studies investigating the gut microbiome, research involving the human lung microbiome is only just emerging. Studies of the lung microbiome present unique challenges compared with those of the gut, including the difficulty in sampling lower airways of human subjects while avoiding upper airway contamination, and a decreased abundance of bacteria that is also more susceptible to contamination from reagents used to obtain and process samples *(179)*. Nonetheless, researchers have begun to study the lung microbiome in diseases such as asthma, COPD, and cystic fibrosis. In patients with asthma, 16S rRNA analysis identified a higher abundance of pathogenic Proteobacteria, such as *Haemophilus* species, in the bronchi of adults and children with asthma as well as patients with COPD compared to controls *(180)*. Additionally, this study found a higher frequency of Bacteroidetes, specifically *Prevotella* species, in controls compared to asthmatics or COPD patients *(180)*. In another study using 16S rRNA analysis, bacterial diversity and the abundance of certain microbial families such as the *Comamonadaceae*, *Sphingomonadaceae*, and *Oxalobacteraceae* correlated with degree of bronchial hyper-responsiveness in poorly controlled asthmatics *(181)*. However, whether a core lung microbiome exists in healthy subjects or in those individuals with specific lung diseases is still under investigation *(179)*.

REFERENCES

1. Staley, J. T.; Konopka, A. *Annu. Rev. Microbiol.* **1985**, *39*, 321–346.
2. Amann, R. I.; Ludwig, W.; Schleifer, K. H. *Microbiol. Rev.* **1995**, *59*, 143–169.
3. Woese, C. R.; Fox, G. E. *Proc. Natl. Acad. Sci. U. S. A.* **1977**, *74*, 5088–5090.
4. Woese, C. R. *Microbiol. Rev.* **1987**, *51*, 221–271.
5. Liu, W. T.; Marsh, T. L.; Cheng, H.; Forney, L. J. *Appl. Environ. Microbiol.* **1997**, *63*, 4516–4522.
6. Anderson, I. C.; Cairney, J. W. G. *Environ. Microbiol.* **2004**, *6*, 769–779.
7. Grice, E. A.; Segre, J. A. *Rev. Genom. Human Genet.* **2012**, *13*, 151–170.
8. Lane, D. J.; Pace, B.; Olsen, G. J.; Stahl, D. A.; Sogin, M. L.; Pace, N. R. *Proc. Natl. Acad. Sci. U. S. A.* **1985**, *82*, 6955–6959.
9. Cole, J. R.; Wang, Q.; Cardenas, E.; Fish, J.; Chai, B.; Farris, R. J.; Kulam-Syed-Mohideen, A. S.; McGarrell, D. M.; Marsh, T.; Garrity, G. M.; Tiedje, J. M. *Nucleic Acids Res.* **2009**, *37*, D141–D145.
10. Hamadyand, M.; Knight, R. *Genome Res.* **2009**, *19*, 1141–1152.
11. Lozupone, C. A.; Knight, R. *Proc. Natl. Acad. Sci. U. S. A.* **2007**, *104*, 11436–11440.
12. García-Martínez, J.; Acinas, S. G.; Antón, A. I.; Rodríguez-Valera, F. *J. Microbiol. Methods* **1999**, *36*, 55–64.
13. Sogin, M. L.; et al. *Proc. Natl. Acad. Sci. U. S. A.* **2006**, *103*, 12115–12120.
14. DeSantis, T. Z.; Brodie, E. L.; Moberg, J. P.; Zubieta, I. X.; Piceno, Y. M.; Andersen, G. L. *Microb. Ecol.* **2007**, *53*, 371–383.
15. Kong, H. H. *Trends Mol. Med.* **2011**, *17*, 320–328.
16. Kolbert, C. P.; Persing, D. H. *Curr. Opin. Microbiol.* **1999**, *2*, 299–305.
17. Kuczynski, J.; Lauber, C. L.; Walters, W. A.; Parfrey, L. W.; Clemente, J. C.; Gevers, D.; Knight, R. *Nat. Rev. Genet.* **2011**, *13*, 47–58.
18. Liu, Z.; Lozupone, C.; Hamady, M.; Bushman, F. D.; Knight, R. *Nucleic Acids Res.* **2007**, *35*, e120.
19. Turnbaugh, P. J.; Hamady, M.; Yatsunenko, T.; Cantarel, B. L.; Duncan, A.; Ley, R. E.; Sogin, M. L.; Jones, W. J.; Roe, B. A.; Affourtit, J. P.; Egholm, M.; Henrissat, B.; Heath, A. C.; Knight, R.; Gordon, J. I. *Nature* **2009**, *457*, 480–484.
20. Petrosino, J. F.; Highlander, S.; Luna, R. A.; Gibbs, R. A.; Versalovic, J. *Clin. Chem.* **2009**, *55*, 856–866.
21. Ehn, M.; Ahmadian, A.; Nilsson, P.; Lundeberg, J.; Hober, S. *Electrophoresis* **2002**, *23*, 3289–3299.

22. Ehn, M.; Nourizad, N.; Bergström, K.; Ahmadian, A.; Nyrén, P.; Lundeberg, J.; Hober, S. *Anal. Biochem.* **2004**, *329*, 11–20.
23. Binladen, J.; Gilbert, M. T. P.; Bollback, J. P.; Panitz, F.; Bendixen, C.; Nielsen, R.; Willerslev, E. *PLoS ONE* **2007**, *2*, e197.
24. Hamady, M.; Walker, J. J.; Harris, J. K.; Gold, N. J.; Knight, R. *Nat. Methods* **2008**, *5*, 235–237.
25. Hoffmann, C.; Minkah, N.; Leipzig, J.; Wang, G.; Arens, M. Q.; Tebas, P.; Bushman, F. D. *Nucleic Acids Res.* **2007**, *35*, e91.
26. Parameswaran, P.; Jalili, R.; Tao, L.; Shokralla, S.; Gharizadeh, B.; Ronaghi, M.; Fire, A. Z. *Nucleic Acids Res.* **2007**, *35*, e130.
27. Meyer, M.; Stenzel, U.; Hofreiter, M. *Nat. Protoc.* **2008**, *3*, 267–278.
28. Wang, Q.; Garrity, G. M.; Tiedje, J. M.; Cole, J. R. *Appl. Environ. Microbiol.* **2007**, *73*, 5261–5267.
29. Wommack, K. E.; Bhavsar, J.; Ravel, J. *Appl. Environ. Microbiol.* **2008**, *74*, 1453–1463.
30. Baker, G. C.; Smith, J. J.; Cowan, D. A. *J. Microbiol. Methods* **2003**, *55*, 541–555.
31. Edwards, R. A.; Rodriguez-Brito, B.; Wegley, L.; Haynes, M.; Breitbart, M.; Peterson, D. M.; Saar, M. O.; Alexander, S.; Alexander, E. C.; Rohwer, F. *BMC Genomics* **2006**, *7*, 57.
32. Huse, S. M.; Huber, J. A.; Morrison, H. G.; Sogin, M. L.; Welch, D. M. *Genome Biol.* **2007**, *8*, R143.
33. Roesch, L. F. W.; Fulthorpe, R. R.; Riva, A.; Casella, G.; Hadwin, A. K. M.; Kent, A. D.; Daroub, S. H.; Camargo, F. A. O.; Farmerie, W. G.; Triplett, E. W. *ISME J.* **2007**, *1*, 283–290.
34. Andersson, A. F.; Lindberg, M.; Jakobsson, H.; Bäckhed, F.; Nyrén, P.; Engstrand, L. *PLoS ONE* **2008**, *3*, e2836.
35. Liu, Z.; DeSantis, T. Z.; Andersen, G. L.; Knight, R. *Nucleic Acids Res.* **2008**, *36*, e120.
36. Frank, J. A.; Reich, C. I.; Sharma, S.; Weisbaum, J. S.; Wilson, B. A.; Olsen, G. J. *Appl. Environ. Microbiol.* **2008**, *74*, 2461–2470.
37. Meyer, A. F.; Lipson, D. A.; Martin, A. P.; Schadt, C. W.; Schmidt, S. K. *Appl. Environ. Microbiol.* **2004**, *70*, 483–489.
38. Harris, J. K.; Kelley, S. T.; Pace, N. R. *Appl. Environ. Microbiol.* **2004**, *70*, 845–849.
39. Marcille, F.; Gomez, A.; Joubert, P.; Ladiré, M.; Veau, G.; Clara, A.; Gavini, F.; Willems, A.; Fons, M. *Appl. Environ. Microbiol.* **2002**, *68*, 3424–3431.
40. Isenbarger, T. A.; Finney, M.; Ríos-Velázquez, C.; Handelsman, J.; Ruvkun, G. *Appl. Environ. Microbiol.* **2008**, *74*, 840–849.
41. Chan, E. Y. *Methods Mol. Biol.* **2009**, *578*, 95–111.
42. Mardis, E. R. *Annu. Rev. Genomics Hum. Genet.* **2008**, *9*, 387–402.
43. Mardis, E. R. *Trends Genet.* **2008**, *24*, 133–141.
44. Ansorge, W. J. *New Biotechnol.* **2009**, *25*, 195–203.
45. Pareek, C. S.; Smoczynski, R.; Tretyn, J. *J. Appl. Genet.* **2011**, *52*, 413–435.
46. Liu, L.; Li, Y.; Li, S.; Hu, N.; He, Y.; Pong, R.; Lin, D.; Lu, L.; Law, M. *J. Biomed. Biotechnol.* **2012**, *251364*, 2012.
47. Zhang, J.; Chiodini, R.; Badr, A.; Zhang, G. *J. Genet. Genom.* **2011**, *38*, 95–109.
48. Metzker, M. L. *Nat. Rev. Genet.* **2010**, *11*, 31–46.
49. Merriman, B.; Rothberg, J. M. *Electrophoresis* **2012**, *33*, 3397–3417.
50. Loman, N. J.; Misra, R. V.; Dallman, T. J.; Constantinidou, C.; Gharbia, S. E.; Wain, J.; Pallen, M. J. *Nat. Biotechnol.* **2012**, *30*, 434–439.
51. Claesson, M. J.; Wang, Q.; O'Sullivan, O.; Greene-Diniz, R.; Cole, J. R.; Ross, R. P.; O'Toole, P. W. *Nucleic Acids Res.* **2010**, *38*, e200.

52. Willner, D.; Daly, J.; Whiley, D.; Grimwood, K.; Wainwright, C. E.; Hugenholtz, P. *PLoS ONE* **2012**, *7*, e34605.

53. Yuan, S.; Cohen, D. B.; Ravel, J.; Abdo, Z.; Forney, L. J. *PLoS ONE* **2012**, *7*, e33865.

54. Zhao, J.; Carmody, L. A.; Kalikin, L. M.; Li, J.; Petrosino, J. F.; Schloss, P. D.; Young, V. B.; LiPuma, J. J. *PLoS ONE* **2012**, *7*, e33127.

55. Vanysacker, L.; Declerck, S. A. J.; Hellemans, B.; De Meester, L.; Vankelecom, I.; Declerck, P. *Appl. Microbiol. Biotechnol.* **2010**, *88*, 299–307.

56. Cuív, P.Ó.; Aguirre de Cárcer, D.; Jones, M.; Klaassens, E. S.; Worthley, D. L.; Whitehall, V. L. J.; Kang, S.; McSweeney, C. S.; Leggett, B. A.; Morrison, M. *Microb. Ecol.* **2011**, *61*, 353–362.

57. Flores, G. E.; Henley, J. B.; Fierer, N. *PLoS ONE* **2012**, *7*, e44563.

58. Zaura, E. *Adv. Dent. Res.* **2012**, *24*, 81–85.

59. Pinard, R.; de Winter, A.; Sarkis, G. J.; Gerstein, M. B.; Tartaro, K. R.; Plant, R. N.; Egholm, M.; Rothberg, J. M.; Leamon, J. H. *BMC Genomics* **2006**, *7*, 216.

60. Human Microbiome Project Consortium. *Nature* **2012**, *486*, 215–221.

61. Liu, B.; Faller, L. L.; Klitgord, N.; Mazumdar, V.; Ghodsi, M.; Sommer, D. D.; Gibbons, T. R.; Treangen, T. J.; Chang, Y.-C.; Li, S.; Stine, O. C.; Hasturk, H.; Kasif, S.; Segrè, D.; Pop, M.; Amar, S. *PLoS ONE* **2012**, *7*, e37919.

62. Hunter, S. J.; Easton, S.; Booth, V.; Henderson, B.; Wade, W. G.; Ward, J. M. *J. Basic Microbiol.* **2011**, *51*, 442–446.

63. Rodrigue, S.; Materna, A. C.; Timberlake, S. C.; Blackburn, M. C.; Malmstrom, R. R.; Alm, E. J.; Chisholm, S. W. *PLoS ONE* **2010**, *5*, e11840.

64. English, A. C.; Richards, S.; Han, Y.; Wang, M.; Vee, V.; Qu, J.; Qin, X.; Muzny, D. M.; Reid, J. G.; Worley, K. C.; Gibbs, R. A. *PLoS ONE* **2012**, *7*, e47768.

65. Kunin, V.; Engelbrektson, A.; Ochman, H.; Hugenholtz, P. *Environ. Microbiol.* **2010**, *12*, 118–123.

66. Haas, B. J.; Gevers, D.; Earl, A. M.; Feldgarden, M.; Ward, D. V.; Giannoukos, G.; Ciulla, D.; Tabbaa, D.; Highlander, S. K.; Sodergren, E.; Methé, B.; DeSantis, T. Z.; Petrosino, J. F.; Knight, R.; Birren, B. W. *Genome Res.* **2011**, *21*, 494–504.

67. Margulies, M.; Egholm, M.; Altman, W. E.; Attiya, S.; Bader, J. S.; Bemben, L. A.; Berka, J.; Braverman, M. S.; Chen, Y.-J.; Chen, Z.; Dewell, S. B.; Du, L.; Fierro, J. M.; Gomes, X. V.; Godwin, B. C.; He, W.; Helgesen, S.; Ho, C. H.; Ho, C. H.; Irzyk, G. P.; Jando, S. C.; Alenquer, M. L. I.; Jarvie, T. P.; Jirage, K. B.; Kim, J.-B.; Knight, J. R.; Lanza, J. R.; Leamon, J. H.; Lefkowitz, S. M.; Lei, M.; Li, J.; Lohman, K. L.; Lu, H.; Makhijani, V. B.; McDade, K. E.; McKenna, M. P.; Myers, E. W.; Nickerson, E.; Nobile, J. R.; Plant, R.; Puc, B. P.; Ronan, M. T.; Roth, G. T.; Sarkis, G. J.; Simons, J. F.; Simpson, J. W.; Srinivasan, M.; Tartaro, K. R.; Tomasz, A.; Vogt, K. A.; Volkmer, G. A.; Wang, S. H.; Wang, Y.; Weiner, M. P.; Yu, P.; Begley, R. F.; Rothberg, J. M. *Nature* **2005**, *437*, 376–380.

68. Quince, C.; Lanzén, A.; Curtis, T. P.; Davenport, R. J.; Hall, N.; Head, I. M.; Read, L. F.; Sloan, W. T. *Nat. Methods* **2009**, *6*, 639–641.

69. Quince, C.; Lanzen, A.; Davenport, R. J.; Turnbaugh, P. J. *BMC Bioinforma.* **2011**, *12*, 38.

70. DeSantis, T. Z.; Hugenholtz, P.; Keller, K.; Brodie, E. L.; Larsen, N.; Piceno, Y. M.; Phan, R.; Andersen, G. L. *Nucleic Acids Res.* **2006**, *34*, W394–W399.

71. Reeder, J.; Knight, R. *Nat. Methods* **2010**, *7*, 668–669.

72. Huse, S. M.; Welch, D. M.; Morrison, H. G.; Sogin, M. L. *Environ. Microbiol.* **2010**, *12*, 1889–1898.

73. Edgar, R. C.; Haas, B. J.; Clemente, J. C.; Quince, C.; Knight, R. *Bioinformatics* **2011**, *27*, 2194–2200.

74. Schloss, P. D.; Westcott, S. L. *Appl. Environ. Microbiol.* **2011**, *77*, 3219–3226.
75. Rappé, M. S.; Giovannoni, S. J. *Annu. Rev. Microbiol.* **2003**, *57*, 369–394.
76. DeSantis, T. Z.; Hugenholtz, P.; Larsen, N.; Rojas, M.; Brodie, E. L.; Keller, K.; Huber, T.; Dalevi, D.; Hu, P.; Andersen, G. L. *Appl. Environ. Microbiol.* **2006**, *72*, 5069–5072.
77. Pruesse, E.; Quast, C.; Knittel, K.; Fuchs, B. M.; Ludwig, W.; Peplies, J.; Glöckner, F. O. *Nucleic Acids Res.* **2007**, *35*, 7188–7196.
78. Saitou, N.; Nei, M. *Mol. Biol. Evol.* **1987**, *4*, 406–425.
79. Yang, Z.; Rannala, B. *Nat. Rev. Genet.* **2012**, *13*, 303–314.
80. Cavalli-Sforza, L. L.; Edwards, A. W. F. *Am. J. Hum. Genet.* **1967**, *19*, 233–257.
81. Gascuel, O.; Steel, M. *Mol. Biol. Evol.* **2006**, *23*, 1997–2000.
82. Price, M. N.; Dehal, P. S.; Arkin, A. P. *PLoS ONE* **2010**, *5*, e9490.
83. Schloss, P. D.; Larget, B. R.; Handelsman, J. *Appl. Environ. Microbiol.* **2004**, *70*, 5485–5492.
84. Schloss, P. D.; Handelsman, J. *Appl. Environ. Microbiol.* **2005**, *71*, 1501–1506.
85. Faith, D. P.; Baker, A. M. *Evol. Bioinformatics Online* **2006**, *2*, 121–128.
86. Lozupone, C.; Hamady, M.; Knight, R. *BMC Bioinforma.* **2006**, *7*, 371.
87. Singleton, D. R.; Rathbun, S. L.; Dyszynski, G. E.; Whitman, W. B. *Mol. Microb. Ecol. Man.* **2004**, *2*, 3263–3274.
88. Martin, A. P. *Appl. Environ. Microbiol.* **2002**, *68*, 3673–3682.
89. Caporaso, J. G.; Kuczynski, J.; Stombaugh, J.; Bittinger, K.; Bushman, F. D.; Costello, E. K.; Fierer, N.; Peña, A. G.; Goodrich, J. K.; Gordon, J. I.; Huttley, G. A.; Kelley, S. T.; Knights, D.; Koenig, J. E.; Ley, R. E.; Lozupone, C. A.; Mcdonald, D.; Muegge, B. D.; Pirrung, M.; Reeder, J.; Sevinsky, J. R.; Turnbaugh, P. J.; Walters, W. A.; Widmann, J.; Yatsunenko, T.; Zaneveld, J.; Knight, R. *Nat. Methods* **2010**, *7*, 335–336.
90. Schloss, P. D.; Westcott, S. L.; Ryabin, T.; Hall, J. R.; Hartmann, M.; Hollister, E. B.; Lesniewski, R. A.; Oakley, B. B.; Parks, D. H.; Robinson, C. J.; Sahl, J. W.; Stres, B.; Thallinger, G. G.; Van Horn, D. J.; Weber, C. F. *Appl. Environ. Microbiol.* **2009**, *75*, 7537–7541.
91. Kuczynski, J.; Stombaugh, J.; Walters, W. A.; González, A.; Caporaso, J. G.; Knight, R. In: *Current Protocols in Bioinformatics;* Baxevanis, A. D., Ed.; 2011, Chapter 10, Unit 10.7.
92. Atherton, J. C.; Blaser, M. J. *J. Clin. Investig.* **2009**, *119*, 2475–2487.
93. McColl, K. E. L. *N. Engl. J. Med.* **2010**, *362*, 1597–1604.
94. El-Serag, H. B.; Sonnenberg, A. *Gut* **1998**, *43*, 327–333.
95. Chen, Y.; Blaser, M. J. *Arch. Intern. Med.* **2007**, *167*, 821–827.
96. Nweneka, C. V.; Prentice, A. M. *BMC Gastroenterol.* **2011**, *11*, 7.
97. Plottel, C. S.; Blaser, M. J. *Cell Host Microbe* **2011**, *10*, 324–335.
98. Hasan, N.; Pollack, A.; Cho, I. *Infect. Dis. Clin. N. Am.* **2010**, *24*, 1019–1039.
99. Reikvam, D. H.; Erofeev, A.; Sandvik, A.; Grcic, V.; Jahnsen, F. L.; Gaustad, P.; McCoy, K. D.; Macpherson, A. J.; Meza-Zepeda, L. A.; Johansen, F.-E. *PLoS ONE* **2011**, *6*, 13.
100. Macfarlane, S.; Macfarlane, G. T. *Proc. Nutr. Soc.* **2003**, *62*, 67–72.
101. Lazarova, D. L.; Bordonaro, M.; Carbone, R.; Sartorelli, A. C. *Int. J. Can.* **2004**, *110*, 523–531.
102. Wang, T.; Cai, G.; Qiu, Y.; Fei, N.; Zhang, M.; Pang, X.; Jia, W.; Cai, S.; Zhao, L. *ISME J.* **2012**, *6*, 320–329.
103. Biarc, J.; Nguyen, I. S.; Pini, A.; Gossé, F.; Richert, S.; Thiersé, D.; Van Dorsselaer, A.; Leize-Wagner, E.; Raul, F.; Klein, J.-P.; Schöller-Guinard, M. *Carcinogenesis* **2004**, *25*, 1477–1484.

104. Huycke, M. M.; Abrams, V.; Moore, D. R. *Carcinogenesis* **2002**, *23*, 529–536.
105. Wu, S.; Rhee, K.-J.; Albesiano, E.; Rabizadeh, S.; Wu, X.; Yen, H.-R.; Huso, D. L.; Brancati, F. L.; Wick, E.; McAllister, F.; Housseau, F.; Pardoll, D. M.; Sears, C. L. *Nat. Med.* **2009**, *15*, 1016–1022.
106. Grivennikov, S. I.; Wang, K.; Mucida, D.; Stewart, C. A.; Schnabl, B.; Jauch, D.; Taniguchi, K.; Yu, G.-Y.; Osterreicher, C. H.; Hung, K. E.; Datz, C.; Feng, Y.; Fearon, E. R.; Oukka, M.; Tessarollo, L.; Coppola, V.; Yarovinsky, F.; Cheroutre, H.; Eckmann, L.; Trinchieri, G.; Karin, M. *Nature* **2012**, *491*, 254–258.
107. Castellarin, M.; Warren, R. L.; Freeman, J. D.; Dreolini, L.; Krzywinski, M.; Strauss, J.; Barnes, R.; Watson, P.; Allen-Vercoe, E.; Moore, R. A.; Holt, R. A. *Genome Res.* **2012**, *22*, 299–306.
108. Kostic, A. D.; Gevers, D.; Pedamallu, C. S.; Michaud, M.; Duke, F.; Earl, A. M.; Ojesina, A. I.; Jung, J.; Bass, A. J.; Tabernero, J.; Baselga, J.; Liu, C.; Shivdasani, R. A.; Ogino, S.; Birren, B. W.; Huttenhower, C.; Garrett, W. S.; Meyerson, M. *Genome Res.* **2012**, *22*, 292–298.
109. Krisanaprakornkit, S.; Kimball, J. R.; Weinberg, A.; Darveau, R. P.; Bainbridge, B. W.; Dale, B. A. *Infect. Immun.* **2000**, *68*, 2907–2915.
110. Signat, B.; Roques, C.; Poulet, P.; Duffaut, D. *Curr. Issues Mol. Biol.* **2011**, *13*, 25–36. http://www.horizonpress.com/cimb/v/v13/25.pdf.
111. Hviid, A.; Svanström, H.; Frisch, M. *Gut* **2011**, *60*, 49–54.
112. Sutherland, L.; Singleton, J.; Sessions, J.; Hanauer, S.; Krawitt, E.; Rankin, G.; Summers, R.; Mekhjian, H.; Greenberger, N.; Kelly, M.; Levine, J.; Thomson, A.; Alpert, E.; Prokipchuk, E. *Lancet* **1991**, *32*, 1071–1075.
113. Arnold, G. L.; Beaves, M. R.; Pryjdun, V. O.; Mook, W. J. *Inflamm. Bowel Dis.* **2002**, *8*, 10–15.
114. Ewaschuk, J. B.; Tejpar, Q. Z.; Soo, I.; Madsen, K.; Fedorak, R. N. *Curr. Gastroenterol. Rep.* **2006**, *8*, 486–498.
115. Lepage, P.; Häsler, R.; Spehlmann, M. E.; Rehman, A.; Zvirbliene, A.; Begun, A.; Ott, S.; Kupcinskas, L.; Doré, J.; Raedler, A.; Schreiber, S. *Gastroenterology* **2011**, *141*, 227–236.
116. Manichanh, C.; Rigottier-Gois, L.; Bonnaud, E.; Gloux, K.; Pelletier, E.; Frangeul, L.; Nalin, R.; Jarrin, C.; Chardon, P.; Marteau, P.; Roca, J.; Dore, J. *Gut* **2006**, *55*, 205–211.
117. Qin, J.; Li, R.; Raes, J.; Arumugam, M.; Burgdorf, K. S.; Manichanh, C.; Nielsen, T.; Pons, N.; Levenez, F.; Yamada, T.; Mende, D. R.; Li, J.; Xu, J.; Li, S.; Li, D.; Cao, J.; Wang, B.; Liang, H.; Zheng, H.; Xie, Y.; Tap, J.; Lepage, P.; Bertalan, M.; Batto, J.-M.; Hansen, T.; Le Paslier, D.; Linneberg, A.; Nielsen, H. B.; Pelletier, E.; Renault, P.; Sicheritz-Ponten, T.; Turner, K.; Zhu, H.; Yu, C.; Li, S.; Jian, M.; Zhou, Y.; Li, Y.; Zhang, X.; Li, S.; Qin, N.; Yang, H.; Wang, J.; Brunak, S.; Doré, J.; Guarner, F.; Kristiansen, K.; Pedersen, O.; Parkhill, J.; Weissenbach, J.; Bork, P.; Ehrlich, S. D.; Wang, J. *Nature* **2010**, *464*, 59–65.
118. Frank, D. N.; St Amand, A. L.; Feldman, R. A.; Boedeker, E. C.; Harpaz, N.; Pace, N. R. *Proc. Natl. Acad. Sci. U. S. A.* **2007**, *104*, 13780–13785.
119. Mondot, S.; Kang, S.; Furet, J. P.; Aguirre De Carcer, D.; McSweeney, C.; Morrison, M.; Marteau, P.; Doré, J.; Leclerc, M. *Inflamm. Bowel Dis.* **2011**, *17*, 185–192.
120. Morgan, X. C.; Tickle, T. L.; Sokol, H.; Gevers, D.; Devaney, K. L.; Ward, D. V.; Reyes, J. A.; Shah, S. A.; LeLeiko, N.; Snapper, S. B.; Bousvaros, A.; Korzenik, J.; Sands, B. E.; Xavier, R. J.; Huttenhower, C. *Genom. Biol.* **2012**, *13*, R79.
121. Sokol, H.; Seksik, P. *Curr. Opin. Gastroenterol.* **2010**, *26*, 327–331.

122. Garrett, W. S.; Gallini, C. A.; Yatsunenko, T.; Michaud, M.; DuBois, A.; Delaney, M. L.; Punit, S.; Karlsson, M.; Bry, L.; Glickman, J. N.; Gordon, J. I.; Onderdonk, A. B.; Glimcher, L. H. *Cell Host Microbe* **2010**, *8*, 292–300.

123. Littman, D. R.; Pamer, E. G. *Cell Host Microbe* **2011**, *10*, 311–323.

124. Ivanov, I. I.; Frutos, R. D. L.; Manel, N.; Yoshinaga, K.; Rifkin, D. B.; Sartor, R. B.; Finlay, B. B.; Littman, D. R. *Cell Host Microbe* **2008**, *4*, 337–349.

125. Raza, A.; Yousaf, W.; Giannella, R.; Shata, M. T. *Expert. Rev. Clin. Immunol.* **2012**, *8*, 161–168.

126. Round, J. L.; Mazmanian, S. K. *Proc. Natl. Acad. Sci. U. S. A.* **2010**, *107*, 12204–12209.

127. Ogura, Y.; Bonen, D. K.; Inohara, N.; Nicolae, D. L.; Chen, F. F.; Ramos, R.; Britton, H.; Moran, T.; Karaliuskas, R.; Duerr, R. H.; Achkar, J. P.; Brant, S. R.; Bayless, T. M.; Kirschner, B. S.; Hanauer, S. B.; Nuñez, G.; Cho, J. H. *Nature* **2001**, *411*, 603–606.

128. Hugot, J. P.; Chamaillard, M.; Zouali, H.; Lesage, S.; Cézard, J. P.; Belaiche, J.; Almer, S.; Tysk, C.; O'Morain, C. A.; Gassull, M.; Binder, V.; Finkel, Y.; Cortot, A.; Modigliani, R.; Laurent-Puig, P.; Gower-Rousseau, C.; Macry, J.; Colombel, J. F.; Sahbatou, M.; Thomas, G. *Nature* **2001**, *411*, 599–603.

129. Franchimont, D.; Vermeire, S.; El Housni, H.; Pierik, M.; Van Steen, K.; Gustot, T.; Quertinmont, E.; Abramowicz, M.; Van Gossum, A.; Deviere, J.; Rutgeerts, P. *Gut* **2004**, *53*, 987–992.

130. Rajilić-Stojanović, M.; Biagi, E.; Heilig, H. G. H. J.; Kajander, K.; Kekkonen, R. A.; Tims, S.; de Vos, W. M. *Gastroenterology* **2011**, *141*, 1792–1801.

131. Jeffery, I. B.; O'Toole, P. W.; Öhman, L.; Claesson, M. J.; Deane, J.; Quigley, E. M. M.; Simrén, M. *Gut* **2012**, *61*, 997–1006.

132. Louis, P.; Young, P.; Holtrop, G.; Flint, H. J. *Environ. Microbiol.* **2010**, *12*, 304–314.

133. Bourdu, S.; et al. *Gastroenterology* **2005**, *128*, 1996–2008.

134. Jeffery, I.; Quigley, E.; Öhman, L. *Gut* **2012**, *3*, 572–576.

135. Abu-Shanab, A.; Quigley, E. M. M. *Nat. Rev. Gastroenterol. Hepatol.* **2010**, *7*, 691–701.

136. Fox, J. G.; Feng, Y.; Theve, E. J.; Raczynski, A. R.; Fiala, J. L. A.; Doernte, A. L.; Williams, M.; McFaline, J. L.; Essigmann, J. M.; Schauer, D. B.; Tannenbaum, S. R.; Dedon, P. C.; Weinman, S. A.; Lemon, S. M.; Fry, R. C.; Rogers, A. B. *Gut* **2010**, *59*, 88–97.

137. Ridlon, J. M.; Alves, J. M.; Hylemon, P. B.; Bajaj, J. S. *Gut Microbes* **2013**, *4*, 1–6.

138. Chen, Y.; Yang, F.; Lu, H.; Wang, B.; Chen, Y.; Lei, D.; Wang, Y.; Zhu, B.; Li, L. *Hepatology* **2011**, *54*, 562–572.

139. Mutlu, E.; Keshavarzian, A.; Engen, P.; Forsyth, C. B.; Sikaroodi, M.; Gillevet, P. *Alcohol. Clin. Exp. Res.* **2009**, *33*, 1836–1846.

140. Yan, A. W.; Fouts, D. E.; Brandl, J.; Stärkel, P.; Torralba, M.; Schott, E.; Tsukamoto, H.; Nelson, K. E.; Brenner, D. A.; Schnabl, B. *Hepatology* **2011**, *53*, 96–105.

141. Bäckhed, F.; Manchester, J. K.; Semenkovich, C. F.; Gordon, J. I. *Proc. Natl. Acad. Sci. U. S. A.* **2007**, *104*, 979–984.

142. Ley, R. E.; Bäckhed, F.; Turnbaugh, P.; Lozupone, C. A.; Knight, R. D.; Gordon, J. I. *Proc. Natl. Acad. Sci. U. S. A.* **2005**, *102*, 11070–11075.

143. Turnbaugh, P. J.; Ley, R. E.; Mahowald, M. A.; Magrini, V.; Mardis, E. R.; Gordon, J. I. *Nature* **2006**, *444*, 1027–1031.

144. Cho, I.; Yamanishi, S.; Cox, L.; Methé, B. A.; Zavadil, J.; Li, K.; Gao, Z.; Mahana, D.; Raju, K.; Teitler, I.; Li, H.; Alekseyenko, A. V.; Blaser, M. J. *Nature* **2012**, *488*, 621–626.

145. Cox, L. M.; Cho, I.; Young, S. A.; Anderson, W. H. K.; Waters, B. J.; Hung, S.-C.; Gao, Z.; Mahana, D.; Bihan, M.; Alekseyenko, A. V.; Methé, B. A.; Blaser, M. J. *FASEB J.* **2013**, *27*, 692–702.

146. Ley, R. E.; Turnbaugh, P. J.; Klein, S.; Gordon, J. I. *Nature* **2006**, *444*, 1022–1023.

147. Schwiertz, A.; Taras, D.; Schäfer, K.; Beijer, S.; Bos, N. A.; Donus, C.; Hardt, P. D. *Obesity (Silver Spring, Md.)* **2010**, *18*, 190–195.

148. Li, J. V.; Ashrafian, H.; Bueter, M.; Kinross, J.; Sands, C.; Le Roux, C. W.; Bloom, S. R.; Darzi, A.; Athanasiou, T.; Marchesi, J. R.; Nicholson, J. K.; Holmes, E. *Gut* **2011**, *60*, 1214–1223.

149. Ajslev, T. A.; Andersen, C. S.; Gamborg, M.; Sørensen, T. I. A.; Jess, T. *Int. J. Obes.* **2005**, *35*, 522–529.

150. Luoto, R.; Kalliomäki, M.; Laitinen, K.; Isolauri, E. *Int. J. Obes.* **2005**, *34*, 1531–1537.

151. Amos, A.; McCarty, D.; Zimmet, P. *Diabet. Med.* **1997**, *14* (Suppl 5), S1–S85.

152. Larsen, N.; Vogensen, F. K.; van den Berg, F. W. J.; Nielsen, D. S.; Andreasen, A. S.; Pedersen, B. K.; Al-Soud, W. A.; Sørensen, S. J.; Hansen, L. H.; Jakobsen, M. *PLoS ONE* **2010**, *5*, e9085.

153. Qin, J.; Li, Y.; Cai, Z.; Li, S.; Zhu, J.; Zhang, F.; Liang, S.; Zhang, W.; Guan, Y.; Shen, D.; Peng, Y.; Zhang, D.; Jie, Z.; Wu, W.; Qin, Y.; Xue, W.; Li, J.; Han, L.; Lu, D.; Wu, P.; Dai, Y.; Sun, X.; Li, Z.; Tang, A.; Zhong, S.; Li, X.; Chen, W.; Xu, R.; Wang, M.; Feng, Q.; Gong, M.; Yu, J.; Zhang, Y.; Zhang, M.; Hansen, T.; Sanchez, G.; Raes, J.; Falony, G.; Okuda, S.; Almeida, M.; LeChatelier, E.; Renault, P.; Pons, N.; Batto, J.-M.; Zhang, Z.; Chen, H.; Yang, R.; Zheng, W.; Li, S.; Yang, H.; Wang, J.; Ehrlich, S. D.; Nielsen, R.; Pedersen, O.; Kristiansen, K.; Wang, J. *Nature* **2012**, *490*, 55–60.

154. Vrieze, A.; van Nood, E.; Holleman, F. *Gastroenterology* **2012**, *143*, 913–916.

155. Karlsson, F. H.; Tremaroli, V.; Nookaew, I.; Bergström, G.; Behre, C. J.; Fagerberg, B.; Nielsen, J.; Bäckhed, F. *Nature* **2013**, *498*, 99–103.

156. Smith, M. I.; Yatsunenko, T.; Manary, M. J.; Trehan, I.; Mkakosya, R.; Cheng, J.; Kau, A. L.; Rich, S. S.; Concannon, P.; Mychaleckyj, J. C.; Liu, J.; Houpt, E.; Li, J. V.; Holmes, E.; Nicholson, J.; Knights, D.; Ursell, L. K.; Knight, R.; Gordon, J. I. *Science (New York, N.Y.)* **2013**, *339*, 548–554.

157. Trehan, I.; Goldbach, H. S.; LaGrone, L. N.; Meuli, G. J.; Wang, R. J.; Maleta, K. M.; Manary, M. J. *N. Engl. J. Med.* **2013**, *368*, 425–435.

158. Wang, Z.; Klipfell, E.; Bennett, B. J.; Koeth, R.; Levison, B. S.; Dugar, B.; Feldstein, A. E.; Britt, E. B.; Fu, X.; Chung, Y.-M.; Wu, Y.; Schauer, P.; Smith, J. D.; Allayee, H.; Tang, W. H. W.; DiDonato, J. A.; Lusis, A. J.; Hazen, S. L. *Nature* **2011**, *472*, 57–63.

159. Koeth, R.; Wang, Z.; Levison, B. S.; Buffa, J. A.; Org, E.; Sheehy, B. T.; Britt, E. B.; Fu, X.; Wu, Y.; Li, L.; Smith, J. D.; DiDonato, J. A.; Chen, J.; Li, H.; Wu, G. D.; Lewis, J. D.; Warrier, M.; Brown, J. M.; Krauss, R. M.; Tang, W. H. W.; Bushman, F. D.; Lusis, A. J.; Hazen, S. L. *Nat. Med.* **2013**, *19*, 576–585.

160. Siva, S.; Barrack, E. R.; Reddy, G. P. V.; Thamilselvan, V.; Thamilselvan, S.; Menon, M.; Bhandari, M. *BJU Int.* **2009**, *103*, 18–21.

161. Sidhu, H.; Holmes, R. P.; Allison, M. J.; Peck, A. B. *J. Clin. Microbiol.* **1999**, *37*, 1503–1509.

162. Magwira, C.; Kullin, B.; Lewandowski, S.; Rodgers, A.; Reid, S. J.; Abratt, V. R. *J. Appl. Microbiol.* **2012**, *113*, 418–428.

163. Aronov, P. A.; Luo, F. J.-G.; Plummer, N. S.; Quan, Z.; Holmes, S.; Hostetter, T. H.; Meyer, T. W. *J. Am. Soc. Nephrol.* **2011**, *22*, 1769–1776.

164. Evenepoel, P.; Meijers, B. K. I.; Bammens, B. R. M.; Verbeke, K. *Kidney Int. Suppl.* **2009**, *76*, S12–S19.

165. Poesen, R.; Meijers, B.; Evenepoel, P. *Semin. Dialysis* **2013**, *26*, 323–332.

166. Scher, J. U.; Abramson, S. B. *Nat. Rev. Rheumatol.* **2011**, *7*, 569–578.

167. Ivanov, I. I.; Atarashi, K.; Manel, N.; Brodie, E. L.; Shima, T.; Karaoz, U.; Wei, D.; Goldfarb, K. C.; Santee, C. A.; Lynch, S. V.; Tanoue, T.; Imaoka, A.; Itoh, K.; Takeda, K.; Umesaki, Y.; Honda, K.; Littman, D. R. *Cell* **2009**, *139*, 485–498.
168. Wu, H.-J.; Ivanov, I. I.; Darce, J.; Hattori, K.; Shima, T.; Umesaki, Y.; Littman, D. R.; Benoist, C.; Mathis, D. *Immunity* **2010**, *32*, 815–827.
169. Abdollahi-Roodsaz, S.; Joosten, L. A.; Koenders, M. I.; Devesa, I.; Roelofs, M. F.; Radstake, T. R.; Heuvelmans-Jacobs, M.; Akira, S.; Nicklin, M. J.; Ribeiro-Dias, F.; Van den Berg, W. B. *J. Clin. Invest.* **2008**, *118*, 205–216.
170. Schön, M. P.; Boehncke, W. H. *N. Engl. J. Med.* **2005**, *352*, 1899–1912.
171. Gao, Z.; Tseng, C.; Strober, B. E.; Peiand, Z.; Blaser, M. J. *PLoS ONE* **2008**, *3*, 9.
172. Fahlén, A.; Engstrand, L.; Baker, B. S.; Powles, A.; Fry, L. *Arch. Dermatol. Res.* **2012**, *304*, 15–22.
173. Ehlers, S.; Kaufmann, S. H. E. *Trends Immunol.* **2010**, *31*, 184–190.
174. Grice, E. A.; Segre, J. A. *Nat. Rev. Microbiol.* **2011**, *9*, 244–253.
175. Price, L. B.; Liu, C. M.; Melendez, J. H.; Frankel, Y. M.; Engelthaler, D.; Aziz, M.; Bowers, J.; Rattray, R.; Ravel, J.; Kingsley, C.; Keim, P. S.; Lazarus, G. S.; Zenilman, J. M. *PLoS ONE* **2009**, *4*, e6462.
176. Grice, E. A.; Snitkin, E. S.; Yockey, L. J.; Bermudez, D. M.; Liechty, K. W.; Segre, J. A.; Mullikin, J.; Blakesley, R.; Young, A.; Chu, G.; Ramsahoye, C.; Lovett, S.; Han, J.; Legaspi, R.; Fuksenko, T.; Reddix-Dugue, N.; Sison, C.; Gregory, M.; Montemayor, C.; Gestole, M.; Hargrove, A.; Johnson, T.; Myrick, J.; Riebow, N.; Schmidt, B.; Novotny, B.; Gupti, J.; Benjamin, B.; Brooks, S.; Coleman, H.; Ho, S.; Schandler, K.; Smith, L.; Stantripop, M.; Maduro, Q.; Bouffard, G.; Dekhtyar, M.; Guan, X.; Masiello, C.; Maskeri, B.; McDowell, J.; Park, M.; Jacques Thomas, P. *Proc. Natl. Acad. Sci. U. S. A.* **2010**, *107*, 14799–14804.
177. Redel, H.; Gao, Z.; Li, H.; Alekseyenko, A. V.; Zhou, Y.; Perez-Perez, G. I.; Weinstock, G.; Sodergren, E.; Blaser, M. J. *J. Infect. Dis.* **2013**, *207*, 1105–1114.
178. McDowell, A.; Gao, A.; Barnard, E.; Fink, C.; Murray, P. I.; Dowson, C. G.; Nagy, I.; Lambert, P. A.; Patrick, S. *Microbiology* **2011**, *157*, 1990–2003.
179. Huang, Y. J.; Charlson, E. S.; Collman, R. G.; Colombini-Hatch, S.; Martinez, F. D.; Senior, R. M. *Am. J. Respir. Crit. Care Med.* **2013**, *187*, 1382–1387.
180. Hilty, M.; Burke, C.; Pedro, H.; Cardenas, P.; Bush, A.; Bossley, C.; Davies, J.; Ervine, A.; Poulter, L.; Pachter, L.; Moffatt, M. F.; Cookson, W. O. C. *PLoS ONE* **2010**, *5*, e8578.
181. Huang, Y. J.; Nelson, C. E.; Brodie, E. L.; Desantis, T. Z.; Baek, M. S.; Liu, J.; Woyke, T.; Allgaier, M.; Bristow, J.; Wiener-Kronish, J. P.; Sutherland, E. R.; King, T. S.; Icitovic, N.; Martin, R. J.; Calhoun, W. J.; Castro, M.; Denlinger, L. C.; Dimango, E.; Kraft, M.; Peters, S. P.; Wasserman, S. I.; Wechsler, M. E.; Boushey, H. A.; Lynch, S. V. *J. Allergy Clin. Immunol.* **2011**, *127*, 372–381.e1–3.

Chapter 5

Emerging RNA-Seq Applications in Food Science

Alberto Valdés, Carolina Simó, Clara Ibáñez and Virginia García-Cañas

Laboratory of Foodomics, Institute of Food Science Research (CIAL), CSIC. Nicolás Cabrera 9, Madrid, Spain

Chapter Outline

1 INTRODUCTION

For the past two decades, gene-expression microarray has been regarded as one of the foremost technological advances in high-throughput analysis. The success of microarray in profiling gene expression has been remarkable in last years, and owing to the extensive optimization and standardization performed in instruments and protocol, microarray has become a mature technology *(1)*. Gene-expression microarray technique has been employed to obtain meaningful insights into the molecular mechanisms underlying complex biological processes relevant to food science. However, gene-expression microarray fails to reach comprehensive and precise characterization of transcriptome

Comprehensive Analytical Chemistry, Vol. 64. http://dx.doi.org/10.1016/B978-0-444-62650-9.00005-1

due to some unsolved technical constraints. As alternative, recent unbiased sequencing methodologies termed RNA-sequencing (RNA-Seq) are now available for genome-wide high-throughput transcriptomics. Such groundbreaking high-throughput technologies are changing the way we investigate the transcriptome landscapes in biological systems.

The transcriptome, defined as the complete set of RNA transcripts produced by the genome at any one time, can be considered as an important link between phenotype and information encoded in the genome *(2)*. Comprehensive and precise characterization of transcriptome encompasses (i) the annotation of all species of transcript, including mRNAs, noncoding RNAs, and small RNAs; (ii) the determination of the transcriptional structure of genes (start sites, 5′- and 3′-ends, splicing variants, etc.); and the quantification of differential expression levels of each transcript under different conditions *(3)*. Revolutionary tools for transcriptomics analysis are providing new opportunities and prospects to investigate previously unanswered questions relevant to food science. In opposition to gene-expression microarray, RNA-Seq technologies are independent of any annotated sequence feature and rely on recent technical advances in high-density microarraying and various sequencing chemistries. They provide extraordinary opportunities to explore many different aspects of entire transcriptomes and also allow the determination of gene-expression levels by robust digital quantitative analysis. Applications of RNA-Seq to the food and nutrition domain are relatively recent compared with their use in basic science applications. This chapter provides insight into recent progress in RNA-Seq technologies, discussing their main advantages and limitations. Innovative applications of RNA-Seq will be discussed in the context of food science to illustrate its extraordinary potential. Finally, some outlooks regarding forthcoming potential applications and technical advances will be drawn.

2 OVERVIEW OF RNA-SEQ TECHNOLOGY

The development of groundbreaking DNA-sequencing strategies known as next-generation sequencing (NGS) technologies has opened a new age in the study of biological systems. Over the last years, the extraordinary progress in NGS technologies has been driven by the strong competition between manufacturers. Thus, different generations of platforms for DNA sequencing have been developed with improved capabilities at costs that were unimaginable few years ago *(4)*. Besides, this novel sequencing technology has been expanded to the analysis of gene expression by specific techniques known as massively parallel sequencing of RNA or RNA-Seq. This technology has the potential to truly embrace the whole transcriptome, offering excellent possibilities for the discovery of new transcripts. In addition, RNA-Seq offers a larger dynamic range to measure RNA abundance with high reproducibility and in a nonrelative quantitative mode. These features, and the upcoming cost reductions, make RNA-Seq an increasingly attractive strategy, even for unknown genome organisms.

The earliest and currently most widely used NGS systems, referred to as second-generation sequencing (SGS) technologies, are characterized by the synchronous-controlled optical detection, together with cyclic reagent washes of a substrate where the extension of thousands DNA templates is carried out in massive and parallel fashion (5). SG sequencers are open platforms designed for sequencing libraries of nucleic acids in a high-throughput and cost-effective way. The three major SG sequencers are Illumina Genome Analyzer (GA), Roche 454 Genome Sequencer FLX system (FLX), and Applied Biosystems SOLiD (SOLiD) platform. Regardless the platform chosen for high-throughput sequencing, the analytical procedure shares common steps: (1) library preparation involving fragmentation of RNA molecules, cDNA synthesis, and ligation to specific adaptors at both ends; (2) clonal amplification of each template since most imaging systems have not been designed to detect single chemiluminescent or fluorescent events; (3) attachment of the amplified DNA templates to a solid support in a flow cell or a reaction chamber; and (4) iterative and synchronized flowing and washing off the reagents for DNA strand extension while signals are acquired by the detection system (6). Following signal acquisition, the resulting raw image data have to be converted into short reads (nucleotidic sequences generated per DNA template) by a process named "base-calling." Depending on the platform, DNA extension (or synthesis) can be attained enzymatically by ligation or polymerization. A brief description of the key features and some technical aspects of the main platforms employed in the food science applications is next provided.

2.1   Library Preparation and Clonal Amplification

The selected method for sample (RNA or DNA) preparation prior to sequencing will depend on the NGS instrument used. Most of the sample preparation methods follow common procedures that are aimed to produce sets of short DNA molecules (library) with adapters ligated in their 3'- and 5'-ends. Owing to the sensitivity limitations of most detection systems in SGS platforms, library amplification is a previous requirement to the sequencing step. As the most widely used methods for amplification are capable to amplify every template from the library in several orders of magnitude in a very controlled way, the process has been commonly termed clonal amplification. Thus, clonal amplification can be performed using either solid-phase bridge amplification (employed in Illumina platforms, formerly Solexa) or emulsion PCR (emPCR, used in Roche and SOLiD platforms). In bridge amplification, an optically transparent surface in a flow cell is derivatized with forward and reverse primers that capture library DNA sequences by hybridizing with the adaptors (7). Each single-stranded DNA molecule, immobilized at one end on the surface, bends over to hybridize with the complementary adapter on the support by its free end, forming a "bridge" structure that serves as template for amplification. The addition of amplification reagents initiates the generation of clusters of thousands of equal DNA fragments in small areas. After several amplification cycles, random clusters of about 1000 copies of single-stranded DNA (termed DNA polonies) are originated on the surface. By contrast, in emPCR, each DNA template from the library is captured onto a bead under conditions that favor one DNA molecule per bead (8). Then, beads are emulsified along with PCR reagents in water-in-oil emulsions. During emPCR, each single template attached to a bead is then clonally amplified to obtain millions of copies of the same sequence. Next, beads can be immobilized in different supports or loaded into individual PicoTiterPlate (PTP) wells, which are made of fiber-optic bundle and is designed to house one single bead per well. This approach has been implemented in 454 Life Technologies/Roche and Applied Biosystems platforms.

2.2 Sequencing Chemistry

The two most commonly used techniques employed in SGS platforms to obtain detectable signals can be classified as sequencing by synthesis and sequencing by ligation. Sequencing by synthesis techniques have as a common feature the use of DNA polymerase enzyme during the sequencing step. Sequencing by synthesis can be divided into cyclic-reversible termination technique and single-nucleotide addition technique.

Cyclic-reversible termination techniques have been implemented in Illumina and Helicos Bioscience sequencers. The general procedure involves the use of nucleotides (reversible terminators) that contain a removable fluorophore that blocks the incorporation of subsequent nucleotides to the newly synthesized chain until the fluorescent group is cleaved off. Thus, a complete cycle encompasses (a) flowing the nucleotides to the sequencing surface, (b) washing-out the unbound nucleotides, (c) fluorescence signals acquisition, and (d) nucleotide deprotection to allow the incorporation of the next reversible terminator. The Illumina technology uses a unique removable fluorescent-dye group for each nucleotide type (Figure 1). In this case, the reaction mixture for the sequencing reactions is supplied onto the amplified DNA created on a surface by bridge amplification. After each terminator nucleotide is incorporated into the growing DNA strand, a fluorescent signal is emitted and recorded by a charge-coupled device (CCD) camera, which detects the position of the fluorescent signal in the support and identifies the nucleotide. By deprotection of blocking groups and washing, a new synthesis cycle starts. Similarly, Helicos Bioscience Technology has implemented this chemistry in the HeliScope sequencer, but using only a single-color fluorescence label. Another particular feature in HeliScope sequencer is the lack of clonal amplification step prior sequencing. This system has been the first commercial single-molecule DNA-sequencing system and its detection

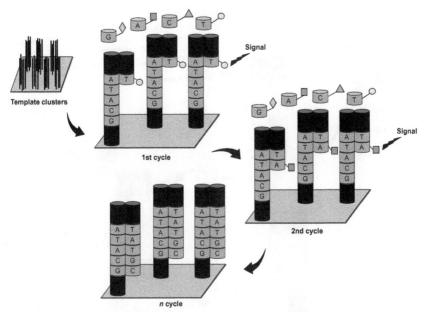

FIGURE 1   Sequencing scheme in Illumina Genome Analyzer platform. (For color version of this figure, the reader is referred to the online version of this chapter.)

system is capable of scanning millions of single molecules of DNA anchored to a glass coverslip flow cell.

Single-nucleotide addition techniques are based on adding limited amounts of an individual dNTP to the reaction media to synthesize DNA and, after washing, sequencing is resumed with the addition of another nucleotide. This technique has been combined with pyrosequencing by Roche and incorporated in the GS and GS FLX sequencers. In brief, after emPCR, the beads positioned onto the PTP wells are subjected to pyrosequencing process. This is initiated by the addition of ATP sulfurylase and luciferase enzymes, adenosine 5' phosphosulfate, and luciferin substrates. Thus, in every sequencing cycle, a single species of dNTPs is flowed into the PTP (Figure 2). The incorporation of a complementary nucleotide results in the release of pyrophosphate (PPi) which eventually leads to a burst of light *(1)*. Individual dNTPs are dispensed in a predetermined sequential order and the chemiluminescence is imaged with a CCD camera. This process results in parallel sequencing of multiple DNA templates attached to a single bead in each well of the PTP. Based on a similar idea, the semiconductor technology, developed by Life Technologies, has developed a novel generation of sequencers (Ion Torrent and Ion Proton PGM) that employ emPCR for clonal amplification in combination with a highly dense microwell array in which each well acts as an individual DNA synthesis reaction chamber. Every time a nucleotide

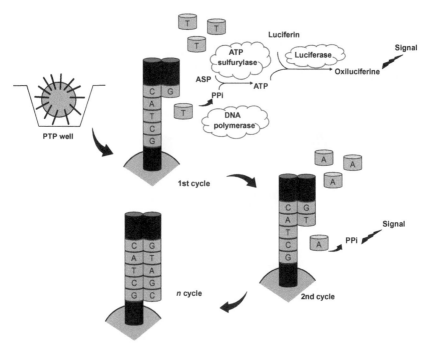

FIGURE 2 Sequencing procedure in Roche 454 Genome sequencer FLX platform. (For color version of this figure, the reader is referred to the online version of this chapter.)

is incorporated in the complementary template, H^+ is released. A layer composed of a highly dense field-effect transistor array is aligned underneath the microwell array to transform the change in pH to a recordable voltage change.

Sequencing by ligation has been implemented in the ABI SOLiD sequencer by Applied Biosystems. This sequencing technique requires the use of a DNA ligase instead of a DNA polymerase to generate the complementary DNA strands *(9)*. The ligation method involves the use of several fluorescent 8-mer oligonucleotides that hybridize with the template. More precisely, amplified templates by emPCR are subjected to a 3′ modification to ensure further attachment to the glass slide. Then, a universal (sequencing) primer hybridizes to the adapter on the templates and sequencing starts by ligation of any of the four competing fluorescent 8-mer oligonucleotides (Figure 3). In each oligonucleotide, the dye color is defined by the first two of the eight bases. A ligation event is detected by the fluorescent label incorporated to the growing strand. After the detection of the fluorescence from the dye, bases 1 and 2 in the sequence can thus be determined. Then, the last three bases and the dye are cleaved to enable a further ligation event. Then, another hybridization–ligation cycle is initiated. After 10 cycles of ligation, the extended primer is removed from the template and the process is

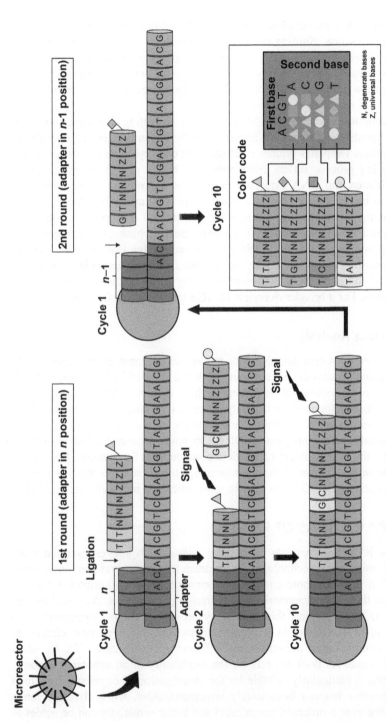

FIGURE 3 Sequencing strategy in Applied Biosystems SOLiD platform. (For color version of this figure, the reader is referred to the online version of this chapter.)

repeated with a universal primer that is shifted one base from the adaptor-fragment position. Shifting the universal primer in five rounds of 10 cycles enables the entire fragment to be sequenced and provides an error correction scheme because each base position is interrogated twice. Color calls from the five ligation rounds are then ordered into a linear sequence and decoded in a process termed "two-base encoding."

A common aspect in these NGS platforms, excepting HeliScope sequencer is that the observed signal is a consensus of the nucleotides or probes added to the identical templates (clones or clusters) in each cycle. In consequence, the sequencing process suffers from numerous biases due to imperfect clonal amplification and DNA extension failure. For example, a strand which has failed to incorporate a base in a given cycle will lag behind the rest of the sequencing run (phasing), whereas the addition of multiple nucleotides or probes in a given cycle results in leading-strand dephasing (prephasing) *(10)*. Phasing and prephasing accumulate during sequencing leading to an increase of base-calling errors toward the end of reads. Table 1 summarizes the comparison of the main NGS platforms including some of their advantages and limitations. For a broader overview of NGS technology, refer to Refs. *(1,3,9)*.

2.3   Data Analysis

Analysis of RNA-Seq data demands tailored bioinformatics strategies able to manage and process the huge amount of data generated by RNA-Seq methods *(11)*. Novel algorithms able to process the huge amount of RNA-Seq data are in continuous development for a variety of applications including sequence alignment, assembly, read annotation, and quantitation, among others. This incessant bioinformatics progress provides improved resources, but in turn, is delaying the establishment of standard practice tools for analysis. For a comprehensive description of the bioinformatics tools and algorithms frequently used for analyzing NGS data refer to some articles in recent literature *(12–14)*.

3   APPLICATIONS OF RNA-SEQ

Current RNA-Seq methods offer some relevant advantages over the more mature microarray technology. In contrast to gene-expression microarray, RNA-Seq provides more complete information about transcriptome because it allows the direct characterization of transcript sequences. As it will be discussed in next sections, this technology provides excellent opportunities to detect point mutations in expressed transcripts, discover new classes of RNA, identify fusion transcripts and unknown splice variants *(15)*. As detection of sequences does not rely on the availability of an annotated genome, RNA-Seq is particularly suitable for the investigation of organisms for which their genome has not been totally sequenced. Also, wider dynamic range (spanning over 5 orders of magnitude) and better sensitivity can be achieved

TABLE 1 Comparison of Main NGS Platforms

	Illumina GA (II)[a] and HiSeq[b]	Roche 454 GS FLX+	Applied Biosystems SOLiD 5500xl	Life Technologies Ion PGM (318)
Amplification	Bridge amplification	Emulsion PCR	Emulsion PCR	Emulsion PCR
Sequencing chemistry	Reversible terminator	Pyrosequencing	Ligation	Proton detection
Method of detection	Fluorescence	Chemiluminescence	Fluorescence	Change in pH
Run time	14 days[a] or 11 days[b]	23 h	8 days	4 h
Read length	75 bp[a] or 100 bp[b]	~800 bp	75 + 35 bp	100 or 200 bp
Millions of reads/run	40[a] or 3000[b]	1	1400	4
Data generation/run	35 Gb[a] or 600 Gb[b]	0.7 Gb	155 Gb	0.86 Gb
Advantage	– Cost effectiveness – Massive throughput – Low hands-on time	– Long-read length improves mapping in repetitive regions – Short-run time	– Low error rate – Massive throughput	– Short-run times – No need for modified DNA bases
Disadvantage	– Long-run time – Short-read lengths	– High reagent costs – High-error rate in homopolymer repeats – High hands-on time	– Short-read lengths – Long-run time	– High-error rate in homopolymers repeats – High hands-on time

[a]Illumina/GA(II) instrument.
[b]Illumina/HiSeq 2000 instrument.

using RNA-Seq, as signal-to noise ratios increase with the sequencing depth (the times that a particular base is sequenced) *(16)*. On the other side, increasing the sequencing depth is directly associated with an increase in the sequencing cost *(2)*. Hence, sequencing costs will vary depending on the sequencing depth required to effectively interrogate a given transcriptome. Also, another issue that complicates the RNA-Seq analysis is the heterogeneity of sequencing depth along the transcript length. This heterogeneity that might be originated during RNA enrichment, fragmentation, ligation, amplification, and sequencing procedures, introduces an important bias in accurate quantification of gene expression. To the contrary, heterogeneity is not a concern in microarray analysis because of the fixed nature of probes that capture the transcripts by hybridization.

3.1 *De Novo* Transcriptome Assembly

Depending on the analytical goals, sequencing reads can be either aligned to a reference genome (or transcriptome) or assembled *de novo* to produce a genome-scale transcription map that consists of both the transcriptional structure and level of expression for each gene *(17)*. The more basic outcome from RNA-Seq analysis is represented by the raw reads that should be subjected to quality evaluation in order to decide which portion of reads are suitable for downstream analysis. After this step, the alignment process is aimed at mapping the reads into the sequence of the reference genome or transcriptome. However, in many sequencing projects, a reference genome or transcriptome is not available. RNA-Seq technologies and potent computational tools can be combined to obtain the *de novo* assembly and annotation of a transcriptome for many organisms. Several strategies have been employed to that aim, including the generation of libraries with different fragmentation degrees; increase the sequencing coverage; use of paired-end reads technology, etc. The recent development of strategies based on sequencing paired-end (mate-paired) reads has improved the efficiency and accuracy of the assembly and mapping process *(18)*. Paired-end read approach involves the sequencing from both ends of the same molecule, thereby generating a larger read with known sequences at either end. Illumina has successfully implemented this technology by using two rounds of sequencing. Thus, once the first strand is sequenced, the template is regenerated to allow a second round of sequencing from the opposite end (complementary strand). On the other side, mate-paired sequencing in SOLiD technology involves joining the fragment ends by recircularization using oligonucleotide adaptors during library production to allow both ends to be determined in a single round of sequencing. The enormous interest on this topic has led to the development of dedicated bioinformatic tools to generate *de novo* assemblies. In spite of the high accuracy of these tools, validation of the generated assemblies is still challenging and requires intensive research.

3.2   Comparative Transcriptomic Analysis/Digital Gene-Expression Profiling

Providing that an already sequenced reference genome or transcriptome is available, reads can be aligned to the reference genome or transcriptome using mapping algorithms. In general, mapping algorithms can deal with single-base differences owing to sequencing errors, mutations, or SNPs; however, these tools cannot accommodate large gaps. In addition, the mapping process is time-consuming and faces several challenges which aggravate when working with complex eukaryote transcriptomes. In such cases, the use of splice-aware mapping tools capable of working with reads that represent alternative splicing patterns improves the mapping efficiency *(12)*. Another important issue that needs to be improved is mapping repetitive sequences (multimapped reads), especially when they match to more than one position in the genome.

After mapping step, the workflow for RNA-Seq data analysis for expression level determination involves the two general steps: calculation of gene-expression levels by counting mapped reads (digital readout) and determination of differential gene expression using statistical tests. Regardless the method selected for quantifying gene expression, either sequencing full RNA molecules or the less expensive method based on sequencing the $3'$-end of each transcript, well-suited bioinformatic tools are required for the estimation of transcript levels *(19)*. The first step for quantifying gene expression involves the conversion of sequence reads into a quantitative value for each transcript. In most cases, the selection of a suitable approach will ultimately depend on the procedure used for library preparation. A common approach when sequencing full RNA molecules involves summarizing the number of reads for each transcript and then normalizing for the length of the transcript. In comparative transcriptomics, the central objective relies on the estimation of differential gene-expression values from different tissues, treatments, conditions, developmental stages, varieties, etc. This can be achieved by an additional transformation of data in order to eliminate differences on sequencing depth between runs or libraries. The most frequently adopted method to that aim is based on the conversion of read counts to reads per kilobase per million mapped reads *(20)*.

Sample preparation methods, specific for sensitive and accurate quantification of each transcript, are also gaining attention *(21)*. For instance, digital gene expression (DGE) tag profiling is a specific method based on Illumina RNA-Seq for accurate quantification of each transcript by sequencing short (20- or 21-bp) cDNA tags rather than the entire transcript. This approach offers enhanced sensitivity without the need for increasing the sequencing depth. The library preparation procedure involves sequential digestion steps of cDNA molecules alternated by enzymatic ligation of adaptors to finally build DNA templates consisting of a single 20- or 21-bp cDNA tag flanked by defined adapters that are further sequenced. The reads delivered by the sequencer are filtered and mapped to a reference genome (or transcriptome), to be subsequently counted and normalized with respect to transcript length and sequencing depth. Since DGE tag profiling does not attempt to sequence the entire length of each transcript, its sensitivity is higher with fewer total reads per run. Then, rare transcript discovery and quantification can be achieved by selecting the depth of coverage.

3.3 Splicing Analysis

In addition to providing DGE profiles, RNA-Seq can also assist on the investigation of many other aspects of the transcriptome such as alternative splicing isoform composition, gene fusion and nucleotide variations, unknown coding or noncoding transcripts, and RNA editing (22). Alternative splicing is a major mechanism of posttranscriptional regulation, by which the immature mRNA of a gene can be spliced into multiple isoforms after the transcription. For instance, in the human, more than 95% of multiexon genes undergo alternative splicing. Considering that the alternative spliced isoforms can have relevant functional meaning and that they are not expressed equal, there has been great interest on the research of alternative splicing and its regulatory mechanisms over the last years. Specific bioinformatic approaches directed to analyze RNA-Seq data have been developed. Thus, junction between exons can be detected in RNA-Seq data even in those cases where the isoform in unknown using splicer aligners such as TopHat and SOAPsplice (23). Those junction reads should be unique to isoforms and may provide information regarding the expression level of the isoform, whereas reads mapped within an exon will be redundant across isoforms sharing that particular exon. As mentioned, RNA-Seq is potentially well suited for alternative splicing analysis, but it is not free of constraints. The success on mapping junctions partially depends on read length and the sequence depth; however, the latter increases the false positives.

3.4 MicroRNA-Seq

In addition to the aforementioned applications, RNA-Seq techniques have the potential to analyze noncoding RNA molecules, such as microRNAs (miRNAs) (24). miRNAs are short (15–25 nucleotides) RNA sequences that play an important role in the regulation of gene expression in a number of biological processes in plants (25). These RNA sequences may act posttranscriptionally by hybridizing to specific 3′-untranslated region in mRNA transcripts, to induce their subsequent degradation, or to inhibit their translation. The RNA-Seq analysis of miRNA requires specialized protocols for preparation of libraries that allow capturing the short miRNA sequences from RNA samples (26).

4   RNA-SEQ IN FOOD SCIENCE

4.1   Production of Food Crops

Global warming and the demand for food of an ever-growing world population are relevant issues attracting much attention worldwide. In this regard, the study of the links existing between gene function and traits relevant to agriculture in food crops is of great interest. Transcriptomic analysis of crops provides valuable information of how genome responds to cellular perturbations and also reveals the expressed genes that control important traits (e.g., yield and tolerance to adverse environmental conditions) (27). Moreover, the detection of differential transcription patterns and identification of novel transcripts at specific stages of development or conditions establishes the foundation for understanding the molecular mechanisms underlying production of proteins and metabolites relevant to food science (e.g., bioactive compounds and nutrients). These insights provide further directions for controlling gene expression to increase or decrease accumulation of the compounds of interest.

In recent years, the genomes of several food crops have been fully sequenced. The progress in NGS techniques has contributed to the generation of comprehensive gene-expression data sets of cell-, tissue-, and developmental-specific gene expression for many food crop species. Data derived from NGS provide the starting point to discover the function of unknown genes and to describe the transcriptome throughout the life cycle for crop species relevant in food production, including rice, soybean, maize, and wheat, to mention few. In crops, adverse growing conditions often result in lower yields that have a negative economic impact for producers and consumers. Understanding the mechanisms involved in the response to unfavorable conditions will help in producing crops with higher tolerance to stress. To this regard, various gene-expression profiling studies have been completed using NGS to investigate a variety of responses to drought, salinity, cold, and diseases. Novel technological advances as those directed to increase the read length and the total number of reads per run combined with novel strategies that utilize paired-end reads have prompted its widespread use to decipher several food crops' transcriptomes. As mentioned above, the generation of longer paired-end reads enables higher levels of mappability, better identification of reads from splice variants, and the assembly of transcriptomes in the absence of a reference genome using de novo assembly approaches (18,28). De novo transcriptome assembly represents an emerging application of RNA-Seq particularly interesting for those food crop species whose genome has not been fully sequenced. Nevertheless, due to the complexity and frequently incomplete representation of transcripts in sequencing libraries, the assembly of high-quality transcriptomes can be challenging. In early RNA-Seq studies, the short 30-bp average read length restricted transcriptome assembly, whereas with the 75 bp or longer reads now available, transcriptomes can be assembled more easily, allowing reads whose ends are anchored in different exons to define splice sites without relying on prior annotations *(28)*. A strategy used in many studies aimed at capturing the most comprehensive transcript representation in the novel assembly relies on pooling samples obtained from different tissues and developmental stages for subsequent sequencing. After transcriptome assembly, validation or quality control for the new assembly output is frequently addressed using bioinformatic tools and searching in databases. This process allows to ascertain the degree of similarity/conservation between the novel assembly and other closely related transcriptomes and also to address transcript and functional annotation. Furthermore, the use of mining tools on *de novo* assemblies provides excellent opportunities for the discovery of novel transcripts, as well as short sequence repeats (SSRs), also known as microsatellites. These microsatellites consist on repetitions of short (two to six) nucleotides and are extremely useful for gene mapping, marker-assisted selection, and comparative genome analysis.

As a general trend, *de novo* transcriptome assembly studies reveal greater transcriptome complexity than expected and provide a blueprint for further studies. *De novo* transcriptome assembly of black pepper is a representative example of the application of SOLiD RNA-Seq technology in nonmodel species *(29)*. In that study, 71 million short reads, representing a sequencing coverage per base (depth) of 62X, were used to assemble a total of 22,363 transcripts in root samples. Transcript and functional annotation was performed based on the sequence homology with other species.

As stated before, paired-end reads method from Illumina technology has been commonly used for *de novo* transcriptome assembly in nonmodel food crop species. For instance, Zhang *et al.* investigated *de novo* assembly of peanut transcriptome with the aim of identifying expressed genes during the fast accumulation period in seeds *(30)*. Prior sequencing step, libraries were prepared from three peanut varieties including low and high oil-content varieties. An average of 26 million paired-end Illumina reads were combined to form longer fragments (i.e., contigs) that, in turn, were used to form longer sequences (i.e., scaffolds), and ultimately unigene sequences. A comparison of the assembly with the NCBI protein database indicated that 42% of the unigenes (24,814 sequences) did not significantly match the mRNA database and were thus considered putative novel transcribed sequences. In addition to these findings, data mining using Perl script MISA enabled the detection of 5883 microsatellites in 4993 unigenes, demonstrating the extraordinary potential of RNA-Seq for the discovery of this type of polymorphisms. The transcriptomic study on sesame samples is another example of RNA-Seq application to study nonmodel species *(31)*. Transcriptome assembly was achieved by sequencing 24 paired-end cDNA libraries using Illumina technology. Also, a survey of the new transcriptome assembly for the presence of SSRs revealed more than 10,000 microsatellites in 42,566 unitranscript sequences, many of them showing an uneven distribution in the transcriptome. In some studies, more than one sequencing platform has been used in parallel to address a novel transcriptome assembly. This strategy has shown to be particularly helpful for improving transcriptome assembly of bread wheat using about 16 million reads provided by Illumina and Roche sequencers *(32)*. The assembly of this complex polyploidy eukaryotic transcriptome was performed following a two-stage approach. First, a rough assembly was produced using the Velvet/Oases assembler, and then, reads in each cluster were reassembled using the high-precision assembler MIRA. Using the FLX platform, the date palm fruit transcriptome has been investigated at seven different developmental stages *(33)*. An average of 1 million reads with a median length of 399 bp was obtained for each sequenced library. Interestingly, date fruit transcriptome showed high homology with grapevine sequences. In a separate report, *de novo* hop transcriptome assembly using Illumina sequencing has provided relevant information regarding the lupulin gland gene expression *(34)*. Transcriptomic data in combination with metabolite analysis provided evidence for the lupulin gland-specific BCAA and isoprenoid metabolism to produce precursors for bitter acids, which are important in brewing industry. Turmeric transcriptome has been assembled using a similar RNA-Seq approach with an Illumina sequencer *(35)*. In that case, pathway annotation of transcripts indicated that a number of expressed genes were related to the biosynthesis of secondary metabolites, some of them have been suggested to exert potential health-promoting effects. Also recently, the complexity of tea transcriptome has been investigated by RNA-Seq *(36)*. In that study, approximate 2.5 Gb were obtained with Illumina technology from different tea plant tissues. Processing and aligning the near 34.5 million 75-bp paired-end reads enabled the construction of 127,094 unigenes, a number 10-fold higher than existing sequences for tea plant in GeneBank. Some of the unigenes were annotated and assigned to putative metabolic pathways that are important to tea quality, such as flavonoid, theanine, and caffeine biosynthesis.

In addition to the unique capability for *de novo* transcriptome assembly and improved transcriptome annotation, RNA-Seq is also useful for comparative transcriptomic analysis. As an example, Ono *et al.* employed Illumina RNA-Seq technology to determine gene-expression levels in fruit peel after accomplishing *de novo* assembly of pomegranate transcriptome *(37)*. In this case, transcript annotation based on homologues grape and *Arabidopsis* genomes, allowed the identification of putative gene sequences involved in the metabolism of terpenoid and phenolic compounds with a role in pigmentation, flavor, and nutritional value of fruits. In a separate report, Feng *et al.* followed a different approach to investigate the transcriptomic changes during development and ripening of Chinese bayberry fruit *(38)*. To achieve that, each of the RNA samples from various tissues and fruit of different development and ripening stages were ligated with a different adaptor and sequenced in the same run using Illumina sequencing. The data produced from the mixed samples were used to construct the whole transcriptome assembly. In the second part of the study, the generated assembly was used as the reference, and data from each separate sample, identified by the adaptor sequences, were used to estimate the global differential gene expression between samples. Moreover, organic acid and sugar profiling data obtained with mass spectrometry-based methods were obtained as complementary information to the transcriptomic profiles in order to gain some insights on the metabolic pathways involved in fruit ripening, color development, and taste quality. Similarly, Illumina RNA-Seq has been used for the study of different aspects in major food crops, such as, for example, the studies on rice including the investigation of seed development (39,40); the transcriptional response to nematode infection (41) and to drought stress (42); differential gene expression between aleurone and starchy wheat endosperm (43); and transcript profiling in maize (44), chickpea (45), and soybean (46). Also, Illumina RNA-Seq technology has been applied to study the miRNA fraction in common bean (47) and rapeseed (48). González-Ibeas et al. used barcoding strategy to prepare 10 RNA libraries from different melon plant tissues for further sequencing using Roche technology (49). This strategy allowed the identification of conserved miRNAs, small interfering RNAs, as well as the discovery of potential melon-specific sequences miRNAs.

In addition to the aforementioned studies, Illumina DGE technology has been successfully applied to identify transcriptome differences in seven tissues on sweet potato (21). Using another approach, Kalavacharla et al. used FLX platform in combination with barcode tagging for the quantitative analysis of gene expression in common bean (50). Moreover, tagging cDNA libraries was very helpful on verifying and validating global gene-expression patterns, and detecting both shared and unique transcripts among the analyzed bean tissues. Also, the study by Garg et al. on chickpea has demonstrated the great potential of FLX sequencer for differential gene-expression analysis (51). In the initial stage of the study, de novo transcriptome assembly of chickpea was obtained from nearly 2 million short reads. For the assembly process, authors assayed eight different programs highlighting the importance of optimizing the assembly procedure.

4.2   Foodborne Pathogenic Microorganisms

One of the main goals for the food industry is the production of safe foods with the desired quality using minimal processing technologies. Foodborne disease, commonly referred to as food poisoning, occurs when food becomes contaminated with harmful species. Although chemical species such as pesticides, among others, can originate important health problems, the vast majority of food poisonings are the direct result of microbiological hazards induced by bacteria, toxigenic molds, and microalgae, viruses, and parasites. Over the past years, the availability of genome sequences of relevant food microorganisms has given rise to extraordinary possibilities for the study the molecular mechanisms in complex biological processes such as food spoilage and biofilm formation (52,53). In this field, RNA-Seq offers great potential to investigate the activities of foodborne microorganisms under strictly controlled conditions in the laboratory as well as in industrial environments or in food products. Despite its good potential, RNA-Seq methods have been applied less frequently to study foodborne pathogens than to investigate food crops. Regarding sample preparation, enrichment for all transcripts other than the abundant rRNA and tRNA species in RNA samples can be challenging, especially for bacterial transcriptomes, lacking mRNAs with poly(A) tail. A frequent solution to this problem involves 16S and 23S rRNA depletion from total RNA fraction isolated from microbial cells. With regard to the investigation on foodborne pathogens, RNA-Seq has been highly valuable in providing global transcriptomic profiles of persistent and nonpersistent *Listeria monocytogenes* isolates (54). This foodborne pathogen is the causative bacterium of serious invasive disease in animals and in humans. The contamination of food-processing facilities and food products with this *L. monocytogenes* is of particular concern because it survives extreme environmental conditions and has the ability to form resistant biofilms. The RNA-Seq study by Fox *et al.* was focused on the comparison of both bacterial strains in response to the treatment with benzethonium chloride, a disinfectant used in food-processing industry. RNA-Seq data suggested that treatment induced a complex peptidoglycan biosynthesis response, which may play a key role in disinfectant resistance. Also, RNA-Seq has provided the information to generate transcription start site maps for pathogenic and nonpathogenic *Listeria* species (55). In that work, the discovery of novel long antisense RNA species using this methodology suggested new mechanisms for the regulation of gene expression in bacteria. Also, the potential of RNA-Seq for the investigation of microorganisms in complex food matrices has been demonstrated in a recent study of *Salmonella* in peanut oil (56). Interestingly, the study revealed that desiccated bacterial cells in peanut oil were in physiologically dormant state with a low portion (<5%) of its genome being transcribed.

4.3 Food Fermentations

A major focus in food biotechnology research is directed to the investigation of cellular and molecular processes involved in industrially relevant microorganisms that are responsible for food fermentations. The acquired information obtained from such studies may ultimately help fermentation-based industries to enhance the quality of the final product, to improve their product yields, and even to develop novel foods. In this research field, high-throughput transcriptomic tools assist in elucidating the molecular mechanisms behind interesting metabolic transformations and functionalities in fermented food ecosystems *(57)*. Over the last years, the complete genomes of significant fermenting microorganisms, including yeasts and bacteria species, have been released into public databases. These genomic data resources have been essential to develop several species-specific microarrays that enable the study of gene expression under different conditions, providing new insight into important metabolic processes. For instance, transcriptome profiling of the yeast *Saccharomyces cerevisiae* and lactic acid bacteria (LAB) has contributed to improve our knowledge about cellular processes and responses of these organisms in different environments. Recent microarray applications in this field include the study of fermentation-related stress factors on the transcriptional response for laboratory or industrial wine, lager brewing and baker's yeast strains *(58–60)*, gene-expression dynamics during different fermentation stages in synthetic media or natural substrates *(61)*, and transcriptional differences between diverse strains and mutants *(62)*. In addition to yeasts, LAB have also industrial relevance since this group of microorganisms has the ability to provide the key flavor, texture, and preservative qualities to variety of fermented foods such as sourdoughs, dairy products, and fermented sausages *(63)*.

Although gene-expression microarray has become a powerful tool in this research field, RNA-Seq has taken gene-expression analysis to a higher level in terms of improving the possibilities for investigating novel aspects of transcriptomes in fermenting microorganisms *(64)*. For instance, many aspects of the transcriptome structure of *S. cerevisiae* have been elucidated using Illumina RNA-Seq method. In their pioneer work, Nagalakshmi *et al.* identified alternative initiation codons, upstream open-reading frames, the presence of several overlapping genes and unexpected 3'-end heterogeneity in the yeast transcriptome *(65)*. Interestingly, RNA-Seq data also indicated that about 75% of the nonrepetitive sequence of the yeast genome was transcribed. In another report, RNA-Seq has also provided remarkable insight into the transcriptome of *Aspergillus oryzae*, a mold used in various oriental fermented foods *(66)*. Among the discoveries using paired-end reads Illumina technology, authors highlighted the identification of novel transcripts, new exons, untranslated regions, alternative splicing isoforms, alternative upstream initiation codons and upstream open-reading frames. Moreover, gene-expression profiling indicated that the mold showed superior protein production grown under solid substrate than in liquid culture. Although the application of RNA-Seq to food metatranscriptomics studies is still lacking, this technique is well suited for the complex uncharacterized microbial ecosystems *(67,68)*. In a near future, it can be expected that RNA-Seq will be applied to investigate the activities involved in global metabolic processes in fermented foods and to decipher the temporal contribution of each species in food ecosystems. Such analyses will constitute the foundation for constructing a system level understanding of microbial activity in complex food ecosystems.

5   FUTURE OUTLOOKS AND CONCLUSIONS

RNA-Seq techniques have demonstrated their impressive analytical potential for gene-expression studies in the context of food science. RNA-Seq has the potential to quickly supersede microarray in many gene-expression studies. However, RNA-Seq technology still remains evolving and several technical and bioinformatics challenges need to be overcome to realize the full potential of this technique in food science. Despite the extraordinary reductions in cost per sequenced base associated with SGS in comparison with more conventional sequencing methods, the application of RNA-Seq to survey the transcriptome (structure and expression levels) is still expensive. Thus, it might be expected that cheaper and faster library preparation methods will be developed to decrease the costs of generating sequencing data in the near future. Also, the development of novel high-throughput enrichment strategies, such as in-solution or in-microarray capture methods, aimed to target specific sequences of interest, will broaden the applicability of RNA-Seq in food science, since such strategies have the potential to improve the sequencing of low-abundance transcripts without the need for increasing the costly sequencing depth (69).

Given the evolution path of RNA-Seq technology, high-end instruments with higher sequencing throughput able to provide longer and accurate reads can be expected in the very near future. For instance, a new generation of sequencers (third-generation sequencing, TGS), based on single-molecule sequencing, is rapidly emerging. An outstanding advantage of these novel approaches is that they do not require routine PCR amplification prior sequencing, thereby avoiding systematic amplification bias occurring in SGS. In addition to their capability for sequencing RNA directly with the corresponding savings in reagents and manpower, novel technologies can also sequence molecules in real-time, decreasing the time of analysis and allowing longer read lengths. TGS systems, including PacBio-, nanopore-sequencing technologies, and direct imaging of individual DNA molecules using advanced microscopy techniques have been recently reviewed (4). In addition to the high-end sequencers, the recent commercialization of bench-top instruments, including the 454 GS Junior (Roche), Ion PGM (Life Technologies), and MiSeq (Illumina), seeking lower cost and time of analysis will bring about RNA-Seq to gain more popularity in food research laboratories in coming years (70).

As it has been discussed in this chapter, transcriptomic profiling of food crops is a hot application for RNA-Seq technologies. Nevertheless, the applicability of RNA-Seq in Food Science and Nutrition is still in its infancy and it is not fully exploited yet. In the near future, it can be anticipated that many relevant topics in Food Science will benefit from RNA-Seq techniques. For instance in Nutrigenomics studies, it can be expected that RNA-Seq studies will improve our limited understanding of the roles of nutritional compounds at molecular level. Also, RNA-Seq may be a suitable profiling tool to characterize and investigate different safety aspects related with genetically modified organisms. Novel aspects related with food pathogens will be addressed such as, for example, the link between pathogen and host, pathogenesis, and virulence factors. For microorganisms that produce toxins in food products, understanding the molecular mechanisms for toxins and their secondary metabolites production is a key point to limit the toxin contamination in food products and food-processing facilities. In consequence, the availability of tools such RNA-Seq, capable to provide detailed information about the transcriptional regulation of the metabolic pathways implicated in the biosynthesis of toxins in toxigenic microorganism in real samples may be very valuable in ongoing and future research.

ACKNOWLEDGMENTS

This work was supported by AGL2011-29857-C03-01 project (Ministerio de Economía y Competitividad, Spain) and CSD2007-00063 FUN-C-FOOD (Programa CONSOLIDER, Ministerio de Educación y Ciencia, Spain). A. V. thanks the Ministerio de Economía y Competitividad for his FPI predoctoral fellowship.

REFERENCES

1. Pareek, C. S.; Smoczynski, R.; Tretyn, A. *J. Appl. Genet.* **2011**, *52*, 413–435.
2. Wang, Z.; Gerstein, M.; Snyder, M. *Nat. Rev. Genet.* **2009**, *10*, 57–63.
3. Malone, J. H.; Oliver, B. *BMC Biol.* **2011**, *9*, 34.
4. Niedringhaus, T. P.; Milanova, D.; Kerby, M. B.; Snyder, M. P.; Barron, A. E. *Anal. Chem.* **2011**, *83*, 4327–4341.
5. Morozova, O.; Marra, M. A. *Genomics* **2008**, *92*, 255–264.
6. Mardis, E. R. *Annu. Rev. Genomics Hum. Genet.* **2008**, *9*, 387–402.
7. Bentley, D. R.; Balasubramanian, S.; Swerdlow, H. P.; Smith, G. P.; Milton, J.; Brown, C. G.; Hall, K. P.; Evers, D. J.; et al. *Nature* **2008**, *456*, 53–59.
8. Margulies, M.; Egholm, M.; Altman, W. E.; Attiya, S.; Bader, J. S.; Bemben, L. A.; Berka, J.; Braverman, M. S.; et al. *Nature* **2005**, *435*, 376–380.
9. Metzker, M. L. *Nat. Rev. Genet.* **2010**, *11*, 31–46.
10. Datta, S.; Datta, S.; Kim, S.; Chakraborty, S.; Gill, R. S. *J. Proteomics Bioinform.* **2010**, *3*, 183–190.
11. McGettigan, P. A. *Curr. Opin. Chem. Biol.* **2013**, *17*, 4–11.
12. Stein, L. D. *Curr. Protoc. Bioinform.* **2011**, *36*, 11.1.1.
13. Hong, H.; Zhang, W.; Shen, J.; Su, Z.; Ning, B.; Han, T.; Perkins, R.; Shi, L.; Tong, W. *Sci. China Life Sci.* **2013**, *56*, 110–118.
14. Scholz, M. B.; Lo, C. C.; Chain, P. S. *Curr. Opin. Biotechnol.* **2012**, *23*, 9–15.
15. Blow, N. *Nature* **2009**, *458*, 239–242.
16. Marguerat, S.; Wilhelm, B. T.; Bähler, J. *Biochem. Soc. Trans.* **2008**, *36*, 1091–1096.
17. Mutz, K.; Heilkenbrinker, A.; Lönne, M.; Walter, J.; Stahl, F. *Curr. Opin. Biotech.* **2013**, *24*, 22–30.
18. Ozsolak, F.; Milos, P. M. *Nat. Rev. Genet.* **2011**, *12*, 87–98.

19. Wilhelm, B. T.; Landry, J. *Methods* **2009**, *48*, 249–257.
20. Mortazavi, A.; Williams, B. A.; McCue, K.; Schaeffer, L.; Wold, B. *Nat. Methods* **2008**, *5*, 621–628.
21. Tao, X.; Gu, Y.; Wang, H.; Zheng, W.; Li, X.; Zhao, C.; Zhang, Y. *PLoS One* **2012**, *7*, e36234.
22. Feng, H.; Qin, Z.; Zhang, X. *Cancer Lett.* **2013**, *340*, 179–191.
23. Malcom, J. W.; Malone, J. H. In: *Fundamentals of Advanced Omics Technologies: From Genes to Metabolites;* Simó, C.; Cifuentes, A.; García-Cañas, V., Eds.; Elsevier: Oxford, UK, 2014; pp 325–354.
24. Zhou, Q.; Gallagher, R.; Ufret-Vincenty, R.; Li, X.; Olson, E. N.; Wang, S. *Proc. Natl. Acad. Sci. U.S.A.* **2011**, *108*, 8287–8292.
25. Edwards, M. A.; Henry, R. J. *J. Cereal Sci.* **2011**, *54*, 395–400.
26. Wegman, D. W.; Krylov, S. N. *Trend Anal. Chem.* **2013**, *44*, 121–130.
27. Rensink, W. A.; Buell, R. *Trends Plant Sci.* **2005**, *10*, 603–609.
28. Iyer, M. K.; Chinnaiyan, A. M. *Nat. Biotechnol.* **2011**, *29*, 599–600.
29. Gordo, S. M.; Pinheiro, D. G.; Moreira, E. C.; Rodrigues, S. M.; Poltronieri, M. C.; de Lemos, O. F.; da Silva, I. T.; Ramos, R. T.; et al. *BMC Plant Biol.* **2012**, *12*, 168.
30. Zhang, H.; Wei, L.; Miao, H.; Zhang, T.; Wang, C. *BMC Genomics* **2012**, *13*, 316.
31. Zhang, J.; Liang, S.; Duan, J.; Wang, J.; Chen, S.; Cheng, Z.; Zhang, Q.; Liang, X.; et al. *BMC Genomics* **2012**, *13*, 90.
32. Schreiber, A. W.; Hayden, M. J.; Forrest, K. L.; Kong, S. L.; Langridge, P.; Baumann, U. *BMC Genomics* **2012**, *13*, 492.
33. Yin, Y.; Zhang, X.; Fang, Y.; Pan, L.; Sun, G.; Xin, C.; Ba Abdullah, M. M.; Yu, X.; et al. *Plant Mol. Biol.* **2012**, *78*, 617–626.
34. Clark, S. M.; Vaitheeswaran, V.; Ambrose, S. J.; Purves, R. W.; Page, J. E. *BMC Plant Biol.* **2013**, *13*, 12.
35. Annadurai, R. S.; Neethiraj, R.; Jayakumar, V.; Damodaran, A. C.; Rao, S. N.; Katta, M. A.; Gopinathan, S.; Sarma, S. P.; Senthilkumar, V.; et al. *PLoS One* **2013**, *8*, e56217.
36. Shi, C.; Yang, H.; Wei, C.; Yu, O.; Zhang, Z.; Jiang, C.; Sun, J.; Li, Y.; et al. *BMC Genomics* **2011**, *12*, 131.
37. Ono, N. N.; Britton, M. T.; Fass, J. N.; Nicolet, C. M.; Lin, D.; Tian, L.; Integr, J. *Plant Biol.* **2011**, *53*, 800–813.
38. Feng, C.; Chen, M.; Xu, C.; Bai, L.; Yin, X.; Li, X.; Allan, A. C.; Ferguson, I. B.; et al. *BMC Genomics* **2012**, *13*, 19.
39. Xu, H.; Gao, Y.; Wang, J. *PLoS One* **2012**, *7*(2), e30646. http://dx.doi.org/10.1371/journal.pone.0030646.
40. Davidson, R. M.; Gowda, M.; Moghe, G.; Lin, H.; Vaillancourt, B.; Shiu, S.-H.; Jiang, N.; Buell, C. R. *Plant J.* **2012**, *71*, 492–502.
41. Kyndt, T.; Denil, S.; Haegeman, A.; Trooskens, G.; Bauters, L.; Van Criekinge, W.; De Meyer, T.; Gheysen, G. *New Phytol.* **2012**, *196*, 887–900.
42. Zong, W.; Zhong, X.; You, J.; Xiong, L. *Plant Mol. Biol.* **2013**, *81*, 175–188.
43. Gillies, S. A.; Futardo, A.; Henry, R. J. *Plant Biotechnol. J.* **2012**, *10*, 668–679.
44. Hui Xu, X.; Chen, H.; Sang, Y. L.; Wang, F.; Ma, J.; Gao, X.; Zhang, X. S. *BMC Genomics* **2012**, *13*, 294.
45. Garg, R.; Patel, R. K.; Jhanwar, S.; Priya, P.; Bhattacharjee, A.; Yadav, G.; Bhatia, S.; Chattopadhyay, D.; Tyagi, A. K.; Jain, M. *Plant Physiol.* **2011**, *156*, 1661–1678.
46. Chen, H.; Wang, F. W.; Dong, Y. Y.; Wang, N.; Sun, Y. P.; Li, X. Y.; Liu, L.; Fan, X. D.; Yin, H. L.; Jing, Y. Y.; Zhang, X. Y.; Li, Y. L.; Chen, G.; Li, H. Y. *BMC Plant Biol.* **2012**, *12*, 122.

47. Peláez, P.; Trejo, M. S.; Iñiguez, L. P.; Estrada-Navarrete, G.; Covarrubias, A. A.; Reyes, J. L.; Sanchez, F. *BMC Genomics* **2012**, *13*, 83.

48. Körbes, A. P.; Machado, R. D.; Guzman, F.; Almerão, M. P.; de Oliveira, L. F. V.; Loss-Morais, G.; Turchetto-Zolet, A. C.; Cagliari, A.; et al. *PLoS One* **2012**, *7*, e50663.

49. Gonzalez-Ibeas, D.; Blanca, J.; Donaire, L.; Saladié, M.; Mascarell-Creus, A.; Cano-Delgado, A.; Garcia-Mas, J.; Llave, C.; et al. *BMC Genomics* **2011**, *12*, 393.

50. Kalavacharla, V.; Liu, Z.; Meyers, B. C.; Thimmapuram, J.; Melmaiee, K. *BMC Plant Biol.* **2011**, *11*, 135.

51. Garg, R.; Patel, R. K.; Jhanwar, S.; Priya, P.; Bhattacharjee, A.; Yadav, G.; Bhatia, S.; Chattopadhyay, D.; et al. *Genome Anal.* **2011**, *156*, 1661–1678.

52. Puttamreddy, S.; Carruthers, M. D.; Madsen, M. L.; Minion, F. C. *Foodborne Pathog. Dis.* **2008**, *5*, 517–529.

53. Andrews-Polymenis, H. L.; Santiviago, C. A.; McClelland, M. *Curr. Opin. Biotech.* **2009**, *20*, 149–157.

54. Fox, E. M.; Leonard, N.; Jordan, K. *Appl. Environ. Microbiol.* **2011**, *77*, 6559–6569.

55. Wurtzel, O.; Sesto, N.; Mellin, J. R.; Karunker, I.; Edelheit, S.; Becavin, C.; Archambaud, C.; Cossart, P.; et al. *Mol. Syst. Biol.* **2012**, *8*, 583.

56. Deng, X.; Li, Z.; Zhang, W. *Food Microbiol.* **2012**, *30*, 311–315.

57. Bokulich, N. A.; Mills, D. A. *BMB Rep.* **2012**, *45*, 377–389.

58. Shima, J.; Kuwazaki, S.; Tanaka, F.; Watanabe, H.; Yamamoto, H.; Nakajima, R.; Tokashiki, T.; Tamura, H. *Int. J. Food Microbiol.* **2005**, *102*, 63–71.

59. Tai, S. L.; Daran-Lapujade, P.; Walsh, M. C.; Pronk, J. T.; Daran, J. *Mol. Biol. Cell* **2007**, *18*, 5100–5112.

60. Rossignol, T.; Postaire, O.; Storaï, J.; Blondin, B. *Appl. Microbiol. Biotechnol.* **2006**, *71*, 699–712.

61. Penacho, V.; Valero, E.; Gonzalez, R. *Int. J. Food Microbiol.* **2012**, *153*, 176–182.

62. Bartra, E.; Casado, M.; Carro, D.; Campama, C.; Piña, B. *J. Appl. Microbiol.* **2010**, *109*, 272–281.

63. Klaenhammer, T. R.; Azcarate-Peril, M. A.; Altermann, E.; Barrangou, R. *J. Nutr.* **2007**, *137*, 748S–750S.

64. Solieri, L.; Dakal, T. C.; Giudici, P. *Ann. Microbiol.* **2013**, *63*, 21–37.

65. Nagalakshmi, U.; Wang, Z.; Waern, K.; Shou, C.; Raha, D.; Gerstein, M.; Snyde, M. *Science* **2008**, *320*, 1344–1349.

66. Wang, B.; Guo, G.; Wang, C.; Lin, Y.; Wang, X.; Zhao, M.; Guo, Y.; He, M.; Zhang, Y.; Pan, L. *Nucleic Acids Res.* **2010**, *38*, 5075–5087.

67. Mäder, U.; Nicolas, P.; Richard, H.; Bessieres, P.; Aymerich, S. *Curr. Opin. Biotech.* **2011**, *22*, 32–41.

68. McNulty, N. P.; Yatsunenko, T.; Hsiao, A.; Faith, J. J.; Muegge, B. D.; Goodman, A. L.; Henrissat, B.; Oozeer, R.; et al. *Sci. Transl. Med.* **2011**, *3*, 106ra106.

69. Mercer, T. R.; Gerhardt, D. J.; Dinger, M. E.; Crawford, J.; Trapnell, C.; Jeddeloh, J. A.; Mattick, J. S.; Rinn, J. L. *Nat. Biotechnol.* **2012**, *30*, 99–104.

70. Loman, N. J.; Constantinidou, C.; Chan, J. Z. M.; Halachev, M.; Sergeant, M.; Penn, C. W.; Robinson, E. R.; Pallen, M. J. *Nat. Rev. Microbiol.* **2012**, *10*, 599–606.

Part II

Proteomics and Peptidomics

Making Progress in Plant Proteomics for Improved Food Safety

Pier Giorgio Righetti*, Elisa Fasoli*, Alfonsina D'Amato* and Egisto Boschetti†

*Department of Chemistry, Materials and Chemical Engineering "Giulio Natta", Politecnico di Milano, Milano, Italy

†EB Consulting, JAM-Conseil, Neuilly s/Seine, France

1 INTRODUCTION

While animal proteomics is well advanced after more than a decade of accelerated development, much remains to be discovered in plant proteomics, a fundamental discipline allowing to understand the biological basis of agricultural production and to manage the genomics development related to plant modifications and consequences. After having demonstrated its capability of global approach to discover novel proteins and the recognition of specific patterns in view of detecting biomarkers, proteomics has been applied to the analysis of foods and drinks as well. With the progress of science and education of the population especially in developed countries, the need to have full information on food and beverages has been continuously stimulated not only

Comprehensive Analytical Chemistry, Vol. 64. http://dx.doi.org/10.1016/B978-0-444-62650-9.00006-3

for educational purposes but also for security issues. Although the advances of genomics allowed the advent of many analytical applications, it did not resolve all questions. More recently, the development of proteomics and meta-bolomics opened up another page of analytical possibilities by addressing directly the ingredients of food that are of major importance such as proteins and all other items that would contribute to human health. These powerful capabilities enriched the panoply of tools to check for better food and drink security. The latter domain was timely because of concomitant development of genetically modified organisms (GMO) introduced for resolving economi-cal situations related to the capability of feeding a growing worldwide popu-lation for the decades to come.

Authenticity, quality and security issues can now be addressed with modern technologies, among which the global protein analysis known as pro-teomics (1). This technology itself made strong progresses especially with the capability of detecting very dilute species expressed at trace level, thus allow-ing the detection of frauds and also adventitious agents such as low-abundance allergens.

What is discussed in this chapter is the situation of plant proteomics in terms of low-abundance proteins. Here is especially described a recent technol-ogy, along with dedicated applications, based on a unique sample treatment using combinatorial peptide ligand libraries (CPLL) that have already largely demonstrated their capabilities for novel discovery in animal protein science (2). The situation of both plant-derived food and drinks is reviewed throughout the context of plant proteomics.

Naturally, plant ingredients are not the only source of food and drinks. Actually, in advanced developed countries, food from animal source is domi-nating where milk and meat represent the most concrete examples of drink and food, respectively. Nevertheless, the prevalence of specific applications, such as protein components useful for human health, stimulates the develop-ment of plant consumption. As a counterpart, plants also contain threatening protein components (toxins and allergens) that require attention for an improved food security; therefore, there is a need of detecting low-abundance proteins (the deep proteome) from plant tissues as schematically illustrated in Figure 1. It is within this context that this chapter purposely limits the discus-sion to plant proteomics. The description about the situation of low-abundance animal-derived food proteomics has been made by the same authors in various other documents (3–6).

2 CURRENT TECHNOLOGIES IN PLANT PROTEOMICS SAMPLE TREATMENT

Proteomics studies in plants are not fundamentally different from animal prote-omics: A protein extraction is needed first; then the analytical methods are the standard ones, such as polyacrylamide gel electrophoresis, two-dimensional

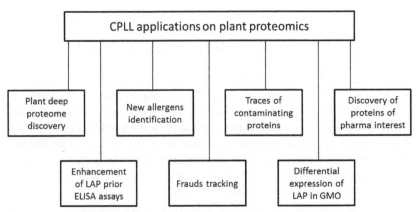

FIGURE 1 Application of deep proteome technologies to food- and drink-based plants. LAP, low-abundance proteins; GMO, genetically modified organisms.

electrophoresis, liquid chromatography, protein breakdown with trypsin or other proteases, and mass spectrometry under various aspects. The dilemma related to the presence of high-abundance proteins mixed to very low-abundance species is also similar and requires similar solutions. The major differences are at the level of the protein content (mainly glycosylated with different glycans compared to animal glycoproteins) that is much lower in plants for a similar amount of tissue and the large amounts of other products such as cellulose, lignin, polyphenols, polysaccharides of various natures, and pigments. All these products need to be eliminated before proteomics studies. A large number of proteins are also part of the cell wall where from their extraction is challenging *(7)*.

In spite of analytical approaches similarities, animal proteomics is more advanced than plant proteomics perhaps due to the much larger interest in human diagnostic and therapy.

Recently, due to food safety implications, plant proteomics gained in importance and progress made is already impressive. Even so, not much is known on plant proteomics details. A number of questions are yet to be answered such as their dynamic concentration, where they are located, and how the proteins are post-translationally modified. The function of many of these proteins is moreover unknown (e.g., their mutual interaction and their response to environmental stimulations).

The fact that the protein content is scarce in plants is not as critical as for animal samples since with plants, most generally, the availability of biological material is quite large. This authorizes a relatively easy identification of proteins that are expressed only in few copies (this can be resolved using the CPLL technology as described in the succeeding text). However, in contrast to animal proteins (especially human serum), the depletion technology is here not easily applicable because of the absence of immunosorbents addressing the most abundant plant proteins. The abundance of a variety of proteins depends on the plant organ in question. Green leaves are, for instance, extremely rich in ribulose-1,5-bisphosphate carboxylase/oxygenase (RuBisCO) constituting about 30–60% of the total protein content; due to its ubiquitous presence in plants, it is considered as the most abundant protein on earth. This major protein interferes with the detection and identification of lower abundance proteins in leaf extracts. RuBisCO is an essential enzyme in the Calvin cycle that allows carbon fixation and plays an important role in photosynthesis. For the depletion of this protein, special immunosorbents have been successfully experimented. Other plant organs may have different concentrated proteins. This is the case, for instance, of tubers that are essentially rich in storage proteins and the same immunodepletion cannot unfortunately be applied.

Common technical approaches to plant proteomics investigations are the extraction and the purification of proteins from crude tissue *(8)*; however, due to the presence of a number of proteases and a low protein charge, a protection against proteolysis is highly recommended by adding selected inhibitors. Then the removal of nonproteinaceous products is operated according to the nature of the plant tissue (see scheme on Figure 2).

The global recovery of proteins from the undesired material mentioned in the preceding text is performed as a second operation just after the initial crude extraction. Precipitation and/or selective extraction is considered, depending on the context. When dealing with seeds with a high lipid content, a delipidation by means of chemical solvents is recommended or even required *(9–12)*. This operation can be performed before or after protein

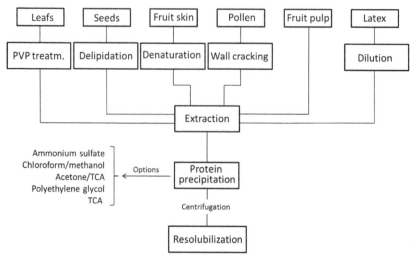

FIGURE 2 Sample pretreatments as a function of plant tissue origin before entering the protein treatment in view of fractionation or depletion or compression of dynamic concentration range. Details and protocols can be found in Boschetti and Righetti *(6)*.

extraction. The elimination of pigments and polyphenols can be operated by a treatment with aqueous phenol, associated or not with some amounts of poly-vinylpyrrolidone *(13)*. The concentration of proteins is obtained by precipita-tion. This can be accomplished by using, for instance, TCA associated with acetone and containing reducing agents. By that procedure, most proteins are collected as pellets, and the supernatant that contains other undesired non-proteinaceous materials is eliminated by centrifugation or filtration *(14)*. Other substances that can interfere with two-dimensional electrophoresis can be separated by protein precipitation with ammonium acetate or ammonium sulfate from a solution of Tris–HCl saturated with phenol. Another approach is the combination of phenol extraction and a precipitation with ammonium acetate in methanol as reported by Faurobert *et al.* *(15)*. Among other possi-bilities, it can also be mentioned a protein precipitation method involving a mixture of chloroform–methanol–water in 1:4:3 proportions *(12)*. However, due to a frequent difficult resolubilization of proteins, the addition of small amounts of chaotropes and zwitterionic detergents can be useful *(16)*. When the extraction of proteins from plant tissues requires some amount of sodium dodecyl sulfate, the latter should be eliminated for further fractionation of analysis by precipitation with hydro-organic mixtures as described in the succeeding text in the examples of plant proteins treated with CPLL.

All of these methods do not address the important question of the enhance-ment of low-abundance proteins to allow detecting rare species or very low-expressed proteins. Protein fractionation by regular ion exchange chro-matography could be one possible way to proceed *(17)*, but it involves a large dilution and yields a large number of fractions with a consequent heavy work-load for protein analysis and identification. Depletion technologies (especially immunodepletion using specific antibodies) largely used for human plasma proteins to remove several high-abundance proteins *(18,19)* are here only marginally applied. In one example, immunodepletion of RuBisCO has been applied to proteomic analysis of rice after cold stress *(20)*. The analysis of the remaining leaf proteins by differential two-dimensional showed 39 differ-ently expressed proteins (19 upregulated and 20 downregulated) out of a total 250–400 protein spots. Compared to a nondepleted leaf extract, the removal of RuBisCO allowed discovering four novel proteins, a modest harvest considering the cost of this operation. In another example, Cellar *et al.* *(21)* demonstrated that immunodepletion was effective on different species such as corn and tobacco with good removal efficiency.

3   THE POWER OF HEXAPEPTIDE LIGAND LIBRARIES

At present, it is current knowledge that a biological sample treated with solid-phase CPLLs yields a much larger number of identified proteins from any biological sample. This is due to the reduction of dynamic concentration range of proteins present in a sample via the large decrement of high-abundance proteins and the concomitant concentration of low-abundance ones. How is this obtained? Although the concept and the theoretical and practical implications are largely described in a dedicated book *(6)*, a summary can here be delivered for noninitiated readers. The material for the treatment of the biological sample is constituted of a mixture of discrete sorbents, each of them carrying a different ligand (here, a hexapeptide). The number of discrete different sorbents is extremely large. Each of them represents an item of a full mathematical combination of 15 amino acids aligned as hexapeptides and obtained by combinatorial synthesis. Each discrete peptide is a potential ligand for one or more proteins from the sample. By loading the combined sorbent mix with a biological sample exceeding the binding capacity, each single sorbent will be saturated with corresponding protein partners, and the excess will be found in the so-called flow-through. Abundant proteins will rapidly saturate the corresponding ligand, while the rare proteins will continue to be adsorbed as long as the loading is increased giving rise to a substantial concentration. Once the sorbent mix is extensively washed to remove the excess of proteins, the captured species are harvested by means of appropriate desorbing agents. All proteins are potentially present but their relative concentration is deeply changed with a strong compression of dynamic concentration range. The consequence of such a situation is the detection of many novel proteins (the low-abundance ones) as a result of (i) the annihilation of the signal suppression due to concentrated species (e.g., albumin in serum) and (ii) the detection of very low-concentration proteins that were below the detectability level prior the sample treatment. These two distinct phenomena coexisting during the CPLL treatments are very powerful when the loading is very large and when all captured proteins are harvested *(2,22)*.

Contingent upon a number of parameters that are all extensively described by Boschetti and Righetti *(6)*, the method of sample processing allows an increase of the number of proteins that can range from 50% to more than five times the number of proteins found in the nontreated sample. Variations are also possible with the use of different loading conditions and elutions as well as analytical methods.

As a comparison, the methods of high-abundance protein removal (specific depletion or immunodepletion) to suppress the signal obscuration due to very concentrated species subtract also associated proteins. However it engenders a very important problem related to the dilution effect of depletion: the rare species that would need to be concentrated to be detected are here diluted, thus aggravating the difficulty for their detection.

In plant proteomics, the depletion by antibodies, very common in animal proteomics processing, is not very popular, and to our best knowledge, it is limited to RuBisCO when leaves are extracted for proteomics investigations (see Section 2). Instead, protein fractionation could be used with well-known advantages and disadvantages.

The treatments of proteins with CPLL could also be partially oriented toward a category of proteins conferring a decisive advantage against immunodepletion. For instance, playing on pH at the capture stage, the cationic or anionic proteins could be selectively enhanced *(23)*; the capture in low ionic strength is somewhat different from the one operated at high ionic strength *(24)*. In addition, the capture of hydrophobic proteins is favored in the presence of lyotropic salts *(25)*. At the elution stage, proteins can be harvested at once all together or under conditions allowing the fractionated desorption for easier analytical determinations for the discovery of novel proteins. All these possibilities are described in a recently published book *(6)* where the practitioner can also find many detailed protocols.

In differentiated plant tissues, soluble proteins are of low concentration *(26)* and are contaminated by undesired material that can prevent the proper function of CPLL; therefore, several points have to be considered. For instance, as described in the previous section, there are methods to remove nonproteinaceous material, but a number of protocols cannot be directly streamlined with CPLL treatment. The main rules to be adopted can be summarized as follows: (a) The aqueous extraction should be performed in relatively low ionic strength to prevent the solubilization of nucleic acids; (b) with highly viscous material, such as lymph exudates, a dilution is recommended; (c) when dealing with proteins that are engaged within the cell wall, such as pollen proteins, some amounts of nonionic detergent (<0.5–1%)—or even 3% sodium dodecyl sulfate—and urea (<3 M) should be used; (d) many plant extracts comprise polyphenols and pigments that can be removed using small amounts of polyvinylpyrrolidone; (e) when dealing with seeds, it is recommended to make an extract using nonpolar solvents first. Figure 3 summarizes several suggestions on how to proceed for protein capture in conjunction with the initial extraction method that may contain products more or less compatible with CPLLs.

All the described pretreatments, prior to CPLL application, contribute to obtain significantly better analytical results especially when using electrophoretic technologies (one-dimensional or two-dimensional electrophoresis). When using a proper extraction, the recovery of proteins can be performed by precipitation with ammonium sulfate at 80–90% saturation. Protein pellets are then desalted (dialysis or diafiltration or gel filtration or centrifugation using appropriate filtration-integrated devices) against a buffer recommended for CPLL treatment.

Nevertheless, in some cases, no pretreatment of the sample prior CPLL is required. This is the case for protein content or contamination in plant-derived beverages such as wines *(30,31)*.

4   EXAMPLES OF PLANT PROTEOMICS STUDIES FOCUSING ON LOW-ABUNDANCE SPECIES

An excellent review on the subject has been recently published by Agrawal *et al. (5)* not only reporting technical issues and solutions around proteomics investigations but also extending the context to food security throughout reducing pathogen risk and food traceability to prevent frauds. Unfortunately,

Protein capture possibilities with CPLL

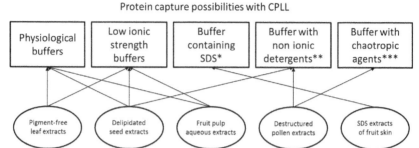

FIGURE 3 Selection of CPLL protein capture conditions as a function of the protein extraction method. With pigment-free leaf extracts, a direct protein adsorption can be performed in physiological conditions or at low ionic strength to increase the amount of harvested proteins per volume unit of beads. Seed extracts that are first submitted to lipid removal can also be treated with CPLL under the same conditions; however, due to a presumed hydrophobic proteins content, the addition of limited amounts of nonionic detergent contributes for a better protein availability to beads. As described in the text, fruit skin and pulp do not behave similarly to CPLL protein capture (see also Figure 4A); therefore, adapted conditions might be required. For pollen proteins, the question is related to the recovery of proteins that are within the pollen globule. Here, a destructuration of the pollen wall can be necessary; this is accomplished by adding some amounts of nonionic detergent and limited amounts of urea that are compatible with the CPLL capture operation (for details, see Shahali *et al. (29)*). *: The sodium dodecyl sulfate concentration used for extraction is around 2–4%; this must be reduced by dilution to 0.1% before CPLL capture. **: The final concentration of nonionic detergents in the capture buffer should be below 0.2%. ***: When the chaotropic agent is urea, its final concentration in the capture buffer should be below 3 M.

despite recent investigations of some research teams, plant proteomics is relatively marginally worked out compared to animal proteomics.

Considering the critical role of plant-based food to cover the future food demand by a rapidly growing worldwide population, a better knowledge of plant proteins (composition, interactions, adverse effects, allergic reactions, and biopharmaceutical discoveries) becomes a strategic challenge.

While it is easy to focus attention to current plant proteins, the low-abundance species that are probably very active in terms of biological performance are at the same time very numerous and present at trace levels precluding a direct detection possibility with current instruments and methods. It is clear that even more than in animal proteomics, the decrease of the concentration of current proteins and the concomitant enhancement of rare species is mandatory to get a complete picture of the situation.

The following example illustrates the capability of CPLL on the potential discovery of plant proteins used either as food or drink origins.

4.1 Plants as Basis for Food

Plants are a relatively large source of food and their use tends to increase compared to meat including in developed countries. Vegetables are very frequently served with meat; pasta is an elaborated food ingredient from wheat; leaves (as salad or spinach) are served with many dishes; seeds represent a large part of human diet as rice and wheat; tubers are also a group of plant products used extensively for cooking; finally, fruits represent also a large part of food taken on daily basis. Besides bringing just food, plants are also a source of many other compounds absent in meat such as fibers, polysaccharides, lipids, vitamins, antioxidants, and several other interesting products.

Indeed, food plant products can have other interesting functions than just being a source of energy such as, for instance, pharmaceutical properties. Compounds with antioxidant properties and with antiaging effects are brought by various vegetables (32).

Conversely, not all plant proteins are good for eating. It is thus essential before ingesting plant-based food to know the composition, which can only be accessed by a global proteomics analytical approach. Correlations between diet and health (living longer, reduction of cardiovascular risks, and reduction of frequency of cancer) have been one of the major concerns of human population for many years. It has been found that different foodstuffs can mitigate inflammation in human tissues, thus attenuating the risk of chronic diseases (33). Some types of diets, such as the Mediterranean one, based on plants containing antioxidants, olive oil, grains, and other vegetable products have stimulated research projects to find scientific information exploitable for health (34,35).

In the past decade, specific attention has been focused on major vegetable products used worldwide in human diets. Rice is one of them. A compelling review has been published in 2011 by Agrawal et al. (36) where rice proteins are discussed under the aspect of proteomics technologies. Their extraction needs special care with a dedicated sample treatment to harvest the largest possible proteome for a more exhaustive analysis of the components. The regulation of seed storage proteins is played by RNA-binding proteins that need to be concentrated. This was accomplished by affinity chromatography using polyU Sepharose (37). Among these proteins in rice seeds, 14-3-3 proteins were identified. Since RNA-binding proteins represent probably more than 200 proteins, many of them of very low abundance, a large amount of work remains to be accomplished. Undoubtedly, CPLLs would contribute to discover many of them, thus allowing functional studies.

Another plant deserving attention for its large importance for human food is potato tubers. In northern countries, potatoes under different aspects constitute a large part of the diet. To attempt covering possible future needs, specific genomic modifications have been tried especially to protect these plants against insect attacks by the expression of protease inhibitors. The impact of genetic manipulations over endogenous proteases has been investigated to detect potential undesirable effects. To this end, potato extracts needed to be treated with polyvinylpyrrolidone followed by a protein precipitation in methanol/chloroform. Proteins were then analyzed by mono-dimensional and two-dimensional electrophoresis separations followed by immunodetection and mass spectrometry. The protein pattern was not significantly modified but several proteins related to the patatin storage protein complex that were not from the transgene appeared differently expressed. Conversely, the protease inhibitor profile appeared very similar to nontransgenic standard tubers suggesting the absence of possible induction of allergic reactions. In spite of this demonstration, one could argue that the analysis was performed using a standard protein extract where the low-abundance proteins expressed thanks to the transgene were at a concentration below the sensitivity of analytical methods and were consequently probably ignored. Perhaps, the picture would have been different if the initial protein sample had been treated with CPLLs. This is the reason why investigations recently started to decipher the deep proteome of key plant organs: leafs, vegetables, and fruits.

The spinach (*Spinacia oleracea*) leaf proteome as a source of food ingredients has been investigated by Fasoli *et al. (38)*. As already indicated, proteins present in leaf extracts are very dilute and comprise minority components; therefore, to analyze their proteomes, it is essential to concentrate them after having eliminated many other components incompatible with proteomics analysis. Leaf proteome includes the high-abundance complex RuBisCO. This protein complex accounts for more than 50% of spinach leaf proteomes. In this situation, it is clear that most low-abundance proteins are masked; therefore, RuBisCO needs to be removed by, for instance, immunodepletion or to be largely decreased in its concentration. To avoid the concomitant depletion of interacting species and also due to nonspecific binding of the immunosorbent, CPLLs appeared more adapted to resolve the situation. This approach largely reduced the concentration of RuBisCO—but preserving a certain amount along with adsorbed species—and at the same time, the low-abundance species were largely enhanced. On the technical point after removal of undesired pigments and polyphenols from the initial aqueous extract, protein treatment with CPLL was performed in physiological reduced conditions and in the presence of protease inhibitors. The elution was operated using a boiling solution of 4% sodium dodecyl sulfate containing 20 mM dithiothreitol. The protein sample thus harvested showed the presence of 322 unique gene products, 114 of which found thanks to CPLL treatment. Mainly two categories of proteins were largely enhanced. One group was about proteins related to chloroplasts: after protein enhancement, among the species found were glucose-1-phosphate adenylyltransferase, glutamate decarboxylase 2, fructose-bisphosphate aldolase 2, inorganic pyrophosphatase 1, phosphoglycerate kinase, 5-methyltetrahydropteroyl-triglutamate-homocysteinemethyl transferase, and ribulose-phosphate 3-epimerase. The second category of protein enrichment was ribosomal proteins such as 30S, 40S, 50S, and 60S.

Another type of leafs used in human food having been investigated for their content in low-abundance proteins are artichoke leaves *(39)*. Globe artichoke (*Cynara scolymus* L.) is a plant belonging to the *Asteraceae* and is consumed within the Mediterranean countries and in the United States. The inflorescence (commonly known as 'head') is consumed immature leaf by leaf. This plant structure is known to comprise compounds of pharmaceutical interest dealing with hepatoprotection *(40)*, protection against free radicals, *(41)* and inhibition of cholesterol biosynthesis *(42)*.

Different extraction techniques have been used for the characterization of the artichoke's proteome by CPLL. The initial extraction buffer contained 50 mM Tris–HCl (pH 7.4), 50 mM sodium chloride, 2% (m/v) CHAPS, 1% (m/v) sodium dodecyl sulfate, and 25 mM dithiothreitol. Protease inhibitor cocktails were added to the extraction buffers to prevent the action of protease. The first extract was then diluted 1:10 (v/v) with the same buffer without sodium dodecyl sulfate to facilitate the protein capture. The extract was separated into four equal aliquots; each of them was titrated to different pH values (4.0, 7.2, and 9.3). The pH of the fourth aliquot was reduced to 2.2 with addition of 0.1% TFA and formic acid. Then individually, each aliquot was added with 100 μL of CPLL beads overnight at room temperature under gentle shaking.

The elution of the captured polypeptides was performed in 4% boiling sodium dodecyl sulfate under reducing conditions. On this protein collection, current analyses of the proteome, such as two-dimensional and SDS electrophoresis coupled to mass spectrometry, were performed. Eight hundred and seventy six unique gene products have been accounted for, 479 of them found exclusively thanks to the contribution of CPLL for the enhancement of low-abundance species. Each single pH of capture contributed to the large body of species: The pH 2.2 capture permitted to reveal a total of 49 novel species, pH 4.0 yielded 136 unique gene products, pH 7.2 detected 114 novel proteins, and pH 9.3 allowed identifying 99 novel species. The synergistic effect of CPLLs was here clearly demonstrated. This study represented the most complete artichoke proteome investigation to date. This has to be compared with a quite elaborated extraction method that allowed evidencing only 70 protein spots reduced to around 40 gene products due to the presence of a number of isoforms *(43)*. This number has been recently increased to 119 by the same authors with a different analytical approach *(44)*; however, this still represents a minor increment compared to what CPLL allowed to discover.

Another plant product of large food interest is avocado (*Persea americana*). This is a tropical fruit largely consumed in various parts of the world as vegetable alone or with salads. It contains a large amount of fats and oils similar to the case of olives for oils. Avocado is also used as a source of products for cosmetics and healthcare *(45)*. Actually, avocado is associated with some important health benefits such as the decrease of the level of cholesterol, low-density lipoproteins, and triglycerides *(46)*. As a result of its oily composition, proteomics studies are quite challenging requiring specific pretreatments. In the context of low-abundance gene product discovery, a paper has been recently published *(27)*. In that case, the protein extraction has been made by using two distinct methods. The first involving a homogenization with 0.125 M Tris–HCl buffer, pH 7.4, comprising a protease inhibitor cocktail (5 g of avocado pulp with 15 mL of buffer).The second method was made using the same buffer without protease inhibitors but containing 3% sodium dodecyl sulfate and 25 mM dithiothreitol. The extraction was performed in a boiling bath for 10 min. From both extracts, proteins were separated by precipitation with a mixture of water/methanol/chloroform *(47)*, and the collected pellets dried at room temperature. From both protein extracts, the protein capture was made at pH 7.4 and pH 2.2 in the presence of 0.1% TFA. While in the control sample (non-CPLL-treated), the total number of gene products found was 193; conversely, the two extractions with and without 3% sodium dodecyl sulfate were more effective yielding, respectively, 762 and 440 gene products (190 proteins being in common) (see Figure 4 left panel). The contribution of pH 2.2 extraction was, however, relatively marginal. Overall, a total number of 1012 unique gene products could be identified against 364 for the untreated control. Among them, the already known avocado allergen Pers a1 was found along with several other potential allergens such as a glucanase, an isoflavone reductase-like protein, a polygalacturonase, a profillin, and a thaumatin-like protein.

Similarly, the pulp and the seed of olives (*Olea europaea*) were extensively analyzed *(48)*. Olives are constituted of a complex matrix for proteomics studies due to their lipidic nature and the large amount of interfering compounds from where the extraction of proteins is difficult *(49)*.This fruit is used for the extraction of oils and as food ingredient in the Mediterranean diet. The olive pulp protein extraction was performed by freezing 5 g of olive

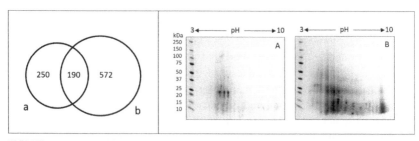

FIGURE 4 Examples of protein treatments results using CPLL. The panel on the left represents the difference of avocado protein capture efficiency by CPLL when the proteins are extracted just in physiological conditions (a) or extracted under denaturing conditions (b) (here, the extraction buffer contained 3% of sodium dodecyl sulfate, which was then diluted to 0.1%). A clear doubling of gene product was accounted. For details, see Esteve *et al. (27)*. Panel on the right: two-dimensional electrophoresis results of mango peel (A) and soft tissue (B) extracts. Peel proteins are clearly difficult to extract without destructuring the fruit skin with a detergent. Both extracts have been here treated with CPLL that increased the harvest of about eight times in each case compared to the respective controls. For details, see Fasoli *et al. (28)*. First dimension, pH 3–10 immobilized pH gradients; second dimension, 8–18% polyacrylamide gel gradient. Colloidal Coomassie Blue staining.

pulp at −20 °C followed by grinding without thawing and homogenization with 15 mL of a buffer containing 0.125 M Tris–HCl (pH 7.4), 5% (v/v) glycerol, and the protease inhibitor cocktail. Afterward, sodium dodecyl sulfate and dithiothreitol, up to concentrations of 3% (m/v) and 25 mM, respectively, were added and the resulting solution was boiled for 10 min. A water/methanol/chloroform precipitation (47) was next applied, obtaining an abundant white precipitate free of sodium dodecyl sulfate. In order to prepare a control sample, a second precipitation with acetone at 4 °C during 1 h was performed. The final precipitate was dried at room temperature.

Proteins were then dissolved in 20 mL of 0.125 M Tris–HCl, pH 7.4 containing 1% (m/v) CHAPS and the protease inhibitor cocktail, and treated with CPLL. With several variations from the standard treatments such as other complementary extractions from seeds, the use of urea-containing extraction buffer, and an additional peptide ligand library, the total number of detected gene products was quite large. Two hundred and thirty one unique proteins were found in the pulp (vs. 56 described previously) and 61 in the seeds (vs. only four reported by the literature). In the latter case, the presence of seed storage proteins, oleosins, and histones were detected; in the pulp, a thaumatin-like protein, an allergenic protein also named Ole e13, was confirmed.

In the domain of fruits, other examples reported on the detection of low-abundance proteins by CPLL dealt with banana and mango.

Banana is mainly grown in tropical countries and represents an important source of energy in the diet of people. Antioxidant and cell antiproliferative properties of banana are associated with a reduced risk in some pathological situations, and it is seen as a model for genomic evolution in relation to biotic and abiotic stresses. The banana pulp contains about 20% carbohydrates and around 1% proteins. Like in many plant organs, proteins are relatively difficult to extract and the signal of proteins of orders-of-magnitude less abundance is obscured by abundant species. To reach this category and thus to enlarge the knowledge of its proteome, the CPLL technology has been applied (50). In conjunction with advanced mass spectrometry techniques, the investigation allowed identifying 1131 proteins. Several enzymes involved in degradation of starch granules and strictly correlated to ripening stage were identified. In addition, the presence of known allergens was confirmed (e.g., musa a1, pectin esterase, and superoxide dismutase) as well as new potential allergens. This study gives starting information for further research to elucidate, for instance, mechanisms of interaction, regulation, and protein expression.

Mango (*Mangifera foetida*) is a tropical fruit native of South Asia. It is a food constituent of a number of desserts (smoothies, ice cream, fruit bars, sweet chili sauce, and many others) or as a component in fresh fruit combinations. In addition to many healthy vitamins, oligoelements, and other health-beneficial compounds, it carries some proteinaceous allergens with symptoms of hypersensitivity reactions that can appear immediately or 48 or 72 h after exposure or ingestion. In spite of the allergenic risk, the proteome of this fruit was almost not known before the analytical determination with CPLL *(28)*. The protein extraction was first made on both the peel and the pulp similarly to other methods described in the preceding text for other fruits. The CPLL treatment of both protein extracts was performed at pH 7.4 and 2.2 as described in the preceding text for avocado. An additional C_{18} resin treatment was made to try harvesting peptides from both extracts. The combined eluates of the mango pulp proteome showed mainly two-dimensional electrophoresis spots located in the relatively low molecular mass region and placed mostly in the acidic pH region (see Figure 4, right panel). In all combined eluates, the number of gene products found was 2693 against 374 for the non-CPLL-treated control, with 212 unique proteins in common. This was accounted for both the peel and the pulp. Most of proteins were found in the pulp except membrane proteins that were a major component of the peel.

As a final example of a proteome from CPLL-treated plant fruit that should be mentioned is of *Lycium barbarum* (goji berries or wolfberries) pulp using the enhancement effect of CPLLs *(51)*. This plant grows mainly in Tibet and China where it has been used in the traditional medicine for its benefits to antiaging *(52)* and stimulation of organ functions such as the kidney and liver *(53)*. The relatively recent introduction of this new food ingredient into the Western diets might generate novel allergic symptoms that in some cases might include anaphylactic reactions *(54,55)*. Although the *L. barbarum* polysaccharides are well known and studied in-depth for their numerous pharmaceutical properties *(53,56)*, very little was known about its proteome. The recent investigation using CPLL evidenced a total of 350 gene products after extraction and protein sample treatment, which represents by far the deepest proteomics investigation of goji berries. The initial crude extraction was operated by addition of up to 3% of sodium dodecyl sulfate in neutral buffer, but the detergent was either diluted to 0.1% or eliminated by precipitation of the proteins with acetone/methanol prior to CPLL capture.

4.2 Plants as Basis for Drinks

Most human drinks are derived from plants and more specifically from fruits. Fruit juice is actually an easy and pleasant way to take liquids for survival. Many examples of plant-derived drinks are available, but the first one is perhaps wine. A first classification can be made between alcoholic beverages and nonalcoholic ones.

Besides water and milk, wine has been one of the oldest plant-derived drinks for humans. Generated by yeast fermentation of grape juice, it became very popular already in the Roman era *(57)*. The global worldwide annual production of wine, with 300 million of hectoliters, represents a significant proportion of the global expanding food market. As a result of fruit extract, it contains some amounts of residual grape proteins that survived the fermentation process. The analysis of this proteome allows to determine the origin of the grape. This being said, a major reason for the proteome analysis of wines is to identify not only the origin but also more importantly the remaining traces of the production process. At the end of the fermentation process, residual grape proteins slowly aggregate and form an insoluble material causing turbidity to wines. To prevent this process that reduces the commercial value of wines, a post-treatment capable of removing the flocculation phenomenon is operated. This treatment implies the addition of flocculating agents, among them casein from bovine milk or egg white albumin; caseins are also known as major food allergens. Although this is an authorized treatment, it is mandatory that all traces of casein after treatment are removed as attested by dedicated ELISA methodologies. According to the current rules, the absence of residual casein is certified by a negative result of ELISA analysis; however, the detection limit of the established ELISA test by the EC is 200 μg/L of casein. Therefore, below this concentration, the wine is considered as being casein-free. In a recent work using a protein 'amplification' by CPLLs, D'Amato et al. (31) and Cereda et al. (30) found the presence of casein in supposedly casein-free wines, thanks to the capture of very dilute proteins from large volumes of wine, down to as little as 5–10 μg/L. A warning is here lightened because it is not clear if, at such concentrations, the allergic effect of casein is fully absent.

In red wine, only one grape protein was found (thaumatin); all other proteins were from *Saccharomyces cerevisiae* and a few others from plant pathogens and fungi (e.g., *Botryotinia fuckeliana*, *Sclerotinia sclerotiorum*, and *Aspergillus aculeatus*) (31).

These findings stimulated the research of grape proteins from wines that are not treated for protein flocculation. Still, by using CPLLs, in a *recioto* wine (a sweet Italian wine prepared from Garganega grapes) (58), 106 unique gene products could be found. The relatively large number of proteinaceous products found is to be compared to 9 gene products found in Champagne (59) and 28 glycoproteins in a chardonnay from Puglia (60).

Another very popular worldwide alcoholic beverage is beer. Most beer proteins are from barley (*Hordeum vulgare*), but others originate from fermentation microorganisms. With a shotgun proteomics analysis, the number of protein found was 33; this comprised 10 proteins from yeast (61). When the beer proteome was explored via the CPLL methodology, no less than 62 proteins (62) could be identified: 20 of them were different barley proteins, 2 were from maize origin, and an additional 40 were essentially from *S. cerevisiae* (except one of them from *Saccharomyces bayanus* and another one from *Saccharomyces pastorianus*). Most if not all these proteins are classified as being of low abundance because their identification was rendered possible only by the use of affinity peptide ligands.

Proteomes of other alcoholic beverages have been explored with the use of CPLLs with the objective of concentrating low-abundance proteins and evidencing thus the plant origin of the beverage. This was the case of the proteome from lemon flavedo (the outer yellow skin) extracts with the aim of not only deciphering the local proteome but also trying to correlate the protein content of a well-known alcoholic beverage called *Limoncello (63)*. The capture with CPLLs of pooled proteins at pH 2.2 and 7.2 revealed the presence of 1011 gene products, 264 of them found also in a genuine homemade *Limoncello* drink, while for other industrial corresponding drinks, the number of proteins found was much lower and in certain cases absent. This is another interesting point to certify the genuineness of these drinks by deep proteomics analysis.

A large variety of nonalcoholic beverages claim the presence of plant and fruit extracts. If so, they should contain some amounts of original proteins even if they are extremely diluted with the addition of water. A number of them have been analyzed for their proteomics profile, but due to large dilution, a preliminary treatment with CPLL was applied. To date, studies on these types of proteomes are very rare; perhaps, the most compelling data are about the detection of peptides as described by Kašička *(64)*.

Recently, the enrichment of proteins by a combinatorial peptide library allowed demonstrating quite easily the presence or the absence of proteins, for instance, in almond's milk and orgeat syrup *(65)*, in coconut milk *(66)*, in a cola drink *(67)*, in ginger ale *(68)*, and even in white wine vinegar that is not really a beverage *(69)* (see in the succeeding text).

As a summary, in almond's milk, 137 unique protein species were found, whereas in the case of orgeat syrup (bitter almond extract), 13 proteins could be identified. Nevertheless, both deep proteome analyses could assess the genuineness of these products. This was not necessarily the case for cheap orgeat extracts sold in supermarkets, suggesting a possible production complemented with synthetic aromas. When present, proteins of this investigation could be easily visible by SDS–polyacrylamide gel electrophoresis after protein capture at pH 7.0 and 9.3 *(65)*.

The proteome investigation of coconut milk showed 307 unique gene products, 200 discovered via CPLL capture and 137 detected in the control. Only 30 gene products were in common between the two sets of data. Compared to a dozen proteins discovered so far in previous studies, this larger proteome could be used as a source of proteins of interest for human health (nutraceuticals) with some unique beneficial effects already described for coconut milk.

It is within the context of finding products that have an effect on human health that a study has been published regarding the gene expression of prostate cancer cells treated with pomegranate fruit juice *(70)* where a comparative proteomics study provided insights for prostate cancer evolution. The involvement of pomegranate fruit products in the molecular mechanism of inducing prostate cancer cell apoptosis has been evidenced. Thus, a possible novel effect of chemopreventive strategy for prostate cancer could be envisioned. Although it is unclear if the origin of the effect was from a polypeptidic fruit product, one could expect additional important insights with the use of CPLL as a mean to 'amplify' the effect in case very low-abundance proteins would be involved.

Concerning the use of CPLL in the detection of protein traces in other types of beverages, an interesting detective story can here be told about the 'invisible' proteome of a Cola drink (67), stated to be produced with a kola nut extract. Indeed, a few proteins in the Mr 15–20 kDa range could be identified from 1 L of drink. The proteins belong to kola nuts and to *Agave* species as this was expected according to what was declared in the commercial label (drink spiked with 6% of agave syrup).

Although not really a nonalcoholic drink, it is interesting to mention the case of white wine vinegar proteome after treatment of this product with CPLLs (69). Globally, 27 gene products could be found, 10 of them belonging to *Vitis vinifera*. Out of the 17 remaining gene products, 13 were found in the general database Uniprot_viridiplantae and 4 by using the entire Swiss-Prot database. The survival of these proteins within a very acidic environment of vinegar suggests that these species could have a tightly packed hydrophobic core conferring a quite good protection against denaturation. These proteins seemed also resistant to proteolytic attack because they have been evidenced by electrophoresis with an apparent molecular mass within 12 and 70 kDa range.

All mentioned reports on proteomics analysis of plant-based beverages could be extended to many other drinks and could be used as a means of detecting traces of proteins justifying the claims and protecting the consumers against potential frauds. In addition, it could be possible to track the presence of other heterologous proteins coming from post-treatments attempting to stabilize some drinks or to confer specific properties. A sort of proteotyping might be useful for a given drink identification against counterfeited products.

5   PROTEOMICS APPLICATIONS IN FOOD SECURITY

Alongside protein biomarkers in human health, markers of food security even for plant proteins might become essential for the determination of food security and authenticity. In the actual world where products travel rapidly throughout continents, one of the main items for security is the traceability of products to try finding the origin of a possible health problem and make corrective actions. The case reported in June 2011 on the contamination of soya germs by a mutant of *E. coli* was symptomatic of what is the current risk for human health requiring a formal traceability of plant products and fully sensitive analysis.

With the advent of GMO that are accepted in certain countries and not in others, except for animal feeding, tracing the presence of such organisms as part of human food, even in trace amounts, is necessary. This can be assessed by the analysis of the genome; however, it is also possible by measuring the expression of genes that not only does result in the presence of novel species but also can modify the normal expression of endogenous proteins. Proteomics profiles are considered as complementary analytical determinations to genomics; they are accomplished by classical two-dimensional electrophoresis and mass spectrometry. These methodologies can be applied not only to animals but also to reengineered plants as described for transgenic maize (71). In this precise case out of 100 proteins that were modified in their expression, 43 of them directly resulted from the single inserted transgenic gene in the corn genome.

With the use of modern very sensitive analytical methods (especially with the possibility to enhance low-expressed proteins), the concept of 'substantial equivalence,' coined years ago, of GMO to conventional counterparts may not have the same impact as at its origin. With proteomics patterns focusing on low-abundance species, the equivalence may never be demonstrated due to the plasticity of protein expression and their interdependence. It is clear today that the effects of transgenesis on transcriptome do not suffice to predict the modifications in proteomes and consequently in metabolomes (72). Needless to say, with CPLLs, many more modifications would be detected due to the capability to reach very low-expressed proteins.

Proteins being at the center of cellular metabolism and development, it is to be expected that they focus a lot of interest in food safety assessment. They actually could negatively impact on human or animal health if they comprise, for example, polypeptidic allergens, proteinaceous toxins, and antagonistic proteins such as special inhibitors. For all these compounds, genome analysis would not bring pertinent information contrary to an in-depth proteome investigation (73).

A number of transgenic plants have been submitted to proteome analysis. For instance, in rice, Wang et al. (74) used classical two-dimensional electrophoresis and mass spectrometry to evidence that more than 50 proteins were differently expressed. Proteomics studies have been performed also in transgenic tomato (75), potato (76), and wheat (77) again by using standard analytical methods.

Beyond the questions around plant transgenicity, there are those regarding the presence of allergens. Plants are known to comprise proteins generating allergic effects, and as such, they should be absent from the current diets; however, when they are part of food compositions with a number of other ingredients, they escape the attention of consumers. Allergens represent clear risks for already sensitized individuals. To prevent accidents, it is in fact not unusual to find information on food labels that the product could contain traces of peanuts, a powerful reservoir of allergens. The question of allergy being a major point for industrially processed food, a specific dedicated section is offered in the succeeding text.

6   THE INNOVATIVE INVOLVEMENT OF CPLL IN PLANT ALLERGEN DISCOVERY

Allergies of plant origin are gradually increasing over the years in all countries. Beyond pollen allergies, among which cypress represents probably the most widespread and highly invasive source in many locations around the Mediterranean basin, there is a quite great increase of food allergies of plant origin. A concrete example is given by the growing number of children who suffer from food allergies *(78)*. The health risk is very high and can go up to life-threatening even if the allergen is present in trace amounts depending on the individual level of sensitivity. To date, many allergens have been identified and well codified, thus helping the medical community to develop detection techniques. Identifying the allergic origin of a disease is an important step; however, another more important issue is to detect the presence of allergen traces in food. The most known allergens from edible plants are from wheat flour *(79)*, peanuts *(80)*, hazelnuts *(81)*, fenugreek *(82)*, and soybeans *(83)*, just to mention a few. The actual list of plant allergens is extending quite rapidly due to advanced proteomics studies. These technologies tend to detect novel allergens present in trace amounts but of strong effect on the immunosystem. A specific attention is to be made to peanut allergens due to the widespread presence in a number of food preparations and the seriousness of the biological reactions. That is the reason why methodologies have been devised to detect the presence of peanut traces in a variety of food products such as chocolate-based food *(84)* and ice cream *(85)*.

Plant allergens contained in food are generally relatively stable against proteolysis and therefore act at the level of the intestinal tract with dramatic effects with systemic allergy reactions *(86)*. Unfortunately, the prevalence of these reactions is significant in terms of cases since allergenic proteins are brought by various types of food not only from plants but also from bacteria and from animal origin (e.g., fish, eggs, and milk). Considering that the origin of allergens is very diverse and that several food components are present in a same preparation, tracing the component responsible for the reaction is quite challenging.

A special attention has been focused toward the transgenic plants as possible carriers of allergens, but this is far from being formally demonstrated and remains still speculative at least at the level of easily detectable proteins. In this respect, it appears essential to develop sensitive methods, targeting very low-expressed proteins, capable to identify trace amounts of known and novel allergens.

One of the first reviews on classical methods for proteomics of allergens was made by Ticha *et al. (87)* with the description of methodologies adapted for these studies. Current tools to separate allergens for identification, such as affinity chromatography ligands, have been reviewed. Among them, there are lectins such as concanavalin A, Cibacron Blue, and polyproline ligands. As affinity collection of combinatorial ligands, CPLLs might positively contribute to this enterprise. Although the detection of plant allergens is clearly not specifically dependent on CPLL technology, when one wants to discover very low-abundance allergens, this enrichment method is a very useful tool as it is for many rare and dilute proteins.

Actually, the demonstration of the detection of low-abundance, never-described allergens subsequent to the CPLL treatment of a biological plant extract has been repeatedly demonstrated. Most compelling examples are the work performed on *Hevea brasiliensis* exudates (rubber material source) *(88)* on *Cupressus sempervirens* pollen extracts *(29)* and *Arachis hypogea* nut extracts *(89)*. In the latter case, the question was not to discover new allergens but rather to find traces of the major peanut allergens Ara h1, Ara h2, and Ara h3 in baked cookies. The general approach after the treatment of the sample was *shotgun* proteomics for the detection of specific peptides from targeted proteolysis via SRM. The allergen detection was successful with a detection limit of around 10 µg peanut/g of matrix.

The first two cases were performed with the aim of trying to find the presence of novel allergens. The treatment of proteins from Hevea latex allowed to discover among 300 unique gene products identified several new allergen candidates. This was possible after having blotted and confronted, by two-dimensional electrophoresis, CPLL-captured proteins with the sera of 18 positive patients. They were hevamine A (43 kDa), a protease inhibitor (8 kDa), a proteasome subunit (30 kDa), glyceraldehyde-3-phosphate dehydrogenase (37 kDa), and a heat shock protein (81 kDa). It is known that the number of Hevea allergens is large, but diagnostic tests are generally limited to the first 13 allergens from the official standard list.

In another different case, the treatment of protein extracts from cypress pollen allowed discovering a total of 108 unique gene products, and among them, 12 IgE-binding proteins were characterized as allergens by two-dimensional electrophoresis followed by blotting against sera from allergic patients. Besides major known allergens, several IgE-binding proteins not previously reported as allergens were characterized (e.g., malate dehydrogenase, glyoxalase I, rab-like protein, glucanase-like protein, and triosephosphate isomerase). These results additionally demonstrated that the immunosystem of allergic patients can raise a diversified IgE response against over- and also under-represented allergenic proteins rendering current cypress tests incomplete.

To complete this short review about the interest of using CPLLs in allergen discovery, a note is to be made for other published results of fungal allergen discovered in blood subsequent to invasive aspergillosis *(90)*, bovine milk allergens *(91)*, and hen's egg allergens *(92)*.

As a summary, the use of CPLL demonstrated its capability not only to amplify the signal of existing allergens but also to discover new allergens that are currently below the sensitivity level of detection. This process could be extended to a large number of allergens potentially present in risk-related plants extracts and also in food components.

7   TOWARD POSSIBLE NEW STEPS

The use of CPLLs as a sample pretreatment to enter the domain of deep proteome and thus to reach the detection of very low-abundance gene products is not to be demonstrated any longer. Hundreds if not thousands of projects in proteomics and other applications are based on the use of CPLLs, and the development continues at great speed. If there is one side that would need to be supported is the proper utilization of the methodology because a number of authors do not use the technology as it should. The capture phase can easily and largely be improved; the elution of captured proteins can be much more effective in order to reduce the possible losses of rare species; more acidic and more alkaline proteins could be captured and hydrophobic proteins could be better targeted. All of these contributes to lower the losses (even if they are limited to a few percent of what is already detectable without pretreatment) and to simplify the process. To speed out all these possibilities, a book has been published in 2013 *(6)* with many scientific and practical information and recipes to guarantee the best results from any possible biological sample supposed to contain proteins, even if in traces.

At this stage, what could be expected is the increase of the number of applications especially in the domain of contamination by foreign proteins in food and drinks. Kits with all chemical ingredients (buffers, chemicals, and expendables) in the same box and a protocol that anyone can implement are a concrete expectation. The reproducibility of CPLL treatment having been demonstrated repeatedly, the results could be guaranteed providing all instructions are followed. Such kits could be protein-specific or group-specific or of general exploration for proteins. It can be also envisioned that the implementation of kits would be performed with a manual or an automatic device, thus saving manipulation time and giving gains on reliability of the tests.

There could be applications that could address specific groups of proteins as those described with the stabilization of white wines (presence of low traces of casein). In these cases, assay kits could comprise also all ingredients for an ELISA test after having treated the sample with CPLL.

All of these applications are virtually ready with the current CPLL products that are commercially available. Beyond the aforementioned possibilities and hypotheses, another field is open for development: this is the advent of more targeted combinatorial libraries and also cheaper libraries to decrease the cost of routine applications, which is of outmost importance especially within the domain of food science.

REFERENCES

1. Agrawal, G. K.; Timperio, A. M.; Zolla, L.; Bansal, V.; Shukla, R.; Rakwal, R. *J. Proteomics* **2013**, *32*, 335–365.
2. Righetti, P. G.; Boschetti, E. *Amino Acids* **2013**, *45*, 219–229.
3. Righetti, P. G.; Boschetti, E. *Electrophoresis* **2012**, *33*, 2228–2239.
4. Boggess, M.; Lippolis, J.; Hurkman, W.; Fagerquist, C.; Briggs, S.; Gomes, A.; Righetti, P. G.; Bala, K. *J. Proteomics* **2013**, *93*, 20–39.
5. Agrawal, G. M.; Sarkar, A.; Righetti, P. G.; Pedreschi, R.; Carpentier, S.; Wang, T.; Barkla, B. J.; Kohli, A.; Ndimba, B. K.; Bykova, N.; Rampitsch, C.; Zolla, L.; Rafudeen, M. S.; Cramer, R.; Bindschedler, L. V.; Tsakirpaloglou, N.; Ndimba, R. J.; Farrant, J. M.; Renaut, J.; Job, D.; Kikuchi, S.; Rakwal, R. *Mass Spectrom. Rev.* **2013**, *32*, 335–365.
6. Boschetti, E.; Righetti, P. G. *Low-Abundance Protein Discovery: State of the Art and Protocols;* Elsevier: Amsterdam, 2013, ISBN: 978-0-12-401734-4.
7. Borderies, G.; Jamet, E.; Lafitte, C.; Rossignol, M.; Jauneau, A.; Boudart, G.; Monsarrat, B.; Esquerre-Tugaye, M. T.; Boudet, A.; Pont-Lezica, R. *Electrophoresis* **2003**, *24*, 3421–3432.
8. Carpentier, S. C.; Panis, B.; Vertommen, A.; Swennen, R.; Sergeant, K.; Renaut, J.; Laukens, K.; Witter, E.; Samyn, B.; Devreese, B. *Mass Spectrom. Rev.* **2008**, *27*, 354–377.
9. Menke, W.; Koenig, F. *Methods Enzymol.* **1980**, *69*, 446–452.
10. Radin, N. S. *Methods Enzymol.* **1981**, *72*, 5–7.
11. Van Renswoude, J.; Kemps, C. *Methods Enzymol.* **1984**, *104*, 329–339.
12. Wessel, D.; Flugge, U. I. *Anal. Biochem.* **1984**, *138*, 141–143.
13. Wigand, P.; Tenzer, S.; Schild, H.; Decker, H. *J. Agric. Food Chem.* **2009**, *57*, 4328–4333.
14. Méchin, V.; Damervaland, C.; Zivy, M. *Methods Mol. Biol.* **2007**, *355*, 1–8.
15. Faurobert, M.; Pelpoir, E.; Chaïb, J. *Methods Mol. Biol.* **2007**, *355*, 9–14.
16. Isaacson, T.; Damasceno, C. M.; Saravanan, R. S.; He, Y.; Catalá, C.; Saladié, M.; Rose, J. K. *Nat. Protoc.* **2006**, *1*, 769–774.
17. Yuan, F.; Robinson, D. P.; Foster, L. J. *J. Proteome Res.* **2010**, *9*, 1902–1912.
18. Echan, L. A.; Tang, H. Y.; Ali-Khan, N.; Lee, K.; Speicher, D. W. *Proteomics* **2005**, *5*, 3292–3303.
19. Shi, T.; Zhou, J. Y.; Gritsenko, M. A.; Hossain, M.; Camp, D. G.; Smith, R. D.; Qian, W. J. *Methods* **2012**, *56*, 246–253.
20. Hashimoto, M.; Komatsu, S. *Proteomics* **2007**, *7*, 1293–1302.
21. Cellar, N. A.; Kuppannan, K.; Langhorst, M. L.; Ni, W.; Xu, P.; Young, S. A. *J. Chromatogr. B* **2008**, *861*, 29–39.
22. Thulasiraman, V.; Lin, S.; Gheorghiu, L.; Lathrop, J.; Lomas, L.; Hammond, D.; Boschetti, E. *Electrophoresis* **2005**, *26*, 3561–3571.
23. Fasoli, E.; Farinazzo, A.; Sun, C. J.; Kravchuck, A. V.; Guerrier, L.; Fortis, F.; Boschetti, E.; Righetti, P. G. *J. Proteomics* **2010**, *73*, 733–742.
24. Di Girolamo, F.; Boschetti, E.; Chung, M. C.; Guadagni, F.; Righetti, P. G. *J. Proteomics* **2011**, *74*, 589–594.
25. Santucci, L.; Candiano, G.; Petretto, A.; Lavarello, C.; Bruschi, M.; Ghiggeri, G. M.; Citterio, A.; Righetti, P. G. *J. Chromatogr. A* **2013**, *1297*, 106–112.
26. Rose, J. K.; Bashir, S.; Giovannoni, J. J.; Jahn, M. M.; Saravanan, R. S. *Plant J.* **2004**, *39*, 715–733.
27. Esteve, C.; D'Amato, A.; Marina, M. L.; García, M. C.; Righetti, P. G. *Electrophoresis* **2012**, *33*, 2799–2805.

28. Fasoli, E.; Righetti, P. G. *Biochim. Biophys. Acta* **2013**, *1834*, 2539–2545.
29. Shahali, Y.; Sutra, J. P.; Fasoli, E.; D'Amato, A.; Righetti, P. G.; Futamura, N.; Boschetti, E.; Senechal, H.; Poncet, P. *J. Proteomics* **2012**, *77*, 101–110.
30. Cereda, A.; Kravchuk, A. V.; D'Amato, A.; Bachi, A.; Righetti, P. G. *J. Proteomics* **2010**, *73*, 1732–1739.
31. D'Amato, A.; Kravchuk, A. V.; Bachi, A.; Righetti, P. G. *J. Proteomics* **2010**, *73*, 2370–2377.
32. Ningappa, M. B.; Srinivas, L. *Toxicol. In Vitro* **2008**, *22*, 699–709.
33. Wu, X.; Schauss, A. G. *J. Agric. Food Chem.* **2012**, *60*, 6703–6717.
34. Seiberand, J. N.; Kleinschmidt, L. *J. Agric. Food Chem.* **2012**, *60*, 6644–6647.
35. Tomás-Barberán, F. A.; Somoza, V.; Finley, J. *J. Agric. Food Chem.* **2012**, *60*, 6641–6643.
36. Agrawal, G. K.; Rakwal, R. *Proteomics* **2011**, *11*, 1630–1649.
37. Crofts, A. J.; Crofts, N.; Whitelegge, J. P.; Okita, T. W. *Planta* **2010**, *231*, 1261–1276.
38. Fasoli, E.; D'Amato, A.; Kravchuk, A. V.; Boschetti, E.; Bachi, A.; Righetti, P. G. *J. Proteomics* **2011**, *74*, 127–136.
39. Saez, V.; Fasoli, E.; D'Amato, A.; Simó-Alfonso, E.; Righetti, P. G. *Biochim. Biophys. Acta* **2013**, *1834*, 119–126.
40. Aktay, G.; Deliorman, D.; Ergun, E.; Ergun, F.; Yesilada, E.; Cevik, C. *J. Ethnopharmacol.* **2000**, *73*, 121–129.
41. Gebhardt, R. *Toxicol. Appl. Pharmacol.* **1997**, *144*, 279–286.
42. Quiang, Z.; Lee, S.; Ye, Z.; Wu, X.; Hendrich, S. *Phytother. Res.* **2011**. http://dx.doi.org/10.1002/ptr.3698.
43. Acquadro, A.; Falvo, S.; Mila, S.; Giuliano Albo, A.; Moglia, A.; Lanteri, S. *Electrophoresis* **2009**, *30*, 1594–1602.
44. Falvo, S.; Di Carli, M.; Desiderio, A.; Benvenuto, E.; Moglia, A.; America, T.; Lanteri, S.; Acquadro, A. *Proteomics* **2012**, *12*, 448–460.
45. Swisher, H. E. *J. Am. Oil Chem. Soc.* **1988**, *65*, 1704–1706.
46. Lopez-Ledesma, R.; Frati-Munari, A. C.; Hernandez-Dominguez, B. C.; Cervantes-Montalvo, S.; Hernandez-Luna, M. H.; Juarez, C.; Morant-Lira, S. *Arch. Med. Res.* **1996**, *27*, 519–523.
47. Mirza, S. P.; Halligan, B. D.; Greene, A. S.; Olivier, M. *Physiol. Genomics* **2007**, *30*, 89–94.
48. Esteve, C.; D'Amato, A.; Marina, M. L.; Garcia, M. C.; Citterio, A.; Righetti, P. G. *J. Proteomics* **2012**, *75*, 2396–2403.
49. Esteve, C.; Del Rio, C.; Marina, M. L.; Garcia, M. C. *Anal. Chim. Acta.* **2011**, *690*, 129–134.
50. Esteve, C.; D'Amato, A.; Marina, M. L.; García, M. C.; Righetti, P. G. *Electrophoresis* **2013**, *34*, 207–214.
51. D'Amato, A.; Esteve, C.; Fasoli, E.; Citterio, A.; Righetti, P. G. *Electrophoresis* **2013**, *34*, 1729–1736.
52. Chang, R. C. C.; So, K. F. *Cell. Mol. Neurobiol.* **2008**, *28*, 643–652.
53. Amagase, H.; Farnsworth, N. R. *Food Res. Int.* **2011**, *44*, 1702–1717.
54. Larramendi, C. H.; Garcia-Abujeta, J. L.; Vicario, S.; Garcia-Endrino, A.; Lopez-Matas, M. A.; Garcia-Sedeno, M. D.; Carnes, J. *J. Investig. Allergol. Clin. Immunol.* **2012**, *22*, 345–350.
55. Carnes, J.; de Larramendi, C. H.; Ferrer, A.; Huertas, A. J.; Lopez-Matas, M. A.; Pagan, J. A.; Navarro, L. A.; Garcia-Abujeta, J. L.; Vicario, S.; Pena, M. *Food Chem.* **2013**, *137*, 130–135.
56. Potterat, O. *Planta Med.* **2010**, *76*, 7–19.
57. Righetti, P. G.; D'Amato, A.; Fasoli, E.; Boschetti, E. *Food Technol. Biotechnol.* **2012**, *50*, 253–260.
58. D'Amato, A.; Fasoli, E.; Kravchuk, A. V.; Righetti, P. G. *J. Proteome Res.* **2011**, *10*, 3789–3801.

59. Cilindre, C.; Jegou, S.; Hovasse, A.; Schaeffer, C.; Castro, A. J.; Clément, C.; Van Dorsselaer, A.; Jeandet, P.; Marchal, R. *J. Proteome Res.* **2008**, *7*, 1199–1208.

60. Palmisano, G.; Antonacci, D.; Larsen, M. R. *J. Proteome Res.* **2010**, *9*, 6148–6159.

61. Picariello, G.; Mamone, G.; Nitride, C.; Addeo, F.; Camarca, A.; Vocca, I.; Gianfrani, C.; Ferranti, P. *J. Proteomics* **2012**, *75*, 5872–5882.

62. Fasoli, E.; Aldini, G.; Regazzoni, L.; Kravchuk, A. V.; Citterio, A.; Righetti, P. G. *J. Proteome Res.* **2010**, *9*, 5262–5269.

63. Fasoli, E.; Colzani, M.; Aldini, G.; Citterio, A.; Righetti, P. G. *Biochim. Biophys. Acta* **2013**, *1834*, 1484–1491.

64. Kašička, V. *Electrophoresis* **2012**, *33*, 48–73.

65. Fasoli, E.; D'Amato, A.; Kravchuk, A. V.; Citterio, A.; Righetti, P. G. *J. Proteomics* **2011**, *74*, 1080–1090.

66. D'Amato, A.; Fasoli, E.; Righetti, P. G. *J. Proteomics* **2012**, *75*, 914–920.

67. D'Amato, A.; Fasoli, E.; Kravchuk, A. V.; Righetti, P. G. *J. Proteome Res.* **2011**, *10*, 2684–2686.

68. Fasoli, E.; D'Amato, A.; Citterio, A.; Righetti, P. G. *J. Proteomics* **2012**, *75*, 1960–1965.

69. Di Girolamo, F.; D'Amato, A.; Righetti, P. G. *J. Proteomics* **2011**, *75*, 718–724.

70. Lee, S. T.; Wu, Y. L.; Chien, L. H.; Chen, S. T.; Tzeng, Y. K.; Wu, T. F. *Proteomics* **2012**, *12*, 3251–3262.

71. Zolla, L.; Rinalducci, S.; Antonioli, P.; Righetti, P. G. *J. Proteome Res.* **2008**, *7*, 1850–1861.

72. Chassy, B. M. *Regul. Toxicol. Pharmacol.* **2010**, *58*, S62–S70.

73. Lovegrove, A.; Salt, L.; Shewry, P. R. In: *Transgenic Wheat, Barley and Oats;* Jones, H. D.; Shewry, P. R., Eds.; *Methods in Molecular Biology*; Vol. 478; Humana Press: New York, 2009.

74. Wang, W. W.; Meng, B.; Ge, X. M.; Song, S. H.; Yang, Y.; Yu, X. M.; Wang, L.; Hu, S. N.; Liu, S. Q.; Yu, J. *Proteomics* **2008**, *8*, 4808–4821.

75. Corpillo, D.; Gardini, G.; Vaira, A. M.; Basso, M.; Aime, S.; Accotto, G. R.; Fasano, M. *Proteomics* **2004**, *4*, 193–200.

76. Lehesranta, S. J.; Davies, H. V.; Shepherd, L. V. T.; Nunan, N.; McNicol, J. W.; Auriola, S.; Koistinen, K. M.; Suomalainen, S.; Kokko, H. I.; Karenlampi, S. O. *Plant Physiol.* **2005**, *138*, 1690–1699.

77. Di Luccia, A.; Lamacchia, C.; Fares, C.; Padalin, L.; Mamone, G.; La Gatta, B.; Gambacorta, G.; Faccia, M.; Di Fonzo, N.; La Notte, E. *Ann. Chim.* **2005**, *95*, 405–414.

78. Hadley, C. *EMBO Rep.* **2006**, *7*, 1080–1083.

79. Akagawa, M.; Handoyo, T.; Ishii, T.; Kumazawa, S.; Morita, N.; Suyama, K. *J.Agric. Food Chem.* **2007**, *55*, 6863–6870.

80. Chassaigne, H.; Tregoat, V.; Nörgaard, J. V.; Maleki, S. J.; van Hengel, A. J. *J. Proteomics* **2009**, *72*, 511–526.

81. Weber, D.; Polenta, G.; Lau, B. P. Y.; Godefroy, S. B. *Food Feed* **2009**, *1020*, 153–182.

82. Fæste, C. K.; Christians, U.; Egaas, E.; Jonscher, K. R. *J. Proteomics* **2010**, *73*, 1321–1333.

83. Houston, N. L.; Lee, D. G.; Stevenson, S. E.; Ladics, G. S.; Bannon, G. A.; McClain, S.; Privalle, L.; Stagg, N.; Herouet-Guicheney, C.; MacIntosh, S. C.; Thelen, J. J. *J. Proteome Res.* **2011**, *10*, 763–773.

84. Shefcheck, K. J.; Callahan, J. H.; Musser, S. M. *J. Agric. Food Chem.* **2006**, *54*, 7953–7959.

85. Shefcheck, K. J.; Musser, S. M. *J. Agric. Food Chem.* **2004**, *52*, 2785–2790.

86. Astwood, J. D.; Leach, J. N.; Fuchs, R. L. *Nat. Biotechnol.* **1996**, *14*, 1269–1273.

87. Ticha, M.; Pacakova, V.; Stulik, K. *J. Chromatogr. B* **2002**, *771*, 343–353.

88. D'Amato, A.; Bachi, A.; Fasoli, E.; Boschetti, E.; Peltre, P.; Sénéchal, H.; Citterio, A.; Righetti, P. G. *J. Proteomics* **2010**, *73*, 1368–1380.

89. Pedreschi, R.; Nørgaard, J.; Maquet, A. *Nutrients* **2012**, *4*, 132–150.

90. Fekkar, A.; Pionneau, C.; Brossas, J. Y.; Marinach-Patrice, C.; Snounou, G.; Brock, M.; Ibrahim-Granet, O.; Mazier, D. *J. Proteomics* **2012**, *75*, 2536–2549.

91. D'Amato, A.; Bachi, A.; Fasoli, E.; Boschetti, E.; Peltre, G.; Senechal, H.; Righetti, P. G. *J. Proteome Res.* **2009**, *8*, 3925–3936.

92. Martos, G.; Lopez-Fandino, R.; Molina, E. *Food Chem.* **2013**, *136*, 775–781.

Advantages and Applications of Gel-Free Proteomic Approaches in the Study of Prokaryotes

John P. Bowman

Food Safety Centre, Tasmania Institute of Agriculture, Hobart, Tasmania, Australia

Comprehensive Analytical Chemistry, Vol. 64. http://dx.doi.org/10.1016/B978-0-444-62650-9.00007-5

1 INTRODUCTION

Proteomics involves the study of the proteins usually in the context of living systems. As proteins are integral to life, their study is crucial for our understanding of the mechanisms underpinning cell biology, physiology, metabolism, and regulation. Integrated with genomics (the study of genomes), transcriptomics (the study of message RNA and gene expression), and metabolomics (the study of cell metabolites), proteomics forms a key component of the science of *systems biology*. System biologists seek to conceive and measure biological functions at a molecular and mathematical level such that cell processes can be computationally predicted and modeled. As part of this grand concept, proteomics involves measuring protein levels either in absolute terms, deriving a concentration or copy number per cell, or in a relative sense where protein abundance is estimated between one situation and another situation or a control. These approaches allow detailed understanding of functions within a life-form with the data translated to assess metabolic pathway enzyme levels, arrangement and responses of regulatory systems, and responses and interactions of protein complexes that represent the apparatuses making up a living entity. Global proteomics essentially measures as many as possible of these proteins that can be extracted and detected. Global studies provide a broad scope to understand phenotypes as a predictable suite of protein abundance changes. The relatively low number of proteins, lower complexity in terms of compartmentalization, and generally direct regulatory and signaling systems make proteomics very amenable for the study of prokaryote systems biology. This is especially the case for global studies in which a large proportion of a proteome is potentially analyzable with modern techniques, thus providing a comprehensive means to assess microbial systems within a myriad of scenarios.

This chapter focuses on the application of quantitative proteomics in microbial systems and covers recent methodological developments in the field and includes an indication of how it is impacting the study of prokaryotes. In particular, this chapter examines gel-free, label-free-based proteomic analysis and its advantages and disadvantages. The review also covers new developments in mass-spectrometric analysis of proteins and summarizes the means available for downstream data processing to quantify proteins. The chapter also briefly examines the assessment of functionally and regulatory allied groups of proteins via data visualization and also looks at the current trends in proteomic data being integrated into metabolic models. Numerous reviews and continual new method developments mentioned in this chapter indicate that proteomics is a highly active area that is riding the information deluge wave created with next-generation sequencing-driven analysis of genomes and metagenomes.

1.1   Proteomics and Integration with Genomics and Transcriptomics

Proteomics represents a highly important facet in the broader field of systems biology connecting the genomics of a system with metabolism, physiology, and metabolites *(1)*. Whereas transcriptomes represent an intermediary stage of the process translating the genetic code to mRNA, proteomes effectively represent the end points in the expressed genome. Global analysis of transcriptomes via microarray analysis or via deep RNA sequencing has been well established for many years and has been widely used in microbiology to study prokaryote responses and to assess and identify gene function and regulatory networks *(2)*. In recent years, the complexity of transcriptional and posttranscriptional processes in prokaryotes has become apparent, owing to the presence of small RNAs, antisense RNAs, and other forms of RNA-based transcripts *(3,4)*. This complexity is also compounded by differences in turnover rates of proteins (half-lives ranging from 2 to 60 h) and RNA (range 0.2–30 min). This results in low-to-modest correlations between mRNAs *(5–7)* and seems to mainly owe to mRNA transcripts for most genes being generally at low copy number in each given cell (one to three copies, *(8,9)*), thus contributing to a high-level noise thresholds. In mammalian cells, by comparison, the cellular abundance of proteins is largely controlled at the level of translation; thus, transcript and protein show reasonably good correspondence *(10)*.

Proteomic analysis thus has a strong place to provide complementary and more exacting insights into cellular-level responses and provides a means to more clearly realize the basis of molecular interactions, metabolite fluxes, and outputs. Unlike mRNA, proteins cannot be amplified, are far more chemically and structurally diverse, are often compartmentalized, and have chemical features that differentially affect extraction. Thus, comprehensive and high-throughput proteomics has been technically much more challenging to implement, and the technology for rapid, accurate, and quantitative analysis of proteins *en masse* has lagged behind nucleic acid analysis-related technologies. New advances in instruments, proteomic workflows, peptide mass spectrometry, and bioinformatic pipeline-related software have in recent years substantially advanced proteomics to the point that the capacity to rapidly, comprehensively, and quantitatively analyze microbial proteomes is possible *(11)* even though the "cutting-edge" analytical methods are limited to a few pioneering laboratories.

1.2   Accessibility of Quantitative Proteomics

Analysis requires access to liquid chromatography coupled to high-resolution, high-sensitivity mass spectrometers plus an associated software platform for peptide spectral analysis. This usually entails employment of one or more expert scientists. As described further in the succeeding text, many basic workflows can be established to provide a certain level of proteomics capabilities in laboratories that are not dedicated specifically to proteomics research. Much of the leading advances are still limited to a very small number of large laboratories within core proteomics facilities. These laboratories focus on proteomics technology development that still requires translation to streamlined forms more accessible to the wider scientific community be it in the form of in-house benchtop instrument systems or via service providers. Much the same has been seen already with next-generation-based sequencing technologies. The technology is readily accessible at least at certain levels if not completely, and in this review, example data will be provided by the author, analyzed within a small proteomics facility in his own institution. The data are derived from a streamlined and relatively simple workflow that focuses on discovery-mode data-dependent acquisition with protein abundance measured on a relative level. This is not the existing cutting edge but provides an effective indication of the type and depth of information that can be yielded as well as provides some insights into inherent limitations that are now being solved via with reconfiguration of existing workflows, new technology, and software.

1.3   Gel-Based Proteomics

Before examining contemporary proteomic workflows, it must be said that the older gel-based technology still remains the largest contributor of proteomic-level information. For many years, the main means to analyze complex protein samples involves separation of proteins on the basis of size and isoelectric point (pI) using two-dimensional polyacrylamide gel electrophoresis (2-DE). The densitometric aspects of gel protein spots provide a direct means to visualize and quantify proteins based on the experimental conditions applied to the source organism. Image analysis provides a means to directly estimate relative changes in protein abundances between treatments. Usually, the data only allow for relative comparisons; however, workflows have been developed to allow for absolute quantification using targeted mass spectrometry by analysis of ion intensities of selected peptides (12). 2-DE analysis was revolutionized with the invention of the means to directly identify proteins from gel spots via MALDI-type MS. Indeed, this discovery led to the birth of the proteomics field itself (13,14). The spot cutting process of peptide analysis has led to the development of other technologies such as bacterial biotyping, which allows rapid identification of microorganisms via a MALDI-MS-based approach (15,16) that is now widely used in diagnostic laboratories. Some refinements in proteomic gel apparatus, automated spot cutting devices, and phosphoimagers have attempted to improve the rather laborious nature of gel-based proteomics. However, it might be considered that this technology has matured close to an end point. Some other innovations, for example, differential gel electrophoresis (DIGE) (17), have improved analysis that depends normally on many replicate gels to achieve consistent data. With DIGE, different protein-binding fluorescent dyes allow one to compare multiple samples simultaneously (up to 3) in a single gel, based on comparisons of the fluorescent signals in different channels. Since 2-DE quantification is based on image analysis, the use of false coloration in an image analysis program (such as Adobe Photoshop) also allows for direct comparison of conventional 2-DE gels via overlaying images. This approach provides a better basis for making decisions of what spots to cut and process for identification of peptides. In the end, 2-DE still has substantial limitations in terms of cost, labor time, consistency of the workflow, limited sensitivity, difficulty in resolving all proteins in one gel due to their wide ranges of pI, translational modifications resulting in multiple spots of the same protein, high-abundance proteins overshadowing low-abundance proteins, extraction biases against certain types of proteins unless fractionation is also applied, and increasing time and cost. A consequence is that global studies are very constrained and tend to be shallow with certain high-abundance "boring" proteins often being observed time and again. Nevertheless, gel-based proteomics when combined with specific protein fractionation approaches or targeted peptide analysis still remains a very powerful means to assess directly abundance changes and very useful for studies focused on protein complexes and posttranslational modification (18,19).

2   PROTEOMIC WORKFLOWS

Gel-free approaches employ liquid chromatography followed by mass spectrometry to analyze peptides in order to assess protein levels in complex samples such as those derived from microbial extracts or environmental samples. This approach allows for quantitative analysis of identified proteins, either in a relative or an absolute sense at a relatively deep level. This power-ful and versatile approach is progressively replacing 2-DE-based analysis. Furthermore, proteome analysis workflows increasingly include label-free approaches that provide cost and labor time reduction and reduce complexity of analysis. Eventually, it is expected that MS-based peptide sequencing will also shift from a data-dependent to a data-independent mode of acquisition allowing for rapid, deeper, and cost-effective absolute quantitative analysis.

One clear aspect of the current trends in the proteomic field is the plethora of ways to achieve informational outcomes. Research developments over recent years have focused on all facets of the proteomic workflow including improvements in the comprehensiveness of analysis, efficiency of workflows, and development of specialized methods. This is aided by several mass spec-trometers being available on the market that have the necessary accuracy, sen-sitivity, speed, and dynamic ranges. The baseline workflow for proteomics however remains essentially the same and involves a number of required steps (summarized in the succeeding text and in Figure 1):

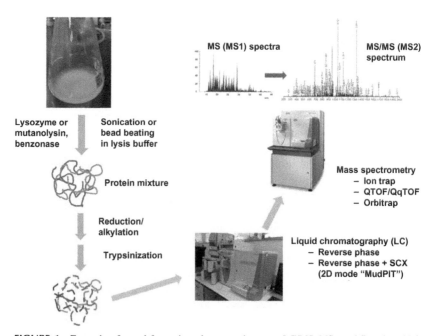

FIGURE 1 Example of a gel-free-oriented proteomics nano-LC/MS–MS workflow in which bacterial culture proteins digested to tryptic peptides are separated via LC and peptides subse-quently analyzed by mass spectrometry. In the process, the spectrometer rapidly cycles every few seconds and examines a size window in which peptide-derived MS1 ions are analyzed to define MS/MS (MS2) spectra. The MS/MS (MS2) spectrum generated for each peptide then enters a bioinformatic pipeline for sequence identification, statistical validation, and quantification.

(1) Protein extraction:
- Subproteome analysis via shaving or direct extraction
- Fractionation on gels
- Affinity and/or size exclusion chromatography

(2) Alkylation, reduction, and digestion

(3) Liquid chromatography

(4) Tandem mass spectrometry

(5) Peptide identification, statistical validation, and quantification

The subsequent downstream processing section, which includes visualization of quantified proteins, statistical validation of differences in treatments or samples, and biological interpretation, is much less defined in terms of work-flow regimens and is discussed toward the end of this chapter. In the succeeding text, various relevant aspects of proteomic workflows that impinge on the data obtained from proteomic analysis of prokaryotes assuming that a gel-free approach is used are discussed.

2.1   Protein Extraction and Immediate Postanalysis

Obtaining proteins is an easy process, since they are rich in cells at all times; however, proteins in microbes exist in different compartments including the cytosol, cell wall, and cell membrane, and this has a strong effect on to what degree they can be retrieved for analysis. Some proteins are secreted outside of the cell into the extracellular milieu, while other proteins can be membrane-bound to different extents (such as transmembrane proteins) or have membrane-associated domains and peptidoglycan covalent linkage domains that expose them on the cell surface and locate them in the periplasm or within inner membrane regions. Membrane proteins are often poorly soluble as indicated by their large grand average of hydropathy (GRAVY) score values or by the proportion of the protein-containing transmembrane helical (TMH) domains. The more TMH domains, the poorer their presence in protein extracts, which indicates that physical separability and lack of inherent solubility can act in tandem. Peptidoglycan-bound proteins, common in Gram-positive bacteria, are often lost in the discarded cell wall debris and also tend to be not overly abundant in the first place. Secreted proteins obviously can be underrepresented and what is examined is usually the precursor that is in the process of being exported by the cell.

2.2   Use of Surfactants

A fundamental part of protein extraction is solubilization. Without some form of fractionation or enrichment processing step, several proteins tend to be underrepresented in datasets. Even with specialized membrane protein selective detergents, some form of bias may occur, as the abundance of these proteins still can be much less than major cytosolic proteins. Extraction of membrane-bound and other hydrophobic proteins can be enhanced via the use of common and more specialized detergents. Unfortunately, numerous popular detergents are incompatible with downstream mass-spectrometric analysis due to their ionic properties including Triton X-100, NP-40, all derivatives of Igepal, Brij derivatives, Tweens, *n*-octyl-β-D-thioglucopyranoside, SDS, CHAPS, and CHAPSO. Several comparatively MS-friendly surfactants are available that are nonionic variants or are cleavable such that the products can be easily removed within the workflow, including *N*-octyl-β-glucopyranoside; Triton X-114 *(20)*; Deoxy Big CHAP—a nonionic version of CHAPS/CHAPSO (EMD Millipore Inc.); and acid or base cleavable detergents (PPS Silent Surfactant, Expedeon Inc.; ProteoMAX, Promega; RapiGest SF, Waters Corp; AASL I and ZALS, Protea Biosciences Inc.) that comprise an uncharged lipophilic compound and a zwitterionic compound, the latter of which can be removed by solid-phase C18 extraction *(21,22)*. Sodium deoxycholate and other surfactants can also be used for membrane protein enrichment but must be removed by precipitation or solvent extraction within the workflow *(23)*. For example, SDS perhaps the most commonly used detergent in proteomics overall can be removed via electrophoresis *(24)*. MS analysis compatible commercial kits have been assessed for their ability to isolate membrane proteins and cytosolic proteins. The cross-contamination of proteins in either the cytosolic or membrane-associated proteins pool was also assessed. After Western blotting analysis of proteins derived from colorectal carcinoma tissue culture cells, the data suggested that the Calbiochem ProteoExtract Kit was the most effective overall in the extraction of membrane proteins with good yields and low cross-contamination, while better yields were found to be obtained using the Qproteome Cell Compartment Kit (Qiagen) at the expense of greater presence of cytosolic protein contaminants *(25)*.

2.3 Subproteome Analysis via Shaving

A strategy for studying subproteomes from membrane or cell walls of microbes involves "shaving" whole cells using peptidases (trypsin or chymotrypsin) or lithium chloride. This can be extended to digestion of separated cell walls or membrane fractions followed by concentration by dialysis. Shaving membrane protein fractions yields a high proportion of peptides associated specifically with transmembrane proteins, thus providing a means to cover this compartment effectively, and can be combined with other approaches to comprehensively analyze a proteome *(26,27)*. Whole cells can be readily analyzed by shaving *(28)*, especially Gram-positive bacteria that have cell walls rich in sorted peptidoglycan-bound proteins. Furthermore, proteins from exogenous sources such as host systems that interact with a bacterial surface can be analyzed in combination using such an approach *(29)*.

2.4   Moonlighting Proteins

Shaved cell wall subproteomes usually include, besides known or predicted cell wall proteins, high-abundance cytosolic proteins such as enzymes of the glycolysis pathway and central metabolism, chaperones, stress proteins, elongation factors, and even ribosomal proteins (30,31). These "moonlighting" proteins or surface-layer-associated proteins that are exposed on the cell surface have been considered to have potentially a role different to their normal cytosolic role (32,33). However, it is also believed that moonlighting proteins within the subproteome could simply be contaminants derived from cells that had previously lysed (28,34); thus, special care is needed to interpret cell wall proteome data against appropriate controls and among biological replicates. Solis and colleagues (35) advocate the use of false-positive control proteins to aid in the decision making when assessing cell wall proteomes.

2.5   Proteome Stability

Some organisms, such as algae and yeast, release proteases from vacuoles that rapidly reduce protein yields and produce undesirable cleavage products unless strong inhibitory agents are used to block proteolytic activity. One solution to mitigate this issue involves a rapid "electroseparation" of proteins (36). Alternatively, precipitation of proteins by the addition of trichloroacetic acid or by a brief thermal treatment also can stabilize proteins in protease-rich samples (37).

2.6   Fractionation on Gels or Columns

One-dimensional gel separation of proteins followed by in-gel digestion and LC/MS–MS analysis (GeLC/MS–MS) (38) as well as off-gel separation of peptides following digestion provides means to increase the resolution and coverage of complex protein samples. GeLC/MS–MS is potentially useful if proteins are targeted within a particular size range, though for global studies, the need to analyze many samples contributes to increased cost, time, and complexity of the workflow as well as complicates downstream quantification. Whole gel processing has been shown to speed up the manual processing without compromising data quality (39).

3   AFFINITY CHROMATOGRAPHY APPLICATIONS

Several approaches relying on affinity chromatography allows for selective protein enrichment including posttranslational-modified proteins and membrane proteins and also can be utilized for depleting unwanted high-abundance proteins. These approaches can be readily inserted as an additional preparative step during protein extraction and purification before peptide digestion and mass spectrometry.

3.1 Enrichment of the Phosphoproteome

In the study of phosphoproteins, high-affinity antiphosphotyrosine (and other phosphorylated amino acid) antibodies can be used for phosphoprotein enrichment with peptide mass and sequence analyses are used to identify phosphorylation sites *(40–43)*.

3.2 Glycoproteomics

Similarly, glycoproteins, usually present in bacteria as surface structure and membrane proteins, can be enriched using lectin-based affinity chromatography *(44,45)*. In bacteria, glycoproteins are typically involved in a virulence context being often associated with bacterial–host cell interactions; thus, glycoproteomics can potentially provide key insights into infectious and parasitic bacteria. Such studies could be extended to bacteria that interact with other systems such as algae- or plant-based systems. Lectin-based approaches could be directly adapted from eukaryotic-applied systems. For example, columns coated with concanavalin A will bind mannose-containing *N*-linked glycoproteins that occur in various Gram-positive pathogens such as *Streptococcus pyogenes (46)*. Hybrid chromatographic and mass-spectrometric analyses have been developed that allow determination of the carbohydrate modifications on glycan peptides *(47,48)*. Specific proteomic analysis thus can provide a more complete picture of glycoproteins in bacteria that aid in determining their biological roles.

3.3 Membrane Protein Tagging

More broadly, membrane proteins can also be enriched via labeling with a membrane-impermeable, cleavable biotin reagent sulfosuccinimidyl-2-[biotinamido]ethyl-1,3-dithiopropionate (Sulfo-NHS-SS-Biotin). The reagent can be applied to intact bacteria and interacts with exposed lysine residues of membrane proteins; however, the sulfonated section of the molecule prevents it from entering the cytoplasm, thus providing considerable selectivity. Using strepavidin affinity chromatography, membrane fraction extracts then can be analyzed. Sulfo-NHS-SS-Biotin comes in a variety of commercial kits. This approach has been used to investigate intracellular bacteria *(49,50)*.

3.4 Immunodepletion of Highly Abundant Proteins

A minority of proteins make up a large proportion of the total protein abundance in most samples, and thus, the coverage of a proteome is highly dependent on the sampling depth *(51)*. The abundance distribution of proteins appears to follow a power-law distribution *(52)*. Thus, removal of obscuring high-abundance proteins via affinity chromatography immunodepletion increases the number of identified proteins. Preparatory immunodepletion has been successfully applied to human and animal tissue-derived samples such as serum, cerebrospinal fluid, saliva, and milk *(53)*, and various immuno-depletion kits are available for commonly studied systems. For example, kits for plasma remove the top 20 most abundant proteins, representing 97% of the total protein abundance. In plant tissue, removal of the enzyme RuBisCO, present in very high levels, aided in the detection of proteins that undergo O-nitrosylation since RuBisCO is similarly modified *(54)*. Proteomic study of microbial pathogens in complex protein-containing liquids such as human plasma has been shown to be considerably aided by immunodepletion of the most abundant plasma proteins *(55)*. For prokaryote systems, deliberate immunodepletion could be useful in various circumstances; however, it has yet to be applied. The observability of low-abundance proteins, however, would not be as great for systems in which a few proteins overwhelm most others and, since they have to be specially tailored for the species being investigated, would represent a large initial investment to achieve.

3.5   Interactome Analysis

The network of physical interactions of molecules in a cell has been referred to as interactomes. The study of interactions may yield important new insights into biological processes. In the case of proteins, this has been referred to as protein–protein interaction networks (PPINs). PPINs in prokaryotes have been studied with large-scale protein complex affinity purification and mass spectrometry *(56,57)*, which is considered the gold standard, or by extensive yeast two-hybrid screens (e.g., *(58)*). The number of interactions has been considered to roughly correlate with the genome size, though the numbers for prokaryotes widely vary and perhaps represent methodological limitations due to observations of indirect interactions or false-positives. The interactions are also not considered complete, and higher throughput approaches that take into account three-dimensional structure of proteins may potentially provide the means for more accurate and complete interaction maps *(59)*.

3.6   Combinatorial Libraries

Though yet to be applied to prokaryote systems, solid-phase combinatorial libraries of hexapeptide ligands have been designed to capture low-abundance proteins. The technology relies on beads coated with hexapeptide ligands, which results in millions of beads with different binding properties. By passing a complex protein extract over the beads, the huge dynamic range of proteins is equalized *(60)*, allowing for better detection of lower-abundance proteins in subsequent eluates. This approach is best served for biomarker discovery in various eukaryote systems or as a way to detect protein impurities in protein products *(61,62)*. Using selective reaction monitoring (SRM; see in the succeeding text), the use of combinatorial libraries seems applicable to quantitative proteomics *(63)* but still may have applicability in prokaryote systems for some applications.

3.7   Lab-on-a-Chip Applications

Attempts to innovate the characterization of proteome analytes is a highly active research area. This includes microfabricated devices that are designed to reduce the complexity of the proteomic analysis workflow as well as experimental development of novel in-line chromatography columns or off-line "lab-on-a-chip" devices *(64)* that allow separation of targeted protein or peptide analytes including toxins, pathogens *(65–67)*, phosphopeptides, glycopeptides *(68–70)*, and proteins from environmental samples *(71)*. The general trend suggests a streamlined workflow will eventually become possible for a range of general and specific proteomics applications.

4   GEL-FREE PEPTIDE CHROMATOGRAPHY AND MASS SPECTROMETRY

4.1   Protein Digestion

Most proteomics applications are bottom-up, relying on identification of peptides derived from proteins. Typically, trypsin is used to yield peptides following the required pretreatment of proteins via alkylation and reduction to prevent interaction of reactive thiol residues with the chromatographic separation phase *(72)*. Trypsin cleaves at the carboxyl sides of Arg and Lys residues (except when followed by a proline residue), thus leaving reliably tryptic ends that aid in subsequent mass spectral analysis. Trypsin is not perfect, and certain cleavage sites on proteins are less accessible than others, leading to missed cleavages usually observed as a small series of peptides of progressively longer chain length for the same region of the protein *(73)*. Nonspecific activity of trypsin may also lead to nontryptic cleavage sites that may reduce the observability of proteins if nontryptic peptides are rejected in downstream analytical processes. Finally, a minority of small proteins with a large number of Arg and Lys residues do not yield peptides of insufficient mass (<900–1000 Da) to allow chromatographic separation and subsequent identification. Membrane proteins due to the presence of TMH domains contain a larger proportion of hydrophobic amino acid residues and thus have fewer on average cleavage sites. As a result, alternative enzymes such as chymotrypsin *(74)* may be useful in providing coverage of proteomes, though in general, the complication to the workflow means that the use of nontrypsin approaches tends to be quite targeted.

4.2 Liquid Chromatography of Peptides

The gel-free aspect of proteomics relies on peptide separation via nano-LC, which is performed in either 1-D or 2-D modes utilizing a gradient solvent profile in order to resolve peptides over a wide hydrophobicity and pI range *(75)*. To avoid contaminants and improve resolution of peptides entering the mass spectrometer, peptides are pumped initially onto a C18-packed capillary trap column at a slow rate. Subsequently, peptides are typically separated on a C18 reverse phase column in 1-D mode or on successive reverse phase and strong cation exchange column in 2-D mode. The 2-D mode or "MudPit" approach improves resolution of peptides, and through a longer retention process, it results in more sequenced peptides *(75)*. Combining longer columns and thus higher-resolution separation along with a fast sequencing mass spectrometer (e.g., Q Exactive LTQ Orbitrap and Thermo Fisher), the deeper analysis of peptide spectra is synergized. Thus, from a limited number of runs and using data-dependent acquisition, deep coverage of microbial proteomes is attainable *(76)*. Interestingly, the dynamic concentration range in proteins (up to 10^5–10^6 fold) identified is covered very effectively as shown with experiments on yeast *(77)*. New approaches in LC are being studied to improve sensitivity and comprehensiveness in proteomics. These include adapting transient isotachophoresis stacking and capillary electrophoresis to improve peptide resolution in multidimensional LC. Potentially, such development will allow dilution of high-abundance peptides but increase the concentration of those that are less abundant *(78,79)* and allow connectivity to other separation technologies *(80)*.

4.3 Mass Spectrometers for Gel-Free Proteomics

Global proteomics coupled with quantification requires tandem mass spectrometers that have fast sequencing speeds, very high mass accuracy determination (typically $<1:10^6$ error), high sensitivity, and the ability to resolve many peptide-derived ions at the same time. Peptide identification and proteome coverage are affected by how instruments respond to sample amounts, the speed of ion acquisition, and the data quality of spectra. Studies examining tandem mass spectrometer performance strongly recommend optimization when different workflows are used *(81)*. Primary instruments used are ion trap spectrometers, including linear trap quadrupole (LTQ) Orbitrap *(82)* and single, double, and triple quadrupole time-of-flight (QTOF, QqTOF, and QqQTOF) instruments. New configurations of original systems such as the LTQ Orbitrap evolving to the Orbitrap Velos and the Orbitrap Elite systems (Thermo Fisher Scientific) improve sensitivity, resolution, and sequencing speeds. A field test of the high field resolution Orbitrap Elite system demonstrated that it is able to identify hundreds of proteins in samples down to 0.1 ng of protein extract *(83)*. Other experimental hybrid ion trap spectrometers such as Orbitrap Velos combined with a Fourier transform ion cyclotron resonance mass spectrometer showed higher rates of peptide ion acquisition than the Velos instrument alone *(84)*. For protein quantification using data-independent means of acquisition (discussed in more detail in the succeeding text), instruments capable of performing high-energy collisions to generate reliable fragmentation ion patterns are required *(85)*, such as the QTOF-type tandem mass instruments, which are becoming more commercially available. It is presumed that new configurations and engineering of mass spectrometers as well as the associated LC separation will combine to offer more accurate, high-throughput means for protein quantification. At this stage, much of the developments are experimental, but there seems to be a rapid transition to the market once an instrument is demonstrated to have improved capability over predecessors.

5   LABELING AND LABEL-FREE APPROACHES FOR PROTEIN QUANTIFICATION

For gel-free-based quantification, various approaches have been devised as listed in Table 1. Labeling approaches typically involve the use of stable isotope dilution, which has been shown to be an effective way to measure protein abundance *(86)*. Specific methods that are popularly used are described hereon. Justification for whether labeling should be used or not very much depends on the goals of the experiment, the budget, and the instrumentation available for use. The different approaches all have particular limitations; thus, combined approaches can be powerful for in-depth analysis *(18)*. All methods mentioned are capable of relative comparisons as a means of protein quantification. Absolute determination of protein abundance is more challenging and has been advanced initially using stable isotope-labeled synthetic peptides *(87)*. This was further advanced with SRM *(88)*, described in more detail in the succeeding text). This has further evolved with data-independent acquisition approaches that exploit consistent peptide mass spectral features that can be collated in an atlas (described in more depth later in this chapter) and combined with SRM for global proteomic quantification *(89)*. Using QqTOF instruments, the peptide feature mapping can be further exploited to directly collate time-resolved fragment ion spectra for all the analytes detectable within a defined mass window *(90)*.

5.1   Spectral Counting

The assumption can be made that the more frequently a peptide spectrum is observed, the more abundant the protein is in the sample *(91)*. In the mass-spectrometric analysis of peptides, ions derived from peptides entering

TABLE 1 Proteomic Protocol Capabilities, Advantages, and Limitations

Method	Gel-Free	Absolute Quantification Possible	Interrun Comparisons Possible	Dynamic Range	Specific Advantages	Specific Disadvantages
Gel electrophoresis						
2-DE	–	+	+	3–4	Protein-level quantification	Limited pI range
DIGE	–	–	–	3–4	Protein-level quantification	Limited pI range
LC/MS-MS						
ICAT	+	–	–	~2	Accurate abundance measurement	Only peptides with cysteine residues analyzable
SILAC	+	–	–	1–2	Accurate abundance measurements	Stable isotope with amino acid incorporation limitations
^{15}N-labeling	+	–	–	1–2	Accurate abundance measurements	Mass alteration of peptides dependent on sequence
iTRAQ	+	–	–	2		Sample processing and workflow must be kept constant
TMT	+	–	–	2		Sample processing and workflow must be kept constant
Label-free	+	+	+	3–4	Global, no-label needed	Sample processing and workflow must be kept constant
AQUA (spiked in heavy peptides)	+	+	+	2	Accurate and sensitive	Targeted analysis only; the need of expensive heavy peptides

Adapted from *(18)*.

the mass spectrometer are scanned for a fixed time across a designated mass window (a swath). The most ions detected are then analyzed to yield an MS/MS (MS2) spectrum. For high-throughput analysis pattern, matching against a database may then precede to a sequence identification of the peptide. The resulting peptide identification process is accumulative, and the spectra can then act effectively as units, a "spectral count" (SpC) that represents a subsample of protein information from a complex pool. The process is semirandom in that certain proteins will be more measurable than others due to abundance differences deriving from the extraction protocol and due to the cleavage behavior of the protein. The number of spectra obtained dictates the number of proteins that can be observed and measured. Biological replication is required to control biological noise, while technical replicates allow for statistical depth. The ratio of test versus control SpCs per protein, pooled from the associated peptides, can yield relative differences in protein abundance.

Various quantitative and statistical validation processes have been described, accounting for the fact that SpCs tend to be small numbers and vary due to the partial stochasticity of the process. In the example datasets included in this chapter, the relationship between the SpC and protein abundance obtained experimentally is shown in Figure 2, demonstrating that many proteins in bacterial cells are low in abundance while a small subset are highly abundant. Several studies have compared label-free with labeling methods or assessed its statistical validity (93–95). Overall, SpC alone should not be used as a means for absolute quantification (92,96), but it is quite adequate for

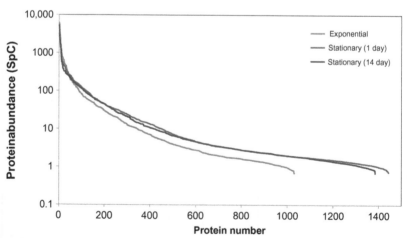

FIGURE 2 Nano-LC/MS–MS-generated spectral counting (SpC) data showing the distribution of protein abundances (corrected for protein length) derived from different growth phases of *L. monocytogenes*. The abundance values follow approximately a power-law relationship with a large fraction of proteins having low SpC values. This results in SpC not being suitable for absolute quantification due to insufficient accuracy across the full dynamic concentration range, which can be up to 10^6-fold (the abovementioned example is approximately 10^4-fold). The collection of SpC has been shown to be quite reproducible and thus is valid for relative quantification (92).

relative quantification. The labeling and label-free approaches also compared well in terms of data output and proteome coverage. SpC analysis possesses the advantage of providing a comparatively simple workflow and downstream data processing. In that respect, it has become one of the most prevalent methods used in gel-free proteomics and is supported by a number of bioinformatic platforms.

A sample dataset is included in this chapter, consisting of proteins extracted from the Gram-positive foodborne pathogenic bacterial species *Listeria monocytogenes* grown to different phases of growth (exponential, early, and late stationary growth phases). Proteins were extracted using a simple procedure involving bead beating with zirconia–silica sand and proteins precipitated from the supernatant before proceeding with the steps described in Figure 1. The peptides were analyzed using a label-free approach on an LTQ Orbitrap mass spectrometer with spectral counting. The recovery of proteins is extensive in relation to pI and GRAVY score (Figure 3) but shows that

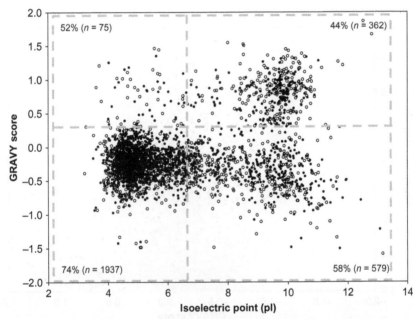

FIGURE 3   A scatterplot showing the distribution of proteins obtained from the foodborne pathogen *L. monocytogenes* strain Scott A. Proteins were extracted without using a surfactant with cells processed using bead beating. Proteins were subsequently analyzed using label-free LC/MS–MS (see *(6,97)* for exact methods). The distribution of observed (closed circles) and unobserved proteins (open circles) divided into four quadrants are shown in relation to isoelectric point (pI) and the GRAVY score for each protein. The observed number of proteins in relation to the total possible is given. The pI indicates overall protein charge, while the GRAVY score indicates overall protein solubility with high GRAVY score proteins usually membrane-associated with natural poor solubility in water. The data clearly show observability is best for proteins falling into the bottom left hand quadrant of the plot consisting of soluble, acidic proteins, which make up most of those located in the cytosol. The remainder quadrants have approximately equal but lower observability, which can be equated to lower abundance at least partly due to extraction bias.

there is a bias toward proteins identified as likely cytosolic rather than in other compartments of the cell. This is clearly shown when abundance (based on SpC) is compared to the solubility of the protein demonstrating the bias against membrane-bound and other hydrophobic proteins (Figure 4).

5.2 MS1 Features and Quantification

The integration of SpC data with mass ion intensity information can provide a more accurate indication of abundance *(98)*. Similarly, methods weighting SpC data with data MS1 feature extracted ion intensities improves reliability *(99)*. To absolutely quantify proteins requires integration of the individual peptide spectral ions generated during mass-spectrometric analysis. To do this requires extraction of peptide MS2 spectra to obtain ion intensity and masses. By correlating ion intensity and accurate mass with internal standard spikes

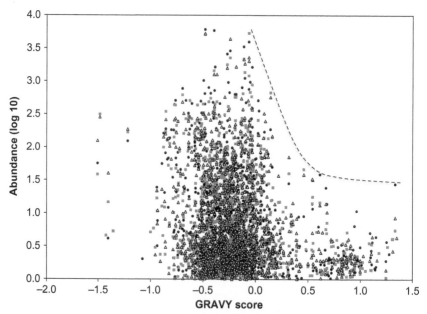

FIGURE 4   A scatter plot showing abundance-dependent biases introduced during protein extraction as also shown in Figures 2 and 3. Abundances of proteins (based on spectral counts and corrected by protein length) are related here to the GRAVY score for each protein. The different represent three different datasets shown by different symbols (exponential; 1-day and 14-day incubated stationary growth phase cultures as indicated in Figure 2). From the distribution, it is clear that most proteins fall within −0.5 to 0 GRAVY score region, reflecting most proteins that were observed are reasonably soluble. The dashed line demarcations are "blind spots" where less-soluble proteins are not observed due to extraction bias. Nevertheless, highly hydrophobic proteins are still observed but have low SpC values (<10 SpC); thus, their relative quantification is quite error-prone.

(e.g., AQUA peptides *(87)*), it is possible to correlate intensity values to con-centrations and copy number of proteins per unit volume or per cell. Various approaches have been developed to allow this computation, and benchmark-ing analysis has suggested that the Top 3 approach, wherein the three most intense ions of a peptide are used for quantification, was the most effective in scaling throughout the dynamic range of observed proteins concentrations *(96)*. Label-free approaches can yield large coverage of proteomes but, though reproducible, can underestimate abundance, reflecting the limitations that many proteins yield only a small number of spectra due to the semirandom sampling process *(100)*. A sample can thus record a null result where a protein is not observed. In order to allow comparisons in SpC-based analyses, a pseudo-SpC is required *(93)*. As SpC values decline for a protein, the signal-to-noise ratio increases, meaning that global proteomic analysis is lim-ited by sampling effort *(101)*. Since there is still good dynamic range from logs 4 to 6 depending on what LC mode is used *(76)*, in the end, the shortcom-ings tend to cancel out, compared to the inherent limitations of labeling approaches (see in the succeeding text). Overall, label-free, gel-free proteo-mics is an attractive approach that is effective for global and more targeted proteomics studies, especially in laboratories that do not have an access to a core proteomics facility. The ability to extract MS1 data provides absolute quantification data, and with data-independent acquisition approaches (see in the succeeding text), limitations associated with the semirandom aspect of analysis are overcome. From the aforementioned *L. monocytogenes* data-sets obtained from a 2 h nano-LC/MS–MS analysis LTQ Orbitrap-based runs, SpC data represented a dynamic range of approximately 10^4 with $\sim$50% of the proteome covered (*(97)*, Figure 3).

5.3 Stable Isotope Labeling with Amino Acids

Stable isotope labeling with amino acids (SILAC) employs amino acids typi-cally containing ^{13}C or ^{15}N atoms, shifting the mass of a peptide upward by a specified amount. Test samples where SILAC have been applied to are com-bined with a corresponding unlabeled control to reduce experimental error. Comparative ratios can be thus made between "heavy" peptides and "light" peptides obtained from mixtures of test and control samples, respectively. Deuterium (^2H) has also been used but, due to its interaction with LC col-umns, shows substantial labeled versus unlabeled peak separation unlike the other isotope labels. This comparison allows assessment of relative changes in protein abundance by comparing the average ion intensities of the pairs of heavy and light peptide pairs detected mass spectrometrically *(102,103)*. SILAC is designed on the basis that they are readily incorporated into cells, thus typically those that are essential for growth. As a result, media conditions and the SILAC peptide incorporation rate and level have to be optimized for different treatment conditions and organisms. Arginine, leucine, and lysine

tend to be favored as they are essential in eukaryotes and occur more frequently proteins than other required amino acids. Since most bacteria are generally not auxotrophic for these amino acids, this has resulted in SILAC
generally being much more favored for research on eukaryotes than prokaryotes. Nevertheless, in a minimal medium, bacteria such as *Escherichia coli*
will readily take up labeled amino acids, and thus, SILAC is a viable means
of analysis in specific contexts *(104)*. SILAC peptides can also be added to
media to "pulse-label" proteins, thus allowing for *de novo* protein synthesis
to be assessed *(105)*. This has been exploited to study proteins newly
expressed in intracellular bacterial cells. In this application, cells are initially
pulse-labeled using SILAC and, after internalization, retrieved by cell sorting
by virtue of being GFP-labeled *(106)*. A limitation of SILAC is its poor
dynamic range (usually no more than 2-log), as the stable isotope-labeled peptide usually has to be relatively abundant to appear in both test and control
samples to allow comparison.

5.4 Isotope-Coded Affinity and Covalent Tagging Methods

Isobaric tags that are bound to protein via affinity (isotope-coded affinity
tagging or ICAT) or via covalent addition (isobaric tags for relative and
absolute quantitation, iTRAQ) represent strategies for relative and absolute
protein abundance analysis. ICAT involves chemically labeling cysteinyl
reactive groups using iodoacetamide followed by the attachment of a biotin
tag and an isotope-coded tag *(107,108)*. In the iTRAQ approach, an isotope-
coded linker *(109)* is instead directly covalently linked to reactive amine
residues of the peptide to which an identifiable tag is added *(110,111)*.
Alternatives to these procedures include tandem mass tagging (TMT)
(112,113), chemical dimethylation of leucines, and isobaric peptide termini
labeling *(114)*. To avoid error, the tagged and untagged proteins or peptides,
depending on the procedure from both the control and treatments, are
combined before subsequent tandem mass-spectrometric analysis. Like
SILAC, the mass differences of the peptide pairs are used to define mass
ions associated with the control or treated samples, and from them, average summed ion intensity of peptide ions are used to calculate an abundance ratio. Though complicating the workflow, isobaric tags provide
accurate estimates of abundance though the ability to cover a broad
dynamic range is restricted. However, one advantage is that isotope-coded
tags used in ICAT and iTRAQ allow for multiplexing of several samples
in the same analysis run. The iTRAQ system comes in a kit form sold by
AB SCIEX Pte Ltd. including 4-plex, 6-plex, and 8-plex tags allowing for
different degrees of multiplexing. Various software have been developed
to aid in the at-times complicated task of assessing abundance, including
those identified as posttranslationally modified *(115)*.

5.5   Metabolic Labeling

Proteins can also be labeled by adding the available nitrogen source as an ^{15}N-labeled version, such as ^{15}N-NH_4^+. This leads to most if not all nitrogen residues in amino acids in a peptide being heavier but showing highly similar LC chromatographic separation. Similarly, ^{18}O labels can be chemically incorporated to peptides by performing trypsin digestion in ^{18}O-enriched water *(116)*. Labeled test and unlabeled control samples are mixed and resolved via LC/MS–MS analysis allowing heavy/light peptide pairs to be compared. The accuracy of quantification was found to rival that of the isobaric tagging approaches *(117)*.

6   PEPTIDE IDENTIFICATION

6.1   Data-Dependent Acquisition

Several database-searching algorithms compatible with shotgun-type proteomics have been developed that act to match a peptide tandem mass spectrum to a predicted mass spectrum, derived typically from genome sequence data. Above a set statistical criterion, the peptide sequence can then be assigned to the peptide spectrum in question. This discovery-based approach has advantages in that it is a rapid and readily integrable process that makes up part of the bioinformatic pipeline in analyzing peptide mass spectral data. The proviso is that prior peptide sequence data are available, supplied as a database. Several probability-based pattern matching software packages are available and have been in continual use since the beginning of the proteomics era, the mainstays being Mascot *(118)*, Sequest *(119)*, and X!Tandem (www.thegpm.org). The performance of the software has been evaluated, but no-one software appear clearly superior *(120,121)*. Many other software programs with specific and broad LC/MS–MS applications are available, well beyond the scope of this chapter to cover.

Various statistical approaches have been developed to identify and thus reduce the inevitable false identification of peptide identification, which increases as spectra become weaker and noisier *(122)*. Decoy databases where peptide spectra are matched against peptides in reverse order or other configurations are designed to acknowledge and reduce the degree of misassignation of identified peptides to proteins. Improvement of this approach for applications in complex proteomic datasets (human proteome and metaproteomes) in which several homologous protein sequences can occur due to the presence of conserved domain regions is an active endeavor *(123–127)*. This situation is less an issue in prokaryotic systems due to fewer repeat proteins. A recent study performed in *E. coli* showed that the posterior error probability distribution of novel peptide spectral hit is almost identical to that of reversed (decoy) hits. This means the inferred proteome is for all attempts and purposes complete annotated entity though it inspires caution on the level of false-discovery rates that can occur *(128)*.

6.2 Data-Independent Acquisition

The gold standard for mass-spectrometric-based quantification of peptides, which assumes prior knowledge of peptide spectral features, is selective (multiple) reaction monitoring (SRM or MRM) and its descendant procedures. SRM was previously used to specifically target compounds in small molecule drug metabolism and pharmacokinetic-type research utilizing triple quadropole mass spectrometers. This approach has been adapted to peptides *(88,129)* and has been used for targeted peptide quantification in several studies. The number of proteins that can be quantified is usually restricted for a single run; however, multiple runs conceivably can be used to quantify whole proteomes *(130,131)*. New peptide-based SRM approaches have expanded the capabilities of quantifying a proteome efficiently *(89,131)* and are summarized in Table 2. The comprehensive *a priori* analysis of peptide mass spectra in a given organism requires the development of a peptide atlas, which compiles high-quality information available on individual protein peptide ion retention time, fragmentation patterns, and detectability. Such atlases are appearing for highly studied proteomes *(132,133)* but require substantial input of time and resources. This slow part of the process potentially could be sped up with sufficient understanding of peptide ion characteristics such that it could be eventually replaced by *in silico* analysis *(134)*. By mapping or imaging peptides, the subsequent analysis of peptides allows for more comprehensive analysis, and depending on the configurations used in the workflow, absolute quantitation is available. The end goal of these developments is for a deep analysis of a proteome within a single injection that is also comparable from run to run. The development of DIA methods has followed a series of increments culminating so far in SWATH-MS (Table 2, *(90)*). Different scan modes have been trialed over sequential series of mass windows ranging from as low as 2.5 Da to 800 Da. SWATH-MS uses a window of 25 Da, which seems to be a good compromise between speed and resolution.

7   PROTEOMIC DATA POSTANALYSIS

7.1   Bioinformatic Pipelines

A number of automated bioinformatic pipelines for processing large proteomic datasets have been developed, and a generalized schematic is included in Figure 5. This has been encouraged since a multitude of software packages for peptide and mass spectra analyses only focus on individual stages of the analysis process, thus the need for integration of a suite of software forming an analysis pipeline *(135)*. Within such pipelines, the essential goals are to

TABLE 2 Mass Spectrometric Approaches for Peptide Identification After Separation By Liquid Chromatography Including Relative Performance Comparisons

Method	Instrument	Scan modes Used	SWATHs	Cycle (s)	Dynamic Range (Logs)	% Peptides Observable
Original peptide SRM*	Q-TOF, QqTOF, or Orbitrap	MS1/MS2 scan	800 Da single	2 (400–1200 m/z)	3–4	Low
Data-independent acquisition	Ion trap	Sequential window scans	10–20 Da window (200 ms/swath)	4 (200 m/z)	>2	Low–moderate
Precursor acquisition independent of ion count (PAcIFIC)	Ion trap	Sequential window scans	2.5 Da window (1.5 Da overlap, 250 ms/swath)	5–6 (45 m/z)	7	High
SWATH-MS	QqTOF	Sequential window scans	25 Da window (1 Da overlap, 100 ms/swath)	3.2 (400–1200 m/z)	4	Very high

SRM, selective reaction monitoring.

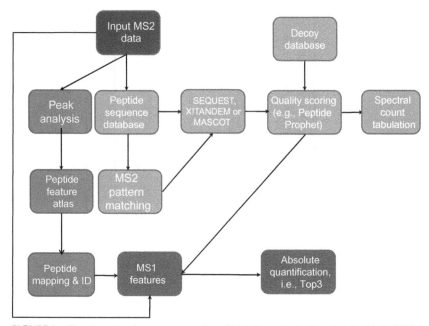

FIGURE 5 Flowchart showing a representation of bioinformatic pipelines involved in LC/MS–MS analysis of peptides. Sections in green relate to a pipeline schematic using a data-dependent (discovery mode) approach, which utilizes a preexisting peptide sequence database or relies on searching a larger database (in the case of metaproteomic-type analysis). Spectral counts for each peptide identified are accumulated to enable relative quantification between samples in the form of spectral counts. The decoy database is a custom designed set of peptides that tests for false assignation of peptides to proteins. The sections in blue relate to a pipeline using a data-independent acquisition approach in which peptide spectral features have been predetermined (peptide atlas), and thus, MS1 features including ion intensity extracted from MS2 spectral data can thus be used to enable quantification (see *(11,90)* for detailed explanations).

process peptide spectral (MS2) data in comparison with peptide datasets (either peptide sequences inferred from genomes or peptide atlas data), calculate SpC, and/or extract MS1 features in order to calculate ion intensities for quantification. MaxQuant was one of the first pipelines to integrate peptide identification with protein quantification *(136)*. An upgraded open-source pipeline includes TOPP, which integrates OpenMS applications *(137,138)*. Commercial software such as Progenesis LC-MS provided means to extract MS1 data and allowed quantification on many platforms, now being adapted beyond its original native applications *(139)*. Integration with web-based tools is important not only allowing analysis but also managing and storing the large volumes of data that tend to be generated. A popular pipeline for this purpose is the Trans-Proteomic Pipeline *(140)*. Based on recent publications, the software integration continues to evolve in terms of integration with data-independent peptide analysis *(138,141)*.

7.2   Statistical Treatment of Quantitative Proteomic Data

For quality control in treatment of quantitative proteomic data, efforts have gone into assessing and developing guidelines for statistical treatment of data. Several considerations are involved with the assessment of proteomic data. Typically, the data involve limited biological replicates, several hundred to thousands of proteins depending on the system being explored (thus increasing the potential for type I error), high dynamic range, and often small numerical values for many proteins. For methods that rely on measurement of ion intensity (MS1 data), namely, the various stable isotope labeling approaches, mass spectrometer type is influential since this affects sensitivity. A benchmarking study by Jones *et al.* *(142)* on immunodepleted serum proteins, a good test bed due to the high dynamic range of proteins present therein, found that the Thermo Fisher LTQ Orbitrap Velos/Q Exactive instrument outperformed TripleTOF instruments in terms of protein detection and suggested that system optimization was important to maximize results. The examination of peptide MS1 data derived from backgrounds of different levels of sample complexity indicated that a variety of strategies can be utilized to assess accurately quantification based on MS1 data derived from a range of experimental approaches *(143–145)*.

In the treatment of label-free data that are based on SpCs (only MS2 data), there are a number of important considerations including sampling depth and the presence of zero values due to either low protein abundance or treatment effects. Since the number of SpC in a given LC/MS–MS run can vary due to small run-to-run variations (typically less than 20%) or deliberate application of multiple technical repeats, some normalization is usually needed. Gokce and colleagues *(146)* indicated that a total SpC count approach for normalization compared very well with internal standards and also with invariant marker proteins, such as certain housekeeping proteins. The actual difference between SpC values of a given protein has been assessed with a wide range of statistical tests; in that respect, Lundgren and colleagues *(93)* had reviewed the limitations of SpC quantification and the types of statistical treatments useful in treating the data. An intersample and between treatment normalization and statistical assessment of significance were proposed by Pham and Jimenez *(147)* based on a negative beta-binomial distribution analysis, which assumes that the distribution of SpC abundances is an accumulation of a finite series of presences or absences, with the results of each test being random. This approach was found to be favorable in reducing false-positives and allowed treatment of zero SpC values. The G-test utilizing a pseudo-SpC value and constrained using a correction factor also performs well *(148)*. Gregori *et al.* *(149)* suggested that minimum absolute log 2-fold change of 0.8 and a minimum signal of 2–4 SpC aid in substantially reducing false-positive results. The integration of MS1 feature data with SpC data *(98)* or MS1 feature extraction seems to get around some limitations of SpC analysis including zero values and has been shown to be more accurate. This requires access to software able to extract MS1 data by processing accumulated MS2 data. Various publically available platforms are available as well as commercial software to do the processing steps, an example of which is SuperHirn *(150)*.

7.3   Functional Profiling and Data Visualization Using Proteomic Data

Functionally based postanalysis of proteomic data is very useful for biological interpretation of complex proteomic data. For this purpose, the use of gene enrichment analysis to assess transcriptome profiles has been adapted to assess global proteomes in a number of studies. Changes in a functionally allied set of proteins can be assessed statistically and typically involve determining whether the protein subset exhibits a significant relative abundance change in comparison with the whole dataset. This form of analysis is thus useful as an *a priori* means to focus on functionally meaningful changes in protein abundance relevant to treatment conditions rather than to examine all individual proteins individually. Experimental design and postanalysis statistical testing of proteomic data has been reviewed by Vaudel *et al. (126)*. The best means to do this form of analysis is somewhat equivocal and requires further research, owing to the diversity of ways proteomic data can be handled. Assignation of functionally allied datasets is subject to interpretation; however, most studies rely on a preordained system such as Gene Ontology (GO) terms or functional terms devised by TIGR *(151)*. Numerous software-driven methods have been devised to test data statistically, but no consensus has been reached about what is "best" or "not the best" way to do this type of analysis. A recent software package has been designed that combines several approaches, and the authors suggest a multipronged approach in doing such analysis *(152)*.

Complex proteomic datasets like many datasets that comprise accumulated abundance-type information are amenable to information visualization and associated statistical analysis to examine broad trends in datasets, such as treatment effects. Alternatively, matching different forms of "-omics" information is also possible with the adaptation of software tools *(153–156)*. Long-standing popular methods in analyzing complex data such as ordination analysis (multidimensional scaling and principal coordinate analysis), heat maps, tree maps, and network diagrams can be readily applied to proteomic data *(157)*. These approaches are increasingly being used to present complex multiomic or protein topological data in research papers or forming parts of specific databases *(158)*.

An example of a functional and visualization analysis is provided here based on the author-generated *L. monocytogenes* growth phase datasets (shown in Figures 2–4). The overall data including individual biological replicates, generated in two separate sessions, are visualized as a multidimensional scaling plot (Figure 6). For this analysis, SpC data were log-transformed, with individual samples compared to calculate a resemblance value in PRIMER 6 (PRIMER-E, Plymouth, the United Kingdom), which was then plotted as a 2-D representation of a 3-D dataset. From this nonparametric visualization, the variation of the experiments and replicates can be assessed, and their relative relation can be statistically examined to determine what treatments or dataset components impart the most variance in the overall grouped dataset. Proteins from this species were grouped on the basis of function based on data from the KEGG database (www.genome.jp/kegg/), MetaCyc database (www.metacyc.org), and conserved domain information collected from NCBI blastp searches and specific research papers. Here, the contributions of each of these protein sets are compared on the basis of abundance (Figure 7), demonstrating that certain sets are predominate though some sets containing transporter proteins are underrepresented due to the inclusion of many membrane proteins. Trends in changes in abundance of these protein sets between different treatments were estimated by a simple penalized *t*-test procedure described by Boorsma *et al. (159)*, which can be implemented in Microsoft Excel. The procedure was originally designed for transcriptomic

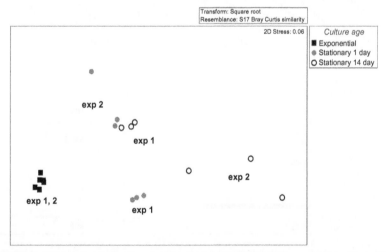

FIGURE 6   Statistical visualization of proteomic datasets distributed in a nonparametric multidimensional scaling (MDS) plot. The MDS plot was generated using the program PRIMER 6 (PRIMER-E, Plymouth, the United Kingdom). The label "exp" denotes different experiments in which sets of biological replicates (representing each point on the plot) were generated. The data used follow those of the previous examples shown in Figures 2–4. The MDS plot is a useful and rapid way to *a priori* establish relationships between treatment effects and assess variability within datasets. The data suggest that while exponential phase samples are very consistent, stationary growth phase data are more variable between experiments and within experiments. Since all samples were processed through the same workflow and analyzed within the same LC/MS–MS run session, the variability is likely derived from biological noise within what is a dynamic and complex system.

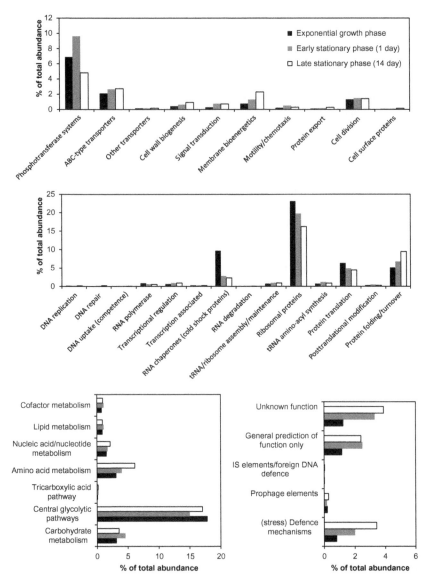

FIGURE 7 Bar graphs showing the contribution of functionally allied protein sets to protein abundance. The data are derived from *L. monocytogenes* SpC datasets shown in the preceding figures. The predominance of certain protein sets are clearly shown and can be assumed to have a central role in cell functionality including central glycolytic pathway, phosphotransferase system, protein translation, protein folding/turnover, ribosomal proteins, and RNA chaperone (cold shock) proteins. Differences in abundance contribution between the datasets suggest physiological adjustments occurring between different growth phases in the model organism demonstrated here.

data but essentially examines trends in abundance changes; importantly, it is flexible and provides meaningful positive and negative values and is not substantially biased by the size of inputted values. The *T*-values that are the output from this analysis are visualized here in a heat map (Figure 8). Overall, the trend analysis simplifies the data and offers a pragmatic guide to biological interpretation. Visualization by hierarchical clustering, heat map analysis, and the equivalent offers a succinct means to display these data. Alternative means such as Voronoi tree maps *(27)* have also been advocated

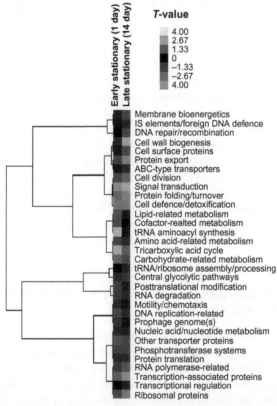

FIGURE 8   A way to represent proteomic data involving treatment comparisons is through heat maps typically used for gene expression data but adapted here to examine differential protein trends in functionally allied protein sets (similar to that done in *(6)*). The data used are from *L. monocytogenes* Scott A as shown in the preceding figures. The differential abundance trends in functional protein sets between growth phases were calculated using a penalized *t*-test method called "T-profiler" *(159)*. The hierarchical clustering (using default parameters) was performed in Cluster 3.0 *(160)*, while the heat map was visualized using Java Treeview *(161)*. The color scale shows the strength of increasing and decreasing trends. *T*-values of >2 and <2 are usually significant ($p < 0.05$) depending on the set size (the smaller the set, the larger the response needed to be significant). In the analysis, only proteins detected and passing filtration criteria (and excluding singleton peptide detections) were considered.

that provide detailed and esthetically pleasing visualization of proteomic data coded to function or to regulatory units.

7.4   Metabolic Modeling and Proteomics

Global and targeted proteomics offers the means to directly and quantitatively contribute to metabolic reconstruction, mapping, and modeling efforts. Using abundance data, the importance of different reactions under different circumstances can be expressed clearly in metabolic network diagrams. Coupled with metabolite profiles, fluxes of reactions can also be included and thus allow generation of more accurate maps and models *(162)*. Thus, combining a genomic sequencing, quantitative proteomics, and absolute quantification of metabolites, possible now with LC-based and stable isotope dilution techniques *(163,164)*, a tangibly reliable and comprehensive metabolic model of a biological system can be built in detail under scenarios of interest. Approaches for quantifying elements of metabolic pathways using proteomics are being developed to a higher level *(165)*; however, many studies have taken the opportunity to map proteomic data to pathways to examine trends caused by different treatment effects such as physicochemical stress *(97,166,167)*. The integration of such data likely will be increasingly made easier and more rapidly via online tools. The developing KBase system (kbase.science.energy.gov), for example, will allow computation of a preliminary metabolic model purely from genome data, which then can be refined with proteomic and metabolite data. Proteomics has also been used extensively to model protein interactions, thus providing means to predicting protein function *(56)*.

8   STATE OF AFFAIRS IN THE STUDY OF PROKARYOTES VIA GEL-FREE PROTEOMICS

Quantitative proteomics studies on bacteria associated with various ecosystems have broad prospects to provide both fundamental knowledge and new concepts for applied utilization, monitoring, and control. A listing of studies utilizing gel-free proteomics studies involving the study of cultured prokaryotes reveals the increasing use (double the number of papers in 2013 compared to 2011) and a diversity of applications (Table 3). The focus of global proteome analysis is most often for studying aspects of metabolism, physiology, and regulation; however, specific fractions of the proteome including cell wall, cell membrane, and secreted, phosphorylated, and glycosylated proteins are also major targets. Label-free and iTRAQ seem to be the most popular analytical approaches most likely because they have straightforward workflows, with iTRAQ reagents available in a kit.

The increased availability of metagenomes increases the prospective application of mass-spectrometric approaches to investigate proteins directly in

TABLE 3 A Selection of Recent Gel-Free Quantitative or Semiquantitative Studies on Proteomes and Subproteomes in Bacteria and Archaea

Bacteria	Nature	Proteomics Applications	Purpose	Reference
Bacillus subtilis	Soil microbe	Label-free	Protein–lysine acetylation	(168)
		Metabolic labeling	Secreted proteins	(169)
		Multiple	Global analysis of protein	(170)
		Multiple	Stress-related physiology	(27)
Bartonella quintana	Pathogen	Label-free	Global analysis of protein	(171)
Cand. "*Blochmannia*" sp.	Insect symbiont	Label-free	Global analysis of proteins	(172)
Clostridium difficile	Pathogen	iTRAQ	Stress-related physiology; toxin formation	(173,174)
Clostridium thermocellum	Anaerobe	Metabolic labeling	Metabolic pathway analysis	(175)
Escherichia coli	Other pathogens	Label-free	Exoproteome, secreted proteins	(176,177)
		ICAT	Proteins subject to peroxidation	(178)
		Label-free	Quantification comparisons	(179)
		SILAC	Phosphoproteome	(104)
		Label-free	Stress-related physiology	(6)
Fusobacterium nucleatum	Oral cavity microbe	Label-free	Metabolic pathway analysis	(167)
Lactobacillus spp.	Probiotic, fermentation	Label-free; iTRAQ; MALDI-based	Global stress/physiology	(166,180,181)
		CID	Glycoproteomics	(182)

Continued

TABLE 3 A Selection of Recent Gel-Free Quantitative or Semiquantitative Studies on Proteomes and Subproteomes in Bacteria and Archaea—Cont'd

Bacteria	Nature	Proteomics Applications	Purpose	Reference
Listeria monocytogenes	Pathogen	Cell wall fractionation; label-free; SRM	Cell wall proteome and associated functions	(183–185)
		iTRAQ	Define role alternate RNA polymerase σ subunit regulators and small RNA	(186,187)
		Label-free	Global stress/physiology	(97,188,189)
		Label-free; iTRAQ	Global and targeted phosphoproteome analysis	(190)
		2-D LC/MS–MS	Proteome in host cells	(191)
Methyloversatilis universalis	Methylotroph	Label-free	Metabolic pathway analysis	(192)
Mycobacterium tuberculosis, *Mycobacterium leprae*	Pathogens	Label-free	Cell wall and membrane subproteomes	(193–195)
		Label-free	Immunological-related analysis	(196)
Nitrosomonas europaea	Autotroph	Metabolic labeling	Metabolic pathway analysis	(197)
Cand. "Pelagibacter ubique"	Marine microbe	Label-free	Global analysis of protein	(198)
Cand. "Phytoplasma aurantifolia"	Plant pathogen	Label-free	Global analysis of proteins	(199)
Porphyromonas gingivalis	Pathogen	Label-free	Metabolic pathway analysis	(167)
Pseudomonas aeruginosa	Opportunistic pathogen	Label-free	Effect of zero gravity	(200)
Psychroflexus torquis	Psychrophile	Label-free	Targeted proteomics, physiology	(201)

Species	Niche/type	Method	Application	Ref.
Roseobacter clade species	Marine taxa	Label-free	Exoproteome	(202)
Serratia marcescens	Opp. pathogen	Label-free	Exoproteome	(203)
Staphylococcus aureus	Pathogen	Metabolic labeling; label-free; iTRAQ	Response to antibiotics	(204–206)
		iTRAQ	Global regulation and virulence	(207)
		iTRAQ	Interactive protein map	(208)
Starkeya novella	Soil microbe	Label-free	Metabolic pathway analysis	(209)
Streptococcus mutans, S. gordonii	Oral cavity microbes	Label-free	Metabolic pathway analysis; stress responses	(167,210)
Streptomyces coelicolor	Soil microbe	Label-free	Metabolic pathway analysis	(211)
Synechocystis sp. PCC6803	Phototroph	iTRAQ	Stress responses; metabolic pathway analysis	(212,213)
Wolbachia	Intracellular microbe	Label-free	Response to antibiotics	(214)
Archaea				
Haloarcula marismortui	Haloarchaeon	Label-free	Metabolic pathway analysis	(215)
Methanococcoides burtonii	Methanogen	iTRAQ	Temperature-dependent stress	(216)
Methanococcus maripaludis	Methanogen	iTRAQ	Metabolic pathway analysis	(217)
Pyrococcus furiosus	Thermophile	Multimethod, multilab	Workflow and quantification validation	(218)
Sulfolobus solfataricus	Thermophile	Label-free	Phosphoproteomics	(219)
Thermoplasma acidophilum	Acidophile	Label-free	Metabolic pathway analysis	(220)

environmental samples. The goal is to determine in greater depth the functionality of organisms *in situ* as well as examine community-level adaptations
and also responses to change. The first metaproteomic analysis utilized gel-
based technology *(221)*. Gel-free, shotgun-type proteomics offers an obvious
advantage for such analysis by increasing the possibility to identify large numbers of proteins and potentially even begin to do quantitative assessments *(222)*.
Given the enormous diversity of life, much of it still unavailable as cultures—
described recently as "microbial dark matter" *(223,224)*, the challenge presented
to the young field of metaproteomics is tremendous. To put that in perspective,
a simple mixed culture consisting of three bacterial species was experimentally
tested using both gel-based and gel-free proteomic analyses to track the populations of bacteria and examine their interactions. The gel-free analysis showed
that 270 proteins from just one of the strains present were considered to be
induced or repressed due to microbial interactions such as production of siderophores *(225)*. In a natural system, the community complexity is likely 2 or more
orders greater than this simple system, with interactions and population
dynamics even more complicated providing a tremendous challenge for the
technology available today. Besides complexity, more prosaically lysis and
macromolecule separation approaches are an additional consideration in metaproteomics. A study recently indicated that extraction methods used for obtaining both DNA and protein heavily affected metaproteome profiles especially
proteins with TMH domains *(226)*. Likely, this problem is worsened by the
shear scale of the metaproteome with many diverse organisms contributing proteins in the dataset; thus, different methods likely differ in their capacity to lyse
cells and separate proteins from other biological materials, especially highly
anionically charged DNA and polysaccharides. Fundamentally, the workflow
approaches used need to be highly consistent, and for interlaboratory comparisons, a standardized approach is obviously required *(227)*. Standardization
schemes are important for comparison between datasets and have been insti-
gated in large proteomic projects, such as the Human Proteomics Organization
(228), which is examining the proteome in different tissues and fluids. Several
reviews have examined the workflows and data so far obtained in pioneering
metaproteomic analysis in a number of different contexts *(229–234)*. The
general consensus is that metaproteomics to be most effective with currently
available technology needs to be used in a complementary and/or targeted
way with other -omics approaches *(235)*.

9 CONCLUSIONS AND FUTURE PROSPECTS

Proteomics as field of technologically based inquiry is rapidly advancing
within the postgenomic era, and its importance as a tool in high-level to routine experimentation and analysis will continue to expand. As expounded in a
recent review, proteomics technology can now be considered robust, accurate,
and reliable means to a degree not imagined a decade ago *(11)*. Developments

of this sort are however also needed at other stages of the proteomic workflow including the initial protein extraction, fractionation, and peptide separation. These areas still require development to create processes that are not only truly high throughput but also cost-effective, reproducible, and unbiased. Despite excellent advances in data-independent acquisition, the sample inputs into such workflows still possess the biases observed with gel-based technology though this can be solved with more complicated workflows incorporating sample fractionation. Nevertheless, other technologies such as the ability to rapidly identify microbes via MALDI-based approaches and its rapid adoption into diagnostic laboratories are examples of the impact of proteomics technology that seems to be far from exhausted. The sophisticated technology required for exact quantitative proteomics at this stage is restricted to few laboratories focused purely on proteomic-based research and development pursuits; however, it can be assumed that commercial systems coupled with streamlining of the instruments and software will eventually produce increasingly accessible and cost-effective platforms for high-quality proteomics applications. Such instrumentation either could exist as benchtop instruments or be available in service provider laboratories if the former is not fully feasible. There is also the possibility that DNA/RNA, protein, and metabolite-based instrumentation technologies could be fused together since there is active research in adapting electrophoretic techniques to analysis of nucleic acids *(236,237)*. Within the big picture, proteomics represents a tool to be used to understand biological systems, and thus, it is most effective when used alongside other approaches, be it -omics-related or orthogonal experiments that provide support and credulity to research conclusions.

ACKNOWLEDGMENTS

The author would like to thank the following people for their excellent support in research pursuits relevant to proteomics: Richard Wilson, Chawalit Kocharunchitt, Rolf Nilsson, Marcia Mata, Maxim Sheludchenko, Ali Al-Naseri, Shi Feng, Bianca Porteus, Mona Al-Harbi, and Reham Mohammed.

REFERENCES

1. Joyceand, A. R.; Palsson, B. *Nat. Rev. Mol. Cell Biol.* **2006**, *7*, 198–210.
2. Pinto, A. C.; Melo-Barbosa, H. P.; Miyoshi, A.; Silva, A.; Azevedo, V. *Genet. Mol. Res.* **2011**, *10*, 1707–1718.
3. Cho, S.; Cho, Y.; Lee, S.; Kim, J.; Yum, H.; Kim, S. C.; Cho, B. K. *Genomics Inform.* **2013**, *11*, 76–82.
4. Filiatrault, M. J. *Curr. Opin. Microbiol.* **2011**, *14*, 579–586.
5. de Sousa Abreu, R.; Penalva, L. O.; Marcotte, E. M.; Vogel, C. *Mol. Biosyst.* **2009**, *5*, 1512–1526.
6. Kocharunchitt, C.; King, T.; Gobius, K.; Bowman, J. P.; Ross, T. *Mol. Cell. Proteomics* **2012**, *11*, M111.009019.

7. Marguerat, S.; Schmidt, A.; Codlin, S.; Chen, W.; Aebersold, R.; Bähler, J. *Cell* **2012**, *151*, 671–683.

8. Maier, T.; Güell, M.; Serrano, L. *FEBS Lett.* **2009**, *583*, 3966–3973.

9. Maier, T.; Schmidt, A.; Güell, M.; Kühner, S.; Gavin, A. C.; Aebersold, R.; Serrano, L. *Mol. Syst. Biol.* **2011**, *7*, 511.

10. Schwanhäusser, B.; Busse, D.; Li, N.; Dittmar, G.; Schuchhardt, J.; Wolf, J.; Chen, W.; Selbach, M. *Nature* **2011**, *473*, 337–342.

11. Picotti, P.; Bodenmiller, B.; Aebersold, R. *Nat. Methods* **2013**, *10*, 24–27.

12. Maass, S.; Sievers, S.; Zühlke, D.; Kuzinski, J.; Sappa, P. K.; Muntel, J.; Hessling, B.; Bernhardt, J.; Sietmann, R.; Völker, U.; Hecker, M.; Becher, D. *Anal. Chem.* **2011**, *83*, 2677–2684.

13. James, P. *Q. Rev. Biophys.* **1997**, *30*, 279–331.

14. Anderson, N. L.; Anderson, N. G. *Electrophoresis* **1998**, *19*, 1853–1861.

15. Carbonnelle, E.; Grohs, P.; Jacquier, H.; Day, N.; Tenza, S.; Dewailly, A.; Vissouarn, O.; Rottman, M.; Herrmann, J. L.; Podglajen, I.; Raskine, L.; Microbiol, J. *J. Microbiol. Methods* **2012**, *89*, 133–136.

16. Khot, P. D.; Couturier, M. R.; Wilson, A.; Croft, A.; Fisher, M. A. *J. Clin. Microbiol.* **2012**, *50*, 3845–3852.

17. Minden, J. S. *Methods Mol. Biol.* **2012**, *854*, 3–8.

18. Otto, A.; Bernhardt, J.; Hecker, M.; Becher, D. *Curr. Opin. Microbiol.* **2012**, *15*, 364–372.

19. Yoshigi, M.; Pronovost, S. M.; Kadrmas, J. L. *Proteome Sci.* **2013**, *11*, 21.

20. Målen, H.; Berven, F. S.; Søfteland, T.; Arntzen, M.Ø.; D'Santos, C. S.; De Souza, G. A.; Wiker, H. G. *Proteomics* **2008**, *8*, 1859–1870.

21. Everberg, H.; Gustavasson, N.; Tjerned, F. *Methods Mol. Biol.* **2008**, *424*, 403–412.

22. Guo, P.; Guan, Z.; Wang, W.; Chen, B.; Huang, Y. *Talanta* **2011**, *84*, 587–592.

23. Lin, Y.; Liu, Y.; Li, J.; Zhao, Y.; He, Q.; Han, W.; Chen, P.; Wang, X.; Liang, S. *Electrophoresis* **2010**, *31*, 2705–2713.

24. Liu, Y.; Lin, Y.; Yan, Y.; Li, J.; He, Q.; Chen, P.; Wang, X.; Liang, S. *Electrophoresis* **2012**, *33*, 316–324.

25. Bünger, S.; Roblick, U. J.; Habermann, J. K. *Cytotechnology* **2009**, *61*, 153–159.

26. Wolff, S.; Hahne, H.; Hecker, M.; Becher, D. *Mol. Cell. Proteomics* **2008**, *7*, 1460–1468.

27. Otto, A.; Bernhardt, J.; Meyer, H.; Schaffer, M.; Herbst, F. A.; Siebourg, J.; Mäder, U.; Lalk, M.; Hecker, M.; Becher, D. *Nat. Commun.* **2010**, *1*, 137.

28. Olaya-Abril, A.; Jiménez-Munguía, I.; Gómez-Gascón, L.; Rodríguez-Ortega, M. J. *J. Proteomics* **2014**, *97*, 164–176.

29. Dreisbach, A.; van der Kooi-Pol, M. M.; Otto, A.; Gronau, K.; Bonarius, H. P.; Westra, H.; Groen, H.; Becher, D.; Hecker, M.; van Dijl, J. M. *Proteomics* **2011**, *11*, 2921–2930.

30. Tjalsma, H.; Lambooy, L.; Hermans, P. W.; Swinkels, D. W. *Proteomics* **2008**, *8*, 1415–1428.

31. Johnson, B.; Selle, K.; O'Flaherty, S.; Goh, Y. J.; Klaenhammer, T. *Microbiology* **2013**, *159*, 2269–2282.

32. Henderson, B.; Martin, A. *Infect. Immun.* **2011**, *79*, 3476–3491.

33. Wang, G.; Xia, Y.; Cui, J.; Gu, Z.; Song, Y.; Chen, Y. Q.; Chen, H.; Zhang, H.; Chen, W. *Curr. Issues Mol. Biol.* **2013**, *16*, 15–22.

34. Olaya-Abril, A.; Gómez-Gascón, L.; Jiménez-Munguía, I.; Obando, I.; Rodríguez-Ortega, M. J. *J. Proteomics* **2012**, *75*, 3733–3746.

35. Solis, N.; Larsen, M. R.; Cordwell, S. J. *Proteomics* **2010**, *10*, 2037–2049.

36. Coustets, M.; Al-Karablieh, N.; Thomsen, C.; Teissié, J. *J. Membr. Biol.* **2013**, *246*, 751–760.

37. Grassl, J.; Westbrook, J. A.; Robinson, A.; Borén, M.; Dunn, M. J.; Clyne, R. K. *Proteomics* **2009**, *9*, 4616–4626.
38. Rezaul, K.; Wu, L.; Mayya, V.; Hwang, S. I.; Han, D. *Mol. Cell. Proteomics* **2005**, *4*, 169–181.
39. Piersma, S. R.; Warmoes, M. O.; de Wit, M.; de Reus, I.; Knol, J. C.; Jiménez, C. R. *Proteome Sci.* **2013**, *11*, 17.
40. Engholm-Keller, K.; Larsen, M. R. *Proteomics* **2013**, *13*, 910–931.
41. Loroch, S.; Dickhut, C.; Zahedi, R. P.; Sickmann, A. *Electrophoresis* **2013**, *34*, 1483–1492.
42. Oda, Y.; Nagasu, T.; Chait, B. T. *Nat. Biotechnol.* **2001**, *19*, 379–382.
43. Roux, P. P.; Thibault, P. *Mol. Cell. Proteomics* **2013** (Sep 13).
44. Hitchen, P. G.; Twigger, K.; Valiente, E.; Langdon, R. H.; Wren, B. W.; Dell, A. *Biochem. Soc. Trans.* **2010**, *38*, 1307–1313.
45. Longwell, S. A.; Dube, D. H. *Curr. Opin. Chem. Biol.* **2013**, *17*, 41–48.
46. Kang, D.; Ji, E. S.; Moon, M. H.; Yoo, J. S. *J. Proteome Res.* **2010**, *9*, 2855–2862.
47. Scott, N. E.; Parker, B. L.; Connolly, A. M.; Paulech, J.; Edwards, A. V.; Crossett, B.; Falconer, L.; Kolarich, D.; Djordjevic, S. P.; Højrup, P.; Packer, N. H.; Larsen, M. R.; Cordwell, S. J. *Mol. Cell. Proteomics* **2011**, *10*, M000031-MCP201.
48. Parker, B. L.; Thaysen-Andersen, M.; Solis, N.; Scott, N. E.; Larsen, M. R.; Graham, M. E.; Packer, N. H.; Cordwell, S. J. *J. Proteome Res.* **2013**, *12*, 5781–5800.
49. Ge, Y.; Rikihisa, Y. *J. Bacteriol.* **2007**, *189*, 7819–7828.
50. Sears, K. T.; Ceraul, S. M.; Gillespie, J. J.; Allen, E. D., Jr.; Popov, V. L.; Ammerman, N. C.; Rahman, M. S.; Azad, A. F. *PLoS Pathog.* **2012**, *8*, e1002856.
51. Koziol, J.; Griffin, N.; Long, F.; Li, Y.; Latterich, M.; Schnitzer, J. *Proteome Sci.* **2013**, *11*, 5.
52. Lu, C.; King, R. D. *Bioinformatics* **2009**, *25*, 2020–2027.
53. Brewis, I. A.; Brennan, P. *Adv. Protein Chem. Struct. Biol.* **2010**, *80*, 1–44.
54. Sehrawat, A.; Abat, J. K.; Deswal, R. *Front Plant Sci.* **2013**, *4*, 342.
55. Eyford, B. A.; Ahmad, R.; Enyaru, J. C.; Carr, S. A.; Pearson, T. W. *PLoS One* **2013**, *8*, e71463.
56. Hu, P.; Janga, S. C.; Babu, M.; Díaz-Mejía, J. J.; Butland, G.; Yang, W.; Pogoutse, O.; Guo, X.; Phanse, S.; Wong, P.; Chandran, S.; Christopoulos, C.; Nazarians-Armavil, A.; Nasseri, N. K.; Musso, G.; Ali, M.; Nazemof, N.; Eroukova, V.; Golshani, A.; Paccanaro, A.; Greenblatt, J. F.; Moreno-Hagelsieb, G.; Emili, A. *PLoS Biol.* **2009**, *7*, e96.
57. Kühner, S.; van Noort, V.; Betts, M. J.; Leo-Macias, A.; Batisse, C.; Rode, M.; Yamada, T.; Maier, T.; Bader, S.; Beltran-Alvarez, P.; Castaño-Diez, D.; Chen, W. H.; Devos, D.; Güell, M.; Norambuena, T.; Racke, I.; Rybin, V.; Schmidt, A.; Yus, E.; Aebersold, R.; Herrmann, R.; Böttcher, B.; Frangakis, A. S.; Russell, R. B.; Serrano, L.; Bork, P.; Gavin, A. C. *Science* **2009**, *326*, 1235–1240.
58. Wang, Y.; Cui, T.; Zhang, C.; Yang, M.; Huang, Y.; Li, W.; Zhang, L.; Gao, C.; He, Y.; Li, Y.; Huang, F.; Zeng, J.; Huang, C.; Yang, Q.; Tian, Y.; Zhao, C.; Chen, H.; Zhang, H.; He, Z. G. *J. Proteome Res.* **2010**, *9*, 6665–6677.
59. Lu, H. C.; Fornili, A.; Fraternali, F. *Expert Rev. Proteomics* **2013**, *10*, 511–520.
60. Righetti, P. G.; Boschetti, E.; Lomas, L.; Citterio, A. *Proteomics* **2006**, *6*, 3980–3992.
61. Fortis, F.; Guerrier, L.; Areces, B. L.; Antonioli, P.; Hayes, T.; Carrick, K.; Hammond, D.; Boschetti, E.; Righetti, G. P. *J. Proteome Res.* **2006**, *5*, 2577–2585.
62. Guerrier, L.; Claverol, S.; Fortis, F.; Rinalducci, S.; Timperio, A. M.; Antonioli, P.; Jandrot-Perrus, M.; Boschetti, E.; Righetti, P. G. *J. Proteome Res.* **2007**, *6*, 4290–4303.
63. Di Girolamo, F.; Righetti, P. G.; Soste, M.; Feng, Y.; Picotti, P. *J. Proteomics* **2013**, *89*, 215–226.

64. Chao, T. C.; Hansmeier, N. *Proteomics* **2013**, *13*, 467–479.
65. Demirev, P. A.; Fenselau, C. *J. Mass Spectrom.* **2008**, *43*, 1441–1457.
66. Ye, Y.; Mar, E. C.; Tong, S.; Sammons, S.; Fang, S.; Anderson, L. J.; Wang, D.; Virol, J. *Methods* **2010**, *163*, 87–95.
67. Schieltz, D. M.; McGrath, S. C.; McWilliams, L. G.; Rees, J.; Bowen, M. D.; Kools, J. J.; Dauphin, L. A.; Gomez-Saladin, E.; Newton, B. N.; Stang, H. L.; Vick, M. J.; Thomas, J.; Pirkle, J. L.; Barr, J. R. *Forensic Sci. Int.* **2011**, *209*, 70–79.
68. Kim, J. Y.; Lee, S. Y.; Kim, S. K.; Park, S. R.; Kang, D.; Moon, M. H. *Anal. Chem.* **2013**, *85*, 5506–5513.
69. Rogeberg, M.; Malerod, H.; Roberg-Larsen, H.; Aass, C.; Wilson, S. R. *J. Pharm. Biomed. Anal.* **2013**, pii: S0731–7085(13)00199–4, 2013.
70. Rudin, T.; Tsougeni, K.; Gogolides, E.; Pratsinis, S. E. *Microelectron. Eng.* **2012**, *97*, 341–344.
71. García-Otero, N.; Barciela-Alonso, M. C.; Moreda-Piñeiro, A.; Bermejo-Barrera, P. *Talanta* **2013**, *115*, 631–641.
72. le Coutre, J.; Whitelegge, J. P.; Gross, A.; Turk, E.; Wright, E. M.; Kaback, H. R.; Faull, K. F. *Biochemistry* **2000**, *39*, 4237–4242.
73. Vandermarliere, E.; Mueller, M.; Martens, L. *Mass Spectrom. Rev.* **2013**, *32*, 453–465.
74. Kalli, A.; Håkansson, K. *J. Proteome Res.* **2008**, *7*, 2834–2844.
75. Yates, J. R., 3rd.; Ruse, C. I.; Nakorchevsky, A. *Annu. Rev. Biomed. Eng.* **2009**, *11*, 49–79.
76. Thakur, S. S.; Geiger, T.; Chatterjee, B.; Bandilla, P.; Frohlich, F.; Cox, J.; Mann, M. *Mol. Cell. Proteomics* **2011**, *10*, M110.003699.
77. Nagaraj, N.; Kulak, N. A.; Cox, J.; Neuhauser, N.; Mayr, K.; Hoerning, O.; Vorm, O.; Mann, M. *Mol. Cell. Proteomics* **2012**, *11*, M111.013722.
78. Wang, Y.; Fonslow, B. R.; Wong, C. C.; Nakorchevsky, A.; Yates, J. R., 3rd. *Anal. Chem.* **2012**, *84*, 8505–8513.
79. Fang, X.; Wang, C.; Lee, C. S. *Methods Mol. Biol.* **2013**, *919*, 181–187.
80. Kler, P. A.; Posch, T. N.; Pattky, M.; Tiggelaar, R. M.; Huhn, C. *J. Chromatogr. A* **2013**, *1297*, 204–212.
81. Kelstrup, C. D.; Young, C.; Lavallee, R.; Nielsen, M. L.; Olsen, J. V. *J. Proteome Res.* **2012**, *11*, 3487–3497.
82. Hu, Q.; Noll, R. J.; Li, H.; Makarov, A.; Hardman, M.; Graham Cooks, R. *J. Mass Spectrom.* **2005**, *40*, 430–443.
83. Pachl, F.; Ruprecht, B.; Lemeer, S.; Kuster, B. *Proteomics* **2013**, *13*, 2552–2562.
84. Weisbrod, C. R.; Hoopmann, M. R.; Senko, M. W.; Bruce, J. E. *J. Proteomics* **2013**, *88*, 109–119.
85. Cotter, R. J. *J. Am. Soc. Mass Spectrom.* **2013**, *24*, 657–674.
86. Mayya, V.; Han, D. K. *Expert Rev. Proteomics* **2006**, *3*, 597–610.
87. Kirkpatrick, D. S.; Gerber, S. A.; Gygi, S. P. *Methods* **2005**, *35*, 265–273.
88. Lange, V.; Picotti, P.; Domon, B.; Aebersold, R. *Mol. Syst. Biol.* **2008**, *4*, 222.
89. Malmström, J.; Beck, M.; Schmidt, A.; Lange, V.; Deutsch, E. W.; Aebersold, R. *Nature* **2009**, *460*, 762–765.
90. Gillet, L. C.; Navarro, P.; Tate, S.; Röst, H.; Selevsek, N.; Reiter, L.; Bonner, R.; Aebersold, R. *Mol. Cell. Proteomics* **2012**, *11*, O111.016717.
91. Old, W. M.; Meyer-Arendt, K.; Aveline-Wolf, L.; Pierce, K. G.; Mendoza, A.; Sevinsky, J. R.; Resing, K. A.; Ahn, N. G. *Mol. Cell. Proteomics* **2005**, *4*, 1487–1502.
92. Cooper, B.; Feng, J.; Garrett, W. M. *J. Am. Soc. Mass Spectrom.* **2010**, *21*, 1534–1546.
93. Lundgren, D. H.; Hwang, S. I.; Wu, L.; Han, D. K. *Expert Rev. Proteomics* **2010**, *7*, 39–53.

94. Collier, T. S.; Sarkar, P.; Franck, W. L.; Rao, B. M.; Dean, R. A.; Muddiman, D. C. *Anal. Chem.* **2010**, *82*, 8696–8702.
95. Grossmann, J.; Roschitzki, B.; Panse, C.; Fortes, C.; Barkow-Oesterreicher, S.; Rutishauser, D.; Schlapbach, R. *J. Proteomics* **2010**, *73*, 1740–1746.
96. Ahrné, E.; Molzahn, L.; Glatter, T.; Schmidt, A. *Proteomics* **2013**, *13*, 2567–2578.
97. Bowman, J. P.; Hages, E.; Nilsson, R. E.; Kocharunchitt, C.; Ross, T. *J. Proteome Res.* **2012**, *11*, 2409–2426.
98. Griffin, N. M.; Yu, J.; Long, F.; Oh, P.; Shore, S.; Li, Y.; Koziol, J. A.; Schnitzer, J. E. *Nat. Biotechnol.* **2010**, *28*, 83–89.
99. Vogel, C.; Marcotte, E. M. *Methods Mol. Biol.* **2012**, *893*, 321–341.
100. Megger, D. A.; Pott, L. L.; Ahrens, M.; Padden, J.; Bracht, T.; Kuhlmann, K.; Eisenacher, M.; Meyer, H. E.; Sitek, B. *Biochim. Biophys. Acta* **2013**, pii: S1570–9639 (13)00289–6.
101. Hendrickson, E. L.; Xia, Q.; Wang, T.; Leigh, J. A.; Hackett, M. *Analyst* **2006**, *131*, 1335–1341.
102. Ong, S. E.; Mann, M. *Nat. Protoc.* **2006**, *1*, 2650–2660.
103. Ong, S. E.; Mann, M. *Methods Mol. Biol.* **2007**, *359*, 37–52.
104. Soares, N. C.; Spät, P.; Krug, K.; Macek, B. *J. Proteome Res.* **2013**, *12*, 2611–2621.
105. Claydon, A. J.; Beynon, R. *J. Methods Mol. Biol.* **2011**, *759*, 179–195.
106. Pförtner, H.; Wagner, J.; Surmann, K.; Hildebrandt, P.; Ernst, S.; Bernhardt, J.; Schurmann, C.; Gutjahr, M.; Depke, M.; Jehmlich, U.; Dhople, V.; Hammer, E.; Steil, L.; Völker, U.; Schmidt, F. *Methods* **2013**, *61*, 244–250.
107. Gygi, S. P.; Rist, B.; Gerber, S. A.; Turecek, F.; Gelb, M. H.; Aebersold, R. *Nat. Biotechnol.* **1999**, *17*, 994–999.
108. Yi, E. C.; Li, X. J.; Cooke, K.; Lee, H.; Raught, B.; Page, A.; Aneliunas, V.; Hieter, P.; Goodlett, D. R.; Aebersold, R. *Proteomics* **2005**, *5*, 380–387.
109. Pappin, D. J. C., Bartlet-Jones, M. U.S. Patent No. 7,195,751, 2007.
110. Ross, P. L.; Huang, Y. N.; Marchese, J. N.; Williamson, B.; Parker, K.; Hattan, S.; Khainovski, N.; Pillai, S.; Dey, S.; Daniels, S.; Purkayastha, S.; Juhasz, P.; Martin, S.; Bartlet-Jones, M.; He, F.; Jacobson, A.; Pappin, D. J. *Mol. Cell. Proteomics* **2004**, *3*, 1154–1169.
111. Zieske, L. R. *J. Exp. Bot.* **2006**, *57*, 1501–1508.
112. Dayon, L.; Hainard, A.; Licker, V.; Turck, N.; Kuhn, K.; Hochstrasser, D. F.; Burkhard, P. R.; Sanchez, J. C. *Anal. Chem.* **2008**, *80*, 2921–2931.
113. Rauniyar, N.; Gao, B.; McClatchy, D. B.; Yates, J. R., 3rd. *J. Proteome Res.* **2013**, *12*, 1031–1039.
114. Treumann, A.; Thiede, B. *Expert Rev. Proteomics* **2010**, *7*, 647–653.
115. Breitwieser, F. P.; Colinge, J. *J. Proteomics* **2013**, *90*, 77–84.
116. Stewart, I. I.; Thomson, T.; Figeys, D. *Rapid Commun. Mass Spectrom.* **2001**, *15*, 2456–2465.
117. Sakai, J.; Kojima, S.; Yanagi, K.; Kanaoka, M. *Proteomics* **2005**, *5*, 16–23.
118. Perkins, D. N.; Pappin, D. J.; Creasy, D. M.; Cottrell, J. S. *Electrophoresis* **1999**, *20*, 3551–3567.
119. MacCoss, M. J.; Wu, C. C.; Yates, J. R., 3rd. *Anal. Chem.* **2002**, *74*, 5593–5599.
120. Kapp, E. A.; Schütz, F.; Connolly, L. M.; Chakel, J. A.; Meza, J. E.; Miller, C. A.; Fenyo, D.; Eng, J. K.; Adkins, J. N.; Omenn, G. S.; Simpson, R. J. *Proteomics* **2005**, *5*, 3475–3490.
121. Dagda, R. K.; Sultana, T.; Lyons-Weiler, J. *J. Proteomics Bioinform.* **2010**, *3*, 39–47.

122. Jeong, K.; Kim, S.; Bandeira, N. *BMC Bioinforma.* **2012**, *13*, S2.

123. Wang, G.; Wu, W. W.; Zhang, Z.; Masilamani, S.; Shen, R. F. *Anal. Chem.* **2009**, *81*, 146–159.

124. Ahrné, E.; Ohta, Y.; Nikitin, F.; Scherl, A.; Lisacek, F.; Müller, M. *Proteomics* **2011**, *11*, 4085–4095.

125. Bern, M. W.; Kil, Y. J. *J. Proteome Res.* **2011**, *10*, 5296–5301.

126. Vaudel, M.; Burkhart, J. M.; Breiter, D.; Zahedi, R. P.; Sickmann, A.; Martens, L. *J. Proteome Res.* **2012**, *11*, 5065–5071.

127. Yadav, A. K.; Kumar, D.; Dash, D. *PLoS One* **2012**, *7*, e50651.

128. Krug, K.; Carpy, A.; Behrends, G.; Matic, K.; Soares, N. C.; Macek, B. *Mol. Cell. Proteomics* **2013**, *12*, 3420–3430.

129. Panchaud, A.; Scherl, A.; Shaffer, S. A.; von Haller, P. D.; Kulasekara, H. D.; Miller, S. I.; Goodlett, D. R. *Anal. Chem.* **2009**, *81*, 6481–6488.

130. Picotti, P.; Bodenmiller, B.; Mueller, L. N.; Domon, B.; Aebersold, R. *Cell* **2009**, *138*, 795–806.

131. Picotti, P.; Aebersold, R. *Nat. Methods* **2012**, *9*, 555–566.

132. Deutsch, E. W. *Methods Mol. Biol.* **2010**, *604*, 285–296.

133. Deutsch, E. W.; Lam, H.; Aebersold, R. *EMBO Rep.* **2008**, *9*, 429–434.

134. Nahnsen, S.; Kohlbacher, O. *BMC Bioinforma.* **2012**, *13* (Suppl 16), S8.

135. Sun, Y.; Braga-Neto, U.; Dougherty, E. R. *BMC Genomics* **2012**, *13* (Suppl 6), S2.

136. Cox, J.; Mann, M. *Nat. Biotechnol.* **2008**, *26*, 1367–1372.

137. Kohlbacher, O.; Reinert, K.; Gröpl, C.; Lange, E.; Pfeifer, N.; Schulz-Trieglaff, O.; Sturm, M. *Bioinformatics* **2007**, *23*, e191–e197.

138. Weisser, H.; Nahnsen, S.; Grossmann, J.; Nilse, L.; Quandt, A.; Brauer, H.; Sturm, M.; Kenar, E.; Kohlbacher, O.; Aebersold, R.; Malmström, L. *J. Proteome Res.* **2013**, *12*, 1628–1644.

139. Qi, D.; Brownridge, P.; Xia, D.; Mackay, K.; Gonzalez-Galarza, F. F.; Kenyani, J.; Harman, V.; Beynon, R. J.; Jones, A. R. *OMICS* **2012**, *16*, 489–495.

140. Deutsch, E. W.; Mendoza, L.; Shteynberg, D.; Farrah, T.; Lam, H.; Tasman, N.; Sun, Z.; Nilsson, E.; Pratt, B.; Prazen, B.; Eng, J. K.; Martin, D. B.; Nesvizhskii, A. I.; Aebersold, R. *Proteomics* **2010**, *10*, 1150–1159.

141. Neuhauser, N.; Nagaraj, N.; McHardy, P.; Zanivan, S.; Scheltema, R.; Cox, J.; Mann, M. *J. Proteome Res.* **2013**, *12*, 2858–2868.

142. Jones, K. A.; Kim, P. D.; Patel, B. B.; Kelsen, S. G.; Braverman, A.; Swinton, D. J.; Gafken, P. R.; Jones, L. A.; Lane, W. S.; Neveu, J. M.; Leung, H. C.; Shaffer, S. A.; Leszyk, J. D.; Stanley, B. A.; Fox, T. E.; Stanley, A.; Hall, M. J.; Hampel, H.; South, C. D.; de la Chapelle, A.; Burt, R. W.; Jones, D. A.; Kopelovich, L.; Yeung, A. T. *J. Proteome Res.* **2013**, *12*, 4351–4365.

143. Mani, D. R.; Abbatiello, S. E.; Carr, S. A. *BMC Bioinforma.* **2012**, *13* (Suppl 16), S9.

144. Oberg, A. L.; Mahoney, D. W. *BMC Bioinforma.* **2012**, *13* (Suppl 16), S7.

145. Higgs, R. E.; Butler, J. P.; Han, B.; Knierman, M. D. *Int. J. Proteomics* **2013**, *2013*, 674282.

146. Gokce, E.; Shuford, C. M.; Franck, W. L.; Dean, R. A.; Muddiman, D. C. *J. Am. Soc. Mass Spectrom.* **2011**, *22*, 2199–2208.

147. Pham, T. V.; Jimenez, C. R. *Bioinformatics* **2012**, *28*, i596–i602.

148. Zhang, B.; VerBerkmoes, N. C.; Langston, M. A.; Uberbacher, E.; Hettichand, R. L.; Samatova, N. F. *J. Proteome Res.* **2006**, *5*, 2909–2918.

149. Gregori, J.; Villarreal, L.; Sánchez, A.; Baselga, J.; Villanueva, J. *J. Proteomics* **2013**, pii: S1874-3919(13)00306-0.

150. Mueller, L. N.; Rinner, O.; Schmidt, A.; Letarte, S.; Bodenmiller, B.; Brusniak, M. Y.; Vitek, O.; Aebersold, R.; Müller, M. *Proteomics* **2007**, *7*, 3470–3480.

151. Hendrickson, E. L.; Lamont, R. J.; Hackett, M. *J. Dent. Res.* **2008**, *87*, 1004–1015.

152. Väremo, L.; Nielsen, J.; Nookaew, I. *Nucleic Acids Res.* **2013**, *41*, 4378–4391.

153. García-Alcalde, F.; García-López, F.; Dopazo, J.; Conesa, A. *Bioinformatics* **2011**, *27*, 137–139.

154. Hirsch-Hoffmann, M.; Gruissem, W.; Baerenfaller, K. *Front Plant Sci.* **2012**, *3*, 123.

155. May, P.; Christian, N.; Ebenhöh, O.; Weckwerth, W.; Walther, D. *Methods Mol. Biol.* **2011**, *694*, 341–363.

156. Wägele, B.; Witting, M.; Schmitt-Kopplin, P.; Suhre, K. *PLoS One* **2012**, *7*, e39860.

157. Scholz, M.; Selbig, J. *Methods Mol. Biol.* **2007**, *358*, 87–104.

158. Omasits, U.; Ahrens, C. H.; Müller, S.; Wollscheid, B. *Bioinformatics* **2013**.

159. Boorsma, A.; Foat, B. C.; Vis, D.; Klis, F.; Bussemaker, H. J. *Nucleic Acids Res.* **2005**, *33*, W592–W595.

160. de Hoon, M. J.; Imoto, S.; Nolan, J.; Miyano, S. *Bioinformatics* **2004**, *20*, 1453–1454.

161. Saldanha, A. J. *Bioinformatics* **2004**, *20*, 3246–3248.

162. Winter, G.; Krömer, J. O. *Environ. Microbiol.* **2013**, *15*, 1901–1916.

163. Bennett, B. D.; Kimball, E. H.; Gao, M.; Osterhout, R.; Van Dien, S. J.; Rabinowitz, J. D. *Nat. Chem. Biol.* **2009**, *5*, 593–599.

164. Gika, H. G.; Theodoridis, G. A.; Vrhovsek, U.; Mattivi, F. *J. Chromatogr. A* **2012**, *1259*, 121–127.

165. Batth, T. S.; Keasling, J. D.; Petzold, C. J. *Methods Mol. Biol.* **2012**, *944*, 237–249.

166. Al-Naseri, A.; Bowman, J. P.; Wilson, R.; Nilsson, R. E.; Britz, M. L. *J. Proteome Res.* **2013**, *12*, 5313–5322.

167. Hendrickson, E. L.; Wang, T.; Dickinson, B. C.; Whitmore, S. E.; Wright, C. J.; Lamont, R. J.; Hackett, M. *BMC Microbiol.* **2012**, *12*, 211.

168. Kim, D.; Yu, B. J.; Kim, J. A.; Lee, Y. J.; Choi, S. G.; Kang, S.; Pan, J. G. *Proteomics* **2013**, *13*, 1726–1736.

169. Goosens, V. J.; Otto, A.; Glasner, C.; Monteferrante, C. C.; van der Ploeg, R.; Hecker, M.; Becher, D.; van Dijl, J. M. *J. Proteome Res.* **2013**, *12*, 796–807.

170. Becher, D.; Büttner, K.; Moche, M.; Hessling, B.; Hecker, M. *Proteomics* **2011**, *11*, 2971–2980.

171. Fabietti, A.; Gaspari, M.; Krishnan, S.; Quirino, A.; Liberto, M. C.; Cuda, G.; Focà, A. *Proteomics* **2013**, *13*, 1375–1378.

172. Fan, Y.; Thompson, J. W.; Dubois, L. G.; Moseley, M. A.; Wernegreen, J. J. *J. Proteome Res.* **2013**, *12*, 704–718.

173. Jain, S.; Graham, C.; Graham, R. L.; McMullan, G.; Ternan, N. G. *J. Proteome Res.* **2011**, *10*, 3880–3890.

174. Moura, H.; Terilli, R. R.; Woolfitt, A. R.; Williamson, Y. M.; Wagner, G.; Blake, T. A.; Solano, M. I.; Barr, J. R. *Int. J. Proteomics* **2013**, *2013*, 293782.

175. Burton, E.; Martin, V. J. *Can. J. Microbiol.* **2012**, *58*, 1378–1388.

176. Deng, W.; Yu, H. B.; de Hoog, C. L.; Stoynov, N.; Li, Y.; Foster, L. J.; Finlay, B. B. *Mol. Cell. Proteomics* **2012**, *11*, 692–709.

177. Orton, D. J.; Arsenault, D. J.; Thomas, N. A.; Doucette, A. A. *Mol. Cell. Probes* **2013**, *27*, 200–207.

178. Lindemann, C.; Lupilova, N.; Müller, A.; Warscheid, B.; Meyer, H. E.; Kuhlmann, K.; Eisenacher, M.; Leichert, L. I. *J. Biol. Chem.* **2013**, *288*, 19698–19714.

179. Arike, L.; Valgepea, K.; Peil, L.; Nahku, R.; Adamberg, K.; Vilu, R. *J. Proteomics* **2012**, *75*, 5437–5448.

180. Lee, J. Y.; Pajarillo, E. A.; Kim, M. J.; Chae, J. P.; Kang, D. K. *J. Proteome Res.* **2013**, *12*, 432–443.

181. Butorac, A.; Dodig, I.; Bačun-Družina, V.; Tishbee, A.; Mrvčić, J.; Hock, K.; Diminić, J.; Cindrić, M. *Rapid Commun. Mass Spectrom.* **2013**, *27*, 1045–1054.

182. Fredriksen, L.; Moen, A.; Adzhubei, A. A.; Mathiesen, G.; Eijsink, V. G.; Egge-Jacobsen, W. *Glycobiology* **2013**, *23*, 1439–1451.

183. García-del Portillo, F.; Calvo, E.; D'Orazio, V.; Pucciarelli, M. G. *J. Biol. Chem.* **2011**, *286*, 34675–34689.

184. Quereda, J. J.; Pucciarelli, M. G.; Botello-Morte, L.; Calvo, E.; Carvalho, F.; Bouchier, C.; Vieira, A.; Mariscotti, J. F.; Chakraborty, T.; Cossart, P.; Hain, T.; Cabanes, D.; García-Del Portillo, F. *Microbiology* **2013**, *159*, 1328–1339.

185. Zhang, C. X.; Creskey, M. C.; Cyr, T. D.; Brooks, B.; Huang, H.; Pagotto, F.; Lin, M. *Proteomics* **2013**, *13*, 3040–3045.

186. Mujahid, S.; Bergholz, T. M.; Oliver, H. F.; Boor, K. J.; Wiedmann, M. *Int. J. Mol. Sci.* **2012**, *14*, 378–393.

187. Mujahid, S.; Orsi, R. H.; Boor, K. J.; Wiedmann, M. *BMC Microbiol.* **2013**, *13*, 156.

188. Nilsson, R. E.; Latham, R.; Mellefont, L.; Ross, T.; Bowman, J. P. *Food Microbiol.* **2012**, *30*, 187–196.

189. Nilsson, R. E.; Ross, T.; Bowman, J. P.; Britz, M. L. *PLoS One* **2013**, *8*, e54157.

190. Misra, S. K.; Milohanic, E.; Aké, F.; Mijakovic, I.; Deutscher, J.; Monnet, V.; Henry, C. *Proteomics* **2011**, *11*, 4155–4165.

191. Donaldson, J. R.; Nanduri, B.; Pittman, J. R.; Givaruangsawat, S.; Burgess, S. C.; Lawrence, M. L. *J. Proteomics* **2011**, *74*, 1906–1917.

192. Lu, H.; Kalyuzhnaya, M.; Chandran, K. *Environ. Microbiol.* **2012**, *14*, 2935–2945.

193. Målen, H.; De Souza, G. A.; Pathak, S.; Søfteland, T.; Wiker, H. G. *BMC Microbiol.* **2011**, *11*, 18.

194. Wiker, H. G.; Tomazella, G. G.; de Souza, G. A. *J. Proteomics* **2011**, *74*, 1711–1719.

195. Gunawardena, H. P.; Feltcher, M. E.; Wrobel, J. A.; Gu, S.; Braunstein, M.; Chen, X. *J. Proteome Res.* **2013**, *12*, 5463–5474.

196. Cho, Y. S.; Dobos, K. M.; Prenni, J.; Yang, H.; Hess, A.; Rosenkrands, I.; Andersen, P.; Ryoo, S. W.; Bai, G. H.; Brennan, M. J.; Izzo, A.; Bielefeldt-Ohmann, H.; Belisle, J. T. *Proteomics* **2012**, *12*, 979–991.

197. Pellitteri-Hahn, M. C.; Halligan, B. D.; Scalf, M.; Smith, L.; Hickey, W. J. *J. Proteomics* **2011**, *74*, 411–419.

198. Smith, D. P.; Kitner, J. B.; Norbeck, A. D.; Clauss, T. R.; Lipton, M. S.; Schwalbach, M. S.; Steindler, L.; Nicora, C. D.; Smith, R. D.; Giovannoni, S. J. *PLoS One* **2010**, *5*, e10487.

199. Monavarfeshani, A.; Mirzaei, M.; Sarhadi, E.; Amirkhani, A.; Khayam Nekouei, M.; Haynes, P. A.; Mardi, M.; Salekdeh, G. H. *J. Proteome Res.* **2013**, *12*, 785–795.

200. Crabbé, A.; Schurr, M. J.; Monsieurs, P.; Morici, L.; Schurr, J.; Wilson, J. W.; Ott, C. M.; Tsaprailis, G.; Pierson, D. L.; Stefanyshyn-Piper, H.; Nickerson, C. A. *Appl. Environ. Microbiol.* **2011**, *77*, 1221–1230.

201. Feng, S.; Powell, S. M.; Wilson, R.; Bowman, J. P. *ISME J.* **2013**, *7*, 2206–2213.

202. Christie-Oleza, J. A.; Piña-Villalonga, J. M.; Bosch, R.; Nogales, B.; Armengaud, J. *Mol. Cell. Proteomics* **2012**, *11*, M111.013110.

203. Fritsch, M. J.; Trunk, K.; Diniz, J. A.; Guo, M.; Trost, M.; Coulthurst, S. J. *Mol. Cell. Proteomics* **2013**, *12*, 2735–2749.

204. Fischer, A.; Yang, S. J.; Bayer, A. S.; Vaezzadeh, A. R.; Herzig, S.; Stenz, L.; Girard, M.; Sakoulas, G.; Scherl, A.; Yeaman, M. R.; Proctor, R. A.; Schrenzel, J.; François, P. *J. Antimicrob. Chemother.* **2011**, *66*, 1696–1711.

205. Hessling, B.; Bonn, F.; Otto, A.; Herbst, F. A.; Rappen, G. M.; Bernhardt, J.; Hecker, M.; Becher, D. *Int. J. Med. Microbiol.* **2013**, *303*, 624–634.

206. Liu, X.; Hu, Y.; Pai, P. J.; Chen, D.; Lam, H. *J. Proteome Res.* **2014**, *13*, 1223–1233.

207. Rivera, F. E.; Miller, H. K.; Kolar, S. L.; Stevens, S. M., Jr.; Shaw, L. N. *Proteomics* **2012**, *12*, 263–268.

208. Cherkasov, A.; Hsing, M.; Zoraghi, R.; Foster, L. J.; See, R. H.; Stoynov, N.; Jiang, J.; Kaur, S.; Lian, T.; Jackson, L.; Gong, H.; Swayze, R.; Amandoron, E.; Hormozdiari, F.; Dao, P.; Sahinalp, C.; Santos-Filho, O.; Axerio-Cilies, P.; Byler, K.; McMaster, W. R.; Brunham, R. C.; Finlay, B. B.; Reiner, N. E. *J. Proteome Res.* **2011**, *10*, 1139–1150.

209. Kappler, U.; Nouwens, A. S. *Front. Microbiol.* **2013**, *4*, 304.

210. Klein, M. I.; Xiao, J.; Lu, B.; Delahunty, C. M.; Yates, J. R., 3rd.; Koo, H. *PLoS One* **2012**, *7*, e45795.

211. Gubbens, J.; Janus, M.; Florea, B. I.; Overkleeft, H. S.; van Wezel, G. P. *Mol. Microbiol.* **2012**, *86*, 1490–1507.

212. Qiao, J.; Wang, J.; Chen, L.; Tian, X.; Huang, S.; Ren, X.; Zhang, W. *J. Proteome Res.* **2012**, *11*, 5286–5300.

213. Gao, L.; Shen, C.; Liao, L.; Huang, X.; Liu, K.; Wang, W.; Guo, L.; Jin, W.; Huang, F.; Xu, W.; Wang, Y. *Mol. Cell. Proteomics* **2013** (Oct 29).

214. Darby, A. C.; Christina Gill, A.; Armstrong, S. D.; Hartley, C. S.; Xia, D.; Wastling, J. M.; Makepeace, B. L. *ISME J.* **2013**. http://dx.doi.org/10.1038/ismej.2013.192.

215. Chu, L. J.; Yang, H.; Shih, P.; Kao, Y.; Tsai, Y. S.; Chen, J.; Huang, G.; Weng, R. R.; Ting, Y. S.; Fang, X.; von Haller, P. D.; Goodlett, D. R.; Ng, W. V. *J. Proteome Res.* **2011**, *10*, 3261–3273.

216. Williams, T. J.; Lauro, F. M.; Ertan, H.; Burg, D. W.; Poljak, A.; Raftery, M. J.; Cavicchioli, R. *Environ. Microbiol.* **2011**, *13*, 2186–2203.

217. Walker, C. B.; Redding-Johanson, A. M.; Baidoo, E. E.; Rajeev, L.; He, Z.; Hendrickson, E. L.; Joachimiak, M. P.; Stolyar, S.; Arkin, A. P.; Leigh, J. A.; Zhou, J.; Keasling, J. D.; Mukhopadhyay, A.; Stahl, D. A. *ISME J.* **2012**, *6*, 2045–2055.

218. Wong, C. C.; Cociorva, D.; Miller, C. A.; Schmidt, A.; Monell, C.; Aebersold, R.; Yates, J. R., 3rd. *J. Proteome Res.* **2013**, *12*, 763–770.

219. Reimann, J.; Esser, D.; Orell, A.; Amman, F.; Pham, T. K.; Noirel, J.; Lindas, A. C.; Bernander, R.; Wright, P. C.; Siebers, B.; Albers, S. V. *Mol. Cell. Proteomics* **2013** (Sep 27).

220. Sun, N.; Pan, C.; Nickell, S.; Mann, M.; Baumeister, W.; Nagy, I. *J. Proteome Res.* **2009**, *9*, 4839–4850.

221. Wilmes, P.; Bond, P. L. *Environ. Microbiol.* **2004**, *6*, 911–920.

222. Verberkmoes, N. C.; Russell, A. L.; Shah, M.; Godzik, A.; Rosenquist, M.; Halfvarson, J.; Lefsrud, M. G.; Apajalahti, J.; Tysk, C.; Hettich, R. L.; Jansson, J. K. *ISME J.* **2009**, *3*, 179–189.

223. Rinke, C.; Schwientek, P.; Sczyrba, A.; Ivanova, N. N.; Anderson, I. J.; Cheng, J. F.; Darling, A.; Malfatti, S.; Swan, B. K.; Gies, E. A.; Dodsworth, J. A.; Hedlund, B. P.; Tsiamis, G.; Sievert, S. M.; Liu, W. T.; Eisen, J. A.; Hallam, S. J.; Kyrpides, N. C.; Stepanauskas, R.; Rubin, E. M.; Hugenholtz, P.; Woyke, T. *Nature* **2013**, *499*, 431–437.

224. Castelle, C. J.; Hug, L. A.; Wrighton, K. C.; Thomas, B. C.; Williams, K. H.; Wu, D.; Tringe, S. G.; Singer, S. W.; Eisen, J. A.; Banfield, J. F. *Nat. Commun.* **2013**, *4*, 2120.

225. Kluge, S.; Hoffmann, M.; Benndorf, D.; Rapp, E.; Reichl, U. *Proteomics* **2012**, *12*, 1893–1901.
226. Leary, D. H.; Hervey, W. J., 4th.; Deschamps, J. R.; Kusterbeck, A. W.; Vora, G. J. *Mol. Cell. Probes* **2013**, *27*, 193–199.
227. Wöhlbrand, L.; Trautwein, K.; Rabus, R. *Proteomics* **2013**, *13*, 2700–2730.
228. Peng, L.; Kapp, E. A.; Fenyö, D.; Kwon, M. S.; Jiang, P.; Wu, S.; Jiang, Y.; Aguilar, M. I.; Ahmed, N.; Baker, M. S.; Cai, Z.; Chen, Y. J.; Van Chi, P.; Chung, M. C.; He, F.; Len, A. C.; Liao, P. C.; Nakamura, K.; Ngai, S. M.; Paik, Y. K.; Pan, T. L.; Poon, T. C.; Salekdeh, G. H.; Simpson, R. J.; Sirdeshmukh, R.; Srisomsap, C.; Svasti, J.; Tyan, Y. C.; Dreyer, F. S.; McLauchlan, D.; Rawson, P.; Jordan, T. W. *Proteomics* **2010**, *10*, 4142–4148.
229. Kolmeder, C. A.; de Vos, W. M. *J. Proteomics* **2013**, pii: S1874–3919(13)00259–5.
230. Muth, T.; Benndorf, D.; Reichl, U.; Rapp, E.; Martens, L. *Mol. Biosyst.* **2013**, *9*, 578–585.
231. Seifert, J.; Taubert, M.; Jehmlich, N.; Schmidt, F.; Völker, U.; Vogt, C.; Richnow, H. H.; von Bergen, M. *Mass Spectrom. Rev.* **2012**, *31*, 683–697.
232. Siggins, A.; Gunnigle, E.; Abram, F. *FEMS Microbiol. Ecol.* **2012**, *80*, 265–280.
233. von Bergen, M.; Jehmlich, N.; Taubert, M.; Vogt, C.; Bastida, F.; Herbst, F. A.; Schmidt, F.; Richnow, H. H.; Seifert, J. *ISME J.* **2013**, *7*, 1877–1885.
234. Wang, D. Z.; Xie, Z. X.; Zhang, S. F. *J. Proteomics* **2013**, pii: S1874–3919(13)00465-X.
235. de Castro, A. P.; Sartori da Silva, M. R.; Quirino, B. F.; Kruger, R. H. *Curr. Protein Pept. Sci.* **2013**, *14*, 447–458.
236. Garcia-Schwarz, G.; Rogacs, A.; Bahga, S. S.; Santiago, J. G. *J. Vis. Exp.* **2012**, *61*, e389.
237. Phung, S. C.; Nai, Y. H.; Powell, S. M.; Macka, M.; Breadmore, M. C. *Electrophoresis* **2013**, *34*, 1657–1662.

Chapter 8

Proteomics Tools for Food Fingerprints: Addressing New Food Quality and Authenticity Challenges

Mónica Carrera*, Benito Cañas† and José M. Gallardo*

*Spanish National Research Council (CSIC), Marine Research Institute (IIM), Vigo, Pontevedra, Spain
†Complutense University of Madrid, Madrid, Spain

1 INTRODUCTION

Food "fingerprint" can be defined as a specific molecular marker diagnostic for a characteristic food/nutritional state or condition. These biological markers also referred as "biomarkers" are specific indicators that can be objectively measured and evaluated (1). They can be genes, proteins, or metabolites. Biomarker discovery and validation is still a current hot topic for nutrition and food science. Emerging "omics" technologies provide a good discovery-based approach for

Comprehensive Analytical Chemistry, Vol. 64. http://dx.doi.org/10.1016/B978-0-444-62650-9.00008-7

food biomarker development. In this sense, Foodomics has been recently defined as a new discipline that studies food and nutrition domains through the application of advanced -omics technologies to improve consumer's well-being, health, and confidence *(2,3)*. Several high-throughput technological approaches like genomic, transcriptomic, proteomic, metabolomic, and lipidomics are involved in this worldwide established field.

Proteomics, a discipline that up to few years ago was practically being used only in biomedical research, is currently being revealed as a powerful tool also in food science research, as it is helping to meet major challenges faced by food analysts and researches: (i) the development of simple, fast, and inexpensive methodologies for routine use; (ii) the analysis of complex or highly processed food matrices; (iii) multiplexing several targets in one analysis; and (iv) the quantification of analytes at trace levels with a high degree of selectivity.

In particular, in this chapter, the powerful and potential application of proteomics in the discovery and monitoring of food fingerprints applied to food quality and authentication is reviewed. In light of this, two consecutive phases of the proteomics pipeline, used in many research laboratories (discovery and target-driven stages), are described in detail. The potentiality of this workflow is emphasized using different topics concerning food quality and safety. Food authentication, allergen detection, microorganism contamination, and quality changes during the storage and processing are the main topics where proteomics methodologies are being successfully applied. Future directions and new potential perspectives are also provided.

2 PROTEOMICS

The concept of proteome was introduced by Marc Wilkins in 1994 at a conference in Siena, Italy *(4)*. The proteome, the total protein content of one particular biological system, is highly dynamic and constantly changes according to different stimuli. As a discipline, proteomics is defined as the large-scale analysis of proteins in a particular biological system at a particular moment in time *(5)*. Proteomics includes not only the study of the structure and function of proteins but also the analysis of modifications, the interactions between proteins, their intracellular location, and the quantification of their abundance. Proteomics analyses was initially based on protein separation by two-dimensional electrophoresis (2-DE) followed by image analysis which provided the first way of displaying thousands of proteins on a single gel *(6,7)*. This approach, the most frequently used in the early 1980s, was mainly descriptive and the results were lists of the differential presence of spots identified by their pI and M_r. Spot abundance was estimated by intensity and a quantitative comparison was possible between gels. However, nowadays, mass spectrometry (MS), mainly matrix-assisted laser desorption/ionization-time of flight (MALDI-TOF), and electrospray-ion trap (ESI-IT) MS, has been recognized as an indispensable tool for the majority of proteomics studies *(8)*. MS is the present method of choice to detect multiple features of a complex protein sample. Moreover, bioinformatics treatment of the data has increased the potential of proteomics tools, representing a powerful strategy for a high-throughput protein and peptide identification and quantification.

Recent advances in proteomics methodologies make them a promising strategy for food science studies and are used by research institutions, industries, agencies, and regulatory laboratories, combining efforts to acquire the needed knowledge on food composition, quality, and safety. Very useful reviews on proteomics applications in food science have been recently published *(9,10)*.

In view of this, our group has greatly contributed in the application of proteomics methodologies for the assessment of the quality and safety of seafood products *(11–14)*. As an example, Figure 1 summarizes the proteomics pipeline and tools that are currently being used at our laboratory for fish authentication *(12,14,15)*. Two consecutive phases, discovery phase and target-driven phase are described in Sections 2.1 and 2.2.

2.1 Discovery Phase

The goal of the discovery phase is to explore comprehensively a particular proteome, in order to identify potential protein/peptide biomarkers or fingerprints, commonly using the proteomics *bottom-up* approach: proteins of interest are separated, digested using proteases such as trypsin, and the resulting peptides analyzed by MS *(8)* (Figure 1, upper panel). 2-DE is usually the method of choice for the separation of proteins in complex biological samples *(16)*. Protein spots of interest are excised, digested with proteases, and the resulting peptides identified by MS or MS/MS following the already mentioned *bottom-up* approach. 2-DE remains the most widely applied separation method in proteome analysis due to the high protein resolution. In addition, this method is the most powerful option for non-completely sequenced organisms (i.e., fish species), in which the identification of proteins is based on comparing peptides from the proteins of interest with orthologous proteins from other species or by means of *de novo* MS sequencing strategies. However, gel-based methods still have some limitations, such as the separation of hydrophobic and poorly soluble proteins and the limited sensitivity of the available detection methods.

In the gel-free approaches, also referred as "*shotgun proteomics*," proteins in a mixture are digested and the peptides released separated by reverse-phase HPLC, either alone or combined with strong cation exchange chromatography *(17)*. Eluted peptides are analyzed by MS/MS. This approach presents high sensitivity and reproducibility. There are several disadvantages such as the loss of information on protein isoforms and the difficulty on the identification and characterization of proteins from species with poorly sequenced genomes.

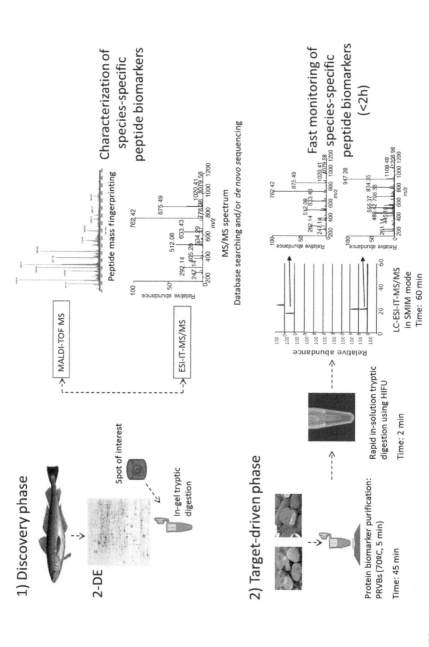

FIGURE 1 Proteomics pipeline used for the identification, characterization, and detection of species-specific peptide biomarkers for fish authentication purposes. (For color version of this figure, the reader is referred to the online version of this chapter.)

Nevertheless, we believe as do other authors that both the gel-based and gel-free methods will continue to prove useful in the long term.

Nowadays, MS is the technique of choice for the characterization and identification of proteins. The general approach consists of the comparison of the MS and/or MS/MS experimental data with calculated mass values obtained from the entries in a sequence database. A search engine, such as Sequest *(18)* or Mascot *(19)*, performs the comparison making a probabilistic calculation to find out if the protein assignation is due to chance. Using peptide mass fingerprinting (PMF) *(20)* approach, the MS experimental data are the masses of the peptides obtained from the enzymatic digestion of the protein. Another approach is that peptide fragmentation fingerprint uses MS/MS fragmentation spectra from one or more peptides instead of the masses of the complete set of peptides from the protein, allowing the sequencing of the peptide and thus yielding the unambiguous identification of the protein *(18)*. For all this approaches, the availability of the corresponding protein sequence in the database is mandatory. If the unknown protein is not present in the database, but highly homologous ones are, the best match is aimed to be the entry with the closest homology, usually proteins from related species. If sequence similarity with database proteins is too low, peptides must be sequenced *de novo*, interpreting MS/MS spectrum either manually or by computer-assisted tools *(21)*.

Finally, the goal of this first phase is to analyze and to compare the resulting proteins and peptides by alignment search tools, such as BLAST (http:// blast.ncbi.nlm.nih.gov/Blast.cgi?CMD=Web&PAGE_TYPE=BlastHome), in order to identify and characterize specific peptide fingerprints, which will be used in the monitoring approach of the next phase of the pipeline.

2.2   Target-Driven Phase

Targeted or hypothesis-driven proteomics approaches are being increasingly used to monitor candidate peptide fingerprints. When these selective and sensitive methods are used, the MS analyzer is centered on analyzing only the compound of interest by selected-reaction monitoring (SRM) *(22,23)*. Monitoring transitions (suitable pairs of precursor and fragment ion *m/z*) constitute a common assay to identify biomarkers. This setup provides high analytical reproducibility, a good signal-to-noise ratio, and an increased dynamic range *(22)*. While SRM performed on a triple-quadrupole is the most sensitive scanning mode (low attomolar) with a broad dynamic range (up to 5 orders of magnitude), optimization for a definite SRM assay is time-consuming. More importantly, using these scanning procedures, complete MS/MS spectra are not registered. The MS/MS spectrum of a molecule is of paramount importance to confirm the structure of the compound detected. To solve this problem, new routines, such as SRM-triggered MS/MS using hybrid Q-IT mass spectrometers, have been explored *(24)*. In such assays, when a significant signal for a specific SRM transition is detected, the instrument switches the third Q automatically to the IT mode, collecting the full MS/MS spectrum. Selected MS/MS ion monitoring (SMIM) in an IT is another scanning mode that allows for a sensitive monitoring of specific molecules, producing complete structural information *(25)*. The high scanning speed attainable in the IT allows for the production of MS/MS spectra in a fraction of a second, registering the information given by the complete spectra. High-confidence MS/MS spectra are recorded due to the possibility of an averaging of the signal during acquisition. In this way, peptide biomarkers identified in the discovery phase are monitored. The utility of this operating mode to assess the quality and safety of food products will be demonstrated in the Section 3.

3 PROTEOMICS FOR ADDRESSING NEW FOOD QUALITY AND AUTHENTICITY CHALLENGES

Currently, consumers demand high-quality, nutritious, and safe products. Thus, food quality and safety are a current founding principle of the international health organizations. In this sense, the current powerful -omics methodologies are having a great impact in the current food science. The applications of the proteomics methodologies for the assessment of quality and authenticity of food products can be divided in four major areas (Figure 2): (i) food authentication, (ii) allergen detection, (iii) monitoring of food spoilage or pathogenic microbes, and (iv) monitoring of changes of food quality during the storage and processing.

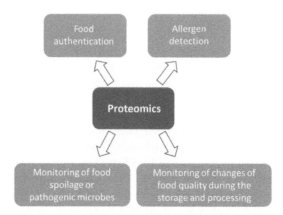

FIGURE 2   Major areas of proteomics applications for the assessment of quality and safety of food products. (For color version of this figure, the reader is referred to the online version of this chapter.)

3.1   Food Authentication

Food authentication is one of the major areas involved in food quality and safety. Several regulations have been implemented to assure correct information and to avoid species substitutions (26). Food species identification has traditionally relied on morphological/anatomical analysis. However, this is a difficult task in the case of closely related species and especially for those products that have been subjected to processing practices. Therefore, there is a strong need for fast and reliable molecular identification methods that provide authorities and food industries the tools needed to comply with labeling and traceability requirements, thus ensuring product quality and protection of the consumer.

The identification of food species traditionally relied on morphological analysis. However, morphological features are particularly difficult to use for species differentiation among certain species, such as seafood species, because of their phenotypic similarities. Moreover, in the case of processed food products, differentiation of the constituents can be difficult, since external features are often removed. Therefore, there is a great need for fast and reliable molecular identification methods that may provide the control authorities and the food industries with the tools to comply with labeling and traceability at both species and origin level, thus assuring product quality and protecting consumer's value.

Over the past two decades, several electrophoretic, immunological, and DNA techniques have been developed for authentication purposes (27). Limitations of these classical protein-based procedures include the lack of stability of some proteins during food processing and the labor and time required for these procedures. Moreover, DNA-based procedures are not exempted from some important limitations. During the processing of food products, disruption of cellular integrity can occur, causing the release of hydrolytic enzymes. A combination of these enzymes with heat treatment and an acidic environment can negatively affect DNA integrity, reducing the length of fragments to be amplified and consequently increasing the chances of having nonspecific identifications.

Due to recent advances on MS, proteomics tools have recently been proposed as faster, sensitive, and high-throughput approaches for the assessment of the authenticity and traceability of species in food products (28,29). MS is being used for both the discovery of species-specific peptide fingerprints in reference samples and for the subsequent detection of those diagnostic peptides in real samples (29,30). Proteomics tools take advantage of the high-throughput capacity of MS to achieve a fast, robust, and sensitive protein and peptide characterization, detection, and quantification. Proteomics-based methodologies may be automated in order to produce fast and reproducible results and to allow a high-throughput analysis of foodstuffs; they can be applied to species that are poorly characterized at the genomic databases, avoiding the time-consuming steps of DNA amplification and sequencing, and overall, the identification and characterization of species-specific diagnostic peptides is the first step toward designing fast and cheap detection analysis, such as antibody-based assays or MRM MS detection methods. A compilation of proteomics-based studies applied to date for food species authentication is resumed in Table 1.

Figure 1 illustrates the proteomics pipeline (discovery phase and target-driven phase) used in our laboratory for seafood authentication (14,48). Regarding the discovery phase, species-specific peptide fingerprints from parvalbumin (PRVB) (15) and arginine kinase (AK) proteins (13) have been demonstrated to be good markers for the identification of fish and shellfish species, respectively. The characterization of these peptide biomarkers is the first step toward their subsequent use as sensitive and diagnostic targets in the next step of the pipeline. In the second stage, a targeted proteomics approach is used to monitor the species-specific peptide biomarkers previously characterized (Figure 1, lower panel). As was described previously, for species without sequenced genomes, the selection of species-specific peptides requires a preceding exhaustive *de novo* MS sequencing analysis (21). Once the species-specific peptides are collected, the MS analyzer is centered on analyzing one or several peptides by SRM, MRM, or SMIM MS modes (23,25). Monitoring transitions (suitable pairs of precursor and fragment ion *m/z*) is a common assay to identify peptide biomarkers. The advantages of this targeted proteomics approach are its high specificity and sensitivity, which permit the quantification of proteins in complex samples at concentrations below the ng/mL range (23). For instance, in one study, species-specific peptides for beef, pork, chicken, and turkey were used to detect less than 0.5% w/w chicken contamination of pork meat, even after cooking. This approach is comparable with other DNA methods (49).

A fast strategy for monitoring species-specific peptides has been recently described by our group (Figure 1). This method combines fast sample preparation using high-intensity focused ultrasound (HIFU) trypsin digestion and the peptide detection ability of MS in SMIM scanning mode (14,25). Because PRVBs are thermostable proteins, the workflow also identifies fish species even in processed and precooked products. This method has been validated using different real commercial samples. This workflow constitutes the fastest method for peptide biomarker monitoring (less than 2 h) and its application to food quality control provides authorities with a rapid and effective method for food authentication and traceability in any foodstuff, which guarantees quality and safety to consumers (14).

3.2   Allergen Detection

Food is one of the most frequent causes of immunoglobulin E-mediated allergy. Its prevalence in regular population has been estimated to be around 0.2–0.6%

TABLE 1 Summary of Proteomics-Based Methods Applied in the Authentication of Species in Food

Main Technique	Species/Food Products	Target Fingerprint	Reference
2-DE	8 Gadoid fishes	Parvalbumins	(11)
	9 Flat fishes	Sarcoplasmic proteins	(31)
	5 Fish species	Myosin light chains	(32)
	3 *Thunnus* species	Triose phosphate isomerase	(33)
IEF+MS	14 Shrimp and prawn species	SCPs	(34)
ESI-MS	Pig, beef, sheep, and horse meats	Hemoglobin/myoglobin	(35)
HPLC-ESI-MS	Cow and goat milk	β-Lactoglobulin	(36)
CE-MS	Cow, goat, and sheep milk	β-Lactoglobulin	(37)
MALDI-TOF protein fingerprinting	Cow, ewe, and buffalo milk	α-Lactalbumin and β-lactoglobulin	(38)
	Mozzarella cheese	α-Lactalbumin and β-lactoglobulin	(39)
	25 Commercial fish species	Sarcoplasmic proteins	(40)
	Honey	–	(41)
	Cow, sheep, goat, and buffalo milk	Caseins	(42)
MALDI-TOF PMF	5 Hake species	Parvalbumins and NDKA	(11)
	3 European mussels	TM	(30)
	10 Merlucciidae hake and grenadier species	Parvalbumins	(12)
	11 Merlucciidae hake and grenadier species	NDKB	(29)
	13 Shrimp and prawn species	AK	(13)
	Pandalus borealis (Northern shrimp)	AK	(43)

Continued

TABLE 1 Summary of Proteomics-Based Methods Applied in the Authentication of Species in Food—Cont'd

Main Technique	Species/Food Products	Target Fingerprint	Reference
MS/MS	5 Hake species	Parvalbumins and NDKA	(11)
	Soybean in meat products	Glycinin, β-conglycinin	(44)
	11 Merlucciidae commercial hakes and grenadiers	NDKB	(29)
	Bovine gelling agent in meat	Fibrinopeptides	(45)
	7 shrimp and prawn species	AK	(13)
	Soy and pea in skimmed-milk powder	Glycinin, β-conglycin, legumin, vicilin	(46)
	Sheep milk in goat and cow cheeses	Casein	(47)
	Pandalus borealis (Northern shrimp)	AK	(43)
HPLC-ESI-MS +XIC	Cow, sheep, goat, and buffalo milk	Caseins	(42)
SIM+MS/MS	3 European mussels	TM	(30)
MRM	11 Merlucciidae hake and grenadier species	NDKB	(29)
	7 Shrimp and prawn species	AK peptides	(48)
	11 Merlucciidae hake and grenadier species	Parvalbumins	(14)
SIM+AQUA	Chicken in meat preparations	Myosin light chain 3	(49)

IEF, isoelectric focusing; MS, mass spectrometry; 2-DE, two-dimensional electrophoresis; ESI, electrospray ionization; HPLC, high-performance liquid chromatography; CE, capillary electrophoresis; MALDI-TOF, matrix-assisted laser desorption/ionization-time of flight; PMF, peptide mass fingerprinting; MS/MS, tandem mass spectrometry; XIC, extracted ion current; SIM, selected ion monitoring; MRM, multiple reaction monitoring; AQUA, absolute quantification; SCPs, sarcoplasmic calcium-binding proteins; AK, arginine kinase; NDKA, nucleoside diphosphate kinase A; NDKB, nucleoside diphosphate kinase B; TM, tropomyosin.

(50). Symptoms appear within 60 min of exposure and include acute and generalized urticaria, nausea, vomiting, abdominal cramps, diarrhea, wheezing, and asthma *(51)*. In the most severe cases, anaphylaxis shock can potentially be life threatening *(52)*. To guarantee the consumers safety, a number of regulations in terms of food allergy have been implemented (Directive 2007/68/EC). In the European Union, these regulations compel the producers to label the 14 food allergens when these have been intentionally introduced in the foodstuffs. However, some products on the market could contain traces of allergens due to cross-contaminations during the food manufacturing processes (Table 2). As consequence, accurate, sensitive, and fast detection methods that

TABLE 2 Summary of the Main Proteomics-Based Methods Applied for Allergen Detection in Food Products

Main Technique	Allergen	Food	Reference
2-DE + WB + MS	8 Major milk proteins	Cow's milk	*(53)*
	20 Major proteins	Wheat	*(54–56)*
	11S Albumin	Hazelnut	*(57)*
	5 Major proteins	Maize	*(58)*
	Ara h 1–4	Peanut	*(59)*
	3 Subunits β-conglycinin	Soybean	*(60)*
	Arginine kinase, tropomyosin	Shrimp	*(61)*
	41 Parvalbumins	Fish	*(15)*
Shotgun proteomics (LC/MS-based strategy)	Casein milk proteins	Cookies	*(62)*
	Caseinates	Wine	*(63)*
	Ara h 1–4	Peanut	*(64,65)*
	Ovalbumin	Hen's egg white	*(66)*
2DE DIGE	Fra a 1	Strawberry	*(67)*
Top-down proteomics	Cow's milk allergens	Mixed fruit-juice	*(68)*
LC-SRM	Peanut Ara h 2 and Ara h 3/4	Several food products	*(69)*
AQUA MRM	8 Allergens	Soybean cultivars	*(70)*
LC-SMIM	165 Parvalbumin	Fish products	*(71)*

2-DE, two-dimensional gel electrophoresis; WB, Western blot; MS, mass spectrometry; LC, liquid chromatography; 2DE DIGE, two-dimensional difference gel electrophoresis, SRM, selected-reaction monitoring; MRM, multiple reaction monitoring; AQUA, absolute protein quantitation; SMIM, selected MS/MS ion monitoring.

permit the direct recognition of allergens in food samples are highly recommendable. A good compilation of proteomics-based studies applied to date for food allergen identification and detection is collected in Picariello *et al. (72,73)*.

Regarding seafood products, parvalbumins beta (β-PRVBs) are considered as the major fish allergens *(74–77)*. These are present in high amounts in the sarcoplasmic fraction of white muscle of fishes. Fish-sensitive patients are commonly allergenic to numerous fish species *(78)*. The clinical cross-reactivity is strongest for taxonomically closely related fish species *(79)*. Recently, a new strategy published by our group allows the rapid direct detection of fish β-PRVBs in any foodstuff *(71)*. The strategy is based on the target-driven phase described previously and basically consists of the use of a fast purification step of β-PRVBs by heat treatment, acceleration of in-solution protein digestion by HIFU, and the monitoring of 19 common β-PRVBs peptide biomarkers by SMIM using LIT mass spectrometer. The results of the sequence alignment of 163 teleostei β-PRVBs available in the UniProtKB database identified the most conserved region, corresponding to the sequence between the residues 46 and 77. This region contains one helix-loop-helix (EF-hand) motif, functional for chelating Ca^{2+} *(80)*. In order to validate this new strategy in commercial sea-foodstuffs, six commercial seafood products were analyzed including processed and even battered pre-cooked products. The methodology allows the direct detection of presence of β-PRVBs in any food product, including processed and precooked, in less than 2 h. To our knowledge, this is the fastest method to achieve the direct detection of this allergen.

3.3 Monitoring of Food Spoilage or Pathogenic Microbes

Food spoilage is a process caused by different biochemical changes due to microbial activities. These alterations depend on inherent and noninherent microflora and on growth conditions such as temperature, pH, and a_w. Contamination incidents during food processing are responsible for significant economic losses for the food industry and serious foodborne diseases. The challenges in food safety have increased due to the development of new products, globalization of the markets, and new processes of preservation of foods. Food intoxications result from the intake of toxins produced in the food by bacteria such as *Staphylococcus aureus* and *Clostridium botulinum*. The poisoning by bacteria can be caused by *E. coli*, *Salmonella spp.*, *Listeria monocytogenes*, *S. aureus*, *B. cereus*, *Shigella spp.*, *Vibrio cholerae*, and *V. parahaemolyticus*, among others. More than 250 known pathogens, mostly microbes and their toxins are known to cause foodborne illness. Although the identification and classification of microorganisms are based on morphological, biochemical, and DNA approaches, proteomic methodologies are being introduced to assist in the identification of foodborne pathogens and microorganisms responsible for food spoilage. For this reason, novel technologies for an accurate and rapid identification and classification of microorganisms, such as the new MS-based proteomics tools, are nowadays complementing the classical and genetic-based identification techniques.

Proteomics technologies have been used in the routine bacterial identification, mainly in clinical microbiology, biodefense, and environmental research *(81)*. However, few works have been done in the field of microbial food MS for the identification of foodborne pathogens and microorganisms responsible for the food spoilage. MALDI-TOF MS of intact bacterial cells was used for the detection and identification of 24 different foodborne pathogens and food spoilage bacteria, including genera such as *Escherichia*, *Yersinia*, *Proteus*, *Morganella*, *Salmonella*, *Staphylococcus*, *Micrococcus*, *Lactococcus*, *Pseudomonas*, *Leuconostoc*, and *Listeria (82)*. These authors were able to distinguish between the pathogenic *E. coli* O157:H7 and the nonpathogenic *E. coli* ATCC 25922. Ochoa and Harrington reported the identification of *E. coli* O157:H7 in ground beef *(83)*. Dieckmann *et al.* identified 126 strains of *Salmonella spp.* in chicken, turkey, swine, and cattle *(84)*. Barbuddhe *et al.* used MALDI to identify 146 strains of *Listeria spp.* in meat, poultry, dairy, and vegetables *(85)*, and Angelakis *et al.* used this methodology for the bacterial identification at the species level in probiotic foods and yogurts *(86)*. MALDI-TOF MS of low-molecular weight proteins extracted from intact bacterial cells was also successfully applied to the safety assessment of fresh and processed seafood (Figure 3). Main species responsible of seafood spoilage and pathogenic Gram-negative bacteria, including *Aeromonas hydrophila*, *Acinetobacter baumannii*, *Pseudomonas spp.*, and *Enterobacter spp.*, were identified *(87,88)*. In a subsequent work, a research on the identification of Gram-positive bacteria in seafood, including *Bacillus spp.*, *Listeria spp.*, *Clostridium spp.*, and *Staphylococcus spp.*, was reported *(89)*. MALDI-TOF MS has been used for the identification of biogenic amine-producing bacteria involved in food poisoning *(90,91)*. *Streptococcus parauberis* is an agent of mastitis in cows and thus a source of economic loss for the dairy industry. Barreiro *et al.* reported the identification of subclinical cow mastitis pathogens in milk using MALDI-TOF MS *(92)*. Recently, the isolation and MALDI-TOF MS identification of *S. parauberis* in vacuum-packed seafood products have been reported *(93)*. Different commercial databases have been created for the bacterial identification by MALDI-TOF MS, such as Spectral Archiving and Microbial Identification System (AnagnosTec GmbH, Zossen, Germany), the Microbelynex bacterial identification system (Waters Corporation, Manchester, UK), and MALDI Biotyper (Bruker Daltonics Inc., Billerica, USA). These private databases are mainly targeted at human pathogens causing infectious diseases, although some bacterial species that play an important role in food spoilage and safety are also included.

A public spectra database containing 24 foodborne bacterial species has been constructed by Mazzeo *et al. (82)* and is freely available on the Web

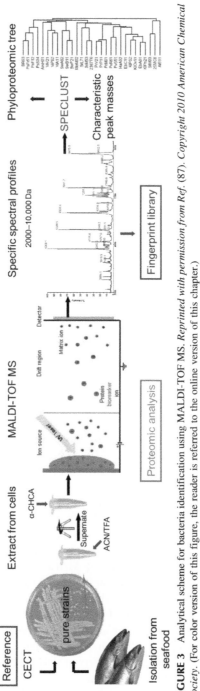

FIGURE 3 Analytical scheme for bacteria identification using MALDI-TOF MS. *Reprinted with permission from Ref. (87). Copyright 2010 American Chemical Society.* (For color version of this figure, the reader is referred to the online version of this chapter.)

(http:/bioinformatica.isa.cnr.it/Descr_Bact_Dbase.htm). Recently, the new public database Spectrabank (http:/www.spectrabank.org) has been created by the Santiago de Compostela University. This library contains the mass spectral fingerprints of the main spoilage-related and pathogenic bacteria species from seafood and includes 120 species of relevant interest in the food sector.

3.4 Monitoring of Changes of Food Quality During the Storage and Processing

Food quality is determined to a large extent not only by the biological and genetic variability of the raw constituents but also by the treatment during production, processing, and storage. During harvest/slaughter and processing, food proteins are subjected to both natural and processing changes that influence key food properties including nutritional value, shelf-life, and quality features such as tenderness, taste, juiciness, and color. Among the different processing treatments that are used in the food manufacturing industry, thermal processing (refrigeration, freezing, cooking, blanching, spray-drying, pasteurization, and sterilization) is one of the most widely used operations. These procedures are employed to improve the safety and organoleptic and nutritional characteristics of food and to extend the shelf-life of foodstuffs. However, due to the variability of products and processes used, chemical reactions can take place among the main components (proteins, lipids, and carbohydrates), and depending on the severity and time of the treatment, detrimental effects on the quality of foodstuffs can occur. A comprehensive overview of proteomics-based studies applied to date for food processing is compiled in Gallardo *et al. (94)*. A brief review concerning the effects of thermal processes on changes and modifications of food proteins using different proteomics methodologies is presented in this chapter.

Oxidation modifications such as carbonylation, thiol oxidation, and aromatic hydroxylation, and Maillard glycation (the reaction of sugars with amino acid side chains) are the protein modifications most frequently reported in foodstuffs that have been subjected to thermal processing. However, condensations and eliminations of side chains or peptide backbone breakdown have also been described *(95)*.

Carbonylation occurs by the oxidation of some amino acid side chains into ketone or aldehyde derivatives by reactions with compounds of lipid oxidation or by glycoxidation with reducing sugars. These protein–carbonyl compounds are markers of protein oxidation, and recently, several carbonylated proteins and protein oxidation sites in milk *(96)*, meat *(97)*, and fishes *(98)* have been identified using a classical *bottom-up* proteomics approach based on 2-DE and MS/MS. Specific labeling of protein carbonyls using fluorescein-5-thiosemicarbazide has been developed and combined with 2-DE and

MALDI-TOF/TOF MS to successfully study the carbonylation of sarcoplasmic and myofibrillar proteins from fish subjected to metal-catalyzed oxidation *(98)*.

Maillard glycation is another most investigated protein modification and occurs mainly during heating and storage of milk and dairy products. Maillard glycation is a nonenzymatic reaction between the amino groups of proteins and reduced sugars (lactose) that produces glycoconjugate condensation products such as lactosylated proteins *(99)*. Diverse proteomics strategies based on bottom-up methods have been used to study Maillard reactions in milk and dairy products *(100)*. These strategies use native size-exclusion chromatography coupled online with ESI-MS/MS for the rapid identification and characterization of lactosylated proteins in thermally processed milk products *(100)*. Recently, the results of new investigations suggest that the ETD MS fragmentation mode is an excellent and powerful technique for the analysis of nonenzymatic glycated peptides. The combination of the selective enrichment of lactosylated peptides, ETD MS, and peptide ligand libraries has been applied for the identification of lactosylated proteins in thermally processed milk *(101)*. Using this methodology, 271 nonredundant modification sites on 33 milk proteins were identified.

Bioinformatics tools play an important role in the identification and prediction of protein modification sites. Within bioinformatics software, there are several programs designed to optimize the detection of protein modifications from MS data, including PTMProphet, ModifiComb, and Peptoscope *(102)*. We expect that the availability of new repositories and databases, such as RedoxDB *(103)*, the use of new specific labeling and enrichment methods, the use of the powerful new MS fragmentation modes, such as ETD and HCD, and the mapping and discovery of protein modifications by top-down proteomics will expand the applications for proteomics in the identification and characterization of protein modifications in further food-related projects.

4   INNOVATIVE PROTEIN-BASED TECHNOLOGIES

Concerning to this, the new developments based on microfluids devices, lab-on-chips, and protein arrays are very promising technologies in the context of modern food science. Analytical microsystems attempts to develop integrated analytical systems at a micro scale to perform in one device all the analytical steps (sample preparation, separation, and detection) needed to carry out a complete analysis *(104)*. Microchips have been used as programmable devices able to measure DNA, protein, and small molecules. Thus, in the context of food science, these new innovative technologies can be implemented for routine control, diagnosis, and monitoring of food products *(105)*.

With the advantage of small size, low cost, ease of integration and automation, microfluidics technology is expected to revolutionize the food industry with portable, inexpensive, and high-performance protein-based devices.

Faster and finer approaches for the analysis of food proteins are necessary for the assurance of food quality and safety *(106)*. Generally, the most recent applications of microchip electrophoresis for the analysis of proteins or peptide biomarkers in foods include the characterization of protein extracts *(107)*, the detection of high-quality value products adulterated with products of inferior quality *(108)*, and the evaluation of the quality due to food storage or different technological process *(109)*.

The lab-on-chip method used by Butikofer *et al.* *(110)* allowed the determination of the percentage of α-lactalbumin + β-lactoglobulin of total protein, in raw, pasteurized, and UHT-treated milk, in order to detect alterations of the natural proportion of the two whey proteins resulting from new or incorrect technological processes. Lab-on-chip electrophoresis provided important information about meat species identity in raw and processed foods *(111)*. In food microbiology, the method for separating cell-wall proteins was applied, for example, to several isolates from olive phylloplane and brine. The high sensitivity of this technique enables discrimination of isolates and the identification of closely related strains within the same species *(112)*. Lab-on-chip technology has also been used to detect pathogens such as *E. coli (113)* and other foodborne pathogens *(114)*. Thus, nowadays different commercially available chip-based pathogen sensing systems are available *(115)*. Herwig *et al.* *(116)* combined immunoprecipitation with automated SDS-microchip capillary gel electrophoresis for the sensitive and selective quantitative detection of *E. coli* in food products. Moreover, a multiplex biochip-based immunoassay has been exploited to investigate the allergen profile of different apple cultivars *(117)*.

5 CONCLUDING REMARKS AND FUTURE DIRECTIONS

Proteomics is contributing enormously to the development of the new modern food science. In this sense, proteomics technologies are helping to address the major challenges in food authentication thanks to the development of cheap, fast, and easy-to-use methodologies for routine use. MS/MS-based proteomics methodologies such as the detection and quantification of species-specific peptides by MRM constitute a great promise as high-throughput and sensitive multiplexed species detection, identification, and quantification in complex matrices are possible using these techniques. In addition, proteomics is contributing to detect allergen proteins and to understand both, postmortem and technological-induced changes, and the relationship of these changes with quality traits. Efforts have been made in the discovery of protein markers for such traits. Thus, the use of fast sample preparation methods coupled to sensitive and reliable MS, for both the discovery and the monitoring of food quality biomarkers, will improve the quality control and assessment in food industry and will greatly help in the optimization and development of food technological processes. Concerning to this, the new developments based on microfluids portable devices, lab-on-chips, and protein arrays are very promising technologies in the context of modern food science.

Proteomics routines, for both, discovery and monitoring, can be implemented for routine control, diagnosis, and monitoring of food products. We anticipate that, within the next decade, these new platforms will be essential components in most food control laboratories and will provide diagnostic information driving decision making by authorities.

On the other hand, the recent advent of -omic high-throughput analytical platforms, coupled with the computational data management using bioinformatics and mathematical models has opened up new vistas in our understanding of complex biological systems inside the field of systems biology. The complex interplay that occurs between the individual in terms of genetics, physiology, health, diet, and environment requires tools to carry out comparative genetic, genomic, proteomic, and metabolomic analyses. In addition, tools for data integration, data mining, and interpretation are also needed. Such global vision offers enormous potential in the elucidation of diseases as well as defining key pathways and networks involved in optimal human health and nutrition. The tools and technologies now available in systems biology offer exciting opportunities to develop the emerging area of individual nutritional profiling.

REFERENCES

1. Mayeux, R. *Am. Soc. Exp. Neurother.* **2004**, *1*, 182–188.
2. Cifuentes, A. *J. Chromatogr. A* **2009**, *1216*, 7109–7110.
3. Herrero, M.; Simó, C.; García-Cañas, V.; Ibáñez, E.; Cifuentes, A. *Mass Spectrom. Rev.* **2012**, *31*, 49–69.
4. First Siena conference. *2D Electrophoresis: From Protein Maps to Genomes*, 5–7 September 1994.
5. Pandey, A.; Mann, M. *Nature* **2000**, *405*, 837–846.
6. Klose, J. *Humangenetik* **1975**, *26*, 231–243.
7. O'Farrell, P. H. *J. Biol. Chem.* **1975**, *250*, 4007–4021.
8. Aebersold, R.; Mann, M. *Nature* **2003**, *422*, 198–207.
9. D'Alessandro, A.; Zolla, L. *J. Proteome Res.* **2012**, *11*, 26–36.
10. Rodrigues, P. M.; Silva, T. S.; Dias, J.; Jessen, F. *J. Proteomics* **2012**, *75*, 4325–4345.
11. Piñeiro, C.; Vázquez, J.; Marina, A. I.; Barros-Velázquez, J.; Gallardo, J. M. *Electrophoresis* **2001**, *22*, 1545–1552.
12. Carrera, M.; Cañas, B.; Piñeiro, C.; Vázquez, J.; Gallardo, J. M. *Proteomics* **2006**, *6*, 5278–5287.
13. Ortea, I.; Cañas, B.; Gallardo, J. M. *J. Proteome Res.* **2009**, *8*, 5356–5362.
14. Carrera, M.; Cañas, B.; López-Ferrer, D.; Piñeiro, C.; Vázquez, J.; Gallardo, J. M. *Anal. Chem.* **2011**, *83*, 5688–5695.
15. Carrera, M.; Cañas, B.; Vázquez, J.; Gallardo, J. M. *J. Proteome Res.* **2010**, *9*, 4393–4406.
16. Miller, I.; Crawford, J.; Gianazza, E. *Proteomics* **2006**, *6*, 5385–5408.
17. Motoyama, A.; Yates, J. R., III. *Anal. Chem.* **2008**, *80*, 7187–7193.

18. Eng, J. K.; McCormack, A. L.; Yates, J. R. I. I. I. *J. Am. Soc. Mass Spectrom.* **1994**, *5*, 976–989.

19. Perkins, D. N.; Pappin, D. J. C.; Creasy, D. M.; Cottrell, J. S. *Electrophoresis* **1999**, *20*, 3551–3567.

20. Pappin, D. J. C.; Hojrup, P.; Bleasby, A. *Curr. Biol.* **1993**, *3*, 327–332.

21. Shevchenko, A.; Wilm, M.; Mann, M. *J. Protein Chem.* **1997**, *16*, 481–490.

22. Lange, V.; Picotti, P.; Domon, B.; Aebersold, R. *Mol. Syst. Biol.* **2008**, *4*, 1–14.

23. Gallien, S.; Duriez, E.; Domon, B. *J. Mass Spectrom.* **2011**, *46*, 298–312.

24. Unwin, R. D.; Griffiths, J. R.; Whetton, A. D. *Nat. Protoc.* **2009**, *4*, 870–877.

25. Jorge, I.; Casas, E.; Villar, M.; Ortega-Pérez, I.; López-Ferrer, D.; Martínez-Ruíz, A.; Carrera, M.; Marina, A.; Martínez, P.; Serrano, H.; Cañas, B.; Were, F.; Gallardo, J. M.; Lamas, S.; Redondo, J. M.; García-Dorado, D.; Vázquez, J. *J. Mass Spectrom.* **2007**, *42*, 1391–1403.

26. Council Regulation (EC) No. 178/2002. Official Journal of European Communities 31, pp 1, 2002.

27. Mafra, I.; Ferreira, I.; Oliveira, M. B. *Eur. Food Res. Technol.* **2008**, *227*, 649–665.

28. Piñeiro, C.; Barros-Velázquez, J.; Vázquez, J.; Figueras, A.; Gallardo, J. M. *J. Proteome Res.* **2003**, *2*, 127–135.

29. Carrera, M.; Cañas, B.; Piñeiro, C.; Vázquez, J.; Gallardo, J. M. *J. Proteome Res.* **2007**, *6*, 3070–3080.

30. López, J. L.; Marina, A.; Álvarez, G.; Vázquez, J. *Proteomics* **2002**, *2*, 1658–1665.

31. Piñeiro, C.; Barros-Velázquez, J.; Sotelo, C. G.; Pérez-Martín, R. I.; Gallardo, J. M. *J. Agric. Food Chem.* **1998**, *46*, 3991–3997.

32. Piñeiro, C.; Barros-Velázquez, J.; Sotelo, C. G.; Gallardo, J. M. *Z. Lebensm. Unters. Forsch. A* **1999**, *208*, 342–348.

33. Martinez, I.; Jakobsen Friis, T.; Careche, M. *J. Sci. Food Agric.* **2001**, *81*, 1199–1208.

34. Pepe, T.; Ceruso, M.; Carpentieri, A.; Ventrone, I.; Amoresano, A.; et al. *Vet. Res. Commun.* **2010**, *34*, S153–S155.

35. Ortea, I.; Rodríguez, A.; Tabilo-Munizaga, G.; Pérez-Won, M.; Aubourg, S. P. *Eur. Food Res. Technol.* **2010**, *230*, 925–934.

36. Taylor, A. J.; Linforth, R.; Weir, O.; Hutton, T.; Green, B. *Meat Sci.* **1993**, *33*, 75–83.

37. Chen, R. K.; Chang, L. W.; Chung, Y. Y.; Lee, M. H.; Ling, Y. C. *Rapid Commun. Mass Spectrom.* **2004**, *18*, 1167–1171.

38. Müller, L.; Barták, P.; Bednár, P.; Frysová, I.; Sevcik, J.; Lemr, K. *Electrophoresis* **2008**, *29*, 2088–2093.

39. Cozzolino, R.; Passalacqua, S.; Salemi, S.; Malvagna, P.; Spina, E.; Garozzo, D. *J. Mass Spectrom.* **2001**, *36*, 1031–1037.

40. Cozzolino, R.; Passalacqua, S.; Salemi, S.; Garozzo, D. *J. Mass Spectrom.* **2002**, *37*, 985–991.

41. Mazzeo, M. F.; de Giulio, B.; Guerriero, G.; Ciarcia, G.; Malorni, A.; Russo, G. L.; et al. *J. Agric. Food Chem.* **2008**, *56*, 11071–11076.

42. Wang, J.; Kliks, M. M.; Qu, W.; Jun, S.; Shi, G.; Li, Q. X. *J. Agric. Food Chem.* **2009**, *57*, 10081–10088.

43. Cuollo, M.; Caira, S.; Fierro, O.; Pinto, G.; Picariello, G.; Addeo, F. *Rapid Commun. Mass Spectrom.* **2010**, *24*, 1687–1696.

44. Pascoal, A.; Ortea, I.; Gallardo, J. M.; Cañas, B.; Barros-Velázquez, J.; Calo-Mata, P. *Anal. Biochem.* **2012**, *421*, 56–67.

45. Leitner, A.; Castro-Rubio, F.; Marina, M. L.; Lindner, W. *J. Proteome Res.* **2006**, *5*, 2424–2430.

46. Grundy, H. H.; Reece, P.; Sykes, M. D.; Clough, J. A.; Audsley, N.; Stones, R. *Rapid Commun. Mass Spectrom.* **2007**, *21*, 2919–2925.
47. Cordawener, J. H. G.; Luykx, D. M. A. M.; Frankhuizen, R.; Bremer, M. G. E. G.; Hooijerink, H.; America, A. H. P. *J. Sep. Sci.* **2009**, *32*, 1216–1223.
48. Ortea, I.; Cañas, B.; Gallardo, J. M. *J. Chromatogr. A* **2011**, *1218*, 4445–4451.
49. Sentandreu, M. A.; Fraser, P. D.; Halket, J.; Patel, R.; Bramley, P. M. *J. Proteome Res.* **2010**, *9*, 3374–3383.
50. Sicherer, S. H. *J. Allergy Clin. Immunol.* **2011**, *127*, 594–602.
51. Jeebhay, M. F.; Robins, T. G.; Lehrer, S. B.; Lopata, A. L. *Occup. Environ. Med.* **2001**, *58*, 553–562.
52. Sampson, H. A. *Pediatrics* **2003**, *111*, 1601–1608.
53. Natale, M.; Bisson, C.; Monti, G.; Peltran, A.; Garoffo, L. P.; Valentini, S.; Fabris, C.; Bertino, E.; Coscia, A.; Conti, A. *Mol. Nutr. Food Res.* **2004**, *48*, 363–369.
54. Maleki, S. J.; Viquez, O.; Jacks, T.; Dodo, H.; Champagne, E. T.; Chung, S. Y.; Landry, S. J. *J. Allergy Clin. Immunol.* **2003**, *112*, 190–195.
55. Sander, I.; Raulf-Heimsoth, M.; Duser, M.; Flagge, A.; Czuppon, A. B.; Baur, X. *Int. Arch. Allergy Immunol.* **1997**, *112*, 378–385.
56. Sotkovský, P.; Hubalek, M.; Hernychova, L.; Novak, P.; Havranova, M.; Setinova, I.; Kitanovicova, A.; Fuchs, M.; Stulik, J.; Tuckova, L. *Proteomics* **2008**, *8*, 1677–1691.
57. Beyer, K.; Grishina, G.; Bardina, L.; Grishin, A.; Sampson, H. A. *J. Allergy Clin. Immunol.* **2002**, *110*, 517–523.
58. Pastorello, E. A.; Farioli, L.; Pravettoni, V.; Scibilia, J.; Conti, A.; Fortunato, D.; Borgonovo, L.; Bonomi, S.; Primavesi, L.; Ballmer-Weber, B. *Anal. Bioanal. Chem.* **2009**, *395*, 93–102.
59. Chassaigne, H.; Tregoat, V.; Nørgaard, J. V.; Maleki, S. J.; van Hengel, A. J. *J. Proteomics* **2009**, *72*, 511–526.
60. Krishnan, H. B.; Kim, W. S.; Jang, S.; Kerley, M. S. *J. Agric. Food Chem.* **2009**, *57*, 938–943.
61. Yu, C. J.; Lin, Y. F.; Chiang, B. L.; Chow, L. P. *J. Immunol.* **2003**, *170*, 445–453.
62. Weber, D.; Raymond, P.; Ben-Rejeb, S.; Lau, B. *J. Agric. Food Chem.* **2006**, *54*, 1604–1610.
63. Monaci, L.; Losito, I.; Palmisano, F.; Visconti, A. *J. Chromatogr. A* **2010**, *1217*, 4300–4305.
64. Chassaigne, H.; Nørgaard, J. V.; Hengel, A. J. *J. Agric. Food Chem.* **2007**, *55*, 4461–4473.
65. Careri, M.; Elviri, L.; Lagos, J. B.; Mangia, A.; Speroni, F.; Terenghi, M. *J. Chromatogr. A* **2008**, *1206*, 89–94.
66. Mann, K.; Mann, M. *Proteome Sci.* **2011**, *9*, 7.
67. Alm, R.; Ekefjard, A.; Krogh, M.; Hakkinen, J.; Emanuelsson, C. *J. Proteome Res.* **2007**, *6*, 3011–3020.
68. Kuppannan, K.; Albers, D. R.; Schafer, B. W.; Dielman, D.; Young, S. A. *Anal. Chem.* **2011**, *83*, 516–524.
69. Careri, M.; Costa, A.; Elviri, L.; Lagos, J. B.; Mangia, A.; Terenghi, M.; Cereti, A.; Garoffo, L. P. *Anal. Bioanal. Chem.* **2007**, *389*, 1901–1907.
70. Houston, N. L.; Lee, D. G.; Stevenson, S. E.; Ladics, G. S.; Bannon, G. A.; McClain, S.; Privalle, L.; Stagg, N.; Herouet-Guicheney, C.; MacIntosh, S. C.; Thelen, J. J. *J. Proteome Res.* **2011**, *10*, 763–773.
71. Carrera, M.; Cañas, B.; Gallardo, J. M. *J. Proteomics* **2012**, *75*, 3211–3220.
72. Picariello, G.; Mamone, G.; Addeo, F.; Nitride, C.; Ferranti, P. *J. Chromatogr. A* **2011**, *1218*, 7386–7398.

73. Picariello, G.; Mamone, G.; Addeo, F.; Nitride, C.; Ferranti, P. In: *Foodomics;* Cifuentes, A., Ed.; John Wiley & Sons Inc.: Hoboken, New Jersey, 2013; pp 69–99

74. Elsayed, S.; Bennich, H. *Scand. J. Immunol.* **1975**, *4*, 203–208.

75. Lindstrøm, C. D.; van Dô, T.; Hordvik, I.; Endresen, C.; Elsayed, S. *Scand. J. Immunol.* **1996**, *44*, 335–344.

76. Swoboda, I.; Bugajska-Schretter, A.; Verdino, P.; Keller, W.; Sperr, W. R.; Valent, P.; Valenta, R.; Spitzauer, S. *J. Immunol.* **2002**, *168*, 4576–4584.

77. Van Do, T.; Hordvik, I.; Endresen, C.; Elsayed, S. *Mol. Immunol.* **2003**, *39*, 595–602.

78. Hansen, T. K.; Bindslev-Jensen, C.; Skov, P. S.; Poulsen, L. K. *Ann. Allergy Asthma Immunol.* **1997**, *78*, 187–194.

79. Bernhiesel-Broadbent, J.; Scanlon, S. M.; Sampson, H. *J. Allergy Clin. Immunol.* **1992**, *89*, 730–737.

80. Ikura, M. *Trends Biochem. Sci.* **1996**, *21*, 14–17.

81. Wilkins, C. L.; Lay, J. O., Eds.; *Identification of Microorganisms by Mass Spectrometry;* John Wiley & Sons Ltd.: Hoboken, 2006.

82. Mazzeo, M. F.; Sorrentino, A.; Gaita, M.; Cacace, G.; Di Stasio, M.; Facchiano, A.; et al. *Appl. Environ. Microbiol.* **2006**, *72*, 1180–1189.

83. Ochoa, M. L.; Harrington, P. B. *Anal. Chem.* **2005**, *77*, 5258–5267.

84. Dieckmann, R.; Helmuth, R.; Erhard, M.; Malorny, B. *Appl. Environ. Microbiol.* **2008**, *74*, 7767–7778.

85. Barbuddhe, S. B.; Maier, T.; Schwarz, G.; Kostrzewa, M.; Hof, H.; Domann, E.; Chakraborty, T.; Hain, T. *Appl. Environ. Microbiol.* **2008**, *74*, 5402–5407.

86. Angelakis, E.; Million, M.; Henry, M.; Raoult, D. *J. Food Sci.* **2011**, *76*, M568–M572.

87. Böhme, K.; Fernández-No, I.; Barros-Velázquez, J.; Gallardo, J. M.; Calo-Mata, P.; Cañas, B. *J. Proteome Res.* **2010**, *9*, 3169–3183.

88. Böhme, K.; Fernández-No, I.; Gallardo, J. M.; Cañas, B.; Calo-Mata, P. *Food Bioprocess Technol.* **2011**, *4*, 907–918.

89. Böhme, K.; Fernández-No, I.; Barros-Velázquez, J.; Gallardo, J. M.; Cañas, B.; Calo-Mata, P. *Electrophoresis* **2011**, *32*, 2951–2965.

90. Fernández-No, I. C.; Böhme, K.; Gallardo, J. M.; Barros-Velázquez, J.; Cañas, B.; Calo-Mata, P. *Electrophoresis* **2010**, *31*, 1116–1127.

91. Fernández-No, I. C.; Böhme, K.; Calo-Mata, P.; Barros-Velázquez, J. *Int. J. Food Microbiol.* **2011**, *151*, 182–189.

92. Barreiro, J. R.; Ferreira, C. R.; Sanvido, G. B.; Kostrzewa, M.; Maier, T.; Wegemann, B.; Böttcher, V.; Eberlin, M. N.; dos Santos, M. V. *J. Dairy Sci.* **2010**, *93*, 5661–5667.

93. Fernández-No, I. C.; Böhme, K.; Calo-Mata, P.; Cañas, B.; Gallardo, J. M.; Barros-Velázquez, J. *Food Microbiol.* **2012**, *30*, 91–97.

94. Gallardo, J. M.; Carrera, M.; Ortea, I. In: *Foodomics: Advances in Mass Spectrometry in Modern Food Science and Nutrition;* Cifuentes, A., Ed.; John Wiley & Sons, Inc.: Hoboken, New Jersey, 2013; pp 125–165

95. Pischetsrieder, M.; Baeuerlein, R. *Chem. Soc. Rev.* **2009**, *38*, 2600–2608.

96. Meyer, B.; Baum, F.; Vollmer, G.; Pischetsrieder, M. *J. Agric. Food Chem.* **2012**, *60*, 7306–7311.

97. Promeyrat, A.; Sayd, T.; Laville, E.; Chambon, C.; Lebret, B.; Gatellier, Ph. *Food Chem.* **2011**, *127*, 1097–1104.

98. Pazos, M.; Pereira da Rocha, A.; Roepstorff, P.; Rogowska-Wrzesinska, A. *J. Agric. Food Chem.* **2011**, *59*, 7962–7977.

99. Martins, S. I. F. S.; Jongen, W. M. F.; van Boekel, M. A. J. S. *Trends Food Sci. Technol.* **2001**, *11*, 364–373.

100. Johnson, P.; Philo, M.; Watson, A.; Clare Mills, E. N. *J. Agric. Food Chem.* **2011**, *59*, 12420–12427.

101. Arena, S.; Renzone, G.; Novi, G.; Paffetti, A.; Bernardini, G.; Santucci, A.; Scaloni, A. *Proteomics* **2010**, *10*, 3414–3434.

102. Savitski, M. M.; Nielsen, M. L.; Zubarev, R. A. *Mol. Cell. Proteomics* **2006**, *5*(5), 935–948.

103. Sun, M. A.; Wang, Y.; Cheng, H.; Zhang, Q.; Ge, W.; Guo, D. *Bioinformatics* **2012**, *28*, 2551–2552.

104. Ríos, A.; Escarpa, A.; Escarpa, M. C.; Crevillen, A. G. *Trends Anal. Chem.* **2006**, *25*, 467–479.

105. Dooley, J. J.; Sage, H. D.; Clarke, M. A.; Brown, H. M.; Garrett, S. D. *J. Agric. Food Chem.* **2005**, *53*, 3348–3357.

106. Nazzaro, F.; Orlando, P.; Fratianni, F.; Di Luccia, A.; Coppola, R. *Nutrients* **2012**, *4*, 1475–1489.

107. García-Cañas, V.; Cifuentes, A. *Electrophoresis* **2008**, *29*, 294–309.

108. Piergiovanni, A. R.; Taranto, G. *J. Agric. Food Chem.* **2005**, *53*, 6593–6597.

109. Miralles, B.; Ramos, M.; Amigo, L. *Milchwissenschaft* **2005**, *60*, 278–282.

110. Butikofer, U.; Meyer, J.; Rehberger, B. *Milchwissenschaft* **2006**, *61*, 263–266.

111. Fratianni, F.; Sada, A.; Orlando, P.; Nazzaro, F. *Open Food Sci. J.* **2008**, *2*, 89–94.

112. Lonigro, S. L.; Valerio, F.; de Angelis, F.; de Bellis, P.; Lavermicocca, P. *Int. J. Food Microbiol.* **2009**, *130*, 6–11.

113. Gehring, A. G.; Patterson, D. L.; Tu, S. I. *Anal. Biochem.* **1998**, *258*, 293–298.

114. Boehm, D. A.; Gottlieb, P. A.; Hua, S. Z. *Sens. Actuators B Chem.* **2007**, *126*, 508–514.

115. Mairhofer, J.; Roppert, K.; Ertl, P. *Sensors* **2009**, *9*, 4804–4823.

116. Herwig, E.; Marchetti-Deschmann, M.; Wenz, C.; Rüfer, A.; Allmaier, G. *Biotechnol. J.* **2011**, *6*, 420–427.

117. Pasquariello, M. S.; Palazzo, P.; Tuppo, L.; Liso, M.; Petriccione, M.; Rega, P.; Tartaglia, A.; Tamburrini, M.; Alessandri, C.; Ciardiello, M. A.; Mari, A. *Food Chem.* **2012**, *135*, 219–227.

Chapter 9

Salivary Peptidomics Targeting Clinical Applications

Rui Vitorino*, Rita Ferreira*, Armando Caseiro*,‡ and
Francisco Amado*,†
*QOPNA, Mass Spectrometry Center, Department of Chemistry, University of Aveiro, Aveiro,
Portugal
†QOPNA, School of Health Sciences, University of Aveiro, Aveiro, Portugal
‡College of Health Technology of Coimbra, Polytechnic Institute of Coimbra, Coimbra, Portugal

1   INTRODUCTION

Peptidomics is an emerging "-omic" technology based on the analysis of peptides in biological samples as body fluids. Among these peptides are small molecules like hormones, cytokines, growth factors, and the products of proteolytic activity *(1)*. As the amount and profile of peptides are modulated by physiological and pathological states, efforts have been made in the application of peptidomics to the detection of specific diagnostic markers and to the identification of cellular pathways modulated by diseases *(2,3)*. Saliva has become an increasingly relevant target for clinically oriented peptidomic analysis. Comparing with other body fluids such as blood and urine *(4)*, saliva offers the benefit of being easily collected, in a noninvasiveness and stress-free way, and involves lower costs related to sample collection and processing.

Comprehensive Analytical Chemistry, Vol. 64. http://dx.doi.org/10.1016/B978-0-444-62650-9.00009-9
223

Moreover, salivary peptidomics involves reduced need for sample preprocessing when compared with serum or urine *(5)*.

Whole saliva composition reflects the secretion of major salivary glands, namely, the parotid gland, submandibular gland, and sublingual gland; numerous minor salivary glands distributed around the oral cavity; and also of plasma, crevicular fluid, and bacteria *(6)*. The resultant large array of proteins and peptides contributes to the multiple functions of saliva, including its protective role against infection and to maintenance of the health of the oral cavity *(5,7,8)*. Despite some overlap between the proteome and the peptidome, approximately 20–30% of total secreted proteins in the whole saliva are peptides, originated from gene expression and posttranslational modifications (PTMs) in addition to the breakdown generated from the proteolysis of proteins and peptides from different sources. There are six structurally related major classes of salivary low-molecular-weight proteins, namely, histatins; basic, acidic, and glycosylated proline-rich proteins (bPRPs, aPRPs, and gPRPs, respectively); statherins; and cystatins. These peptide classes are involved in wound healing, remineralization processes, bacteria aggregation, and antiviral protection. Because of their importance in the homeostasis of oral cavity, several studies have been focusing on the analysis of salivary peptides aiming to better understand their physiological role and disclose the impact of diseases on their profile *(9–14)*.

Traditional biochemical techniques such as liquid chromatography (LC), gel electrophoresis, capillary electrophoresis (CE), and mass spectrometry (MS) have been widely used for the complete analysis of salivary proteins and peptides. The recent advances in these technical approaches applied to peptidomics have allowed a better comprehensive analysis of peptides in human whole saliva, envisioning the identification of potential salivary biomarkers of oral and systemic diseases. Sample preparation is a critical experimental step for the successful identification of peptides using MS-based approaches, for their quantitation and identification of PTMs.

2 SAMPLE PREPARATION FOR SALIVARY PEPTIDOME

The first and most important step in peptidomics is the peptidome extraction from biological samples. Contrary to other bodily fluids such as urine, serum, plasma, or cerebrospinal fluid, saliva is rich in peptides, making easier the separation of these biological molecules from proteins *(15)*. Nevertheless, a wide range of variables can markedly influence peptide profiling including sample collection, freezing conditions, the number of freeze–thaw cycles, pH, salts, proteins/glycoproteins concentration, and bacterial interferences. Attempts have been made aiming to define a standard procedure for the enrichment of all salivary peptides. De Jong *et al. (16)* evaluated some of these variables concluding that salivary peptidome is relatively resistant to fasting versus fed status of donor subjects and room temperature incubation up to 15 min does not induce significant sample peptidome degradation.

Recently, our research group evaluated the influence of centrifugation as well as different extraction procedures showing that, with the exception of ammonium sulfate, all traditional extraction procedures (organic solvent, chaotropic agents, acid, and ultrafiltration (UF)) result in peptide-rich extracts *(17)*. Most of the studies on salivary peptidome characterization have been based on experimental procedures using trifluoroacetic acid for protein precipitation *(18–23)* or UF to extract peptides *(16,24–27)*. Acetonitrile combined with ammonium hydrogen carbonate and UF was also proposed, given the higher yield of extracted salivary peptides *(17)*. Albeit no notorious effect in the yield of salivary peptides, centrifugation might lead to the loss of hydrophobic peptides, which tend to aggregate with high-molecular-weight proteins in saliva. To overcome this methodological limitation, a new procedure was proposed based on the addition of urea to saliva immediately after its collection, followed by a sonication step *(28)*. This procedure allows a better protein solubilization, and, after extraction with acetonitrile–HCl, higher amounts and a higher yield of different peptides might be obtained. With this procedure, a good characterization of salivary peptidome might be achieved with only 400 μL of whole saliva.

3 METHODOLOGIES FOR SALIVARY PEPTIDOME ANALYSIS

In the analysis of peptides, some considerations are note worthy as their low molecular mass, less than ~ 20 kDa, which result in physicochemical properties different from proteins. In addition, peptides tend not to denature irreversibly, and their generally lower hydrophobicity enables them to dissolve easily in aqueous systems without the use of detergents. PTMs such as glycosylation and phosphorylation are the most common in proteins. Considering peptides' properties, the most prominent separation approaches for molecules of up to ~ 20 kDa are based on high-performance liquid chromatography, in particular reversed-phase (RP) and ion-exchange chromatography. Indeed, despite the huge application of one- or two-dimensional sodium dodecyl sulfate polyacrylamide gel electrophoresis (1-D or 2-D SDS-PAGE) for the separation and characterization of salivary proteins, very few works were performed with gel-based approaches for the analysis of salivary peptides. The high amount of glycoproteins, such as amylase, immunoglobulin, and mucin, in saliva mask the low-molecular-weight proteins and peptides, interfering with protein separation through 2-DE. Multidimensional approaches combining different methodologies are typically used to counteract those drawbacks and enlarge the number of peptides identified and quantified (Figure 1).

Since the 1970s, researchers have being pursued the basic knowledge on salivary peptidome, using conventional chromatography and/or modified PAGE (e.g., cationic and tricine). Efforts have resulted in the separation and characterization of peptides from major salivary classes including PRPs, histatins, statherins, and cystatins. These peptides are comprehended in a wide range of molecular weights (from 0.4 to more than 20 kDa) and p*I* (from

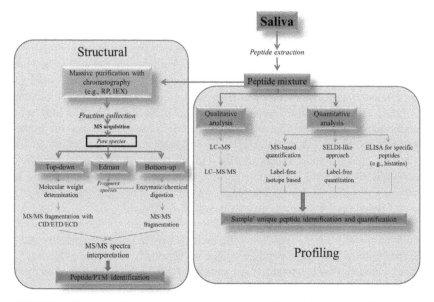

FIGURE 1 Flowchart of the peptidomic platforms typically used for the identification and quantitation of salivary peptides.

3.5 to 12). Thus, at present, and similar to proteomics, there are two basic strategies for the study of salivary peptidomics. In *bottom-up* approach, purified peptides are subjected to chemical or enzymatic cleavage, and the resultant small peptide products are separated by chromatography followed by MS analysis. In *top-down* peptidomics, intact peptide ions or large peptide fragments without any enzymatic digestion are subjected to gas-phase fragmentation for directly analysis by MS. This approach is usually performed on multiple charged ions generated from electrospray ion sources. Fragmentations occur by collision-induced dissociation (CID), by electrons as in electron capture dissociation (ECD), or by electron transfer dissociation. The combination of ECD and CID has been used in the identification of peptides deriving from basic PRPs including P-B, P-D, P-H, and P-E and IB-1 without ambiguity, as well as a new truncated form of peptide P-D and two variants of peptide II-2 *(29)*. Included in *top-down* peptidomics is surface-enhanced laser desorption/ionization time-of-flight (SELDI-TOF) MS, a kind of protein chip array technology using a modified target to achieve biochemical affinity with analytes. This technology has been used in the field of differential proteomics to detect several disease-associated proteins or peptides in a variety of biological tissues and body fluids. An example of its application is the monitoring of obese subjects' inflammatory status using saliva *(30)*. The impact of periodontitis in the salivary peptidome of obese subjects was studied using the same strategy, and defensins were identified in lower levels than in controls *(31)*.

Similar to SELDI-TOF MS profiling, magnetic RP beads and Matrix-assisted laser desorption/ionization time-of-flight (MALDI-TOF) MS have been used for salivary peptidome characterization under pathophysiological conditions, as, for example, in the evaluation of periodontal–orthodontic treatment *(32)*. However, the information retrieved using these strategies is limited requiring complementary *bottom-up* approaches.

Most of the reported findings on salivary peptidomics were obtained with *bottom-up* approaches. One typical experimental methodology is based on the isolation and massive concentration of specific peptides using size-molecular exclusion *(26,33)* or reversed-phase chromatography *(34)*. For endogenous peptides with masses lower than 5 kDa, secondary to the oral cavity's proteolytic activity, LC-ESI-MS/MS and LC-MALDI-MS/MS are usually chosen. The resultant MS/MS spectra can be processed in an automated fashion using nonredundant databases (defining *no enzyme*) where hundreds of peptide sequences are assigned or can be inspected manually using *de novo sequencing* tools to identify new sequences or PTMs *(35–37)*. For instance, spectra manual inspection of tyrosine sulfation, a widespread PTM implicated in the intracellular trafficking of secreted peptides but often mismatched as phosphorylation (same neutral loss of 80 Da), resulted in the identification of four sulfated tyrosines (Tyr 27, 30, 34, and 36) in histatin 1 *(18)*. Cyclo-statherin at Gln-37, a cyclic derivative obtained by the action of transglutaminase 2, was also identified by manual inspection of MS spectra *(20)*. *De novo sequencing* also allowed the detection of *N*- and *O*-glycosylation sites in basic PRPs *(38)*. Recently, the structure of six glycoforms of IB-8a was characterized, evidencing the presence of biantennary *N*-linked glycan fucosylated in the innermost *N*-acetylglucosamine of the core and showing from zero to four additional fucoses in the antenna *(39)*.

The combination of *top-down* and *bottom-up* strategies on the salivary characterization presents the advantages of each one and relied on the identification of more than 2000 different peptides *(5,27,35–37,40–42)*. From those, only about 30–40% identified peptides belong to the major salivary peptide classes referred in the preceding text, the remnant peptides originating from sources such as plasma, squamous cells, and crevicular fluid. Peptides of major classes are usually detected by LC–MS analysis, independently of intra- and interindividual variability. Indeed, a repertoire of 233 masses belonging to these naturally occurring peptides has been consistently detected by LC–ESI-MS analysis of the salivary peptidome *(43)*.

One major issue underlying salivary peptidome characterization is peptide quantification. While some quantitative methods developed for proteome studies were inherited, peptidome analysis has its own quantification traits. Direct profiling with MALDI-MS or SELDI-MS has been one of the most widely used approaches for peptidome despite its limitations. For instance, Zhang et al. *(32,44)* used magnetic beads combined with MALDI-MS for saliva profiling in response to orthodontic procedures, solid foods, and teeth eruption, but only few MS peaks were significantly distinct among the conditions *(27)*. Considering these drawbacks, more accurate quantification methods based on isotopes have also been applied. However, the number of analyzed samples is limited to the distinct isotopes available. The biggest difference between peptidome and proteome quantitative analyses using isotope-labeling techniques is that peptidome requires a full labeling, such as amine and carboxyl group labeling with iTRAQ *(9,45)* and O18 *(46)*, respectively. The group of Castagnola proposed an alternative quantification of peptides based on the extracted ion current (XIC) *(47)*. Three specific *m/z* values corresponding to ions with +3/+5 charged are selected, and the measured XIC peak area is considered proportional to the peptide amount. Independent of the traits underlying each quantification strategy, they are indispensable for the evaluation of the impact of physiological and pathological states on saliva composition based on comparative peptidomic studies.

4   FUNCTIONAL RELEVANCE OF SALIVARY PEPTIDOME

Similar to other bodily fluids, saliva contains several protein species that largely contribute to the oral cavity homeostasis. For instance, statherin allows saliva to keep their state oversaturated of calcium and phosphate salts, contributing to the maintenance of an intact dentition and inhibiting spontaneous precipitation of calcium phosphate. Histatins, small cationic histidine-rich peptides, possess antifungal activity *(25,48,49)*. Glycoproteins such as mucins or glycosylated PRPs are involved either in preventing dehydration and lubricating the oral cavity or in binding toxins and bacterial agglutination. Additionally, they participate in the acquired pellicle formation, which is a thin layer covering the tooth surface *(48,50)*. Nevertheless, a particularity of salivary peptides is their multifunctionality presenting at the same time more than one functional role in the oral cavity. Histatins, in particular histatins 5, behind the antifungal activity, also exhibit wound healing properties *(51)* (Figure 2).

Taking advantage of the methodologies referred in the preceding text, the systematic study of all salivary secretory peptidome components, PTMs, and protein–peptide complexes has been the initial key step of saliva basic knowledge for further application to clinical applications *(52–56)*. Besides their potential use as biomarkers, salivary peptides have been studied so far mainly in relation to their function in the oral cavity, for example, in the modulation of enamel mineralization *(57,58)*, diurnal variations *(45)*, age-dependent modifications *(11,27,59)*, or diet influence *(60)*.

For instance, Hardt *et al.* *(45)* observed that histatins 1, 3, and 5 show an acrophase in the late afternoon synchronous with the flow rate of whole saliva, whereas the concentration of statherin does not change. The study of age-dependent modifications evidenced that the concentration of salivary acidic PRPs, histatin 5, histatin 6, and monophosphorylated and diphosphorylated cystatin S showed a minimum between 6 and 9 years of age.

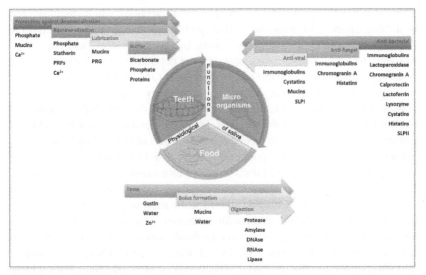

FIGURE 2 Illustration of major salivary protein/peptide constituents and their functional role in health of the oral cavity.

Interestingly, basic PRPs, almost absent in the saliva of children, reached adult levels only after puberty *(11)*, suggesting a potential role of these peptides in the modulation of taste perception. Indeed, it is widely assumed that the molecular origin of astringency of many beverages and foods is the precipitation of basic PRPs following polyphenol binding and the consequent change to the mucous layer in the mouth *(61)*. Specifically, Cabras *et al.* *(60)* showed that responsiveness to bitter taste is associated with salivary levels of II-2 peptide and Ps-1 protein. Overall, these aspects must be taken in consideration for the individual analysis as well as the collection time in the aim of any peptidomic study.

5 THE MISSING LINK: CONNECTING PEPTIDES TO PEPTIDASES

The systematic analysis of salivary peptidome also provided new insights on the proteolytic cleavage as the most relevant PTM of salivary proteins. The proteolytic fragmentation could be related to physiological processes of oral and dental microenvironment and might be relevant for oral cavity defense from different pathogens and for the modulation of oral flora growth. Enzymes responsible for in-mouth proteolysis may be originated from the oral microflora or from the host, produced within the salivary glands by mucosa epithelial cells or serum proteases transported to the oral cavity via the crevicular fluid. Despite all attempts to achieve their identification, the proteolytic cleavage is characterized as a complex and not well-known process involving a set of endo- and exoproteinases that generate a multitude of small peptides, which might be seen as a "fingerprint" *(6)*. Previous studies performed on saliva showed that the majority of proteolytic fragments belong to PRPs, histatins, and statherins. Through the analysis of cleavage sites on these peptides, generation of small salivary peptides has been attributed to cathepsin D *(35)* or enzymes with activity resembling those of the gastrointestinal tract, namely, trypsin, chymotrypsin, or elastase *(41)*. Aiming to identify the proteases involved in PRP, histatin, and statherin fragmentation, Robinson *et al. (46)* developed an approach combining isotopic labeling and *ex vivo* incubation of parotid saliva yielding lysine as the predominant cleavage site, potentially attributed to kallikrein-like protease activity. Later, Sun *et al. (62)* used histatin 5 as a substrate in zymography and identified a total of 13 proteases with cleavage specificities toward arginine and lysine residues. Nevertheless, the identification of glutamine endoproteinase that recognizes KPQ↓ as the main consensus sequence remains inconclusive, though it is being suggested that it probably derives from dental plaque and it is likely of microbial origin *(63)*. In addition, a consistent fragmentation pattern was noticed by Morzel *et al. (27)* between infants and adults, highlighting the exacerbated activity of endoglutamine protease in infants' saliva promoted by solid foods, possibly due to an increased load and/or diversity of the bacterial flora brought about by foods.

Because of the challenges of protease identification, clinical outcomes have been analyzed from salivary peptidome analysis. Amado *et al. (6)* evaluated the salivary peptidome of 10 patients with head and neck cancer (HNC) and identified 1834 fragments belonging to 289 unique proteins. From these, 158 were only identified in HNC patients, mostly involved in gene expression regulation, extracellular matrix organization, and tissue development. Recently, our research group focused on the evaluation of the effect of more than 12 years of type 1 diabetes mellitus (T1DM) and related complications on the salivary peptidome. Data not only supported a T1DM-related higher susceptibility of PRPs, statherins, and histatins to proteolysis but also evidenced an increased content of some specific protein fragments known to be related to bacterial attachment and the accumulation of phosphopeptides involved in tooth protection against erosion. A predominance of basic PRP1 unique peptides was notorious in diabetics, while a higher number of acidic PRP and statherin unique peptides were noticed in healthy individuals *(9)*. Moreover, peptidome profiling evidenced several fragments of type I collagen as a result of disease-related increased proteolytic activity, which might suggest an association to periodontitis *(64)*. Despite the high intra- and interindividual qualitative and quantitative diversity of the salivary peptidome among subjects *(35,65,66)*, diabetics showed an overall higher amount of protein fragments supported by higher activity in zymography analysis. An integrated perspective of these data can be obtained through the STRING *(67)* analysis of salivary proteome *(9)*, peptidome *(64)*, and proteolytic activities *(68)*. Figure 3 illustrates protein–protein interactions, upon querying the database with the participating proteins, peptides, and proteases relying in three clusters

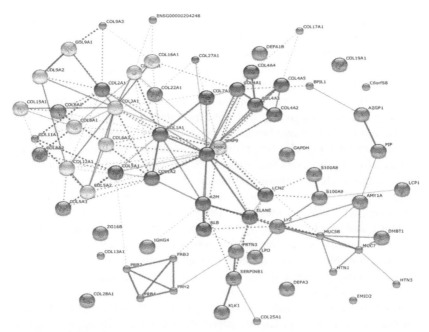

FIGURE 3   STRING protein network evidencing interplay between salivary proteome, peptidome, and proteolytic activities.

(presented with distinct colors in Figure 3) that are evidenced and tightly connected to the functional modules: one cluster comprehending collagen type II, collagen type I subunits interaction, and MMP-9; a second cluster involving MMP-2 and collagen type I; and a third cluster comprehending salivary and inflammatory proteins. According to Kegg pathways, these clusters are mainly associated to extracellular matrix–receptor interaction, focal adhesion, pathways in cancer, and leukocyte transendothelial migration.

From the protease-generated peptides identified in saliva with potential physiological role, GPPPQGGRPQ peptide from C-terminal of acidic PRPs is known to bind Gram-positive *Propionibacterium acnes*, considerably inhibiting bacterial growth *(69)*. The 12-amino acid fragment resultant from histatin 5 degradation is recognized to have an anticandidal activity comparable to that of histatin 5, which can be increased by twofold through amidation of the P-113 C terminus *(70)*. No other salivary peptide resultant from proteolytic activity has a clearly known physiological role.

6   IMPACT OF DISEASES IN THE HUMAN SALIVARY PEPTIDOME

Many recent studies have been focused on the impact of diseases on salivary peptidomics envisioning their clinical diagnosis *(6,71–73)*, including systemic diseases such as breast cancer, rheumatoid arthritis, diabetes mellitus (types I and 2), and Sjögren's syndrome (SS), as well as oral pathologies such as dental caries, periodontitis, cleft palate, and oral cancer (recently reviewed in Amado *et al. (6)*) (Table 1). Vitorino *et al. (74)* compared the salivary peptidome profile of caries-free and caries-susceptible subjects and observed an upregulation of acidic phosphopeptides, histatin 1, and statherin in the absence of caries, suggesting that these peptides might protect from dental caries. Furthermore, high levels of cystatin S were also reported in the acquired enamel pellicle, suggesting an additional protective role against demineralization in caries-free subjects *(75)*. Significant inverse correlations between the levels of statherin and truncated cystatin S and caries and supragingival plaque, total streptococci highlighted the potential risk markers for caries and other oral diseases *(78)*. The phosphorylation levels of statherin, histatin 1, and acid PRPs were reported to be significantly lower in a subgroup of autistic patients, suggesting a relation of salivary peptides' hypophosphorylation to possible asynchronies in the phosphorylation of proteins involved in the development of central nervous system *(12)*. The peptidome profiling from HNC patients showed that basic PRP fragments are more prone to *N*-acetyl hexosamine modification compared with healthy controls *(38)*. Another study focused on salivary peptidome showed significant lower levels of statherins, SMR3B, and histatins and higher concentration of α-defensins 1, 2, and 4 and S100A9 in children and adolescents with T1DM. These data suggest their commitment in the safeguard of the oral cavity in children with T1DM and highlight the major incidence of dental and periodontal diseases in these patients *(79)*.

Aiming to disclose the impact of diseases on salivary peptidome profile, Amado *et al. (6)* performed an integrated analysis, using bioinformatic tools, of all reported salivary peptides whose levels were altered under pathophysiological conditions. Cluster analysis evidenced several peptides, including cystatin S, cystatin C, defensins, and statherins, similarly modulated by diseases like dental caries, periodontitis, SS, HNC, and diabetes mellitus. For instance, cystatin SA-III was found upregulated in pathologies like bleeding oral cavity, SS, and breast cancer *(82)*, whereas acidic PRP1 was downregulated in SS and T1DM *(10,79)*, which invalidates their specificity as disease biomarkers.

To the best of our knowledge, only a peptidomic study was performed targeting the evaluation of therapeutics, specifically on SS, a systemic autoimmune disease manifested by severe impairment of exocrine gland function and focal mononuclear cell infiltrates within the lacrimal and salivary glands *(83)*. The administration of pilocarpine in adult patients restored the levels of several salivary peptides including α-defensin 1 and β-defensin 2 evaluated by LC–MS using a label-free approach *(77)*. Until now, the majority of studies that analyzed the effect of therapeutic strategies have been focused on salivary proteome, either using gel-based as gel-free approaches *(31,76,84–90)*.

TABLE 1 Summary of the Main Findings Retrieved from the Analysis of Salivary Peptidome Under Distinct Pathophysiological Conditions

Pathophysiological Condition	Goal	Sample	Sample Pretreatment	Quantitation	Major Findings	Ref.
Healthy	Peptidome characterization	WS SM/SL P	UF (<5 kDa; <50 kDa), TFA, Guanidine	None	• >2000 peptides identified • Only 26% of total identified peptides correspond to salivary peptides • One hundred and eighty-two of identified peptides are predominantly derived from acidic and basic PRPs, statherins, and histatins • Lys-Pro-Gln as a novel cleavage site	(23,36,40,64)
Healthy	Peptidome characterization	WS	TFA	Label-free	• Several PTMs and isoforms of salivary peptide classes including PRPs, statherins, histatins, cystatins, thymosins, S100	(18–23,33,39,71–75)

Continued

TABLE 1 Summary of the Main Findings Retrieved from the Analysis of Salivary Peptidome Under Distinct Pathophysiological Conditions—Cont'd

Pathophysiological Condition	Goal	Sample	Sample Pretreatment	Quantitation	Major Findings	Ref.
Healthy	Sample collection and handling factors	WS	UF (<10 kDa)	Label-free	• Peptide abundance levels in saliva are rather forgiving toward variations in sample handling and donor nutritional status • Freezing methods may affect peptide abundance	(16)
Healthy	Extraction procedure evaluation for peptidome analysis	WS	UF (<50 kDa), chaotropic, organic, acid	None	• The extraction method with bicarbonate/acetonitrile (ACN) followed by filtration resulted in the higher number of identified peptides • Centrifugation should be critically reconsidered based on the hydrophobic peptides that can be lost	(17)
Healthy	Diurnal variation influence	P	UF (5 kDa)	iTRAQ	• Histatins 1, 3, and 5 were most abundant in the afternoon; leptin peaked at 12 a.m., whereas endothelin levels were relatively constant	(45)

Diet and tasting	Diet effect	WS	TFA, WAX	Label-free	• Responsiveness to bitter taste is associated with salivary levels of basic PRPs • Solid foods promoted overexpression of fragments belonging to acidic and basic PRPs	(27,59)
Type 1 diabetes mellitus	Disease effect	WS	Urea plus acetone and ACN/HCl; TFA	iTRAQ Label-free	• A predominance of bPRP1 peptides in diabetics and a higher number of aPRP and statherin unique peptides were noticed in healthy individuals • Higher levels of α-defensin 3 neutrophil-specific, leukocyte elastase inhibitor, matrix metalloproteinase-9, neutrophil elastase, plastin-2, protein S100-A8, and protein S100-A9 related with microvascular complications as retinopathy and nephropathy	(9,63,76)

Continued

TABLE 1 Summary of the Main Findings Retrieved from the Analysis of Salivary Peptidome Under Distinct Pathophysiological Conditions—Cont'd

Pathophysiological Condition	Goal	Sample	Sample Pretreatment	Quantitation	Major Findings	Ref.
					• Statherin, proline-rich peptide P-B, P-C peptide, and histatins, were significantly less concentrated in saliva of diabetic patients • A higher number of collagen fragments were identified in diabetic patients	
Type 1 diabetes mellitus, head and neck cancer, healthy	Disease effect in salivary PRPs	WS	UF (<10 kDa)	None	• Forty-five new PRP modified residues were identified (including glycosylation, phosphorylation, and conversion of Gln to pyro-Glu) • Predominance of N-acetyl hexosamine modification on bPRPs in head and neck cancer patients	(38)

Periodontal/ orthodontic treatment	Disease effect	WS	WCX magnetic bead	Label-free	• Fragments of proteins including F2 prothrombin precursor, SERPINA1 (PRO2275), FGA isoform 1 of fibrinogen alpha chain precursor, FGA isoform 1 of fibrinogen alpha chain precursor, and VWCE isoform 1 of von Willebrand factor C and EGF domain-containing protein precursor were in higher amounts compared with periodontal and orthodontic treatment	(32,44)
					• Peak intensities at proteins 1817.7, 2010.7, 2744, and 2710.2 Da showed a steady time-dependent increasing trend	
Dental caries	Disease effect	WS	TFA	Label-free	• Strong correlation between large amounts of phosphopeptides (acidic PRP, histatin 1, and statherin) and the absence of dental caries	(77)

Continued

TABLE 1 Summary of the Main Findings Retrieved from the Analysis of Salivary Peptidome Under Distinct Pathophysiological Conditions—Cont'd

Pathophysiological Condition	Goal	Sample	Sample Pretreatment	Quantitation	Major Findings	Ref.
Edentulous	Disease effect	WS	TFA	Label-free	• Levels of α-defensins 1–4 were significantly lower in totally edentulous patients	(78)
Acute graft-versus-host disease	Disease effect	WS	TFA	Label-free	• Variable expression of S100 protein family members (S100A8, S100A9, and S100A7) was detected	(79)
Age-dependent modifications	Influence on salivary gland development during the first months	WS	TFA UF (5 kDa) WAX	Label-free	• Concentrations of the major salivary proteins/peptides showed a minimum between 0- and 6-month-olds and increased with age • The levels of histatin 1 reached a maximum between 7- and 12-month-olds and a minimum in 13- and 24-month-aged babies and increased again in 25- to 36-month-old children	(11,27,58,59,80–83)

				• S-type cystatins were almost undetectable in 0- to 6-month-old babies; P-B peptide concentration increased with age
				• Histatin 3 1/24 and statherin concentrations did not shown any age-related variation
				• Secretion of thymosin increases from about 12 weeks to about 21 weeks of gestational age
				• Adult saliva peptidome is qualitatively comparable to infants revealing similar proteolytic processing of salivary proteins
				• Teeth eruption had a very moderate effect on peptide abundance
Autism	Disease effect	WS	TFA	Label-free	• Phosphorylation levels of salivary phosphopeptides including statherin, histatin 1, and acidic PRPs were significantly lower in autistic patients	(12)

Continued

TABLE 1 Summary of the Main Findings Retrieved from the Analysis of Salivary Peptidome Under Distinct Pathophysiological Conditions—Cont'd

Pathophysiological Condition	Goal	Sample	Sample Pretreatment	Quantitation	Major Findings	Ref.
Sjögren's syndrome (SS)	Therapeutic effect	WS	TFA	Label-free	• Higher levels of α-defensin 1 and the presence of α-defensin 2 were detected in the saliva of patients with primary SS • Pilocarpine partially restored the levels of salivary peptides from patients with primary SS	(84)

WS, whole saliva; P, parotid saliva; SM/SL, submandibular/sublingual saliva; TFA trifluoroacetic acid; UF, ultrafiltration, WAX, weak anion exchange chromatography; ACN, acetonitrile; PTMs, posttranslational modifications.

Regarding the dynamic nature of salivary peptidome, it is time to devote more attention to the comprehension of peptidome adaptation to therapeutics based on peptidomic analysis combined with quantitation strategies.

7   FUTURE PERSPECTIVE

Efforts on salivary peptidomics using state-of-the-art instrumentation are expected to promote a deeper knowledge on saliva peptidome dynamics and the ongoing development of salivary-based diagnosis. In the near future, more studies are predicted to be devoted to the identification of PTMs, including phosphorylation, glycosylation, sulfation, acetylation, and proteolysis, and to the application of quantitative methods to salivary peptide species. The use of bioinformatic tools and the creation and enrichment of dedicated databases will be fundamental for dealing with the predicted huge amount of data. With all the technical and methodological conditions implemented, salivary peptidome analysis will certainly be more straightforward. At the end, the clinical translation of peptidomic data will ideally improve the diagnosis and prognosis of diseases as SS, dental caries, periodontitis, diabetes mellitus, and oral cancer.

ACKNOWLEDGMENTS

This work was supported by the Portuguese Foundation for Science and Technology (FCT), European Union, QREN, FEDER, and COMPETE that funded the QOPNA research unit (project PEst-C/QUI/UI0062/2013), RNEM (Portuguese mass spectrometry network) and the research project (FCT PTDC/EXPL/BBB-BEP/0317/2012; QREN (FCOMP-01-0124-FEDER-027554)).
Conflict of Interest: All authors have read the journal's policy on conflicts of interest and have none to declare.

REFERENCES

1. Diamandis, E. P. *J. Proteome Res.* **2006**, *5*, 2079–2082.
2. Boonen, K.; Landuyt, B.; Baggerman, G.; Husson, S. J.; Huybrechts, J.; Schoofs, L. *J. Sep. Sci.* **2008**, *31*, 427–445.
3. Lee, J. E.; Zamdborg, L.; Southey, B. R.; Atkins, N., Jr.; Mitchell, J. W.; Li, M.; Gillette, M. U.; Kelleher, N. L.; Sweedler, J. V. *J. Proteome Res.* **2013**, *12*, 585–593.
4. Tammen, H.; Peck, A.; Budde, P.; Zucht, H. D. *Expert. Rev. Mol. Diagn.* **2007**, *7*, 605–613.
5. Amado, F.; Lobo, M. J.; Domingues, P.; Duarte, J. A.; Vitorino, R. *Expert Rev. Proteomics* **2010**, *7*, 709–721.
6. Amado, F. M.; Ferreira, R. P.; Vitorino, R. *Clin. Biochem.* **2013**, *46*, 506–517.
7. Helmerhorst, E. J.; Oppenheim, F. G. *J. Dent. Res.* **2007**, *86*, 680–693.
8. Castagnola, M.; Cabras, T.; Vitali, A.; Sanna, M. T.; Messana, I. *Trends Biotechnol.* **2011**, *29*, 409–418.
9. Caseiro, A.; Ferreira, R.; Padrao, A.; Quintaneiro, C.; Pereira, A.; Marinheiro, R.; Vitorino, R.; Amado, F. *J. Proteome Res.* **2013**, *12*, 1700–1709.

10. Cabras, T.; Pisano, E.; Montaldo, C.; Giuca, M. R.; Iavarone, F.; Zampino, G.; Castagnola, M.; Messana, I. *Mol. Cell. Proteomics* **2013**, *12*, 1844–1852.

11. Cabras, T.; Pisano, E.; Boi, R.; Olianas, A.; Manconi, B.; Inzitari, R.; Fanali, C.; Giardina, B.; Castagnola, M.; Messana, I. *J. Proteome Res.* **2009**, *8*, 4126–4134.

12. Castagnola, M.; Messana, I.; Inzitari, R.; Fanali, C.; Cabras, T.; Morelli, A.; Pecoraro, A. M.; Neri, G.; Torrioli, M. G.; Gurrieri, F. *J. Proteome Res.* **2008**, *7*, 5327–5332.

13. Helmerhorst, E. J.; Alagl, A. S.; Siqueira, W. L.; Oppenheim, F. G. *Arch. Oral Biol.* **2006**, *51*, 1061–1070.

14. Campese, M.; Sun, X.; Bosch, J. A.; Oppenheim, F. G.; Helmerhorst, E. J. *Arch. Oral Biol.* **2009**, *54*, 345–353.

15. Amado, F. M.; Vitorino, R. M.; Domingues, P. M.; Lobo, M. J.; Duarte, J. A. *Expert Rev. Proteomics* **2005**, *2*, 521–539.

16. de Jong, E. P.; van Riper, S. K.; Koopmeiners, J. S.; Carlis, J. V.; Griffin, T. J. *Clin. Chim. Acta* **2011**, *412*, 2284–2288.

17. Vitorino, R.; Barros, A. S.; Caseiro, A.; Ferreira, R.; Amado, F. *Talanta* **2012**, *94*, 209–215.

18. Cabras, T.; Fanali, C.; Monteiro, J. A.; Amado, F.; Inzitari, R.; Desiderio, C.; Scarano, E.; Giardina, B.; Castagnola, M.; Messana, I. *J. Proteome Res.* **2007**, *6*, 2472–2480.

19. Inzitari, R.; Cabras, T.; Rossetti, D. V.; Fanali, C.; Vitali, A.; Pellegrini, M.; Paludetti, G.; Manni, A.; Giardina, B.; Messana, I.; Castagnola, M. *Proteomics* **2006**, *6*, 6370–6379.

20. Cabras, T.; Inzitari, R.; Fanali, C.; Scarano, E.; Patamia, M.; Sanna, M. T.; Pisano, E.; Giardina, B.; Castagnola, M.; Messana, I. *J. Sep. Sci.* **2006**, *29*, 2600–2608.

21. Messana, I.; Cabras, T.; Inzitari, R.; Lupi, A.; Zuppi, C.; Olmi, C.; Fadda, M. B.; Cordaro, M.; Giardina, B.; Castagnola, M. *J. Proteome Res.* **2004**, *3*, 792–800.

22. Castagnola, M.; Inzitari, R.; Rossetti, D. V.; Olmi, C.; Cabras, T.; Piras, V.; Nicolussi, P.; Sanna, M. T.; Pellegrini, M.; Giardina, B.; Messana, I. *J. Biol. Chem.* **2004**, *279*, 41436–41443.

23. Castagnola, M.; Cabras, T.; Inzitari, R.; Zuppi, C.; Rossetti, D. V.; Petruzzelli, R.; Vitali, A.; Loy, F.; Conti, G.; Fadda, M. B. *Eur. J. Morphol.* **2003**, *41*, 93–98.

24. Vitorino, R.; Guedes, S.; Tomer, K.; Domingues, P.; Duarte, J.; Amado, F. *Anal. Biochem.* **2008**, *380*, 128–130.

25. Vitorino, R.; Lobo, M. J.; Ferrer-Correira, A. J.; Dubin, J. R.; Tomer, K. B.; Domingues, P. M.; Amado, F. M. *Proteomics* **2004**, *4*, 1109–1115.

26. Hardt, M.; Thomas, L. R.; Dixon, S. E.; Newport, G.; Agabian, N.; Prakobphol, A.; Hall, S. C.; Witkowska, H. E.; Fisher, S. J. *Biochemistry* **2005**, *44*, 2885–2899.

27. Morzel, M.; Jeannin, A.; Lucchi, G.; Truntzer, C.; Pecqueur, D.; Nicklaus, S.; Chambon, C.; Ducoroy, P. *J. Proteomics* **2012**, *75*, 3665–3673.

28. Vitorino, R.; Guedes, S.; Manadas, B.; Ferreira, R.; Amado, F. *J. Proteomics* **2012**, *75*, 5140–5165.

29. Halgand, F.; Zabrouskov, V.; Bassilian, S.; Souda, P.; Loo, J. A.; Faull, K. F.; Wong, D. T.; Whitelegge, J. P. *Anal. Chem.* **2012**, *84*, 4383–4395.

30. Ciavarella, D.; Mastrovincenzo, M.; D'Onofrio, V.; Chimenti, C.; Parziale, V.; Barbato, E.; Lo Muzio, L. *Prog. Orthod.* **2011**, *12*, 126–131.

31. Range, H.; Leger, T.; Huchon, C.; Ciangura, C.; Diallo, D.; Poitou, C.; Meilhac, O.; Bouchard, P.; Chaussain, C. *J. Clin. Periodontol.* **2012**, *39*, 799–806.

32. Zhang, J.; Zhou, S.; Li, R.; Cao, T.; Zheng, H.; Wang, X.; Zhou, Y.; Du, N.; Chen, F.; Lin, J. *Proc. Natl. Acad. Sci. U. S. A.* **2012**, *10*, 63.

33. Inzitari, R.; Cabras, T.; Onnis, G.; Olmi, C.; Mastinu, A.; Sanna, M. T.; Pellegrini, M. G.; Castagnola, M.; Messana, I. *Proteomics* **2005**, *5*, 805–815.

34. Hay, D. I.; Bennick, A.; Schlesinger, D. H.; Minaguchi, K.; Madapallimattam, G.; Schluckebier, S. K. *Biochem. J.* **1988**, *255*, 15–21.
35. Vitorino, R.; Barros, A.; Caseiro, A.; Domingues, P.; Duarte, J.; Amado, F. *Proteomics Clin. Appl.* **2009**, *3*, 528–540.
36. Helmerhorst, E. J.; Sun, X.; Salih, E.; Oppenheim, F. G. *J. Biol. Chem.* **2008**, *283*, 19957–19966.
37. Huq, N. L.; Cross, K. J.; Ung, M.; Myroforidis, H.; Veith, P. D.; Chen, D.; Stanton, D.; He, H.; Ward, B. R.; Reynolds, E. C. *Int. J. Pept. Res. Ther.* **2007**, *13*, 547–564.
38. Vitorino, R.; Alves, R.; Barros, A.; Caseiro, A.; Ferreira, R.; Lobo, M. C.; Bastos, A.; Duarte, J.; Carvalho, D.; Santos, L. L.; Amado, F. L. *Proteomics* **2010**, *10*, 3732–3742.
39. Cabras, T.; Boi, R.; Pisano, E.; Iavarone, F.; Fanali, C.; Nemolato, S.; Faa, G.; Castagnola, M.; Messana, I. *J. Sep. Sci.* **2012**, *35*, 1079–1086.
40. Huang, C. M.; Zhu, W. *Comb. Chem. High Throughput Screen.* **2009**, *12*, 521–531.
41. Lucchi, G.; Chambon, C.; Truntzer, C.; Pecqueur, D.; Ducoroy, P.; Schwartz, C.; Nicklaus, S.; Morzel, M. *Int. J. Pept. Res. Ther.* **2009**, *15*, 177–185.
42. Schipper, R.; Loof, A.; De Groot, J.; Harthoorn, L.; Van Heerde, W.; Dransfield, E. *Ann. N. Y. Acad. Sci.* **2007**, *1098*, 498–503.
43. Castagnola, M.; Cabras, T.; Iavarone, F.; Vincenzoni, F.; Vitali, A.; Pisano, E.; Nemolato, S.; Scarano, E.; Fiorita, A.; Vento, G.; Tirone, C.; Romagnoli, C.; Cordaro, M.; Paludetti, G.; Faa, G.; Messana, I. *J. Matern. Fetal Neonatal Med.* **2012**, *25*, 27–43.
44. Zhang, J.; Zhou, S.; Zheng, H.; Zhou, Y.; Chen, F.; Lin, J. *Biochem. Biophys. Res. Commun.* **2012**, *421*, 844–849.
45. Hardt, M.; Witkowska, H. E.; Webb, S.; Thomas, L. R.; Dixon, S. E.; Hall, S. C.; Fisher, S. J. *Anal. Chem.* **2005**, *77*, 4947–4954.
46. Robinson, S.; Niles, R. K.; Witkowska, H. E.; Rittenbach, K. J.; Nichols, R. J.; Sargent, J. A.; Dixon, S. E.; Prakobphol, A.; Hall, S. C.; Fisher, S. J.; Hardt, M. *Proteomics* **2008**, *8*, 435–445.
47. Messana, I.; Cabras, T.; Pisano, E.; Sanna, M. T.; Olianas, A.; Manconi, B.; Pellegrini, M.; Paludetti, G.; Scarano, E.; Fiorita, A.; Agostino, S.; Contucci, A. M.; Calo, L.; Picciotti, P. M.; Manni, A.; Bennick, A.; Vitali, A.; Fanali, C.; Inzitari, R.; Castagnola, M. *Mol. Cell. Proteomics* **2008**, *7*, 911–926.
48. Dodds, M. W.; Johnson, D. A.; Yeh, C. K. *J. Dent.* **2005**, *33*, 223–233.
49. Vitorino, R.; Calheiros-Lobo, M. J.; Duarte, J. A.; Domingues, P.; Amado, F. *Bull. Group. Int. Rech. Sci. Stomatol. Odontol.* **2006**, *47*, 27–33.
50. Vitorino, R.; Calheiros-Lobo, M. J.; Williams, J.; Ferrer-Correia, A. J.; Tomer, K. B.; Duarte, J. A.; Domingues, P. M.; Amado, F. M. *Biomed. Chromatogr.* **2007**, *21*, 1107–1117.
51. Oudhoff, M. J.; Bolscher, J. G.; Nazmi, K.; Kalay, H.; van 't Hof, W.; Amerongen, A. V.; Veerman, E. C. *FASEB J.* **2008**, *22*, 3805–3812.
52. Granger, D. A.; Kivlighan, K. T.; Fortunato, C.; Harmon, A. G.; Hibel, L. C.; Schwartz, E. B.; Whembolua, G. L. *Physiol. Behav.* **2007**, *92*, 583–590.
53. Castagnola, M.; Picciotti, P. M.; Messana, I.; Fanali, C.; Fiorita, A.; Cabras, T.; Calo, L.; Pisano, E.; Passali, G. C.; Iavarone, F.; Paludetti, G.; Scarano, E. *Acta Otorhinolaryngol. Ital.* **2011**, *31*, 347–357.
54. Pfaffe, T.; Cooper-White, J.; Beyerlein, P.; Kostner, K.; Punyadeera, C. *Clin. Chem.* **2011**, *57*, 675–687.
55. Hu, S.; Loo, J. A.; Wong, D. T. *Expert Rev. Proteomics* **2007**, *4*, 531–538.
56. Brinkmann, O.; Spielmann, N.; Wong, D. T. *Dent. Today* **2012**, *31*, 54, 56–57; quiz 58–59.
57. Vitorino, R.; Calheiros-Lobo, M. J.; Duarte, J. A.; Domingues, P. M.; Amado, F. M. *J. Sep. Sci.* **2008**, *31*, 523–537.

58. Siqueira, W. L.; Custodio, W.; McDonald, E. E. *J. Dent. Res.* **2012**, *91*, 1110–1118.

59. Manconi, B.; Cabras, T.; Pisano, E.; Sanna, M. T.; Olianas, A.; Fanos, V.; Faa, G.; Nemolato, S.; Iavarone, F.; Castagnola, M.; Messana, I. *J. Proteomics* **2013**, *91*, 536–543.

60. Cabras, T.; Melis, M.; Castagnola, M.; Padiglia, A.; Tepper, B. J.; Messana, I.; Tomassini Barbarossa, I. *PLoS One* **2012**, *7*, e30962.

61. Soares, S.; Vitorino, R.; Osorio, H.; Fernandes, A.; Venancio, A.; Mateus, N.; Amado, F.; de Freitas, V. *J. Agric. Food Chem.* **2011**, *59*, 5535–5547.

62. Sun, X.; Salih, E.; Oppenheim, F. G.; Helmerhorst, E. J. *Proteomics Clin. Appl.* **2009**, *3*, 810–820.

63. Thomadaki, K.; Helmerhorst, E. J.; Tian, N.; Sun, X.; Siqueira, W. L.; Walt, D. R.; Oppenheim, F. G. *J. Dent. Res.* **2011**, *90*, 1325–1330.

64. Caseiro, A.; Vitorino, R.; Barros, A. S.; Ferreira, R.; Calheiros-Lobo, M. J.; Carvalho, D.; Duarte, J. A.; Amado, F. *Biomed. Chromatogr.* **2012**, *26*, 571–582.

65. Quintana, M.; Palicki, O.; Lucchi, G.; Ducoroy, P.; Chambon, C.; Salles, C.; Morzel, M. *J. Proteomics* **2009**, *72*, 822–830.

66. Le Yondre, N.; Tammen, H.; Hess, R.; Budde, P.; Jürgens, M. *Open Proteomics J.* **2008**, *1*, 99–105.

67. Szklarczyk, D.; Franceschini, A.; Kuhn, M.; Simonovic, M.; Roth, A.; Minguez, P.; Doerks, T.; Stark, M.; Muller, J.; Bork, P.; Jensen, L. J.; von Mering, C. *Nucleic Acids Res.* **2011**, *39*, D561–D568.

68. Caseiro, A.; Ferreira, R.; Quintaneiro, C.; Pereira, A.; Marinheiro, R.; Vitorino, R.; Amado, F. *Clin. Biochem.* **2012**, *45*, 1613–1619.

69. Huang, C. M.; Torpey, J. W.; Liu, Y. T.; Chen, Y. R.; Williams, K. E.; Komives, E. A.; Gallo, R. L. *Antimicrob. Agents Chemother.* **2008**, *52*, 1834–1836.

70. Rothstein, D. M.; Spacciapoli, P.; Tran, L. T.; Xu, T.; Roberts, F. D.; Dalla Serra, M.; Buxton, D. K.; Oppenheim, F. G.; Friden, P. *Antimicrob. Agents Chemother.* **2001**, *45*, 1367–1373.

71. Liu, J.; Duan, Y. *Oral Oncol.* **2012**, *48*, 569–577.

72. Spielmann, N.; Wong, D. T. *Oral Dis.* **2011**, *17*, 345–354.

73. Al Kawas, S.; Rahim, Z. H.; Ferguson, D. B. *Arch. Oral Biol.* **2012**, *57*, 1–9.

74. Vitorino, R.; Lobo, M. J.; Duarte, J. R.; Ferrer-Correia, A. J.; Domingues, P. M.; Amado, F. M. *Biomed. Chromatogr.* **2005**, *19*, 214–222.

75. Vitorino, R.; de Morais Guedes, S.; Ferreira, R.; Lobo, M. J.; Duarte, J.; Ferrer-Correia, A. J.; Tomer, K. B.; Domingues, P. M.; Amado, F. M. *Eur. J. Oral Sci.* **2006**, *114*, 147–153.

76. Schulz, B. L.; Cooper-White, J.; Punyadeera, C. K. *Crit. Rev. Biotechnol.* **2012**, *33*, 246–259.

77. Peluso, G.; De Santis, M.; Inzitari, R.; Fanali, C.; Cabras, T.; Messana, I.; Castagnola, M.; Ferraccioli, G. F. *Arthritis Rheum.* **2007**, *56*, 2216–2222.

78. Rudney, J. D.; Staikov, R. K.; Johnson, J. D. *Arch. Oral Biol.* **2009**, *54*, 91–100.

79. Cabras, T.; Pisano, E.; Mastinu, A.; Denotti, G.; Pusceddu, P. P.; Inzitari, R.; Fanali, C.; Nemolato, S.; Castagnola, M.; Messana, I. *Mol. Cell. Proteomics* **2010**, *9*, 2099–2108.

80. Castagnola, M.; Inzitari, R.; Fanali, C.; Iavarone, F.; Vitali, A.; Desiderio, C.; Vento, G.; Tirone, C.; Romagnoli, C.; Cabras, T.; Manconi, B.; Sanna, M. T.; Boi, R.; Pisano, E.; Olianas, A.; Pellegrini, M.; Nemolato, S.; Heizmann, C. W.; Faa, G.; Messana, I. *Mol. Cell. Proteomics* **2011**, *10*, M110 003467.

81. Nemolato, S.; Messana, I.; Cabras, T.; Manconi, B.; Inzitari, R.; Fanali, C.; Vento, G.; Tirone, C.; Romagnoli, C.; Riva, A.; Fanni, D.; Di Felice, E.; Faa, G.; Castagnola, M. *PLoS One* **2009**, *4*, e5109.

82. Al-Tarawneh, S. K.; Border, M. B.; Dibble, C. F.; Bencharit, S. *OMICS* **2011**, *15*, 353–361.

83. Delaleu, N.; Jonsson, R.; Koller, M. M. *Eur. J. Oral Sci.* **2005**, *113*, 101–113.

84. Border, M. B.; Schwartz, S.; Carlson, J.; Dibble, C. F.; Kohltfarber, H.; Offenbacher, S.; Buse, J. B.; Bencharit, S. *Mol. Biosyst.* **2012**, *8*, 1304–1310.

85. Thomadaki, K.; Bosch, J.; Oppenheim, F.; Helmerhorst, E. *Oral Dis.* **2013**, *19*, 781–788.

86. Xiao, H.; Zhang, L.; Zhou, H.; Lee, J. M.; Garon, E. B.; Wong, D. T. *Mol. Cell. Proteomics* **2012**, *11*, M111 012112.

87. Rocha Dde, M.; Zenobio, E. G.; Van Dyke, T.; Silva, K. S.; Costa, F. O.; Soares, R. V. *J. Appl. Oral Sci.* **2012**, *20*, 180–185.

88. Jarai, T.; Maasz, G.; Burian, A.; Bona, A.; Jambor, E.; Gerlinger, I.; Mark, L. *Pathol. Oncol. Res.* **2012**, *18*, 623–628.

89. Bencharit, S.; Altarawneh, S. K.; Baxter, S. S.; Carlson, J.; Ross, G. F.; Border, M. B.; Mack, C. R.; Byrd, W. C.; Dibble, C. F.; Barros, S.; Loewy, Z.; Offenbacher, S. *Mol. Biosyst.* **2012**, *8*, 3216–3223.

90. Ambatipudi, K. S.; Swatkoski, S.; Moresco, J. J.; Tu, P. G.; Coca, A.; Anolik, J. H.; Gucek, M.; Sanz, I.; Yates, J. R., 3rd.; Melvin, J. E. *Proteomics* **2012**, *12*, 3113–3120.

Part III

Metabolomics

Chapter 10

Metabolomics in the Study of Alzheimer's Disease

Clara Ibáñez, Alberto Valdés, Virginia García-Cañas and Carolina Simó
Laboratory of Foodomics, Institute of Food Science Research (CIAL), CSIC. Nicola's Cabrera 9, Madrid, Spain

1 INTRODUCTION

Alzheimer's disease (AD) is the most prevalent form of late-life mental failure in humans *(1,2)*. It is expected that AD incidence will triple over the next 50 years *(3)*. AD pathology affects the central nervous system (CNS), leading to a progressive destruction and atrophy of brain cortex especially those regions related to superior mental functions. Namely, neocortex and hippocampus areas are primarily affected in AD *(4)*. Clinically, AD is characterized by the presence of intracellular neurofibrillary tangles composed of hyperphosphorylated tau proteins *(5)* and Aβ peptide aggregates (Aβ) in the form of extracellular amyloid plaques *(6)* and amyloid infiltrates in the brain microvasculature *(7)*. These formations are considered key factors in neuronal dysfunction and cell death *(8–10)*. Other neuropathological hallmarks for AD include synaptic loss and/or dysfunction, diminished neuronal metabolism, and loss of multiple neurotransmitter systems *(1)*. These alterations in the CNS bring about an unspecific symptomatology that begins with cognitive deficits to remember autobiographical events specific to a time and place *(11)*. As the pathology advances to cortical brain areas, the long-term memory becomes impaired and ordinary abilities including semantic memory and attention are affected leading to a dementia syndrome *(12)*.

Comprehensive Analytical Chemistry, Vol. 64. http://dx.doi.org/10.1016/B978-0-444-62650-9.00010-5

AD is preceded by a mild cognitive impairment (MCI) state followed by dementia, and it has been indicated that the annually conversion rate from MCI to dementia is about 15% *(13)* and reach up to 80% in a 6-year follow-up *(14)*. MCI is defined by a cognitive decline greater than expected for certain age and education level, but not severe enough to be considered dementia or to interfere notably with activities of daily life *(13)*. This condition is characterized by memory impairment that can be proved and diagnosed by objective measures *(15)*. Up to date, there is no clinical method to determine which patients with MCI will progress to dementia except for a long clinical follow-up *(16)*. The speed and severity of clinical progression after MCI diagnosis vary and depend on multiple factors, most not well elucidated yet. For instance, depression has been proposed as a risk factor for dementia and cognitive decline *(17)*, postulated as a causal effect by means of hippocampal damage *(18)* or as a possible prodromal state *(19)* of MCI state. There is a great clinical need to identify incipient AD in patients with MCI, and many efforts are being made in the discovery of predictor markers of progression via the use of cognitive tests *(20,21)*, structural and functional neuroimaging *(12,22,23)*, CSF analysis *(24,25)*, and other biomarkers, either isolated or in combination *(26,27)*. On the other hand, the U.S. National Institute of Aging and the Alzheimer's disease Association is working on the determination of the best predictor factors in the progression from normal cognition (presymptomatic or preclinical state) to MCI *(28)*. Some of the proposed research criteria and markers for preclinical AD are summarized in Figure 1 *(28)*. These

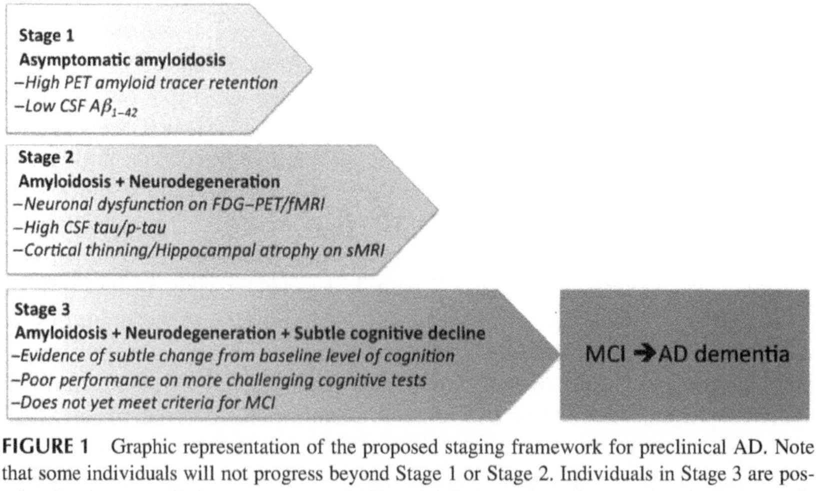

Stage 1
Asymptomatic amyloidosis
−*High PET amyloid tracer retention*
−*Low CSF Aβ_{1-42}*

Stage 2
Amyloidosis + Neurodegeneration
−*Neuronal dysfunction on FDG–PET/fMRI*
−*High CSF tau/p-tau*
−*Cortical thinning/Hippocampal atrophy on sMRI*

Stage 3
Amyloidosis + Neurodegeneration + Subtle cognitive decline
−*Evidence of subtle change from baseline level of cognition*
−*Poor performance on more challenging cognitive tests*
−*Does not yet meet criteria for MCI*

MCI → AD dementia

FIGURE 1   Graphic representation of the proposed staging framework for preclinical AD. Note that some individuals will not progress beyond Stage 1 or Stage 2. Individuals in Stage 3 are postulated to be more likely to progress to MCI and AD dementia. AD, Alzheimer's disease; Aβ, amyloid beta; PET, position emission tomography; CSF, cerebrospinal fluid; FDG, fluorodeoxyglucose, fMRI, functional magnetic resonance imaging, sMRI, structural magnetic resonance imaging. *Reproduced from Ref.* (28). (For the color version of this figure, the reader is referred to the online version of this chapter.)

initiatives are extremely important as there is evidence suggesting AD pathological processes begin years or even decades before dementia onset *(29)*.

Leaving aside the detection and identification of the predementia states, even the diagnosis of patients suffering an advanced AD needs to be improved. AD is routinely diagnosed following the criteria established in 1984 by the National Institute for Neurological and Communicative Disorders and Stroke/Alzheimer's Disease and Related Disorders Association (NINCDS/ADRDA, now known as Alzheimer's Disease Association) criteria *(30)*. These criteria establish that AD can exclusively be diagnosed with certainty at autopsy. A definitive AD diagnostic is only achieved with the detection of NFTs and Aβ in postmortem brain tissue (BT) *(31)*. Nonetheless, CSF is considered a valuable source of AD biomarkers, very promising for the early detection and progress evaluation of AD *(32)*. Several markers, generally protein molecules, have been already linked to AD *(33,34)*. Until date, the ratio between total tau protein and Aβ peptide levels in CSF has been proposed as a high sensitive measure for diagnostic purposes for advanced AD patients, which in combination with brain imaging techniques is able to diagnose up to 90% of advanced AD patients, assuming a high economic burden per evaluated subject (ca. 6000–10,000€ approximately) *(35)*. NINCDS/ADRDA criteria include standard comprehensive assessment protocol including clinical examination, brain imaging, electroencephalography, analyses of blood and CSF (including total tau, phospho-tau, and $A\beta_{1-42}$), and a detailed neuropsychological evaluation to diagnose AD. If after these tests it is considered that the subject suffers mild-to-moderate AD, different therapeutic options are possible. Currently, there is no remedy for AD but symptomatic therapy. Most recommended medications for mild-to-moderate AD include cholinesterase inhibitors (donepezil, rivastigmine, or galantamine) that increase the levels of acetylcholine in the brain. However, controversy about the satisfactory results *(36)* or limited benefits *(37)* of these drugs has been reported *(38)*. A combination of cholinesterase inhibitors with memantine (*N*-methyl-D-aspartate (NMDA) receptor antagonist) seems to produce consistent benefits in moderate-to-severe AD *(39,40)* with respect to a single drug administration. The search for therapeutic treatments at all phases of cognitive decline is imperative since up to date, no drug can reverse the pathological process of this chronic disease. On the other hand, there are evidences suggesting that concurrent pharmacological and behavioral methods may exploit functional benefits for patients suffering from dementia *(41,42)*. Although no specific disease-modifying treatment has yet been shown to be effective for dementias, it is particularly important to develop effective, targeted treatments to halt or delay the onset of cognitive decline from preclinical and MCI states to AD-related dementia. An accurate identification of AD cases at the MCI state or, even better at the preclinical state, might enable early therapeutic interventions to reduce considerably, stop, or even reverse the numerous pathological, emotional, and economic costs of the illness *(43)*.

The distinction of preclinical AD stage from changes of normal aging represents a major demand in AD investigation. Unlike the MCI state, diagnostic criteria for this prodromal phase have not yet been established. In this sense, one of the main focuses of the scientific community is to elucidate the link between the neurodegenerative cascade of AD and the apparition of first clinical symptoms. At the same time, the lack of effective therapeutic treatment for AD entails an ethical conflict in the preclinical stage diagnosis because of the potential emotional repercussion. The discovery of biomarkers suitable to monitor the evolution and progression from the preclinical phases to MCI related to AD is particularly challenging and may offer key information to deepen in the origin, causes, and pathophysiology of AD. The detection of AD biomarkers can be achieved via two different approaches: knowledge-based and unbiased. The traditional one consists of the "knowledge-based" or targeted (deductive method) approach and relies on a direct understanding of the neuropathological processes. Until date, several studies have targeted protein biomarkers related to AD pathology, among which β-amyloid peptide and its precursor (amyloid precursor protein, APP) stand out *(44–46)* followed by the analysis of other proteins such as tau *(47,48)* or BACE (β-site APP cleaving enzyme-1) *(49,50)*. From a genetic point of view, allelic variation of apolipoprotein E (ApoE) has been described as the most influential genetic risk factor to develop sporadic AD *(51)*. To a lesser extent, cholesterol *(52,53)* and homocysteine *(54)* have also been targeted as potential metabolic AD biomarkers. The great lack of knowledge of pathophysiological processes that govern AD makes the alternative "unbiased" approach more suitable than the targeted approach in the generation of new hypothesis *(55)* about the origin, causes, and progression of AD. In contrast to targeted analysis of selected analytes, the goal of untargeted approach is to find biomarkers through the comprehensive study of the maximum number of molecules in a biological system. The use of omics technologies in biomarker discovery has evolved from traditional targeted to nonbiased approaches. Omics technologies refer to a group of advanced analytical technologies used in a high-throughput manner to explore the composition, roles, and relationships of a variety of molecules in a biological system. Omics are able to significantly improve the experimental models that offer only a temporal snapshot of the huge complexity and dynamic nature of biological networks that govern human health and disease *(56)*. Thus, Transcriptomics *(57)*, Proteomics *(33,34)*, and Metabolomics *(58–61)* technologies have been applied in AD research, and they are expected to significantly help in the advance of the investigation of biomarkers implicated in the pathogenesis of AD *(62)*.

2 METABOLOMICS IN AD

The development of new methodologies to discover early and high-confident biomarkers related to AD progression is compulsory. This investigation will contribute to reveal mechanisms by which neurodegeneration starts and progresses, so far unknown (63). Among omics technologies, Metabolomics intensifies variations occurring in both the proteome and the genome, and represents a faithful reflection of the organism phenotype in health and disease (59). Metabolomics, the newest of the omics technologies, deals with the comprehensive study of the metabolome or the entire set of small molecules (about <1500 Da) in a biological system (cell, biofluid, tissue, or organism) at a time, under given conditions (64). The metabolome corresponds to the end stage of all molecular events owing to gene variation and expression (genomics), protein expression and modification (proteomics), and environmental exposures in a biological sample (65). Massive and high-throughput determination of metabolites in complex samples is a powerful tool to characterize organism phenotypes and to discover biomarkers for health/disease states (66). Metabolomics is increasingly applied to AD investigation (67) due to the complex nature of the disease. In fact, increasing evidences suggest that AD is a heterogeneous and multifactorial disorder resulting from combination between the genetic susceptibility and environmental influences (68) including life-style factors (69). Biological processes such as inflammation, cholesterol metabolism, oxidative stress, and homocysteine homeostasis have been observed to be different in AD patients compared to control subjects in "knowledge-based" approaches, but these biomarkers did not achieve sufficient discriminatory power (70). Since the causes of this neurological disease are still not clear, nontargeted Metabolomics opens new frontiers in AD investigation. The unbiased nature of these metabolomic strategies requires special attention on study design, as metabolites have very different molecular structures (lipids, carbohydrates, nucleic acids, amino acids, organic acids, and others) and are present in biofluids and tissues in a wide dynamic range of concentrations (pmol–mmol) (71). Therefore, the complete analysis of entire metabolome in biological systems is still challenging, and no single analytical platform can cover the entire metabolic signature of a given biological sample. Mass spectrometry (MS) and nuclear magnetic resonance (NMR) are by far the two analytical platforms most predominantly used in Metabolomics. NMR permits the analysis of complex mixtures of metabolites with little or no sample preparation in a rapid and nondestructive way (72) and has been largely applied in Metabolomics (73). NMR was firstly used in AD Metabolomics in 1993 (74), 9 years before than the application of the first MS-based metabolomic study (75). This analytical approach has been applied in AD investigation in a variety of applications as can be seen in Table 1. For instance, NMR has been used to uncover AD metabolic biomarkers involved in specific metabolic pathways such as oxidative (76), lipid (77), and cholesterol (78) metabolism.

Compared to NMR, MS technique is highly sensitive enabling broader surveys of the metabolome and provides spectral information for the identification process of metabolites (102). The use of ultra-high-resolution mass

TABLE 1 Metabolomic Applications in AD Research

Sample	Groups of Samples	Metabolomic Approach	Analytical Platform	Refs.
Postmortem CSF	Twenty-three samples from AD patients and healthy controls	Metabolic profiling of 2.4–2.9 NMR region	NMR	(74)
Postmortem human brain	Subjects with no dementia ($n=5$), very mild dementia ($n=3$), mild dementia ($n=4$), moderate dementia ($n=6$), and severe dementia ($n=4$)	Lipid profiling	ESI–QqQ MS	(75)
CSF	Demented patients ($n=12$), healthy subjects ($n=17$), multiple sclerosis patients ($n=19$)	Metabolic profiling of oxidative metabolism	NMR	(76)
Serum	MCI ($n=19$) and control ($n=26$) subjects	Lipid profiling	NMR	(77)
Postmortem human brain	AD patients ($n=9$) and control subjects ($n=10$)	Metabolic profiling of cholesterol metabolism	HPLC–Q/MS NMR	(78)
Postmortem mice brain	Eight brain regions from young mice TgCRND8 ($n=5$) and control young mice ($n=4$), and aged mice TgCRND8 ($n=3$) and control old mice ($n=5$)	Metabolic fingerprinting	NMR	(79)
Postmortem human brain	AD ($n=8$) and AML ($n=11$) patients	Metabolic fingerprinting	NMR	(80)
Rat urine	Control ($n=4$) and transgenic APP/tau rats at three life-time points: 4 ($n=4$), 10 ($n=4$), and 15 ($n=3$) months	Metabolic fingerprinting	NMR	(81)
CSF	Healthy subjects with typical AD biomarkers (Aβ y tau) ($n=10$) and without typical AD biomarkers ($n=34$)	Metabolic fingerprinting	NMR	(82)
Plasma	AD patients ($n=20$), healthy subjects ($n=20$)	Metabolic fingerprinting	UHPLC–QqQ/MS	(83)
CSF	AD patients ($n=79$) and control subjects ($n=51$)	Metabolic fingerprinting	GC–TOF MS LC–MS/MS	(84)

Sample	Subjects	Profiling type	Technique	Ref.
CSF	AD ($n=23$), MCI with progression to AD ($n=9$), MCI without progression to AD ($n=22$), and control ($n=19$) subjects. Validation set ($n=12$)	Metabolic fingerprinting	CE–TOF MS	(85)
Postmortem human brain	AD patients ($n=15$) and control subjects ($n=15$). Validation set ($n=60$)	Metabolic fingerprinting	UHPLC–Q/TOFMS	(86)
CSF, plasma, urine	AD patients ($n=8$), healthy subjects ($n=8$)	Metabolic profiling of amino acids and dipeptides	HPLC–MS^2	(87)
Postmortem CSF	AD patients ($n=15$) and control subjects ($n=15$)	Metabolic fingerprinting	HPLC–ECA	(88)
CSF	Four pool of samples from AD ($n=27$), MCI with progression to AD ($n=13$), MCI without progression to AD ($n=26$), and control ($n=46$) subjects	Metabolic profiling of amino acids	MEKC–LIF	(89)
CSF	AD ($n=21$), MCI with progression to AD ($n=12$), MCI without progression to AD ($n=21$), and control ($n=21$) subjects	Metabolic fingerprinting	UHPLC–TOF MS	(90)
CSF	AD patients ($n=18$), healthy subjects ($n=18$)	Hydrophilic metabolites profiling	nLC–Q/TOF MS	(91)
CSF	AD ($n=40$), MCI ($n=36$), and control ($n=38$) subjects	Metabolic fingerprinting	HPLC–ECA	(92)
CSF	AD patients ($n=40$) and control subjects ($n=38$)	Metabolic fingerprinting	HPLC–ECA GC–TOF MS	(93)
Postmortem mice brain	Control ($n=3$) and transgenic (APP/PS1) ($n=3$) mice	Metabolic profiling of energetic metabolism	GC–TOF MS	(94)
CSF and Plasma	AD ($n=15$), MCI ($n=10$), and control ($n=10$) subjects	Metabolic fingerprinting	UHPLC–TOF MS	(95)
Serum	AD ($n=37$), MCI with progression to AD ($n=52$), MCI without progression to AD ($n=91$), and control ($n=46$) subjects	Lipid profiling Metabolic fingerprinting	UHPLC–MS GC × GC–TOF MS	(96)
Plasma	AD ($n=16$), MCI ($n=12$), and control ($n=10$) subjects	Metabolic profiling of bile acids	UHPLC–MS/MS	(97)

Continued

TABLE 1 Metabolomic Applications in AD Research—Cont'd

Sample	Groups of Samples	Metabolomic Approach	Analytical Platform	Refs.
Plasma	AD ($n=10$), MCI ($n=10$), and control ($n=10$) subjects	Phospholipid profiling	HPLC–LTQ Orbitrap	(98)
Plasma	AD patients ($n=26$) and control subjects ($n=26$)	Lipid profiling	QqQ–MS/MS	(99)
CSF and plasma	Training set (CSF and plasma): AD patients ($n=10$) and control subjects ($n=10$) Validation set (Plasma): AD ($n=41$), MCI ($n=26$), and control ($n=42$) subjects	Steroid profiling	HPLC–Q/MS LTQ Orbitrap	(100)
Postmortem human brain Postmortem mice brain	AD patients ($n=30$) and control subjects ($n=14$) APPswedish mice ($n=5$) and control mice ($n=8$)	Phospholipid profiling	FIA[a]–MS/MS	(101)

[a]FIA, flow injection analysis.

spectrometers (e.g., TOF, FT-ICR MS, Orbitrap®) is essential in MS-based metabolomic approaches, to obtain accurate mass measurements for the determination of elemental compositions of metabolites and for their tentative identification with the help of metabolite databases *(103)*. On the other hand, MS^n experiments provide additional structural information for metabolite identification purposes, especially when product ions are analyzed at high resolution (with Q-TOF, TOF–TOF, or LTQ Orbitrap). Evaluation and comparison of NMR and MS analytical platforms on the application in Metabolomics have been extensively studied *(102–105)*. Both analytical platforms can be used as a standalone or in combination with previous separation techniques such as gas chromatography (GC), liquid chromatography (LC), or capillary electrophoresis (CE). The human metabolome includes a great portion of polar compounds such as polyamines, amino acids, or simple sugars. Compared to GC, both LC and CE can make possible the separation of higher polarity compounds and represent the two separation techniques most useful for medium to highly polar metabolic profiling. On the other hand, GC is especially suitable for the analysis of organic molecules and generally requires sample derivatization of metabolites to create volatile compounds.

Two different and complementary analytical strategies are basically followed in nontargeted Metabolomics: metabolic "fingerprinting" and "profiling." "Metabolic fingerprinting" has been proposed as a means of analyzing the total set of metabolites, avoiding biases against certain classes of compounds, for sample classification *(106,107)*, and "metabolic profiling" is referred to identification and quantification of a predefined group of metabolites (chemically related metabolites or associated with a particular metabolic pathway) *(64)*. Metabolic fingerprinting is gaining extensive interest across a wide variety of disciplines with the emerging main focus on biomarker discovery for disease prognosis, diagnoses, and therapy monitoring *(108)*. This approach offers a rapid biochemical snapshot of the metabolome *(107)* with the possibility of correlating the perturbations from the homeostatic or healthy state with disease. Moreover, metabolic fingerprints reveal characteristic metabolic patterns broadening our understanding of unknown pathological processes *(109)*, which enhance its utility in AD biochemical alterations. In an ideal metabolic fingerprinting approach, metabolite extraction method and analytical platform should not be biased toward any group of molecules. In practice, these issues have not been resolved yet but tried to be minimized. The broad information obtained by metabolic fingerprinting approach makes it the most attractive and suitable strategy for biomarker discovery *(108)*, while at the same time, it represents the most challenging and time-consuming approach in Metabolomics. In AD research, the obtaining of metabolic fingerprints is a recent practice by using both NMR *(79–81)* and MS-based *(82–86)* techniques.

Metabolic profiling approaches are developed for the determination and quantitation of metabolites of a particular pathway or for a class of compounds. This strategy generates quantitative information of metabolites

belonging to known biochemical pathways and physiological interactions
(110). Using metabolic profiling, optimized analytical design (sample treatment, analytical platform, etc.) for a group of compounds is performed, optimizing sensitivity and specificity for the group of molecules of interest. The
information obtained from the metabolic profiles not only describes the metabolic state of a certain sample but in many cases is used for comparison
purposes.

The combination of characteristics from both metabolic profiling (generally, a hypothesis-driven approach) and metabolic fingerprinting
(hypothesis-generating approach) represents a powerful and highly complementary methodology. In fact, it can be especially useful to get the benefit
from the potential to discover novel unforeseen metabolic factors or biomarkers in fingerprinting approaches with the option of demonstrating the
observed changes of known metabolites or confirming the potential biomarkers by means of the metabolic profiling mode *(111)*. A recent study followed
this comprehensive metabolomic approach in the investigation of biomarkers
for AD by the comparison of postmortem BT from AD patients and healthy
control subjects *(112)*. First, a metabolic fingerprint to determine potential
biomarkers and affected pathways in AD pathology was performed. Then
the specific metabolic pathway related to potential AD biomarkers spermine
and spermidine, namely, biogenic polyamine metabolism, was explored in
detail, finding three more molecules altered due to AD pathology *(112)*.

Although Metabolomics is still in its infancy in AD investigation, a number of studies have been published in the last years (Table 1), as this chapter
shows. Most of them are based on the analysis of CSF, blood, or postmortem
BT samples, although in a lesser extent, urine has also been examined in the
search of AD biomarkers *(81,87)*.

2.1 Metabolomics and AD in CSF

Metabolomics works focused on AD investigation are mostly based on the
analysis of CSF. Main reason is that CSF composition is directly affected
by the CNS. Therefore, alterations due to AD pathology will be more probably observed in this sample than in other biofluids. The role of CSF includes a
mechanical protection of CNS against trauma, and transport of nutrients to
ensure the homeostasis of CNS cells *(113)*.

Prior to metabolomic analysis, sample treatment is typically needed, as
CSF contains approximately 0.3 mg/mL protein *(114)* that may hinder metabolite analysis. Consequently, CSF sample treatment is essentially directed to
protein removal by means of organic solvent addition *(84,88)* or by ultrafiltration *(85,89,90)*. The final metabolic extract composition will depend in a great
extent on the sample treatment *(115)*, and it will be selected mostly regarding
the metabolomic approach and the analytical technique that will be afterward
applied.

In some metabolic applications, an additional step of purification is needed. For instance, Myint *et al.* *(91)* removed CSF protein content using 5-kDa centrifuge filters, and then an additional step of solid-phase extraction (SPE) through a polymeric sorbent was carried out to purify hydrophilic metabolites. In that study, analysis of CSF samples from AD-diagnosed subjects ($n = 18$) and healthy controls ($n = 18$) was carried out by the development of a new chromatographic method based on nanoLC–Q/TOF in the positive ionization mode. By using multivariate analysis, satisfactory separation was obtained between two groups of subjects using PCA statistics *(91)*. In that study, an extensive optimization of stationary phases for separation of weak and strong cationic metabolites was required.

When we speak about the type of metabolic species covered by a certain technique, it is also worth mentioning that the analysis of polar metabolites is specially challenging in Metabolomics. CE is a fast and high-resolution separation technique especially suitable for the separation of ionic and highly polar molecules that cannot be easily obtained by LC *(115)* and has been applied in the investigation of early biomarkers related to AD progression, from MCI and healthy states *(85,89)*. According to a previous statement that considered free amino acids as important molecules in neurotransmission receptor function and implicated in neurotoxicity *(116)*, Samakashvili *et al.* determined chiral amino acid alterations in progression to AD. For that purpose, a micellar electrokinetic chromatography (MEKC) with β-cyclodextrin as a chiral selector and laser-induced fluorescence (LIF) detection was developed. This enantioselective method permitted the determination of 11 amino acids in CSF samples from healthy subjects, MCI patients who progressed to AD in a 2-year follow-up period, MCI patients without progression to AD in that follow-up period, and AD patients. Decreased levels of L-arginine, L-glutamic acid, L-aspartic acid, and L-lysine and increased levels of γ-amino butyric acid were related to the progression of MCI patients to AD *(89)* and were suggested as potential early biomarkers. The concept of early diagnostic biomarkers has a long history, with many studies showing that AD biomarkers can be used to predict conversion from MCI to AD. As stated in Section 1, till now the clinical diagnosis of AD requires the presence of MCI state *(117)*, and up to date, there is no clinical method to determine which MCI cases will progress to AD except for a long clinical follow-up period *(43)*. The discovery of markers that indicate progression from MCI to AD is still an unresolved issue, and a topic of high relevance in AD research. Several groups have been addressed this issue using metabolomic approaches in CSF samples. Thus, CSF metabolic fingerprints from patients showing different cognitive status related to AD were obtained by CE–MS *(85)*. In that study, 85 CSF samples were obtained from individuals in four different cognitive states, namely, age-matched controls, patients who at the sample collection time suffered from MCI, and finally, a group of AD patients. After a clinical follow-up period of 2 years, some patients with MCI developed AD and others remained stable.

After CE–MS analysis and extensive data processing (peak detection, filtering, migration time alignment, normalization, etc.), 71 metabolites were detected in the 85 samples. As a result of the multivariate statistical analysis, a 90% of correct sample assignment was achieved with the determination of 10 potential metabolic biomarkers *(85)*. The most similar metabolic fingerprints were observed when MCI patients without development to AD and control subjects were compared. That finding led to the merger of these two groups of samples as one "non-AD" sample group, and other discriminant analysis was performed reaching up to 97% of correct assignment with the determination of 14 potential metabolic markers. Increased levels of choline and a decreased concentration of carnitine and creatine were observed in MCI patients who progressed to AD compared to the other groups of samples under study *(85)*. UHPLC–MS analytical platform was used to improve metabolome coverage of CSF samples *(90)*. Thus, by using RP/UHPLC–MS and HILIC/UHPLC–MS, and a subsequent data processing, a total of 524 high-confident metabolites were detected in CSF samples. After a discriminant analysis, a 98.7% correct assignment was achieved with the determination of 17 significantly different metabolites among the four groups of samples (control, MCI with progression to AD, MCI without progression to AD, and AD). Furthermore, values above 95% for sensitivity and specificity were obtained. In MCI patients who progressed to AD, decreased levels of a neuroprotective metabolite as taurine and increased levels of neurotoxic metabolites such as dopamine-quinone or methylsalsolinol were observed *(90)*. A less common analytical approach in Metabolomics, HPLC with electrochemical array detection (ECA), has been applied to study metabolic signatures from 114 CSF samples from three groups of patients: 40 control individuals (healthy), 36 subjects with MCI, and 38 AD patients *(92)*. ECA is an extremely selective and sensitive detection technique with simpler and cheaper instrumentation, but compared to NMR or MS, less structural information is obtained by LC–ECA. Before analysis, CSF proteins were precipitated by methanol addition, and metabolic fraction was then analyzed by LC–ECA. As a result, 21 of a total of 71 detected metabolites could be identified by the analysis of reference standard mixtures. Patients presenting AD and MCI showed elevated levels of methionine, whereas the methionine/reduced glutathione ratio was decreased, suggesting a glutathione depletion associated with AD pathology *(92)*. In that work, a correlation between metabolite levels and concentration of $A\beta_{1-42}$, total tau, and phospho-tau proteins in CSF was carried out. As a result, an association between the levels of vanillylmandelic acid and xanthine with total tau protein in CSF was found, inferring that norepinephrine pathway and purine pathway may be involved in total tau aggregation and pathology generation *(92)*. Later, the same research group in collaboration with other laboratories combined the metabolic information obtained by HPLC–ECA with that obtained from GC–TOF MS platform *(93)*. In that work, 40 AD patients and 38 healthy control subjects were considered. After the GC–TOF MS analysis and statistical evaluation, higher levels of two metabolites showed the most consistent association with the disease. However, authors declared that unknown identity of these two metabolites hinders biological interpretation *(93)*.

Other study that deepened in the disease phenotype by means of the comparison of AD and healthy subjects was published by Czech *et al.* in 2012 *(84)*. CSF samples from 79 AD patients and 51 age-matched controls were studied. CSF deproteinization and metabolite fractionation (polar and no polar metabolites) were accomplished by using a mixture of ethanol and dichloromethane. Polar and no polar metabolic fractions were analyzed by GC–MS and LC–MS/MS, respectively. More than 340 metabolites were determined. After statistical analysis, an interesting finding was highlighted: by using univariate statistical analysis, female AD patients showed more significant changes than male AD patients, compared to control subjects. Cysteine, uridine, cortisol, 3-methoxy-4-hydroxy phenylglycol, dopamine, noradrenaline, and normetanephrine were significantly different in AD patients compared to control group. Then, discriminant analysis showed cysteine and uridine as the best metabolite pair to predict AD with sensitivity and specificity values of 75% (Figure 2) *(84)*.

The assessment of metabolic changes related to the beginning and progress of AD could improve the diagnosis and monitoring of AD but, in a more challenging scope, metabolic patterns could even help to evaluate the risk of healthy people to develop the disease in the future or the detection of subjects in the prodromal state of the disease. As stated in Section 1, the investigation of early biomarkers of the long preclinical stage of AD could improve our understanding on AD pathophysiology including the development of new therapeutic targets to revert AD pathology. Focused on this idea, Jukarainen *et al.* analyzed CSF samples from 44 neurological controls who did not present any sign of dementia or chronic neurological diseases *(82)*. Subjects were divided into two groups consisting of 10 individuals who presented a typical concentration of AD protein markers, namely, reduced soluble $A\beta_{1-42}$ and increased tau concentrations in CSF, and 34 control subjects with normal levels of AD protein markers in CSF. After NMR analysis of the samples, a curve fitting method based on constrained total line shape method showed that 31 metabolites covered the 85% of the total spectral intensity. However, only one metabolite was significantly different between the two groups of subjects, namely, creatinine was higher in CSF from subjects with typical AD markers concentrations *(82)*.

One of the biochemical processes that have covered much of the research performed on AD over the last years is the oxidative stress mechanisms *(118)*. Mitochondrial dysfunction has been suggested to underlie AD pathophysiology *(94)*. Deposition of heavy metals and alteration of the mitochondrial functionality are some of the facts and effects of oxidative stress observed in AD process *(119)*. Given its importance, a number of reviews have tackled this

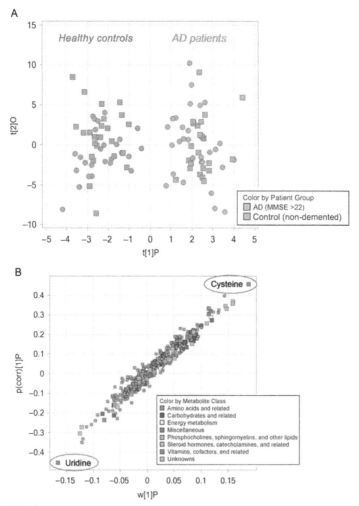

FIGURE 2 Determination of analytes contributing significantly to separation between AD patients and healthy controls by OPLS-DA of the CSF data set. (A) OPLS-DA score plot for light AD patients (MMSE.22) versus controls. (B) OPLS-DA coefficient plot for light AD patients (MMSE.22) versus controls. Cysteine and uridine are the most relevant analytes for separating light AD patients (MMSE.22) from healthy subjects. *Reproduced from Ref.* (84). (For the color version of this figure, the reader is referred to the online version of this chapter.)

complex topic *(119–122)*. To explore this aspect at metabolomic level, oxidative-stress-related metabolites were studied using a profiling approach *(76)*. Thus, CSF samples from 48 individuals (12 AD patients, 19 with multiple sclerosis, and 17 control subjects) were analyzed by ^{1}H NMR. After collection, CSF samples were lyophilized and no protein removal was required. In agreement with brain oxidative metabolism impairment, increased concentrations of several amino acids (alanine, lysine, valine, leucine or iso-leucine, tyrosine, and glutamine), acidic molecules (lactate and pyruvate), and inositol were found in AD samples *(76)*.

2.2   Metabolomics and AD in Plasma and Serum

Lumbar puncture to obtain CSF can produce several side effects *(123)*. Head-ache lasting less than 1 week occurs in up to 40% of patients, representing the most common complication after lumbar puncture. Occasionally, it is accom-panied by nausea, vomiting, vertigo, reduced hearing, and blurred vision. Other complications, including headaches lasting from 8 days to 1 year, cra-nial neuropathies, continued backache, nerve root injury, and meningitis, are rare, following less than 1% of lumbar punctures *(124)*. In addition to these side effects, lumbar puncture is a complicated *(125)* and expensive procedure, so other alternatives such as peripheral blood could be of interest in the search of biomarkers related to AD.

From the clinical point of view, there is a clear trend toward the analysis of minimum invasive samples (e.g., blood or urine) for biomarkers search *(126)*. CSF is constantly exchanged and cleared via the blood *(127)*. This direct contact between CSF and blood encourages investigating this biofluid (plasma, serum, and/or cells), which could reflect pathological changes in the brain *(128)*. However, exchanges between blood and CSF are not well elu-cidated yet, and thus blood biomarkers usefulness is limited owing to the highly selective blood–brain barrier *(70)*. Since 2000s, important efforts to find reliable AD biomarkers in peripheral blood following proteomic approaches have been carried out with moderate success *(129)*. In this sense, Metabolomics can be considered a relatively little explored technology, which may offer valuable information. Recently, Trushina *et al. (95)* established to what extent metabolic changes observed in CSF were or not reflected in plasma obtained from the same individuals. Metabolic fingerprints from CSF and plasma from AD ($n = 15$), MCI ($n = 15$), and control ($n = 15$) indivi-duals was compared to find the correlation between both biofluids and also to uncover biomarkers of AD progression. Integrated information from CSF and plasma samples was used to select and validate reliable plasma biomarkers related to AD *(95)*. As expected, the number of affected pathways in both bio-fluids increased with disease progression. Interestingly, compared to control subjects, nearly 30% and 60% of the metabolic pathways altered in the CSF of MCI and AD patients, respectively, were also affected in plasma samples from the same individuals showing certain correlation between biofluids *(95)*. Amino acid and dipeptide content correlation between CSF and plasma was also accomplished by Fonteh *et al.* including also urine in that case *(87)*. CSF, plasma, and urine from eight AD patients and eight control subjects were submitted to metabolite extraction. Briefly, the extraction of the com-pounds of interest of 200 µL of CSF, 100 µL of plasma, and 200 µL of urine was performed by a SPE and followed by derivatization. Amino acids and dipeptides were then detected by LC–MS/MS, and differences between AD and control groups were determined in the three biofluids. As a result, 23, 37, and 28 metabolites could be detected in CSF, plasma, and urine, respectively, and were compared between control and AD subjects. As can be seen in Figure 3 *(87)*, metabolic variations due to AD were very different among the three biofluids. Altered metabolites due to AD pathology were identified and related to different biochemical processes such as neurotransmission (L-Dopa and dopamine), urea cycle/detoxification or NO formation (arginine, citrulline, ornithine), inhibitory processes (glycine), and antioxidation (carnosine) *(87)*.

With the aim to detect potential metabolite biomarkers of AD in plasma, Li *et al. (83)* compared the metabolic fingerprints obtained by UHPLC–MS, from 20 control subjects and 20 AD patients. The statistical analysis by means of a PCA showed metabolic changes between AD and control samples mainly associated with nine potential biomarkers. Tryptophan, different lysophosphatidylcholines, and sphingosines were highlighted as the metabolites that most differentiated AD from control groups of plasma samples *(83)*. Oresic et al. monitored the progression markers in plasma samples from control, MCI without progression to AD (in a 1- to 4-year follow-up), MCI who progressed

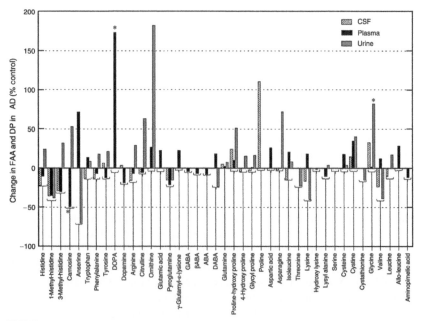

FIGURE 3 Changes in free amino acid (FAA) and dipeptide (DP) concentrations in AD samples. After LC/MS^2 analysis, mole quantities (nmol $=$ dl) of FAAs and DPs in samples from subjects without or with AD were calculated. Then the change in the mean concentration of FAA and DPs in AD ($n=8$) compared to healthy controls ($n=8$) was achieved and expressed as the % change in AD compared to controls ($*p < 0.05$). *Modified from Ref. (87)*.

to AD (in that follow-up period), and AD-diagnosed subjects was carried out *(96)*. In that work, two different analytical platforms were used to widen the metabolome coverage. GC × GC–MS method was developed for the analysis of amino acids, free fatty acids, ketoacids, organic acids, sterols, and sugars, while UHPLC–MS was optimized for the determination of lipids *(96)*. As a result, 683 metabolites (139 lipids and 544 polar compounds) were determined. The high degree of coregulation among the detected molecules was represented by clustering metabolites into subsets following the Bayesian model. Seven and six clusters were obtained for UHPLC- and GC × GC-based metabolomic data, respectively. An alteration of pentose phosphate pathway between the two groups of MCI patients (MCI with and without progression to AD) together with an overall trend toward lower lipid levels in AD patients was observed *(96)*. Those samples with lower lipid contents were associated to processes linked to membrane lipid remodeling in AD patients. Lipids are not only major constituents of cell membrane physiology but also involved in crucial processes constituting one of the main focuses on Metabolomics. The vast majority of AD metabolic profiling studies of peripheral blood samples are focused on the investigation of lipid-related metabolites. Alteration in AD lipid metabolism has been suggested since BT from AD patients display a higher occurrence of "adipose inclusions" or "lipoid granules," already observed by Alois Alzheimer when he first described the disease *(130)*. In fact, a close link between lipid metabolism and AD has been established when the ε4 allele of the ApoE gene was identified as the strongest genetic risk factor for AD *(131,132)*. ApoE encodes a ∼34-kDa protein that serves as a crucial regulator of cholesterol metabolism in the brain and of triglyceride metabolism throughout the body. The processes by which ApoE mediates the uptake of lipoprotein particles in the brain via the low-density lipoprotein receptor-related protein and the very low density family lipoprotein receptor have been already reviewed and described in detail *(133,134)*. Several evidences suggest that systemic lipid levels associate more strongly with the development of AD than previously assumed *(135)* and may vary in the progression to AD. To determine the molecular insights on the potential early lipid biomarkers, serum samples from 19 MCI and 26 control subjects were studied *(77)*. To obtain information on three different groups of metabolites, namely, lipoproteins, polar metabolites, and lipids, three NMR approaches were optimized. Lipoproteins group was analyzed after the application of optimized NMR conditions to detect macromolecules (mainly lipoprotein lipids and albumin). To obtain information about polar metabolites macromolecules, signals were suppressed by the application of a pulse sequence. Finally, lipid signals were detected after the breakdown of lipoprotein particles. Information obtained from the three groups of metabolites was afterward combined and correlated with clinical data reported from the individuals under study. Interestingly, vascular factors, metabolic syndrome, and cognitive decline were shown to be closely interrelated. Furthermore, lower ω-3 fatty acids, sphingomyelin, and phosphatidylcholine appear to be associated with AD risk factors presence, in MCI patients *(77)*.

Bile acids have also been highlighted as possible metabolic markers of AD progression through a metabolic fingerprinting approach *(97)*. Plasma samples from 10 control subjects, 12 MCI, and 16 AD patients were analyzed using UHPLC–MS/MS. After the PCA did not show any grouping of subjects by disease state, a partial least-squares discriminant analysis (PLS-DA) was applied showing separation for the three groups of samples. Increased levels of glycocholate, glycochenodeoxycholate, and glycodeoxycholate were observed in MCI and AD patients. However, the unsuitability in the validation of the PLS-DA model for disease prediction and diagnosis was attributed by the authors to the high levels of inter- and intrasubject variability and the small number of samples. The same problem was considered in another study by Sato *et al. (98)* where plasma from 10 control subjects, 10 MCI, and 10 AD patients was analyzed using LC–MS. Lysophopholipid (18:1) was revealed as a potential early biomarker because it was observed gradually decreased in the progression to AD *(98)*.

Novel approaches such as the nontargeted multidimensional mass spectrometry-based shotgun lipidomics (MDMS-SL) have also been applied to study AD. By using MDMS-SL approach, plasma from 26 AD patients and 26 cognitively normal subjects by ESI–QqQ MS were analyzed in the negative ion mode *(99)*. More than 800 lipid-related molecular species from nine classes of lipids (choline glycerophospholipid, lysophosphatidylcholine, ethanolamine glycerophospholipid, phosphatidylinositol, sphingomyelin, ceramide, triacylglycerol, cholesterol, and cholesterol esters) were detected in plasma samples. Significant reductions of sphingomyelin and significant increase in ceramide content were observed in AD patients, suggesting that the increased ceramide content might result from the accelerated sphingomyelins hydrolysis or increased biosynthesis in the brain *(99)*.

A number of epidemiological studies have shown a potential link between cholesterol lowering compounds, especially statins and a strongly decreased prevalence or incidence for dementia *(136–138)*. These evidences may indicate that targeting lipid metabolism *(139)* and more specifically cholesterol-related metabolites in humans may be a potential strategy for AD prevention. Most predominant pathways related to clearance of key protein AD markers which are affected by cholesterol are described elsewhere *(139)*. For instance, it was observed higher plasma cholesterol levels in individuals with the ε4 allele of ApoE, which elevates the risk for early (<65 years of age) AD onset *(140)*. In agreement with these findings, cholesterol has been highlighted in a recent study as a powerful biomarker for AD *(100)*. After a preliminary discovery and identification of demosterol and cholesterol as potential AD biomarkers, an LC–MS method was optimized to quantify these two metabolites *(100)*. The analysis revealed that desmosterol was found to be decreased in AD versus controls plasma. The developed analytical method to quantify desmosterol and cholesterol revealed that desmosterol and the desmosterol/cholesterol ratio were significantly decreased in AD patients. Results were confirmed with a validation group of 109 plasma samples from clinically well-described patients.

2.3   Metabolomics and AD in Postmortem BT

Postmortem BT has brought about the main theories to explain origin and to design treatments for AD *(141)*. In contrast to plasma or CSF biofluids, the brain area or region of tissue must be carefully selected in the study design, as metabolite composition is not homogeneous. Commonly affected regions in the brain include the association of cortical and limbic areas with especial damage of the cerebral cortex, certain subcortical regions, and the hippocampus. Although amyloid plaques and NFTs are differentially distributed among AD patients, autopsies have shown predominance toward the temporal lobe. Contraction of the cerebral cortex and hippocampus and increase in brain ventricles are other alterations observed in postmortem BT of AD patients. This complexity makes the majority of metabolomic studies to analyze several brain regions. For instance, postmortem BT samples from frontal, parietal, and occipital lobe from 10 control health subjects and 10 AD patients have been studied by using a metabolomic approach *(112)*. Metabolite extraction was performed on frozen-grinded samples by the addition of methanol and Tris–HCl as solvent and followed by ultrafiltration to discard macromolecules. Metabolic fingerprinting by UHPLC–TOF MS analysis was primarily used to evaluate the diversity of low-molecular-weight molecules patterns from the frontal, parietal, and occipital lobes in the positive and negative ion modes. After a multivariate statistical analysis, a total of 431 metabolite species were selected to identify possible biomarker candidates based on all brain regions under study. Five possible biomarkers in the frontal and parietal lobe areas were observed. Among them, identity of spermine and spermidine was confirmed by comparison with chromatograms and MS spectra of commercial standards. With the aim to increase the sensitivity and selectivity toward polyamine analysis, a derivatization procedure was applied to all the BT samples by using of 4-(*N,N*-dimethylaminosulfonyl)-7-fluoro-2,1,3-benzoxadiazole. Then the analysis was performed by UHPLC–MS/MS *(112)*. As a result, other polyamines were uncovered linked to AD pathology: an increase in putrescine, acetylspermidine, and acetylspermine in AD frontal lobes was observed. In that work, authors suggested a new theory shown in Figure 4 *(112)* based on the increase in the activity of ornithine decarboxylase enzyme induced by the amyloid plaques and/or NFT and the effects of polyamines on NMDA receptors in the AD brain. In a different work, UHPLC–MS was also the selected analytical platform *(86)* to study neocortex metabolic fingerprints from postmortem BT of 15 AD patients and 15 control healthy subjects. After an orthogonal projection to latent structures discriminant analysis

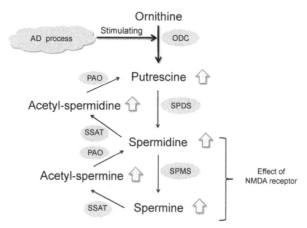

FIGURE 4 Metabolic pathway of polyamines. Increased levels of spermidine, spermine, putrescine, acetylspermidine, and acetylspermine without a change of ornithine in AD pathology were observed. One theory suggests that the NMDA receptor excitotoxicity is caused by an excess of spermidine and spermine due to ornithine decarboxylase activity induced by plaque and/or tangle deposition in specific brain regions. *Reproduced from Ref.* (112). (For the color version of this figure, the reader is referred to the online version of this chapter.)

(Figure 5) and a *t*-test, 34 and 32 candidate biomarkers were observed to be statistically different in the two groups of samples, in the positive and negative ion modes, respectively *(86)*.

Polar biomarkers have also been investigated by NMR *(80)* in postmortem BT samples. These samples were treated with perchloric acid, and metabolic extracts were analyzed to compare AD with amyotrophic lateral sclerosis (ALS) *(80)*. After multivariate statistical analysis of the 8 AD and the 11 ALS samples, a high score in the statistical parameters of OPLS was obtained with the exclusion of 1 AD sample. Increased concentration of branched amino acids, alanine, acetate, glutamine, glutamate, and glycerophosphocholine in AD patients was observed when compared to ALS *(80)*.

One of the possibilities to overcome the difficulty to obtain BT samples is the use of animal models reproducing certain characteristics of the AD pathology. In this sense, numerous transgenic mouse models have been used in AD investigation. Most common mice used in AD investigation express human mutant APP. There are APP mutants mice that develop robust amyloid plaque pathology (e.g., Tg2576 (K670N, M671L) and TgCRND8 (KM670/671NL + V717F)), presenilin protein mutants (e.g., M146L), and biogenic transgenic mice that express not only mutant APP but also mutant presenilin protein (PS) (e.g., APP/PS1) *(142–146)*. Using APP mutants, metabolic fingerprints of eight different brain regions of TgCRND8 mice (five young and three aged mice) and control mice (four young and five aged mice) were obtained by NMR *(79)*. Decreased levels of *N*-acetyl-L-aspartate, glutamate, glutamine, taurine (exception hippocampus), γ-amino butyric acid, choline and

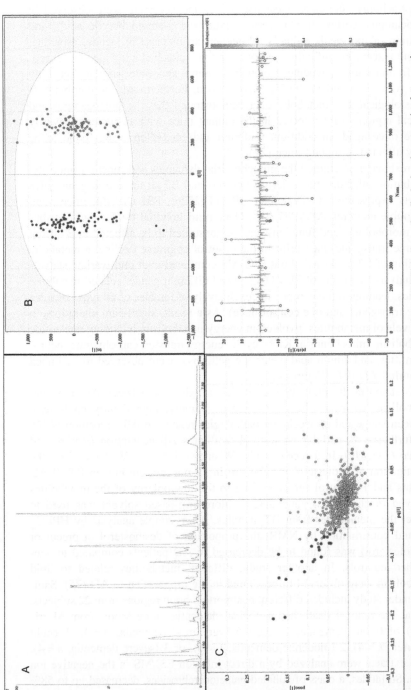

FIGURE 5 (A) UHPLC ESI+chromatogram of the polar extract of postmortem brain tissue. (B) The scores plot displaying the separation between the two sample groups (AD = blue; controls = red). Explained variance (R2) was 97%, predictive ability (Q2) was 93%, and root-mean-square error of validation (RMSEV) was 15%. (C) The loadings plot which correspond to the scores in (B). Indicated in blue (n = 24) and red (n = 17) are metabolites which significantly differ between groups (higher in AD and controls, respectively). These are further emphasized by the S-line-plot (D) and their relative variable importance to the model. *Reproduced from Ref.* (86). (For interpretation of the references to color in this figure legend, the reader is referred to the online version of this chapter.)

phosphocholine, creatine, phosphocreatine, and succinate were observed in hippocampus, cortex, frontal cortex (exception γ-amino butyric acid), and midbrain of transgenic animals. Moreover, an increase in lactate, aspartate, glycine (except in midbrain), and other amino acids including alanine (exception frontal cortex), leucine, isoleucine, valine, and water-soluble free fatty acids were observed in the TgCRND8 mice. Furthermore, the combination of histological and metabolic data demonstrated that the hippocampus and cortical regions were affected in the mutant mice with an increase severity as the mice aged. In addition, midbrain and cerebellum were found to be partly affected in mutant mice *(79)*.

Trushina *et al.* studied the mitochondrial dynamics and function (motility, distribution, ultrastructure, etc.) in neurons and BT of three transgenic mice models expressing mutant human APP (Tg2576), PS1 (M146L mice), and the double mutation APP/PS1 *(94)*. Once mitochondrial trafficking, distribution, morphology, and function were demonstrated to be affected in BT from all three transgenic mice prior to the onset of cognitive decline, a metabolic profiling of BT was carried out. GC–MS combined with multivariate statistical analysis by means of PLS-DA showed different gender-related metabolic profiles, leading to the comparison exclusively of females of all types of mice (three transgenic and one control mice). As a result, significant alterations in the levels of metabolites involved in energy metabolism, including nucleotide metabolism, mitochondrial Krebs cycle, energy transfer, carbohydrate, neurotransmitter, and amino acid metabolic pathways, were observed to be linked to familial AD *(94)* (Figure 6).

As stated in the previous section 2.2, a tight link between AD and lipid metabolism has been already described. Namely, high dietary cholesterol has been suggested to accelerate pathologies related to Aβ deposition *(147)*, and furthermore, cellular cholesterol facilitates Aβ generation *in vitro* and *in vivo (148,149)*. In a recent study, Wisniewski *et al. (78)* focused on the investigation of cholesterol metabolism in 19 postmortem human BT (9 AD patients and 10 control subjects). Due to the low polarity of the metabolites involved in cholesterol pathways, a chloroform-based solvent was used to extract the metabolites from BT samples. After sample analysis by HPLC–MS and confirmation by NMR, the proportion of desmosterol (a precursor of cholesterol) was found to be decreased in AD patients compared to age-matched controls. In another study, different metabolites related to lipid metabolism were observed to be altered in the progression to AD *(75)*. Samples under study included different postmortem brain regions from 22 subjects, who at the time of death had a clinical dementia rating score (from Morris *(150)*) of 0 (no dementia, $n=5$), 0.5 (very mild dementia, $n=3$), 1 (mild dementia, $n=4$), 2 (moderate dementia, $n=6$), or 3 (severe dementia, $n=4$). Lipid extracts were analyzed by a direct infusion ESI/MS in the negative ion mode. Sulfatides, a class of sulfated galactocerebrosides, decreased up to 58% in white matter and were found to be depleted up to 93% in gray matter at

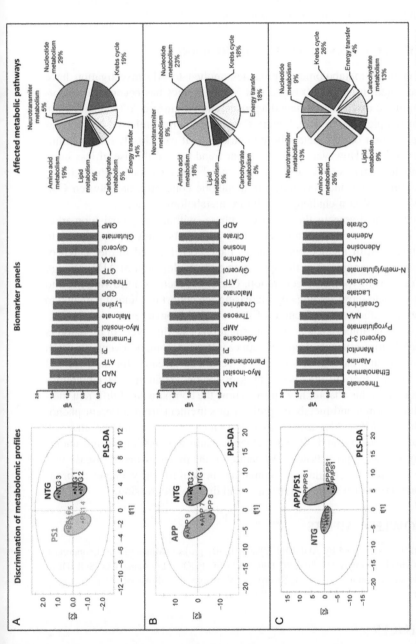

FIGURE 6 Comparison of individualized metabolomic profiles and affected metabolic pathways in FAD mouse models. (A–C, left panels): PLS-DA score plots showing distinct metabolomic profiles of PS1 (A), APP (B), and APP/PS1 (C) female mutant mice (Tg) compared to nontransgenic (NTG) littermates. (A–C, central panels): Panels of specific biomarkers as a plot of variable importance in the projection (VIP) indicating the 15 most significant metabolites in discriminating between metabolomic profiles of NTG and Tg groups in the PLS-DA model. (A–C, right panels): Metabolic pathways specifically affected in each FAD mouse model. *Reproduced from Ref.* (94). (For the color version of this figure, the reader is referred to the online version of this chapter.)

the earliest clinical stage of AD (0.5 dementia score). Authors surmised sulfa-tide deficiency as the earliest clinical stage of AD, suggesting that its deficiency in AD may begin prior to the appearance of clinical symptoms *(75)*.

3 CONCLUSIONS AND FUTURE RESEARCH

In this chapter, metabolomic approaches applied to AD investigation have been presented. It can be observed that Metabolomics has become an impor-tant tool in the discovery of new diagnostic and prognostic AD biomarkers in different samples. Up to date, human CSF has been the preferred biofluid for the investigation of AD metabolic biomarkers as it can closely reflect brain-specific changes. However, there is a clear trend toward the use of less inva-sive sample analysis. Research on the metabolites exchange through the blood–brain barrier and correlation in metabolic concentration between CSF and blood represent a challenge for future metabolomic studies.

To date, the frame in the discovery phase, within the global biomarker finding process, is the common aspect of all metabolomic works linked to AD. Thus, most of metabolomic studies have been focused in the search of new biomarker candidates in the apparition and development of AD with high impact on novel hypotheses generation. Currently, there is a growing need to reliably validate those findings by increasing the cohort of clinical patients, to measure and validate the potential of the revealed biomarkers.

Of particular importance in AD investigation is the discovery of biomarkers expressed before clinical symptoms appear. Presymptomatic events research represents an approach of special relevance in AD research for the potential impact on early therapy and disease progression understanding. Metabolomics works shown in this chapter have proved how the comparison of different met-abolic fingerprints and metabolic profiles procurement from different pheno-types (health, MCI, AD) can be used to detect specific metabolic changes, potential biomarkers, and altered metabolic pathways, leading to the under-standing of disease progression. Deepening in the mechanisms of disease will provide a source of potential new biomarkers helpful for early diagnosis, prog-nosis, prediction of response to therapy, and disease progression.

ACKNOWLEDGMENTS

This work was supported by AGL2011-29857-C03-01 project (Ministerio de Economía y Competitividad, Spain) and CSD2007-00063 FUN-C-FOOD (Programa CONSOLIDER, Ministerio de Educaciín y Ciencia, Spain). A. V. thanks the Ministerio de Economía y Competitividad for his FPI predoctoral fellowship.

REFERENCES

1. Selkoe, D. J. *Physiol. Rev.* **2001**, *81*, 741–766.
2. Blennow, K.; de Leon, M. J.; Zetterberg, H. *Lancet* **2006**, *368*, 387–403.

3. Wimo, A.; Reed, C. C.; Dodel, R.; Belger, M.; Jones, R. W.; Happich, M.; Argimon, J. M.; Bruno, G.; Novick, D.; Vellas, B.; Haro, J. M. *J. Alzheimers Dis.* **2013**, *36*, 385–399.

4. Francis, P. T.; Palmer, A. M.; Snape, M.; Wilcock, G. K. *J. Neurol. Neurosurg. Psychiatry* **1999**, *66*, 137–147.

5. Grundke-Iqbal, I.; Iqbal, K.; Tung, Y. C.; Quinlan, M.; Wisniewski, H. M.; Binder, L. I. *Proc. Natl. Acad. Sci. U.S.A.* **1986**, *83*, 4913–4917.

6. Jarrett, J. T.; Berger, E. P.; Lansbury, P. T., Jr. *Biochemistry* **1993**, *32*, 4693–4697.

7. Suzuki, T.; Oishi, M.; Marshak, D. R.; Czernik, A. J.; Nairn, A. C.; Greengard, P. *EMBO J.* **1994**, *13*, 1114–1122.

8. Carter, J.; Lippa, C. F. *Curr. Mol. Med.* **2001**, *1*, 733–737.

9. Kadowaki, H.; Nishitoh, H.; Urano, F.; Sadamitsu, C.; Matsuzawa, A.; Takeda, K.; Masutani, H.; Yodoi, J.; Urano, Y.; Nagano, T.; Ichijo, H. *Cell Death Differ.* **2005**, *12*, 19–24.

10. Calissano, P.; Matrone, C.; Amadoro, G. *Commun. Integr. Biol.* **2009**, *2*, 163–169.

11. Welsh, K. A.; Butters, N.; Hughes, J. P.; Mohs, R. C.; Heyman, A. *Arch. Neurol.* **1992**, *49*, 448–452.

12. Nestor, P. J.; Scheltens, P.; Hodges, J. R. *Nat. Med.* **2004**, *10*, S34–S41.

13. DeCarli, C. *Lancet Neurol.* **2003**, *2*, 15–21.

14. Petersen, R. C. *J. Intern. Med.* **2004**, *256*, 183–194.

15. Petersen, R. C.; Smith, G. E.; Waring, S. C.; Ivnik, R. J.; Tangalos, E. G.; Kokmen, E. *Arch. Neurol.* **1999**, *56*, 303–308.

16. Blennow, K.; Hampel, H. *Lancet Neurol.* **2003**, *2*, 605–613.

17. Jorm, A. F. *Gerontology* **2000**, *46*, 219–227.

18. Sapolsky, R. M., Ed.; *Stress, the Aging Brain, and the Mechanisms of Neuron Death*; MIT Press: Cambridge, 1992.

19. Reding, M.; Haycox, J.; Blass, J. *Arch. Neurol.* **1985**, *42*, 894–896.

20. Guarch, J.; Marcos, T.; Salamero, M.; Blesa, R. *Int. J. Geriatr. Psychiatry* **2004**, *19*, 352–358.

21. De Jager, C. A.; Hogervorst, E.; Combrinck, M.; Budge, M. M. *Psychol. Med.* **2003**, *33*, 1039–1050.

22. detoledo-Morrell, L.; Stoub, T. R.; Bulgakova, M.; Wilson, R. S.; Bennett, D. A.; Leurgans, S.; Wuu, J.; Turner, D. A. *Neurobiol. Aging* **2004**, *25*, 1197–1203.

23. Jack, C. R., Jr.; Shiung, M. M.; Gunter, J. L.; O'Brien, P. C.; Weigand, S. D.; Knopman, D. S.; Boeve, B. F.; Ivnik, R. J.; Smith, G. E.; Cha, R. H.; Tangalos, E. G.; Petersen, R. C. *Neurology* **2003**, *62*, 591–600.

24. Arai, H.; Nakagawa, T.; Kosaka, Y.; Higuchi, M.; Matsui, T.; Okamura, N.; Tashiro, M.; Sasaki, H. *Alzheimer Dis. Assoc. Disord.* **1997**, *3*, 211–213.

25. Buerger, K.; Teipel, S. J.; Zinkowski, R.; Blennow, K.; Arai, H.; Engel, R.; Hofmann-Kiefer, K.; McCulloch, C.; Ptok, U.; Heun, R.; Andreasen, N.; DeBernardis, J.; Kerkman, D.; Moeller, H.; Davies, P.; Hampel, H. *Neurology* **2002**, *59*, 627–629.

26. Okamura, N.; Arai, H.; Maruyama, M.; Higuchi, M.; Matsui, T.; Tanji, H.; Seki, T.; Hirai, H.; Chiba, H.; Itoh, M.; Sasaki, H. *Am. J. Psychiatry* **2002**, *159*, 474–476.

27. Fakhri El, G.; Kijewski, M. F.; Johnson, K. A.; Syrkin, G.; Killiany, R. J.; Becker, J. A.; Zimmerman, R. E.; Albert, M. S. *Arch. Neurol.* **2003**, *60*, 1066–1072.

28. Sperling, R. A.; Aisen, P. S.; Beckett, L. A.; Bennett, D. A.; Craft, S.; Fagan, A. M.; Iwatsubo, T.; Jack, C. R., Jr.; Kaye, J.; Montine, T. J.; Park, D. C.; Reiman, E. M.; Rowe, C. C.; Siemers, E.; Stern, Y.; Yaffe, K.; Carrillo, M. C.; Thies, B.; Morrison-Bogorad, M.; Wagster, M. V.; Phelps, C. H. *Alzheimers Dement.* **2011**, *7*, 280–292.

29. Morris, J. C. *Alzheimer Dis. Assoc. Disord.* **2005**, *19*, 163–165.

30. McKhann, G.; Drachman, D.; Folstein, M.; Katzman, R.; Price, D.; Stadlan, E. M. *Neurology* **1984**, *34*, 939–944.

31. Tiraboschi, P.; Hansen, L. A.; Thal, L. J.; Corey-Bloom, J. *Neurology* **2004**, *62*, 1984–1989.

32. Anoop, A.; Singh, P. K.; Jacob, R. S.; Maji, S. K. *Int. J. Alzheimers Dis.* **2010**, *2010*, 606802.

33. Shi, M.; Caudle, W. M.; Zhang, J. *Neurobiol. Dis.* **2009**, *35*, 157–164.

34. Lista, S.; Faltraco, F.; Prvulovic, D.; Hampel, H. *Prog. Neurobiol.* **2013**, *101–102*, 1–17.

35. Cedazo-Minguez, A.; Winblad, B. *Exp. Gerontol.* **2010**, *45*, 5–14.

36. Birks, J.; Harvey, R. J. *Cochrane Database Syst. Rev.* **2006**, *25*, CD001190.

37. Kaduszkiewicz, H.; Zimmermann, T.; Beck-Bornholdt, H. P.; van den Bussche, H. *BMJ* **2005**, *331*, 321–327.

38. Dwolatzky, T.; Clarfield, A. M. *Aging Health* **2012**, *8*, 233–237.

39. Gauthier, S.; Molinuevo, J. L. *Alzheimers Dement.* **2013**, *9*, 326–331.

40. Tariot, P. N.; Farlow, M. R.; Grossberg, G. T.; Graham, S. M.; McDonald, S.; Gergel, I. *JAMA* **2004**, *291*, 317–324.

41. Choi, J.; Twamley, E. W. *Neuropsychol. Rev.* **2013**, *23*, 48–62.

42. Buschert, V.; Bokde, A. L. W.; Hampel, H. *Nat. Rev. Neurol.* **2010**, *6*, 508–517.

43. Chong, M. S.; Sahadevan, S. *Lancet Neurol.* **2005**, *4*, 576–579.

44. Portelius, E.; Zetterberg, H.; Gobom, J.; Andreasson, U.; Blennow, K. *Expert Rev. Proteomics* **2008**, *5*, 225–237.

45. Zetterberg, H.; Blennow, K.; Hanse, E. *Exp. Gerontol.* **2010**, *45*, 23–29.

46. Zetterberg, H. *Scand. J. Clin. Lab. Invest.* **2009**, *69*, 18–21.

47. Shi, M.; Sui, Y. T.; Peskind, E. R.; Li, G.; Hwang, H.; Devic, I.; Ginghina, C.; Edgar, J. S.; Pan, C.; Goodlett, D. R.; Furay, A. R.; Gonzalez-Cuyar, L. F.; Zhang, J. *J. Alzheimers Dis.* **2011**, *27*, 299–305.

48. Meredith, J. E., Jr.; Sankaranarayanan, S.; Guss, V.; Lanzetti, A. J.; Berisha, F.; Neely, R. J.; Slemmon, J. R.; Portelius, E.; Zetterberg, H.; Blennow, K.; Soares, H.; Ahlijanian, M.; Albright, C. F. *PLoS One* **2013**, *8*, e76523.

49. Zhong, Z.; Ewers, M.; Teipel, S.; Burger, K.; Wallin, A.; Blennow, K.; He, P.; McAllister, C.; Hampel, H.; Shen, Y. *Arch. Gen. Psychiatry* **2007**, *8*, 718–726.

50. Zetterberg, H.; Andreasson, U.; Hansson, O.; Wu, G.; Sankaranarayanan, S.; Andersson, M. E.; Buchhave, P.; Londos, E.; Umek, R. M.; Minthon, L.; Simon, A. J.; Blennow, K. *Arch. Neurol.* **2008**, *8*, 1102–1107.

51. Raber, J.; Huang, Y.; Ashford, J. W. *Neurobiol. Aging* **2004**, *25*, 641–650.

52. Tan, Z. S.; Seshadri, S.; Beiser, A.; Wilson, P. W.; Kiel, D. P.; Tocco, M.; D'Agostino, R. B.; Wolf, P. A. *Arch. Intern. Med.* **2003**, *163*, 1053–1057.

53. Evans, R. M.; Emsley, C. L.; Gao, S.; Sahota, A.; Hall, K. S.; Farlow, M. R.; Hendrie, H. *Neurology* **2000**, *54*, 240–242.

54. Seshadri, S.; Beiser, A.; Selhub, J.; Jacques, P. F.; Rosenberg, I. H.; D'Agostino, R. B.; Wilson, P. W.; Wolf, P. A. *N. Engl. J. Med.* **2002**, *346*, 476–483.

55. Ghidoni, R.; Benussi, L.; Paterlini, A.; Albertini, V.; Binetti, G.; Emanuele, E. *Neurodegener. Dis.* **2011**, *8*, 413–420.

56. Ozdemir, V.; Suarez-Kurtz, G.; Stenne, R.; Somogyi, A. A.; Kayaalp, S. O.; Kolker, E. *OMICS* **2009**, *13*, 43–62.

57. Sutherland, G. T.; Janitz, M.; Kril, J. J. *J. Neurochem.* **2011**, *116*, 937–946.

58. Xu, X. H.; Huang, Y.; Wang, G.; Chen, S. D. *Neurosci. Bull.* **2012**, *28*, 641–648.

59. Barba, I.; Fernandez-Montesinos, R.; Garcia-Dorado, D.; Pozo, D. *J. Cell. Mol. Med.* **2008**, *12*, 1477–1485.

60. Ibañez, C.; Simo, C.; Cifuentes, A. *Electrophoresis* **2013**, *34*, 2799–2811.

61. Trushina, E.; Mielke, M. M. *Biochim. Biophys. Acta* **2013**. http://dx.doi.org/10.1016/j.bbadis.2013.06.014, (in press).
62. Dunckley, T.; Coon, K. D.; Stephan, D. A. *Drug Discov. Today* **2005**, *10*, 326–334.
63. Bazenet, C.; Lovestone, S. *Biomark. Med* **2012**, *6*, 441–454.
64. Fiehn, O. *Comp. Funct. Genomics* **2001**, *2*, 155–168.
65. Zhang, X.; Wei, D.; Yap, Y.; Li, L.; Guo, S.; Chen, F. *Mass Spectrom. Rev.* **2007**, *26*, 403–431.
66. Oresic, M.; Vidal-Puig, A.; Hänninen, V. *Expert. Rev. Mol. Diagn.* **2006**, *6*, 575–585.
67. Hassan-Smith, G.; Wallace, G. R.; Douglas, M. R.; Sinclair, A. J. *J. Neuroimmunol.* **2012**, *248*, 48–52.
68. Hardy, J.; Selkoe, D. J. *Science* **2002**, *297*, 353–356.
69. Povova, J.; Ambroz, P.; Bar, M.; Pavukova, V.; Sery, O.; Tomaskova, H.; Janout, V. *Biomed. Pap. Med. Fac. Univ. Palacky Olomouc Czech Repub.* **2012**, *156*, 108–114.
70. Irizarry, M. C. *NeuroRx* **2004**, *1*, 226–234.
71. Dunn, W. B.; Ellis, D. I. *Trends Anal. Chem.* **2005**, *24*, 285–293.
72. Zhang, S.; Nagana-Gowda, G. A.; Ye, T.; Raftery, D. *Analyst* **2010**, *135*, 1490–1498.
73. Zhang, B.; Powers, R. *Future Med. Chem.* **2012**, *4*, 1273–1306.
74. Ghauri, F. Y.; Nicholson, J. K.; Sweatman, B. C.; Wood, J.; Beddell, C. R.; Lindon, J. C.; Cairns, N. J. *NMR Biomed.* **1993**, *6*, 163–167.
75. Han, X.; Holtzman, D. M.; McKeel, D. W., Jr.; Kelley, J.; Morris, J. C. *J. Neurochem.* **2002**, *82*, 809–818.
76. Nicoli, F.; VionDury, J.; ConfortGouny, S.; Maillet, S.; Gastaut, J. L.; Cozzone, P. J. *C. R. Acad. Sci. III* **1996**, *319*, 623–631.
77. Tukiainen, T.; Tynkkynen, T.; Mäkinen, V. P.; Jylänki, P.; Kangas, A.; Hokkanen, J.; Vehtari, A.; Gröhn, O.; Hallikainen, M.; Soininen, H.; Kivipelto, M.; Groop, P. H.; Kaski, K.; Laatikainen, R.; Soininen, P.; Pirttilä, T.; Ala-Korpela, M. *Biochem. Biophys. Res. Commun.* **2008**, *375*, 356–361.
78. Wisniewski, T.; Newman, K.; Javitt, N. B. *J. Alzheimers Dis.* **2013**, *33*, 881–888.
79. Salek, R. M.; Xia, J.; Innes, A.; Sweatman, B. C.; Adalbert, R.; Randle, S.; McGowan, E.; Emson, P. C.; Griffin, J. L. *Neurochem. Int.* **2010**, *56*, 937–947.
80. Botosoa, E. P.; Zhu, M.; Marbeuf-Gueye, C.; Triba, M. N.; Dutheil, F.; Duyckäerts, C.; Beaune, P.; Loriotb, M. A.; Le Moyec, L. *IRBM* **2012**, *33*, 281–286.
81. Fukuhara, K.; Ohno, A.; Ota, Y.; Senoo, Y.; Maekawa, K.; Okuda, H.; Kurihara, M.; Okuno, A.; Niida, S.; Saito, Y.; Takikawa, O. *J. Clin. Biochem. Nutr.* **2013**, *52*, 133–138.
82. Jukarainen, N. M.; Korhonen, S. P.; Laakso, M. P.; Korolainen, M. A.; Niemitz, M.; Soininen, P. S.; Tuppurainen, K.; Vepsäläinen, J.; Pirttila, T.; Laatikainen, R. *Metabolomics* **2008**, *4*, 150–160.
83. Li, N. J.; Liu, W. T.; Li, W.; Li, S. Q.; Chen, X. H.; Bi, K. S.; He, P. *Clin. Biochem.* **2010**, *43*, 992–997.
84. Czech, C.; Berndt, P.; Busch, K.; Schmitz, O.; Wiemer, J.; Most, V.; Hampel, H.; Kastler, J.; Senn, H. *PLoS One* **2012**, *7*, e31501.
85. Ibañez, C.; Simo, C.; Martín-Álvarez, P. J.; Kivipelto, M.; Winblad, B.; Cedazo-Mínguez, A.; Cifuentes, A. *Anal. Chem.* **2012**, *84*, 8532–8540.
86. Graham, S. F.; Chevallier, O. P.; Roberts, D.; Hölscher, C.; Elliott, C. T.; Green, B. D. *Anal. Chem.* **2013**, *85*, 1803–1811.
87. Fonteh, A. N.; Harrington, R. J.; Tsai, A.; Liao, P.; Harrington, M. G. *Amino Acids* **2007**, *32*, 213–224.

88. Kaddurah-Daouk, R.; Rozen, S.; Matson, W.; Han, X.; Hulette, C. M.; Burke, J. R.; Doraiswamy, P. M.; Welsh-Bohmer, K. A. *Alzheimers Dement.* **2011**, *7*, 309–317.

89. Samakashvili, S.; Ibañez, C.; Simo, C.; Gil-Bea, F. J.; Winblad, B.; Cedazo-Mínguez, A.; Cifuentes, A. *Electrophoresis* **2011**, *32*, 2757–2764.

90. Ibañez, C.; Simo, C.; Barupal, D. K.; Fiehn, O.; Kivipelto, M.; Cedazo-Mínguez, A.; Cifuentes, A. *J. Chromatogr. A* **2013**, *1302*, 65–71.

91. Myint, K. T.; Aoshima, K.; Tanaka, S.; Nakamura, T.; Oda, Y. *Anal. Chem.* **2009**, *81*, 1121–1129.

92. Kaddurah-Daouk, R.; Zhu, H.; Sharma, S.; Bogdanov, M.; Rozen, S. G.; Matson, W.; Oki, N. O.; Motsinger-Reif, A. A.; Churchill, E.; Lei, Z.; Appleby, D.; Kling, M. A.; Trojanowski, J. Q.; Doraiswamy, P. M.; Arnold, S. E. *Transl. Psychiatry* **2013**, *3*, e244.

93. Motsinger-Reif, A. A.; Zhu, H.; Kling, M. A.; Matson, W.; Sharma, S.; Fiehn, O.; Reif, D. M.; Appleby, D. H.; Doraiswamy, P. M.; Trojanowski, J. Q.; Kaddurah-Daouk, R.; Arnold, S. E. *Acta Neuropathol. Commun.* **2013**, *1*, 28.

94. Trushina, E.; Nemutlu, E.; Zhang, S.; Christensen, T.; Camp, J.; Mesa, J.; Siddiqui, A.; Tamura, Y.; Sesaki, H.; Wengenack, T. M.; Dzeja, P. P.; Poduslo, J. F. *PLoS One* **2012**, *7*, e32737.

95. Trushina, E.; Dutta, T.; Persson, X. M.; Mielke, M. M.; Petersen, R. C. *PLoS One* **2013**, *8*, e63644.

96. Oresic, M.; Hyötyläinen, T.; Herukka, S. K.; Sysi-Aho, M.; Mattila, I.; Seppänan-Laakso, T.; Julkunen, V.; Gopalacharyulu, P. V.; Hallikainen, M.; Koikkalainen, J.; Kivipelto, M.; Helisalmi, S.; Lötjönen, J.; Soininen, H. *Transl. Psychiatry* **2011**, *1*, e57.

97. Greenberg, N.; Grassano, A.; Thambisetty, M.; Lovestone, S.; Legido-Quigley, C. *Electrophoresis* **2009**, *30*, 1235–1239.

98. Sato, Y.; Nakamura, T.; Aoshima, K.; Oda, Y. *Anal. Chem.* **2010**, *82*, 9858–9864.

99. Han, X.; Rozen, S.; Boyle, S. H.; Hellegers, C.; Cheng, H.; Burke, J. R.; Welsh-Bohmer, K. A.; Doraiswamy, P. M.; Kaddurah-Daouk, R. *PLoS One* **2011**, *6*, e21643.

100. Sato, Y.; Suzuki, I.; Nakamura, T.; Bernier, F.; Aoshima, K.; Oda, Y. *J. Lipid Res.* **2012**, *53*, 567–576.

101. Grimm, M. O.; Grösgen, S.; Riemenschneider, M.; Tanila, H.; Grimm, H. S.; Hartmann, T. *J. Chromatogr. A* **2011**, *1218*, 7713–7722.

102. Lei, Z.; Huhman, D. V.; Sumner, L. W. *J. Biol. Chem.* **2011**, *286*, 25435–25442.

103. Dettmer, K.; Aronov, P. A.; Hammock, B. D. *Mass Spectrom. Rev.* **2007**, *26*, 51–78.

104. Barding, G. A., Jr.; Salditos, R.; Larive, C. K. *Anal. Bioanal. Chem.* **2012**, *404*, 1165–1179.

105. Smolinska, A.; Blanchet, L.; Buydens, L. M.; Wijmenga, S. S. *Anal. Chim. Acta.* **2012**, *750*, 82–97.

106. Fiehn, O. *Plant Mol. Biol.* **2002**, *48*, 155–171.

107. Kell, D. B.; Brown, M.; Davey, H. M.; Dunn, W. B.; Spasic, I.; Oliver, S. G. *Nat. Rev. Microbiol.* **2005**, *3*, 557–565.

108. Ellis, D. I.; Dunn, W. B.; Griffin, J. L.; Allwood, J. W.; Goodacre, R. *Pharmacogenomics* **2007**, *8*, 1243–1266.

109. Barderas, M. G.; Laborde, C. M.; Posada, M.; de la Cuesta, F.; Zubiri, I.; Vivanco, F.; Alvarez-Llamas, G. *J. Biomed. Biotechnol.* **2011**, *2011*, 790132.

110. Dettmer, K.; Hammock, B. D. *Environ. Health Perspect.* **2004**, *112*, A396–A397.

111. Luedemann, A.; von Malotky, L.; Erban, A.; Kopka, J. *Methods Mol. Biol.* **2012**, *860*, 255–286.

112. Inoue, K.; Tsutsui, H.; Akatsu, H.; Hashizume, Y.; Matsukawa, N.; Yamamoto, T.; Toyo'oka, T. *Sci. Rep.* **2013**, *3*, 2364.

113. Roche, S.; Gabelle, A.; Lehman, S. *Proteomics Clin. Appl.* **2008**, *2*, 428–436.
114. Ramström, M.; Zuberovic, A.; Grönwall, C.; Hanrieder, J.; Bergquist, J.; Hober, S. *Biotechnol. Appl. Biochem.* **2009**, *52*, 159–166.
115. Simo, C.; Ibañez, C.; Gómez-Martínez, A.; Ferragut, J. A.; Cifuentes, A. *Electrophoresis* **2011**, *32*, 1765–1777.
116. Advokat, C.; Pellegrin, A. I. *Neurosci. Biobehav. Rev.* **1992**, *16*, 13–24.
117. Jack, C. R., Jr.; Knopman, D. S.; Jagust, W. J.; Shaw, L. M.; Aisen, P. S.; Weiner, M. W.; Petersen, R. C.; Trojanowski, J. Q. *Lancet Neurol.* **2010**, *9*, 119–128.
118. Perry, G.; Cash, A. D.; Smith, M. A. *J. Biomed. Biotechnol.* **2002**, *2*, 120–123.
119. Pohanka, M. *Curr. Med. Chem.* **2013**, *21*, 356–364.
120. Jomova, K.; Vondrakova, D.; Lawson, M.; Valko, M. *Mol. Cell. Biochem.* **2010**, *345*, 91–104.
121. Butterfield, D. A. *Free Radic. Res.* **2002**, *36*, 1307–1313.
122. Smith, D. G.; Cappai, R.; Barnham, K. J. *Biochim. Biophys. Acta* **2007**, *1768*, 1976–1990.
123. Wang, L. P.; Schmidt, J. F. *Dan. Med. Bull.* **1997**, *44*, 79–81.
124. Evans, R. W. *Neurol. Clin.* **1998**, *16*, 83–105.
125. Boon, J. M.; Abrahams, P. H.; Meiring, J. H.; Welch, T. *Clin. Anat.* **2004**, *17*, 544–553.
126. Aftab, M. F.; Waraich, R. S. *Am. J. Neurosci.* **2012**, *3*, 54–62.
127. Hye, A.; Lynham, S.; Thambisetty, M.; Causevic, M.; Campbell, J.; Byers, H. L.; Hooper, C.; Rijsdijk, F.; Tabrizi, S. J.; Banner, S.; Shaw, C. E.; Foy, C.; Poppe, M.; Archer, N.; Hamilton, G.; Powell, J.; Brown, R. G.; Sham, P.; Ward, M.; Lovestone, S. *Brain* **2006**, *129*, 3042–3050.
128. Humpel, C.; Hochstrasser, T. *World J. Psychiatry* **2011**, *1*, 8–18.
129. Blennow, K.; Hampel, H.; Weiner, M.; Zetterberg, H. *Nat. Rev. Neurol.* **2010**, *6*, 131–144.
130. Foley, P. *Biochim. Biophys. Acta* **2010**, *1801*, 750–753.
131. Corder, E. H.; Saunders, A. M.; Strittmatter, W. J.; Schmechel, D. E.; Gaskell, P. C.; Small, G. W.; Roses, A. D.; Haines, J. L.; Pericak-Vance, M. A. *Science* **1993**, *261*, 921–923.
132. Bertram, L.; Tanzi, R. E. *Nat. Rev. Neurosci.* **2008**, *9*, 768–778.
133. Kim, J.; Basak, J. M.; Holtzman, D. M. *Neuron* **2009**, *63*, 287–303.
134. Hartmann, T.; Kuchenbecker, J.; Grimm, M. O. *J. Neurochem.* **2007**, *103*, 159–170.
135. Sagin, F. G.; Sozmen, E. Y. *Curr. Alzheimer Res.* **2008**, *5*, 4–14.
136. Sparks, D. L.; Sabbagh, M. N.; Connor, D. J.; Lopez, J.; Launer, L. J.; Browne, P.; Wasser, D.; Johnson-Traver, S.; Lochhead, J.; Ziolwolski, C. *Arch. Neurol.* **2005**, *62*, 753–757.
137. Buxbaum, J. D.; Cullen, E. I.; Friedhoff, L. T. *Front. Biosci.* **2002**, *7*, a50–a59.
138. Simons, M.; Schwärzler, F.; Lütjohann, D.; von Bergmann, K.; Beyreuther, K.; Dichgans, J.; Wormstall, H.; Hartmann, T.; Schulz, J. B. *Ann. Neurol.* **2002**, *52*, 346–350.
139. Grosgen, S.; Grimm, M. O.; Friess, P.; Hartmann, T. *Biochim. Biophys. Acta* **2010**, *1801*, 966–974.
140. Sing, C. F.; Davignon, J. *Am. J. Hum. Genet.* **1985**, *37*, 268–285.
141. Beach, T. G. *J. Alzheimers Dis.* **2013**, *33*, S219–S233.
142. Chishti, M. A.; Yang, D. S.; Janus, C.; Phinney, A. L.; Horne, P.; Pearson, J.; Strome, R.; Zuker, N.; Loukides, J.; French, J.; Turner, S.; Lozza, G.; Grilli, M.; Kunicki, S.; Morissette, C.; Paquette, J.; Gervais, F.; Bergeron, C.; Fraser, P. E.; Carlson, G. A.; George-Hyslop, P. S.; Westaway, D. *J. Biol. Chem.* **2001**, *276*, 21562–21570.
143. Games, D.; Adams, D.; Alessandrini, R.; Barbour, R.; Berthelette, P.; Blackwell, C.; Carr, T.; Clemens, J.; Donaldson, T.; Gillespie, F.; Guido, T.; Hagopian, S.;

Johnsonwood, K.; Khan, K.; Lee, M.; Leibowitz, P.; Lieberburg, I.; Little, S.; Masliah, E.; McConlogue, L.; Montoyazavala, M.; Mucke, L.; Paganini, L.; Penniman, E.; Power, M.; Schenk, D.; Seubert, P.; Snyder, B.; Soriano, F.; Tan, H.; Vitale, J.; Wadsworth, S.; Wolozin, B.; Zhao, J. *Nature* **1995**, *373*, 523–527.

144. Hsiao, K.; Chapman, P.; Nilsen, S.; Eckman, C.; Harigaya, Y.; Younkin, S.; Yang, F.; Cole, G. *Science* **1996**, *274*, 99–102.

145. Janus, C.; Phinney, A. L.; Chishti, M. A.; Westaway, D. *Curr. Neurol. Neurosci. Rep.* **2001**, *1*, 451–457.

146. Mucke, L.; Masliah, E.; Yu, G. Q.; Mallory, M.; Rockenstein, E. M.; Tatsuno, G.; Hu, K.; Kholodenko, D.; Johnson-Wood, K.; McConlogue, L. *J. Neurosci.* **2000**, *20*, 4050–4058.

147. Refolo, L. M.; Malester, B.; LaFrancois, J.; Bryant-Thomas, T.; Wang, R.; Tint, G. S.; Sambamurti, K.; Duff, K.; Pappolla, M. A. *Neurobiol. Dis.* **2000**, *7*, 321–331.

148. Simons, M.; Keller, P.; De Strooper, B.; Beyreuther, K.; Dotti, C. G.; Simons, K. *Proc. Natl. Acad. Sci. U.S.A.* **1998**, *95*, 6460–6464.

149. Frears, E. R.; Stephens, D. J.; Walters, C. E.; Davies, H.; Austen, B. M. *Neuroreport* **1999**, *10*, 1699–1705.

150. Morris, J. C. *Neurology* **1993**, *43*, 2412–2414.

Chapter 11

Application of Metabolomics to Cardiovascular and Renal Disease Biomarker Discovery

Gloria Alvarez-Llamas*, Maria G. Barderas†, Maria Posada-Ayala*, Marta Martin-Lorenzo* and Fernando Vivanco*,‡

*Department of Immunology, IIS-Fundación Jiménez Díaz, Madrid, Spain
†Department of Vascular Physiopathology, Hospital Nacional de Parapléjicos, SESCAM, Toledo, Spain
‡Department of Biochemistry and Molecular Biology I, UCM, Madrid, Spain

1 INTRODUCTION

1.1 Cardiorenal Syndrome

Chronic kidney disease (CKD) is increasingly recognized as a major public health problem. Defined as an irreversible progressive loss of renal function for 3 months or longer, CKD shows a glomerular filtration rate (GFR) less

Comprehensive Analytical Chemistry, Vol. 64. http://dx.doi.org/10.1016/B978-0-444-62650-9.00011-7

than 60 mL/min/1.73 m^2 or the presence of other markers of kidney damage as albuminuria or proteinuria (albumin to creatinine ratio (ACR) equal or higher than 30 mg/g). CKD is categorized into five stages of increasing severity (CKD1–5) and its prevalence is continuously rising. In early stages, it develops in a silent and asymptomatic progression, which enormously impedes early diagnosis and intervention. In general terms, clinicians are alerted by the presence of CKD in the basis of serum creatinine data and estimated GFR. However, serum creatinine is a late marker and it is affected by muscle mass, age, and race; assessment of renal function based on GFR is less reliable when this value is higher than 60 mL/min/1.73 m^2. Addition of cystatin C to the combination of creatinine and ACR measures has been recently proposed to improve the predictive accuracy for all-cause mortality and end-stage renal disease (ESRD) *(1)*, pointing to the continuous demand of candidate markers to better define disease progression. In a parallel scenario, cardiovascular disease (CVD) continues to be the leading cause of morbidity and mortality worldwide, and, particularly, rates of cardiovascular (CV) events and death consistently increase as kidney function worsens. In other words, CKD acts as a risk multiplier and, independent from other risk factors, patients with stage 4 or 5 CKD have a death risk for CV complications that is two- to fourfold higher than in the general population *(2,3)*. In particular, subjects with mild-to-moderate renal dysfunction have a higher probability of dying by a CV event than dying due to the kidney disease itself before reaching ESRD. In fact, the American Heart Association stated that CKD patients should be regarded as the highest risk group for subsequent CVD *(4)*. The risk of unfavorable CV prognosis increases as renal insufficiency progresses toward ESRD, reaches its maximum, and persists following successful renal transplantation.

The cross talk between the heart and the kidneys in the systems biology context is clearly evidenced but not fully understood. Observational and clinical data showed that acute/chronic cardiac disease directly contributes to acute/chronic worsening of kidney function and vice versa, constituting the cardiorenal syndrome *(5)*.

1.2 CV Risk and On-Time Action: The Real Challenge

One of the main questions clinical cardiologists are concerned with is how to prevent a fatal event or, at least, how to improve current monitoring of people at high risk to act on time. Clinical nephrologists are concerned about how to predict whether CKD will progress or remain stable and how to accurately evaluate the risk to suffer a CV event in renal disease patients, that is, identification of patients who will die prematurely from CVD instead of progressing to CKD stage 5 *(6)*. The same question in two different but interrelated contexts, with not necessarily the same answer. The discovery and use of novel markers of CV risk complementing the best existing ones will allow identification of individuals at risk for a CV event and who could be targeted for preventive measures based on a preventive lifestyle or timely pharmacological treatment.

This is a multidisciplinary task, demanding for deep collaboration between clinicians and basic researchers to pursue the final goal of a personalized CV medicine. Different approaches can be raised. In terms of defining a panel of markers able to provide with information related to CV risk stratification, prospective studies are a good option aimed to, collect biological samples taken when individuals are in a healthy condition (at different levels of risk) and further classify them at "risk" or "no risk," later in time and once the information on whom of those individuals has suffered an acute event is available. To identify markers of CVD itself, comparison of healthy and pathological biological fluids, circulating cells, microvesicles, or tissues can be approached. Pathological could refer to arterial atheromatous plaques or urine, serum, and plasma collected at the time of an acute event (myocardial infarction (MI), angina, ictus, etc.). Identification of novel markers of prognosis, recovery, and response to treatment would imply a different strategy, based on serial sample collection at consecutive time points. Sample type of choice (fluid or tissue) will mark a key difference in the pursued information: tissue studies will be helpful to discover and better understand molecular mechanisms underlying the physiopathology but could not be the option of choice in the search for CV markers of diagnosis or clinical evolution.

This chapter does not intend to thoroughly review and compile all studies published so far, but to discuss novel aspects in the search for novel markers related to CVD, risk, and progression markers, in and out of the renal disease context. Note that separate consideration is given to these two different scenarios: CV risk in the healthy population and in the chronic renal disease patients as explained in the following sections.

2   WHAT TO EXPECT FROM THE "OMICS": CHANGING THE PERSPECTIVE

Clinical proteomics is a discipline based on the application of the proteomics technologies to diagnosis, prognosis, treatment, and course of human diseases. Isoforms and posttranslational modifications are detectable only by studying the proteins directly and can be indicative of specific protein functions. Furthermore, there is evidence that only one-third of proteins with altered expression display a concomitant change in mRNA expression, so proteomics reflect the actual cellular processes more accurately than do genomics or transcriptomics. Among the most recent of the three "omics" in systems biology, metabolomics accounts for great impact in the scientific community and raising interest, able to translate a particular phenotype into a metabolic signature.

Classical focus investigates one or various molecules for which there is evidence or proved connection with the disease under research. These can be molecules potentially involved in development and progression of CKD and related CV events discovered in post hoc studies. Most of these studies are, therefore, based on the preselection of potential targets whose potential implication had been previously evidenced. In all cases, it was known what to look for and what to follow up, and this is a huge limitation considering the complexity of the interactions and underlying mechanisms operating in the cross talk among the different organs. The omics strategy is somehow "blind" at the discovery phase. This means that no potential marker and no key target are preselected, but all the proteins and/or metabolites' sets are investigated as a whole. In this way, not only particular pathways or responding molecules commonly measured in routine biochemical patient's analysis are being investigated, but also those whose relationship with the pathophysiological processes taking place are still unknown.

The proteomics/metabolomics approach allows focusing to the whole map and to the whole picture in a first discovery phase where any subjacent molecule is discarded and any information is being missed *a priori*. In the validation step, the discovery phase makes all the sense and it is then when candidate makers are further investigated, confirming previous results in wider cohorts and setting valid conclusions for the clinical practice. Metabolomics approach applied to the study of CVD *(7)* or renal pathology *(8)* has been recently reviewed, pointing to a limited list of metabolomics studies and candidate markers of a reduced number of kidney diseases, as referred in more detail in the following sections. Although scarce and not sufficient, these studies constitute a clear proof of concept.

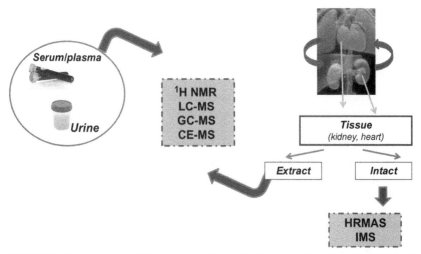

FIGURE 1 Overview of metabolomics approaches in the discovery of cardiovascular and renal disease biomarkers.

3   METABOLOMICS IN CVD

CVDs remain the leading cause of death in developed countries and are expected to become so in emerging countries in the next years *(9,10)*. However, operating mechanisms are not fully understood. One of the main problems in clinical practice is that symptoms occur late in the course of disease. For this reason, early diagnosis will provide the opportunity to institute a preventive lifestyle and pharmacological treatments. In this sense, metabolomics research can be aimed to discover a characteristic pattern to be used with diagnostic purpose, in substitution of traditional clinical approaches of higher cost (i.e., angiography). In the succeeding text, relevant findings in CVD are discussed and compiled in Table 1.

3.1   Metabolic Fingerprinting in CVD

One of the few metabolomics studies in the field of CVD comprised the comparative study of metabolomics fingerprinting of plasma samples from non-ST elevation acute coronary syndrome (NSTEACS) patients, stable atherosclerosis patients, and healthy patients by gas chromatography-mass spectrometry (GC-MS) *(35)*. Citric acid, 4-hydroxyproline (4OH-Pro), aspartic acid, and fructose were found to be decreased in NSTEACS patients whereas lactate, urea, glucose, and valine were found to be increased. Decreased 4OH-Pro finding was especially interesting because circulating 4OH-Pro is thought to prevent the binding of LDL to lipoprotein previously deposited in the vascular wall and release already deposited LDL from the atherosclerotic lesions. It is also a component of collagen, which confers stability to the atherosclerotic plaque. Using the same technique, Laborde *et al.* *(11)* performed a metabolomics analysis in plasma from 35 patients and 35 healthy donors, finding 15 metabolites with statistical differences ($p < 0.05$) between experimental groups. Additionally, validation by GC-MS and liquid chromatography-mass spectrometry (LC-MS) in an independent cohort of patients (15 NSTEACS patients and 15 healthy controls used for GC-MS validation and 10 NSTEACS patients and 10 healthy controls used for LC-MS validation) permitted to identify a potential panel of biomarkers formed by 5-OH-tryptophan, 2-OH-butyric acid, and 3-OH-butyric acid. This panel of biomarkers reflects the oxidative stress and the hypoxic state that suffers the myocardial cells and consequently constitutes a metabolomics signature of the atherogenesis process that could be used for early diagnosis of ACS.

The high-resolution power of capillary electrophoresis-mass spectrometry (CE-MS) makes it a powerful technique for the separation and analysis of charged metabolites; however, only a few metabolomics applications have been published to date. Polypeptide fraction isolated from urine or plasma was analyzed by CE-MS and used to discriminate between coronary artery

TABLE 1 Selected Metabolomics Approaches in the Study of Cardiovascular Disease

Sample	Pathology	No. of Subjects	Methodology	Relevant Metabolites	References
Human plasma	NSTEACS	45	GC-MS	Citric acid ↓ 4-Hydroxyproline ↓ Aspartic acid ↓ Fructose ↑ Lactate ↑ Urea ↑ Glucose ↑ Valine ↑	(11)
Human urine and plasma	Atherosclerosis		CE-MS	Collagen α1	(12)
Human aneurysm wall	Human intramural thrombus	8	LC-QTOF-MS	Hippuric acid	(13)
Rat urine	Myocardial infarction	25	LC-QTOF-MS	Creatine Uridine Glutamate Pantothenic acid Oxalosuccinic acid Nicotinamide mononucleotide Phenylacetylglycine Xanthosine Shexiang Baoxin Pill	(14)
Human blood samples	Myocardial infarction	36	MS-based metabolite profiling platform	B: Alanine, aminoisobutyric acid, hypoxanthine, isoleucine/leucine, malonic acid, threonine, and trimethylamine N-oxide A: 1-Methylhistamine, choline, inosine, serine, proline, and xanthine C: Taurine, ribose-5-phosphate DMPC, lactic acid, AMP, malic acid, succinic acid, and glycerate-2-phosphate	(15)

Sample	Condition	n	Technique	Metabolites	Ref.
Rat plasma and urine	Atherosclerosis		UFLC/MS-IT-TOF	Plasma: leucine↓, phenylalanine ↓, tryptophan ↓, acetylcarnitine ↓, butyrylcarnitine ↓, propionylcarnitine ↓, spermine ↓ Ursodeoxycholic acid ↑, chenodeoxycholic acid ↑ Urine: 3-*O*-methyl-dopa ↓, ethyl *N2*-acetyl argininate ↓, and leucylproline ↓ Glucuronate ↓, *N*(6)-(*N*-threnylcarbonyl)-adenosine ↓, methyl-hippuric acid ↓, and hippuric acid ↑	(16)
Mice liver	Atherosclerosis		LC-MS/MS		(17)
Rabbit and rat myocardial cells and tissue	Atherosclerosis		ESI-MS	Plasmalogens	(18)
Human plasma	Cardiovascular status in healthy voluntaries	864	NMR	3-Hydroxybutyrate α-Ketoglutarate Threonine Dimethylglycine	(19)
	Atherosclerosis		NMR GC-MS	Glutamate Ketoglutarate Succinyl-CoA 4-OH-ʟ-proline Creatine Pyruvate Malate Glycolate	(20)

Continued

TABLE 1 Selected Metabolomics Approaches in the Study of Cardiovascular Disease—Cont'd

Sample	Pathology	No. of Subjects	Methodology	Relevant Metabolites	References
Human urine	Cardiovascular risk	4680	¹H NMR	Isoleucine Valine Methylguanidine Hippurate 2-Hydroxibutirate Succine Scyllo-inositol Trans-aconitate Proline betaine	(21)
	Left ventricular dysfunction			Glucose ↓ Lactate ↓ Free fatty acids ↓ Ketones ↓ 3-Hydroxibutyrate ↓ Leucine/isoleucine ↓ Glutamate ↓ Alanine ↑	(22)
Peripheral blood	CAD	117	Quantitative mass spectrometry-based metabolic profiling	Arginine Ornithine Alanine Proline Leucine/isoleucine Valine Glutamate/glutamine Phenylalanine Glycine	(23)

Sample	Disease	N	Technique	Metabolites	Reference
	Human atrial fibrillation		NMR	β-Hydroxybutyrate Ketogenic amino acids Glycine	(24)
Human plasma	Human heart failure		LC-MS/MS	Hexanoic acid Succinate Hypoxanthine L-Lysine Uric acid L-Arginine Dodecanoic acid Palmitic acid Linoleic acid Stearic acid	(25)
Human plasma	Diabetes	2422	LC-MS/MS	Isoleucine Leucine Valine Tyrosine Phenylalanine	(26)
Rat myocardial tissue	Myocardial infarction		MSI	Phospholopase-A2 Arachidonic acid	(27)
Mouse cardiac myocyte-specific RAMP2	Heart failure		MSI	Cardiolipin ↓ cAMP ↓	(28)
ApoE mice Human femoral arteries	Atherosclerosis		MSI	Mouse and human: Phosphatidylcholine Cholesteryl linoleate Cholesterol oleate Human: triacylglycerol	(29)
	Atherosclerosis	15	TOF-SIMS (MSI)		(30)

Continued

TABLE 1 Selected Metabolomics Approaches in the Study of Cardiovascular Disease—Cont'd

Sample	Pathology	No. of Subjects	Methodology	Relevant Metabolites	References
Atherosclerosic aortic sinuses from ApoE knockout mice				Cholesterol esters (cholesteryl palmitate and cholesteryl oleate): markers of atherosclerosis progression	
Human atherosclerotic carotid from diabetic/nondiabetic	Atherosclerosis	4 diabetic, 4 nondiabetic	TOF-SIMS (MSI)	Nonesterified fatty acids. Differences in concentration for diabetic and non diabetic	(31)
Rat aorta: control Atherosclerotic aorta: human	Atherosclerosis	1 + 1	TOF-SIMS (MSI)	Cholesterol, oxysterol, and diacylglycerols detected and localized	(32)
Hearts. Rat model of infracted myocardium	Coronary-heart disease. Myocardial infraction	—	MALDI-MSI	Phospholipid: intact and lysophospholipids	(27)
Carotids from rat	Carotid artery ligation	—	MALDI-MSI	Triacylglycerides accumulation in response to cilostazol	(33)
Human carotids	Atherosclerosis	4 patients	TOF-SIMS	Distribution of fatty acids, cholesterol, vitamin. E, phosphatidic acids, phosphatidylinositols, and triglycerides	(34)

disease (CAD) and non-CAD patients who had presented with clinical symptoms and had been subjected to coronary angiography *(12)*. The stability of urine samples and their resistance to oxidizing processes or precipitation demonstrate the advantages of this biological fluid for proteomic analysis. Polypeptide profiling in urine resulted to be more reproducible than in plasma with no significant loss of polypeptides over time when performing consecutive analyses in a 24 h time period, which also demonstrates the reproducibility of the CE-MS. In total, 200 of the most abundant polypeptides were detected and a set of 17 urinary polypeptides allowed separation of CAD and non-CAD groups. Among them, collagen α1 types I and III were found to be increased in CAD samples, which was corroborated by their increased expression found in atherosclerotic plaques. This fact points to an important role of collagen in atherosclerosis development.

Ciborwoski *et al.* used the LC-MS-based metabolomics fingerprinting, as a tool to obtain a unique pattern characterizing metabolism as an application that may extend knowledge about process responsible for the development and progression of abdominal aortic aneurysm (AAA). Liquid chromatography coupled with accurate mass quadrupole time-of-flight MS (LC-QTOF-MS) was used as a metabolomics approach to analyze metabolites released by human intraluminal thrombus (ILT) (divided into luminal and abluminal layers, $n=9$), aneurysm wall ($n=8$), and healthy wall ($n=4$). Statistical analysis was used to compare luminal with abluminal ILT layers, ILT with aneurysm wall, and aneurysm wall with healthy wall to select the metabolites exchanged between tissue and external medium. Identified compounds are related to inflammation and oxidative stress and indicate the possible role of fatty acids amides in AAA. Some metabolites such as hippuric acid had not been previously described in aneurysm; others have arisen, indicating a very promising line of research *(13)*.

The same methodology was applied to obtain a systematic view of the development and progression of MI. By combination with partial least squares discriminant analysis (PLS-DA), 16 biomarkers in rat urine ($n=25$) were identified, and 8 of them were related to the pathway of energy metabolism (creatine, uridine, glutamate, pantothenic acid, oxalosuccinic acid, nicotinamide mononucleotide, phenylacetylglycine, and xanthosine). The developed method was also used to analyze the therapeutic effects of traditional Chinese medicine named Shexiang Baoxin Pill (SBP). The results showed that SBP administration could provide satisfactory effects on MI through partially regulating the perturbed pathway of energy metabolism *(14)*.

3.2   Metabolic Profiling in CVD

Metabolite profiling research has been used in CVD in order to identify and quantify metabolites, which could be used as new biomarkers (Figure 1). A metabolite profiling of peripheral blood obtained from individuals undergoing planned myocardial infarction (PMI) was established *(15)*. Serial blood samples of 36 patients were obtained at various intervals after PMI and changes in circulating levels of metabolites were identified by an MS-based metabolite profiling platform. Most of the alterations were observed in tricarboxylic acid cycle, purine and pyrimidine catabolism, and pentose phosphate pathway. Ten minutes after the onset of the myocardial injury, seven metabolites were significantly changed ($p < 0.005$): alanine, aminoisobutyric acid, hypoxanthine, isoleucine/leucine, malonic acid, threonine, and trimethylamine *N*-oxide (TMANO). All these alterations were especially interesting, since they were observed before any significant rise in the clinically available biomarkers in plasma (CKMB and troponin T) was detectable. After 60 min, six new metabolites were also significantly changed ($p < 0.005$): 1-methylhistamine, choline, inosine, serine, proline, and xanthine. The anatomic origin of the early metabolic changes observed was further explored in a subgroup of 13 patients by comparison of metabolite levels in samples obtained, simultaneously, from peripheral blood and from a catheter placed in the coronary sinus (CS). Eight additional metabolites, taurine, ribose-5-phosphate, DCMP, lactic acid, AMP, malic acid, glutamine, and glutamic acid, and six additional metabolites, glycerol-3-phosphate, orotic acid, succinic acid, glycerate-2-phosphate, taurine, and malic acid, showed greater than 1.3-fold transmyocardial enrichment 10 and 60 min after PMI, respectively.

Plasma and urine samples from atherosclerotic and control rats were comparatively analyzed by ultrafast liquid chromatography coupled with ion trap-time-of-flight (IT-TOF) MS (UFLC-IT/TOF-MS) *(16)*. They identified 12 metabolites in rat plasma and 8 metabolites in rat urine as potential biomarkers. Concentrations of leucine, phenylalanine, tryptophan, acetylcarnitine, butyrylcarnitine, propionylcarnitine, and spermine in plasma and 3-*O*-methyl-dopa, ethyl *N*2-acetyl-L-argininate, leucylproline, glucuronate, *N*(6)-(*N*-threonylcarbonyl)-adenosine, and methyl-hippuric acid in urine were decreased in atherosclerosis rats; ursodeoxycholic acid, chenodeoxycholic acid, LPC (C16:0), LPC (C18:0), and LPC (C18:1) in plasma and hippuric acid in urine were increased in atherosclerosis rats. The altered metabolites demonstrated abnormal metabolism of phenylalanine, tryptophan, bile acids, and amino acids. Lysophosphatidylcholine (LPC) plays an important role in inflammation and cell proliferation, which shows a relationship between LPC in the progress of atherosclerosis and other inflammatory diseases.

The lipidome profile of mice liver homogenates of free cholesterol, low cholesterol, and high cholesterol diets showed the influence between dietary cholesterol intake and atherosclerosis *(17)*. To get individual metabolite fingerprints, they measured near 300 metabolites such as di- and triglycerides, phosphatidylcholines, LPCs, and cholesterol esters in plasma samples by LC-MS/MS. It was observed that when dietary cholesterol intake was increased, the liver compensated for elevations in plasma cholesterol by adjusting metabolic and transport processes related to lipid metabolism, which leads to an inflammatory, proatherosclerotic state. Free cholesterol diet did not induce early atherosclerosis, the low cholesterol diet did so, but only mildly, and the high cholesterol diet induced it very strongly as an increase of dietary cholesterol stress induces proinflammatory gene expression. Therefore, they hypothesize that a relationship between cholesterol intake (measured as cholesterol plasma levels) and atherosclerotic lesion size exists.

Lipidomes of membrane cells and tissues were investigated by measuring plasmalogens contained in rabbit and rat myocardial nuclei by ESI-MS *(18)*. Plasmalogen is an ether lipid where the first position of glycerol binds a vinyl residue with the double bond next to the ether bond. The second carbon has a typical ester-linked fatty acid, and the third carbon usually has a phospholipid head group, which can protect cells against the damaging effect of singlet oxygen. This seems to be the reason for the membrane of myocardial cells to be highly enriched in plasmalogens as found.

In a different study, the nuclear magnetic resonance (NMR) metabolomics profiles of 864 healthy voluntaries were examined. The aim was to investigate whether NMR of plasma contains a fingerprint of the CV status of healthy individual and whether this fingerprint is constituted also by metabolites not previously associated with CV risk. They analyzed healthy individuals with increased CV risk parameters (low HDL, high LDL, high total cholesterol, high triglycerides, and in general high Framingham score), and they compared them with the subjects with a low-risk pattern using a multivariate statistics. They found that those subjects that are classified as at high or low risk using the common clinical markers can also be discriminated using NMR metabolomics signature. After Bonferroni correction, other metabolites not associated with CVDs appeared (3-hydroxybutyrate, α-ketoglutarate, threonine, and dimethylglycine) *(19)*.

Metabolic changes associated with atherosclerosis were also investigated through metabolite profiling by NMR and GC-MS *(20)*. Authors have also given a biochemical explanation to their findings and have related these metabolite alterations to different disorders, which cause the final atherosclerotic lesion. For instance, they found that in diabetic patients, insulin resistance increases the activity of transaminases, which are critical enzymes in the metabolic pathways of amino acids. This means that if there is a lack of insulin response, these pathways of amino acids will be altered and many other metabolites as glutamate, ketoglutarate, succinyl-CoA, 4-OH-L-proline (4OHPro), 2-hydroxybutyrate (2OHB), creatinine, pyruvate, oxaloacetate, malate, glycolate, and 2,3,4-trihydroxybutyrate will be affected and could indicate damage into tissue from intima artery walls.

Yap *et al.* investigated a population of biomarkers that might relate the difference in CVD risk between southern and northern Chinese population by ^{1}H NMR spectroscopy-based metabolome-wide association approach *(21)*. The study was carried out among 4680 men and women ages 40–59 years from 17 diverse population samples in China, Japan, United Kingdom, and United

States. NMR spectra were acquired from 24 hour urine from 523 northern and 243 southern Chinese participants. Discriminating metabolites were identified using orthogonal partial least square discriminant analysis (OPLS-DA) and assessed for statistical significance with conservative familywise error rate <0.01 to minimize false-positive findings. Urinary metabolites significantly ($p < 1.2 \times 10^{-16}$–2.9×10^{-69}) higher in northern (higher CV risk) than in southern Chinese populations included dimethylglycine, alanine, lactate, branched-chain amino acids (isoleucine, leucine, and valine), N-acetyls of glycoprotein fragments, N-acetyl neuraminic acid, pentanoic/heptanoic acid, and methylguanidine; metabolites significantly ($p < 1.1 \times 10^{-12}$–2×10^{-12}) higher in the south (lower CV risk) were gut microbial cometabolites (hippurate, 4-cresyl sulfate, phenylacetylglutamine, and 2-hydroxybutyrate), succine, creatine, scyllo-inositol, proline betaine, and *trans*-aconitate. These findings indicate the importance of environmental influences, endogenous metabolism, and mammalian-gut microbial cometabolism, which may help explain north–south China differences in CVD risk.

The myocardial metabolic response was investigated by Turer *et al. (22)* in CAD and left ventricular dysfunction (LVD) patients, at baseline and following ischemia–reperfusion (I/R). The study revealed that, in a preischemia state, glucose, lactate, free fatty acids, total ketones, 3-hydroxybutyrate, pyruvate, leucine/isoleucine, and glutamate were present at lower concentrations in CS than in arterial samples (reflecting myocardial uptake); on the contrary, alanine concentration was in higher concentration (reflecting release). Principal component analysis showed several metabolic changes as potentially important in postoperative clinical course. The dysfunctional ventricle was associated with global suppression of metabolic fuel uptake and limited myocardial metabolic reserve and flexibility following global I/R stress is associated with cardiac surgery.

Zhao *et al. (36)* found that some citric acid metabolites are decreased in acute ischemia and acute myocardial disease. The citric acid plays an important role in oxidative phosphorylation and ATP production in the cardiomyocytes in which levels of citric acid cycle intermediates are supplied by glycolysis and β-oxidation of fatty acids.

Shah *et al. (23)* used metabolomics profile to study the heritability of premature coronary disease by performing quantitative MS-based metabolic profiling in 117 individuals unaffected by CAD but with a family member affected. They found high heritabilities for amino acids, such as arginine, ornithine, alanine, proline, leucine/isoleucine, valine, glutamate/glutamine, phenylalanine, and glycine, and free fatty acids, such as arachidonic, palmitic, linoleic, and acylcarnitines. They concluded that there is a high heritability of metabolic changes in coronary artery disease and a strong relationship with age. This may indicate that metabolic processes could be controlled genetically, implying a correlation between genotype and phenotype in families with CAD.

Metabolism changes in human atrial fibrillation (AF) were investigated by NMR performing quantitative analysis of 24 metabolites selected beforehand. Significant differences were found for β-hydroxybutyrate, ketogenic amino acids, and glycine, all of them raised in AF patients versus controls, pointing to a role for ketone bodies. Metabolic profiles allowed classification of more than 80% of patients at risk of AF at the time of coronary artery bypass grafting as a discordant regulation of energy metabolites was found to precede postoperative AF *(24)*.

An optimized and refined protocol applicable to human plasma was used by Chan *et al.* to study heart failure. This study contributes to methodology development with respect to deproteinization, incubation, reconstitution, and detection with MS. Nineteen metabolites were chosen based on their demonstrated functions in the relevant biological process to heart failure, specifically participating in the regulation of extracellular matrix, energy metabolism, inflammation, insulin resistance, renal dysfunction, and cardioprotection. With these newly developed platforms, they extract in an effective way 13 of 19 of the targeted metabolites. This workflow in conjunction with LC-MS/MS will enable them to establish a high-throughput metabolomics platform to characterize and validate these target metabolites as a potential biomarker in human heart failure *(25)*.

Wang *et al.* performed a metabolite profiling in participants from two longitudinal studies, with the goal of identifying early pathophysiological changes that might also serve as a novel predictor of future diabetes. The study included 2422 normoglycemic individuals followed for 12 years. Amino acids, amines, and other polar metabolites were profiled in baseline specimens using LC-MS/MS. Cases and controls were matched by age, body mass index, and fasting glucose. Five branched-chain and aromatic amino acids had highly significant associations with future diabetes: isoleucine, leucine, valine, tyrosine, and phenylalanine. A combination of three amino acids predicted future diabetes (>5-fold higher cohort). These findings underscore the potential importance of amino acid metabolism early in the pathogenesis of diabetes and suggest that amino acid profiles could aid in diabetes risk assessment *(26)*.

In other study, Shah *et al.* performed a quantitative profiling of 69 metabolites with the goal of assessing the capability of metabolite profiles to discriminate the presence of CVD and to explore the relation of metabolite profiles with subsequent CV events. The study was developed using MS in peripheral blood of subjects from CATHGEN biorepository. Two groups were profiled: 174 CAD cases and 174 controls and 140 CAD cases and 140 controls' replication. A third independent group was profiled, $n = 63$ cases with CV events and 66 controls. Analysis included principal components, linear regression, and Cox proportional hazards. Several metabolites were associated with CAD: branched-chain amino acid metabolites and urea cycle metabolites. Dicarboxylacylcarnitines predicted death/MI and was associated with CV events *(37)*.

3.3   Metabolomics Directly in Tissue

Mass spectrometry imaging (MSI) constitutes a novel technical approach with increasing popularity. It is a microprobe technique that generates chemically selective images from thin tissue sections *(38,39)*. It generates molecular profiles directly from tissue containing hundreds of distinct biomolecular ions; at the same time, it provides their original spatial distribution *(40)*. To date, MALDI-MSI analyses have identified the spatial distribution of endogenous and exogenous compounds including proteins, peptides, metabolites, lipids, and drugs *(41)*. One advantage of this technique is the ability to elucidate relative intensity changes and spatial distributions resulting from external stimuli such as administration of an exogenous drug or injury models. Chaurand *et al.* *(42)* explored some of the requirements necessary for the successful analysis and imaging of the phospholipids from thin tissue sections of various dimensions. They addressed methodology issues relative to the imaging of whole-body sections such as those cut from model laboratory animals, sections of intermediate dimensions typically prepared from individual organs, and the requirements for imaging areas of interest from these sections at cellular scale. An MSI study was conducted on cardiac tissue by Menger *et al.* *(27)*, in order to analyze multiple compound classes. First, the spatial distribution of a small metabolite, creatine, was used to identify areas of infracted myocardium. Second, multivariate data analysis and MS/MS were used to identify phospholipid (PL) markers of MI. A number of lysophospholipids demonstrated increased ion signal in the area of infarction. In contrast, selected intact PLs demonstrated decreased ion signal on the area of infarction. The complementary nature of these two lipid classes suggests increased activity of phospholipase A_2 (PAL$_2$), as this enzyme hydrolyzes PLs yielding free lysophospholipids among other metabolites. Furthermore, PAL$_2$ is an enzyme that has been implicated in coronary heart disease inflammation. With knockout receptor activity-modifying protein (RAMP) mice, Yoshizawa *et al.* *(28)* analyzed the embryonic and adult cardiac myocytes metabolome by MSI. This analysis reveals early downregulation of cardiolipin, a mitochondrial membrane-specific lipid. Furthermore, primary cultured cardiac myocytes from inducible RAMP2$^{-/-}$ (C-RAMP2$^{-/-}$) showed reduced mitochondrial membrane potential and enhanced reactive oxygen species production in RMP2. C-RAMP2$^{-/-}$ showed downregulated activation of the cAMP response element-binding protein, one of the main regulators of mitochondria-related genes. These data demonstrated that adrenomedullin–RAMP2 system is essential for cardiac metabolism and homeostasis.

Using MALDI-MSI-based histopathologic examination, Zaima *et al.* *(29)* identified and visualized specific markers for aortic atherosclerosis lesions in mice ApoE and in human. They found the same distribution in mice and human samples, in several molecules: cholesterol linolate, cholesterol oleate, and phosphatidylcholine. The triacylglycerol only appeared in human samples. With these data, they proposed MSI-based pathologic approach as a new histopathology examination. Atherosclerotic lesions from ApoE knockout mice at different stages of progression have been related to lipid composition and distribution inside the plaque by TOF-SIMS, pointing to the regional ratio of cholesterol esters as a marker of atherosclerosis progression *(30)*. The same technique evidenced a significant increase in the quantity of several nonesterified fatty acids in diabetic versus nondiabetic atheroma plaques *(31)* and different localization for lipids when comparing healthy rat aorta with atherosclerotic human artery *(32)* and allowed to map fatty acid, cholesterol, vitamin E, phosphatidic acid, phosphatidylinositol, and triglyceride distribution on human atherosclerotic plaques *(34)*. In an *in vivo* rat model of MI, MSI of cardiac tissue allowed correlating creatine, specific lysophospholipids, and intact phospholipid with areas of infarcted myocardium *(27)*.

4   METABOLOMICS IN RENAL DISEASE

The application of metabolomics analysis to investigate renal pathology has been recently reviewed *(43)* pointing to the relevant contribution that omics may offer to clinical medicine, in a continuous upgrade. In Table 2, a compilation of most relevant studies exemplifies the applicability of metabolomics in the study of kidney diseases. Techniques of choice mainly include NMR and LC-MS/MS or GC-MS/MS. Although usually employed to analyze low-molecular-weight protein fragments (peptidomics), CE-MS offers increased separation efficiency for soluble molecules. CE-MS was the technique of choice, that is, for the analysis of uremic solutes as a simple and accurate method to assess renal function, finding 22 cations and 30 anions that accumulated significantly as the estimated glomerular filtration rate (eGFR) decreased *(56)*. In the attempt to discover additional markers of CKD and its evolution (i.e., progressive reduction of eGFR), serum metabolite analysis has been approached in patients at different stages *(55)* and in the general population *(45)*, in a way to complement serum creatinine and urinary albumin to improve diagnosis or prediction. Focusing on early stages of disease (CKD1–2), carnosine, hypotaurine, adenosine, and guanine (in urine), and hypotaurine and taurine (in serum) were found altered *(44)*. In urine, a discovery phase approached by NMR revealed a panel of seven metabolites, which were confirmed by SRM (LC-MS/MS) in an independent cohort, to be significantly different between CKD and non-CKD: 5-oxoproline, glutamate, guanidoacetate, α-phenylacetylglutamine, taurine, citrate, and trimethylamine *N*-oxide (TMANO) *(54)*. Diabetic nephropathy (DN) is the leading cause of chronic renal disease and a major cause of CV mortality. In a prospective study, worsening of albuminuria was investigated by classification of normoalbuminuric DN patients into normo–micro- or macroalbuminuric condition after 5.5 years of follow-up. Urinary metabolites were analyzed by

TABLE 2 Selected Metabolomics Approaches in the Study of Kidney Diseases

Sample	Pathology	No. of Subjects	Methodology	Relevant Metabolites	References
Plasma	Diabetes without nephropathy, diabetic nephropathy III, IV, and V	Healthy subjects 50; patients 88	HPLC-UV-MS/MS	Adenosine, inosine, uric acid, xanthine, and creatinine	(44)
Urine	Hyperalbuminuria	Healthy subjects 21; patients 29	¹H NMR LC-MS/MS	Lysine, glycine, alanine, valine, n-butyrate, 3-hydroxybutyrate, betaine, dimethylglycine, kynurenic acid, and xanthurenic acid	(45)
Kidney urine	Lithium-induced nephrogenic diabetes insipidus	Rat model	¹H NMR	Acetate, lactate, allantoin, TMA, and creatinine	(46)
Plasma	ESRD hemodialysis	Healthy subjects 16; patients 10	UPLC-MS	1-Methylinosine	(47)
Serum	ESRD hemodialysis and peritoneal dialysis	Healthy subjects 18; hemodialysis 18; peritoneal dialysis 18	¹H NMR	Hypoxanthine, inosine (hemodialysis) Lactate, glucose, maltose, pyruvate, succinate, alanine, and glutamate (peritoneal dialysis)	(48)
Plasma	CKD uremic toxicity	Patients 41	CE-MS	>500 metabolites	(31,49)
Plasma	Chronic glomerulonephritis	Healthy subjects 18; glomerulonephritis 18; chronic renal failure 17	LC-MS	Nineteen phospholipid species (C18:0–C18:2, C18:0–C18:1, C18:0–C18:0, and C18:1–C20:4)	(50)
Serum urine	CKD	Healthy subjects 7; patients 15	CE-TOF MS	Carnosine, hypotaurine, adenosine, guanine (urine) hypotaurine, and taurine (serum)	(27)

Urine	CKD	Healthy subjects 30; patients 30	NMR LC-MRM MS	5-Oxoproline, glutamate, guanidoacetate, α-phenylacetylglutamine, taurine, citrate, and TMANO	(43)
Serum	AKI	Healthy subjects 17; patients 17	LC-MS	Acylcarnitines, methionine, homocysteine, pyroglutamate, ADMA, phenylalanine, arginine, and lysophosphatidylcholines	(51)
Serum	CKD	Healthy subjects 28; patients 80	1H NMR	Glucose, lactate, valine, alanine, glutamate, glycine, betaine, myoinositol, taurine, and glycerolphosphocholine	(32)
Urine	Tubulointerstitial lesions	Glomerulonephritis 77	1H NMR	Citrate, hippurate, glycine, creatinine, lactate, acetate, and trimethylamine-N-oxide	(52)
Urine	Nephrolithiatic children	Healthy subjects 74; patients 73	UPLC Q-TOF MS	Hypoxanthine, proline, and 5C-aglycone	(53)
Serum	Diabetic nephropathy (DN)	Healthy subjects 25; DN patients 8; type 2 diabetes mellitus patients 33	UPLC Q-TOF MS	Leucine, dihydrosphingosine, and phytosphingosine	(54)
Serum	DN with macroalbuminuria and diabetic patients without albuminuria	DN patients 78	CE-TOF MS	Creatinine, aspartic acid, γ-butyrobetaine, citrulline, symmetric dimethylarginine, kynurenine, azelaic acid, and galactaric acid	(55)
Urine	Progressive form of albuminuria from the nonprogressive form of albuminuria	Type 1 diabetic patients normoalbuminuric56. After 5.5 years: micro/macroalbuminuric 26; normoalbuminuric 26	GC-MS LC-MS	Acylcarnitines, acyl-glycines, and metabolites related to the tryptophan metabolism	(56)

Continued

TABLE 2 Selected Metabolomics Approaches in the Study of Kidney Diseases—Cont'd

Sample	Pathology	No. of Subjects	Methodology	Relevant Metabolites	References
Serum	Uremic syndrome	Control rats (6/5)	LC-MS	Indoxyl sulfate, phenyl sulfate, hippuric acid, and p-cresyl-sulfate	(57)
Plasma	ESRD	10 age-matched, at-risk fasting control subjects; fasting subjects with ESRD, before and after hemodialysis 44	LC-MS	Adipate, malonate, methylmalonate, and maleate, biogenic amines, nucleotide derivatives, phenols, and sphingomyelins	(58)
Kidney	Drug toxicity affecting kidney	Rat model	MALDI-MSI NMR-HRMAS LC-MS	Identification and mapping of renal crystalline deposits Identification of molecular changes in pathological areas	(59)
Kidney	Imaging of lipids	Mouse model	MALDI-MSI	Unique distribution of specific lipids in the cortex, medulla, pelvic, and perinephric tissue	(60)
Kidney	Renal RAS	Mouse model	MALDI-TOF/ TOF-MS	Molecular imaging of renal RAS Localization of AngII and enzymes activity in kidney regions	(61)
Kidney	Olanzapine distribution (OLZ)	Rat model	MALDI-MSI MS/MS	Different distribution of OLZ: intact drug along outer cortex, oxidized metabolites along the medulla	(62)
Kidney urine	IgA nephropathy	Mouse model (Blab/c and hyper IgA vs control)	MALDI-QIT-TOF MS LC-LTQ MS	Specific lipid signature in the tubular lesion of hyper-IgA mice	(63)

GC-MS and LC-MS revealing 8 and 10 metabolites, respectively, significantly varied between progressive and nonprogressive normoalbuminuric patients, including acylcarnitines, acyl-glycines, and compounds related to the tryptophan metabolism *(64)*. By CE-MS, a serum signature composed of creatinine, aspartic acid, γ-butyrobetaine, citrulline, symmetric dimethylarginine, kynurenine, azelaic acid, and galactaric acid allowed to distinguish between DN patients with and without albuminuria *(46)*. In the same line, the capacity to correlate a potential increase in urinary albumin with a urinary metabolic pattern was investigated by NMR and LC-MS, showing a potential cost-effective solution of high-throughput analysis with pattern recognition tools to be applied in the real clinical situation. Alterations at metabolome level related to tryptophan metabolism, citric acid cycle, and amino acid metabolism, among others, were found in the patients' group *(65)*. Plasma metabolomics profile was also shown to be altered in DN showing, in particular, adenosine, inosine, uric acid, xanthine, and creatinine as altered metabolites *(58)*. Similarly, differences in serum metabolic pattern were investigated for DN in comparison with type 2 diabetes mellitus (DM) itself and control levels, finding altered values for leucine, dihydrosphingosine, and phytosphingosine, both in DM and in DN patients compared to the control group *(47)*.

Therapy effects at metabolome level were also investigated in DN by LC-MS, in adenine-induced CKD *(48)* by LC-MS, and in chronic lithium treatment inducing nephrogenic diabetes insipidus by NMR *(66)*. In particular, rosiglitazone normalized levels of about one-third of the urinary metabolites shown to be altered in diabetes and reduced reactive oxygen species accumulation to prevent type 1 diabetic mice from developing DN *(67)*. The effect of dialysis was investigated by LC-MS/MS by measuring changes of uremia markers in the plasma of ESRD patients before and after dialysis. Plasma profiles were also compared to those of healthy subjects finding dicarboxylic acids (adipate, malonate, methylmalonate, and maleate), biogenic amines, nucleotide derivatives, phenols, and sphingomyelins as markers of ESRD. Lipid pattern was also altered in this clinical condition pointing to a disturbance in triglyceride catabolism and/or oxidation. The dialysis procedure resulted in the expected decrease of many polar analytes but, more interestingly, reflected a potential activation of glycolysis, lipolysis, ketosis, and nucleotide breakdown. In this case, authors point to the utility of a simultaneous multimolecular analysis able to reveal unknown mechanisms *(52)*. A different study of plasma metabolome from hemodialysis (HD) patients by LC-MS revealed 1-methylinosine as an effective candidate biomarker to estimate adequate HD dose *(51)*. In a different study, serum metabolites were profiled in the search for differences in subjects receiving HD or peritoneal dialysis (PD) *(68)*.

Classical strategies do not always offer sufficient specificity, sensitivity, or accuracy to allow appropriate and timely interventions. Investigating alterations at metabolite level will provide extra information related to, that is,

I/R injury, assessment of immunosuppressive drug toxicity and organ function, and localization of organ damage *(69)*. Metabolomics has also been proposed as a valuable tool to monitor success of kidney transplantation. In this study, metabolites in urine samples from transplant recipients were analyzed by MALDI-FTMS, as a noninvasive method to predict acute tubular injury of renal allografts in replacement of biopsy. Loading graphs were constructed so that the position of an individual urine sample in the graph may reflect the tubular injury status *(50)*. Histopathologically assessed tubulointerstitial lesions were found to correlate with urinary metabolite profile, in particular with alterations of citrate, hippurate, glycine, creatinine, lactate, acetate, and trimethylamine-*N*-oxide *(70)*. Acute kidney injury (AKI) is a frequently encountered complication in renal transplantation. In a pilot study by LC-MS, serum levels of acylcarnitines and amino acids (methionine, homocysteine, pyroglutamate, asymmetric dimethylarginine (ADMA), and phenylalanine) increased, while arginine and several LPCs were found decreased in patients with AKI compared to healthy subjects *(71)*.

Less commonly investigated by metabolomics approaches are renal pathologies as polycystic kidney disease or lupus nephritis (LN). In a rat model, 2-ketoglutaric acid, allantoin, uric acid, and hippuric acid were identified by GC-MS as potential markers of cystic disease *(72)*. In view of current difficulties in clinical practice to distinguish between class III/IV and class V LN and focal segmental glomerulosclerosis, metabolic profiling of patient's urine samples was analyzed by NMR in an attempt to ameliorate invasiveness of diagnosis, finding taurine, citrate, and hippurate as differentially excreted *(73)*. Plasma phospholipids have been proposed as potential markers on the progress of primary chronic glomerulonephritis by LC-MS *(74)*. In a different study, spontaneous hypertensive rats at early stage of hypertension showed citrate, α-ketoglutarate, and hippurate as main metabolites decreased in urine with creatinine-based normalization data *(75)*.

Although most of the studies investigate the metabolome of biological fluids (urine, serum, and plasma), it is worthwhile considering that relevant information can be obtained from direct study of the tissue in a similar way as proteomics does *(76)*. If thinking in terms of diagnosis, one may think more in the limitations this kind of approaches may have. However, changes occurring at tissue level may have a potential reflection in urine or plasma—changes that may have not been discovered directly in these complex sample sources where the ultimate response for the whole body to a pathological or biological disturbance occurs. The renal protective effects of the Chinese herb *Cordyceps sinensis* in rats were evaluated by analyzing kidney aqueous extracts by NMR, showing a potential amelioration effect of CKD *(77)*. Furthermore, investigating directly in tissue by HRMAS (tissue NMR) or MSI will reveal underlying mechanisms of action taking place in the kidney allowing access to a focused knowledge. HRMAS NMR spectroscopy of tissues has shown to be valuable in assessing tissue metabolite profiles nondestructively *(60)*.

Very few studies applied this technique to investigate kidney diseases *(63)*, and, as one of the few examples found in the literature, metabolite changes at hepatic and renal cortical tissue in diabetic db/db mice were investigated *(62)*. MSI has recently begun to be applied to investigate kidney diseases, opening a new field of research with promising future *(59,61)*. Lipid distribution *(78,79)*, drugs metabolism *(41)*, mapping of molecular changes in pathological areas *(80)*, and *in situ* profiling of a particular physiopathological mechanism *(81)* are some of the studies approached by MSI.

5   APPROACHING THE CARDIORENAL PUZZLE

One may think that risk factors and misregulated molecules known to be acting in CVD in the general population can be directly translated into the renal disease frame. However, the relationship between risk markers and CV events in kidney pathology often differs from that in the general population. One of the main reasons is that traditional CV risk factors, that is, age, male gender, hypertension, diabetes, dyslipidemia, smoking, overweight, and hyperhomocystinemia, are highly prevalent in CKD; additionally, contribution of risk factors specifically associated with kidney disease should be added, that is, anemia, calcium-phosphate disorders, electrolyte imbalances, chronic inflammation, oxidative stress, hypercatabolism, uremic condition, and vitamin D deficiency *(82)*. Some therapy trials (pharmaceuticals testing) are focused on known drugs acting to control abovementioned risk factors (blood pressure management, lipid lowering, antiplatelet therapy, etc.). However, the results are usually unexpected or not convincing, finding a nonclear response in the adequate direction. On the other hand, most of studies about CVD in CKD are carried out from a post hoc subgroup analysis of larger trials that were not specifically designed to investigate effects on CKD population so that final conclusions should be postponed until further evidence is found *(83)*. In other cases, CV trials even excluded CKD patients from enrollment. Thus, despite research efforts, active cardioprotective treatment is underemployed in renal disease patients, probably due to the uncompleted knowledge of the cardiorenal puzzle.

In this context, documented evidence exists to consistently prove that CV complications should be considered a key, unsolved, and state-of-the art issue in kidney disease, in terms of early on-time diagnosis, accurate prognosis prediction, and in-depth understanding of underlying mechanisms. Currently, potential markers have been described in the literature to predict the CV risk and associated mortality in CKD, that is, brain natriuretic peptide *(84)*, fibroblast growth factor 23 (FGF-23) *(85)*, fetuin A *(86)*, osteoprotegerin *(87)*, C-reactive protein *(88)*, uric acid *(89)*, pentraxin 3 *(90)*, vascular endothelial growth factor *(91)*, fibrinogen *(92)*, ADMA *(93)*, and TNF-related apoptosis-inducing ligand *(94)*, among others *(95)*. Recently, Fassett *et al.* reviewed all data related to biomarkers described in the diagnosis and progression of CKD *(96)*. Among all them, significantly highlighted are neutrophil gelatinase-associated lipocalin, as CKD progression marker, and cystatin C as renal function, disease progression, and CV risk marker. It is key to note that these data result from hypothesis-driven studies where specific molecules, beforehand selected, are investigated. Metabolomics plasma signatures associated with mild and severe CVD in CKD stage 5D patients have been investigated by CE-MS. CVD was assessed according to either arterial calcification or aortic stiffness finding a panel of metabolites able to distinguish mild and severe CVD with sensitivity of 89% and specificity of 67%; among them, fragments of collagen α1 types I and III and one fragment of apolipoprotein CIII were identified as markers of vascular disease *(97)*.

6   MICROVESICLES: A NOVEL SOURCE OF BIOMARKERS

6.1   Metabolomics in Exosomes Research

Exosomes are membranous vesicles 40–100 nm sized, released by cells in different extracellular fluids. Blood exosomes, despite being previously considered inert debris without specific function, have been demonstrated to be important in the regulation of different mechanisms in vascular pathologies *(98)*. On the other hand, urinary exosomes are released from every renal epithelial cell type facing the urinary space, and therefore, they may carry molecular markers of renal dysfunction and structural injury. Additionally, urinary exosomes can shuttle information between nonrenal cells, providing relevant information to a variety of disease processes and constituting a promising source of biomarkers. In this context, exosomes represent only 3% of the whole urine proteome, and, although its isolation and characterization is extremely challenging, exosomes otherwise constitute an enriched subproteome with reduced complexity compared to the whole urine. Low-abundance molecules that may have pathophysiological significance can be in this way enriched against high-abundance ones.

By LC-MS/MS, lysophosphatidylethanolamine has been tentatively associated with a renal cell carcinoma signature in urinary exosomes, although further validation in wider and independent cohort of samples is yet pendant *(99)*. Exosome-mediated transport is among the mechanisms postulated for cell-to-cell communication, both in close environment and at longer distances (organ-to-organ). Virtually, every type of molecule could be transported through these microvesicles, and, interestingly, anticancer drugs may be included in this context. This mechanism could at least partially explain a potential role of exosomes in their mediated expulsion of intracellular drugs, constituting a barrier in the proper action of most of the commonly used therapeutics, targeted agents, and their intracellular metabolites and playing a role in cancer resistant. In a different study, Beloribi *et al.* hypothesized whether exosomal lipids were key elements for cell death through exosome interactions with tumor cells via cholesterol-rich membrane microdomains. Synthetic exosome-like nanoparticles with ratios framing that of human SOJ-6 pancreatic tumor cell exosomes resulted to decrease tumor cell survival, due to activation of cell death with inhibition of Notch pathway *(100)*. Exosomes from macrophages and dendritic cells have been proposed to participate in inflammation through the involvement of some of their functional enzymes in granulocyte recruitment and leukotrienes biosynthesis, a potent proinflammatory lipid mediator with key roles in the pathogenesis of asthma and inflammation *(101)*. The presence of drug-metabolizing enzymes in microvesicles secreted to the extracellular medium by hepatocytes has been also proved, pointing to a potential implication of these extracellular microparticles (MPs) in xenobiotic metabolism and development of drug resistance. Although some issues have to be addressed, this observation indicates that these hepatic microvesicles may play a role in the metabolism of endogenous and xenobiotic compounds and help to control the homeostasis of the body *(102)*.

Although studies focused on the alterations at the metabolome level in exosomes are scarce, there is enough evidence to prove their potential role as "key messengers" of organism disturbances, which will nicely translate into a more accessible fluid (i.e., urine) those changes occurring at tissue (i.e., kidney) level.

6.2   Metabolomics in MP Research

MPs (also called microvesicles) are present in the blood of healthy individuals and are increased in various diseases including CVD. They are small membrane vesicles derived from activated and apoptotic cells. Importantly, MPs have been proposed to play roles in thrombosis, inflammation, and angiogenesis *(103)*. MPs are also released in the circulation *(104)*, and ever since their potent procoagulatory properties were first recognized in the field of homeostasis *(105)*, the interest in their potential pathophysiological importance has increased *(106)*. MP metabolome may be a useful and reliable source of biologically relevant disease biomarkers.

Mayr *et al.* performed a metabolomics study of MPs in order to obtain insights into the role of MPs in atherogenesis. Metabolite profiles were determined by high-resolution NMR spectroscopy. In this work, they compared MPs ($n = 26$) with carotid endarterectomy (CE) samples ($n = 13$). Strikingly, taurine, the most abundant free organic acid in human neutrophils, which is implicate in the feedback inhibition of neutrophil/macrophage respiratory burst by scavenging myeloperoxidase-catalyzed free radicals, was highly enriched in MPs compared with CE samples. They observed that MPs also contained high concentrations of lactate and a low concentration of glycerol phosphocholine, which was increased in CE in this case *(107)*.

REFERENCES

1. Peralta, C. A.; Shlipak, M. G.; Judd, S.; Cushman, M.; McClellan, W.; Zakai, N. A.; Safford, M. M.; Zhang, X.; Muntner, P.; Warnock, D. *JAMA* **2011**, *305*, 1545–1552.
2. Baigent, C.; Burbury, K.; Wheeler, D. *Lancet* **2000**, *356*, 147–152.
3. Go, A. S.; Chertow, G. M.; Fan, D.; McCulloch, C. E.; Hsu, C. Y. *N. Engl. J. Med.* **2004**, *351*, 1296–1305.
4. Sarnak, M. J.; Levey, A. S.; Schoolwerth, A. C.; Coresh, J.; Culleton, B.; Hamm, L. L.; McCullough, P. A.; Kasiske, B. L.; Kelepouris, E.; Klag, M. J.; Parfrey, P.; Pfeffer, M.; Raij, L.; Spinosa, D. J.; Wilson, P. W. *Circulation* **2003**, *108*, 2154–2169.
5. Cruz, D. N. *Adv. Chronic Kidney Dis.* **2013**, *20*, 56–66.
6. Spasovski, G.; Ortiz, A.; Vanholder, R.; El Nahas, M. *Proteomics Clin. Appl.* **2011**, *5*, 233–240.
7. Rhee, E. P.; Gerszten, R. E. *Clin. Chem.* **2012**, *58*, 139–147.
8. Weiss, R. H.; Kim, K. *Nat. Rev. Nephrol.* **2011**, *8*, 22–33.
9. Roger, V. L. Circ. Cardiovasc. Qual. Outcomes. *4*, 257–259.
10. Bassand, J. P.; Hamm, C. *Pol. Arch. Med. Wewn.* **2007**, *117*, 391–393.
11. Laborde, Carlos M.; Mourino-Alvarez, Laura; Posada-Ayala, María; Alvarez-Llamas, Gloria; Serranillos-Reus, Manuel Gómez; Moreu, José; Vivanco, Fernando; Padial, Luis R.; Barderas, María G. *Metabolomics*, in press.
12. Muhlen, C.; Schiffer, E.; Peter, K. *J. Proteome Res.* **2009**, *8*, 335–345.
13. Ciborowski, M.; Martin-Ventura, J. L.; Meilhac, O.; Michel, J. B.; Ruperez, F. J.; Tuñon, J.; Egido, J.; Barbas, C. *J. Proteome Res.* **2011**, *10*, 1374–1382.
14. Jiang, P.; Dai, W.; Yan, S.; Chen, Z.; Xu, R.; Ding, J.; Xiang, L.; Wang, S.; Liu, R.; Zhang, W. *Mol. Biosyst.* **2011**, *7*, 824–831.
15. Lewis, G.; Wei, R.; Liu, E.; Sabatime, M.; Gerszten, R. *J. Clin. Invest.* **2008**, *118*, 3503–3512.
16. Zhang, F.; Jia, Z.; Xu, G. *Talanta* **2009**, *79*, 836–844.
17. Kleeman, R.; Verschuren, L.; Kooistra, T. *Genome Biol.* **2007**, *8*, R200.
18. Albert, C.; Eckelkamp, J. *Front. Biosci.* **2009**, *12*, 2750–2760.
19. Bernini, P.; Bertini, I.; Luchinat, C.; Tenori, L.; Tognaccini, A. *J. Proteome Res.* **2011**, *10*, 4983–4992.
20. Teul, J.; Martín Ventura, J. L.; Blanco Colio, L. M.; Egido, J.; Barbas, C. *J. Proteome Res.* **2009**, *8*, 5580–5589.
21. Yap, I. K.; Brown, I. J.; Chan, Q.; Wijeyesekera, A.; Garcia-Perez, I.; Bictash, M.; Loo, R. L.; Chadeau-Hyam, M.; Ebbels, T.; De Iorio, M.; Maibaum, E.; Zhao, L.; Kesteloot, H.; Daviglus, M. L.; Stamler, J.; Nicholson, J. K.; Elliott, P.; Holmes, E. *J. Proteome Res.* **2010**, *9*, 6647–6654.
22. Turer, A.; Stevens, R.; Podgoreanu, M. *Circulation* **2009**, *119*, 1736–1746.
23. Shah, S. H.; Hauser, E. R.; Granger, C. B.; Kraus, W. E. *Mol. Syst. Biol.* **2009**, *5*, 258.
24. Mayr, M.; Yusuf, S.; Weir, G.; Chung, Y. L.; Mayr, U.; Yin, X.; Ladroue, C.; Madhu, B.; Roberts, N.; de Souza, A.; Fredericks, S.; Stubbs, M.; Griffiths, J. R.; Jahangiri, M.; Xu, Q.; Camm, A. J. *J. Am. Coll. Cardiol.* **2008**, *51*, 585–594.
25. Chan, C. X.; Khan, A. A.; Choi, J. H.; Ng, C. D.; Cadeiras, M.; Deng, M.; Ping, P. *Clin. Proteomics* **2013**, *10*, 7.
26. Wang, T. J.; Larson, M. G.; Vasan, R. S.; Cheng, S.; Rhee, E. P.; McCabe, E.; Lewis, G. D.; Fox, C. S.; Jacques, P. F.; Fernandez, C.; O'Donnell, C. J.; Carr, S. A.; Mootha, V. K.; Florez, J. C.; Souza, A.; Melander, O.; Clish, C. B.; Gerszten, R. E. *Nat. Med.* **2011**, *17*, 448–453.

27. Menger, R. F.; Sttus, W. L.; Anbukumar, D. S.; Bowden, J. A.; Ford, D. A.; Yost, R. A. *Anal. Chem.* **2012**, *84*, 1117–1125.

28. Yoshizawa, T.; Sakurai, T.; Kamiyoshi, A.; Ichikawa-Shindo, Y.; Kawate, H.; Iesato, Y.; Koyama, T.; Uetake, R.; Yang, L.; Yamauchi, A.; Tanaka, M.; Toriyama, Y.; Igarashi, K.; Nakada, T.; Kashihara, T.; Yamada, M.; Kawakami, H.; Nakanishi, H.; Taguchi, R.; Nakanishi, T.; Akazawa, H.; Shindo, T. *Hypertension* **2013**, *61* (2), 341–351.

29. Zaima, N.; Sasaki, T.; Tanaka, H.; Cheng, X. W.; Onoue, K.; Hayasaka, T.; Goto-Inoue, N.; Enomoto, H.; Unno, N.; Kuzuya, M.; Setou, M. *Atherosclerosis* **2011**, *217*, 427–432.

30. Lee, E. S.; Shon, H. K.; Lee, T. G.; Kim, S. H.; Moon, D. W. *Atherosclerosis* **2013**, *226*, 378–384.

31. Mas, S.; Martínez-Pinna, R.; Martín-Ventura, J. L.; Pérez, R.; Gomez-Garre, D.; Ortiz, A.; Fernandez-Cruz, A.; Vivanco, F.; Egido, J. *Diabetes* **2010**, *59*, 1292–1301.

32. Malmberg, P.; Börner, K.; Chen, Y.; Friberg, P.; Hagenhoff, B.; Månsson, J. E.; Nygren, H. *Biochim. Biophys. Acta* **2007**, *1771*, 185–195.

33. Tanaka, H.; Zaima, N.; Ito, H.; Hattori, K.; Yamamoto, N.; Konno, H.; Setou, M.; Unno, N. *J. Vasc. Surg.* **2013**, PII: S0741-5214(13)00196-1.

34. Mas, S.; Touboul, D.; Brunelle, A.; Aragoncillo, P.; Egido, J.; Laprévote, O.; Vivanco, F. *Analyst* **2007**, *132*, 24–26.

35. Vallejo, M.; Martín Ventura, J. L.; Egido, J.; Barbas, C. *Anal. Bioanal. Chem.* **2009**, *394*, 1517–1524.

36. Zhao, G. *Am. J. Physiol. Heart Circ. Physiol.* **2008**, *294*, H936.

37. Shah, S. H.; Bain, J. R.; Muehlbauer, M. J.; Stevens, R. D.; Crosslin, D. R.; Haynes, C.; Dungan, J.; Newby, L. K.; Hauser, E. R.; Ginsburg, G. S.; Newgard, C. B.; Kraus, W. E. *Circ. Cardiovasc. Genet.* **2010**, *3* (2), 207–214.

38. Caprioli, R. M.; Farmer, T. B.; Gile, J. *Anal. Chem.* **1997**, *69*, 4751–4760.

39. March, R. E.; Todd, J. F. J., Eds.; *Practical Aspects of Trapped Ion Mass Spectrometry*; CRC Press: Canada, 2009.

40. McDonnell, L. A.; Heeren, R. M.; Andrén, P. E.; Stoeckli, M.; Corthals, G. L. *J. Proteomics* **2012**, *75*, 5113–5121.

41. Cornett, D. S.; Frappier, S. L.; Caprioli, R. M. *Anal. Chem.* **2008**, *80*, 5648–5653.

42. Chaurand, P.; Cornett, D. S.; Angel, P. M.; Capriolo, R. M. *Mol. Cell. Proteomics* **2011**, *10*, O110.004259.

43. Weiss, R. H.; Kim, K. *Nat. Rev. Nephrol.* **2012**, *8*, 22–33.

44. Hayashi, K.; Sasamura, H.; Hishiki, T.; Suematsu, M.; Ikeda, S.; Soga, T.; Itoh, H. *Nephrourol. Mon.* **2011**, *3*, 164–171.

45. Goek, O. N.; Döring, A.; Gieger, C.; Heier, M.; Koenig, W.; Prehn, C.; Römisch-Margl, W.; Wang-Sattler, R.; Illig, T.; Suhre, K.; Sekula, P.; Zhai, G.; Adamski, J.; Köttgen, A.; Meisinger, C. *Am. J. Kidney Dis.* **2012**, *60*, 197–206.

46. Hirayama, A.; Nakashima, E.; Sugimoto, M.; Akiyama, S.; Sato, W.; Maruyama, S.; Matsuo, S.; Tomita, M.; Yuzawa, Y.; Soga, T. *Anal. Bioanal. Chem.* **2012**, *404*, 3101–3109.

47. Zhang, J.; Yan, L.; Chen, W.; Lin, L.; Song, X.; Yan, X.; Hang, W.; Huang, B. *Anal. Chim. Acta.* **2009**, *650*, 16–22.

48. Zhao, Y. Y.; Feng, Y. L.; Bai, X.; Tan, X. J.; Lin, R. C.; Mei, Q. *PLoS One* **2013**, *8*, e59617.

49. Toyohara, T.; Suzuki, T.; Morimoto, R.; Akiyama, Y.; Souma, T.; Shiwaku, H. O.; Yoichi Takeuchi, E. M.; Abe, M.; Tanemoto, M.; Masuda, S.; Kawano, H.; Maemura, K.; Nakayama, M.; Sato, H.; Mikkaichi, T.; Yamaguchi, H.; Fukui, S.; Fukumoto, Y.; Shimokawa, H.; Inui, K.-i.; Terasaki, T.; Goto, J.; Sadayoshi Ito, T. H.; Rubera, I.;

Tauc, M.; Yoshiaki Fujii-Kuriyama, H. Y.; Moriyama, Y.; Soga, T.; Abe, T. *J. Am. Soc. Nephrol.* **2009**, *20*, 2546–2555.

50. Wang, J.; Zhou, Y.; Xu, M.; Rong, R.; Guo, Y.; Zhu, T. *Transplant. Proc.* **2011**, *43* (10), 3738–3742.

51. Sato, E.; Kohno, M.; Yamamoto, M.; Fujisawa, T.; Fujiwara, K.; Tanaka, N. *Eur. J. Clin. Invest.* **2011**, *41*, 241–255.

52. Rhee, E. P.; Souza, A.; Farrell, L.; Pollak, M. R.; Lewis, G. D.; Steele, D. J. R.; Thadhani, R.; Clish, C. B.; Greka, A.; Gerszten, R. E. *J. Am. Soc. Nephrol.* **2010**, *21*, 1041–1051.

53. Duan, H.; Guan, N.; Wu, Y.; Zhang, J.; Ding, J.; Shao, B. *J. Chromatogr. B Analyt. Technol. Biomed. Life Sci.* **2011**, *879*, 3544–3550.

54. Posada-Ayala, M.; Zubiri, I.; Martin-Lorenzo, M.; Sanz-Maroto, A.; Molero, D.; Gonzalez-Calero, L.; Fernandez-Fernandez, B.; Cuesta, F. D.; Laborde, C. M.; Barderas, M. G.; Ortiz, A.; Vivanco, F.; Alvarez-Llamas, G. *Kidney Int.* **2013**, *85*, 103–111. http://dx.doi.org/10.1038/ki.2013.328.

55. Qi, S.; Ouyang, X.; Wang, L.; Peng, W.; Wen, J.; Dai, Y. *Clin. Transl. Sci.* **2012**, *5*, 379–385.

56. Toyohara, T.; Akiyama, Y.; Suzuki, T.; Takeuchi, Y.; Mishima, E.; Tanemoto, M.; Momose, A.; Toki, N.; Sato, H.; Nakayama, M.; Hozawa, A.; Tsuji, I.; Ito, S.; Soga, T.; Abe, T. *Hypertens. Res.* **2010**, *33*, 944–952.

57. Kikuchi, K.; Itoh, Y.; Tateoka, R.; Ezawa, A.; Murakami, K.; Niwa, T. *J. Chromatogr. B Analyt. Technol. Biomed. Life Sci.* **2010**, *878*, 1662–1668.

58. Xia, J.-F.; Liang, Q.-L.; Liang, X.-P.; Wang, Y.-M.; Hu, P.; Li, P.; Luo, G.-A. *J. Chromatogr. B Analyt. Technol. Biomed. Life Sci.* **2009**, *877*, 1930–1936.

59. Herring, K. D.; Oppenheimer, S. R.; Caprioli, R. M. *Semin. Nephrol.* **2007**, *27* (6), 597–608.

60. Lindon, J. C.; Beckonert, O. P.; Holmes, E.; Nicholson, J. K. *Prog. Nucl. Magn. Reson. Spectrosc.* **2009**, *55*, 79–100.

61. Magni, F.; Lalowski, M.; Mainini, V.; Marchetti-Deschmann, M.; Chinello, C.; Urbani, A.; Baumann, M. *J. Nephrol.* **2013**, *26* (3), 430–436.

62. Xu, J.; Zhang, J.; Cai, S.; Dong, J.; Yang, J. Y.; Chen, Z. *Anal. Bioanal. Chem.* **2009**, *393*, 1657–1668.

63. Garrod, S.; Humpfer, E.; Spraul, M.; Connor, S. C.; Polley, S.; Connelly, J.; Lindon, J. C.; Nicholson, J. K.; Holmes, E. *Magn. Reson. Med.* **1999**, *41*, 1108–1118.

64. van der Kloet, F. M.; Tempels, F. W. A.; Ismail, N.; van der Heijden, R.; Kasper, P. T.; Rojas-Cherto, M.; van Doorn, R.; Spijksma, G.; Koek, M.; van der Greef, J.; Mäkinen, V. P.; Forsblom, C.; Holthöfer, H.; Groop, P. H.; Reijmers, T. H.; Hankemeier, T. *Metabolomics* **2012**, *8*, 109–119.

65. Law, W. S.; Huang, P. Y.; Ong, E. S.; Sethi, S. K.; Saw, S.; Ong, C. N.; Fong, S.; Li, S. F. *J. Proteome Res.* **2009**, *8*, 1828–1837.

66. Hwang, G.-S.; Yang, J.-Y.; Ryu, D. H.; Kwon, T.-H. *Am. J. Physiol. Renal Physiol.* **2010**, *298*, 461–470.

67. Zhang, H.; Saha, J.; Byun, J.; Lee Schin, M.; Lorenz, M.; Kennedy, R. T.; Kretzler, M.; Feldman, E. L.; Pennathur, S.; Brosius, F. C., III *Am. J. Physiol. Renal Physiol.* **2008**, *295*, 1071–1081.

68. Choi, J.-Y.; Yoon, Y. J.; Choi, H.-J.; Park, S.-H.; Kim, C.-D.; Kim, I.-S.; Kwon, T.-H.; Do, J.-Y.; Kim, S.-H.; Ryu, D. H.; Hwang, G.-S.; Kim, Y.-L. *Nephrol. Dial. Transplant.* **2011**, *26*, 1304–1313.

69. Wishart, D. S. *Curr. Opin. Nephrol. Hypertens.* **2006**, *15*, 637–642.

70. Psihogios, N. G.; Kalaitzidis, R. G.; Dimou, S.; Seferiadis, K. I.; Siamopoulos, K. C.; Bairaktari, E. T. *J. Proteome Res.* **2007**, *6*, 3760–3770.

71. Sun, J.; Shannon, M.; Ando, Y.; Schnackenberg, L. K.; Khan, N. A.; Portilla, D.; Beger, R. D. *J. Chromatogr. B Analyt. Technol. Biomed. Life Sci.* **2012**, *893–894*, 107–113.

72. Abbiss, H.; Maker, G. L.; Gummer, J.; Sharman, M. J.; Phillips, J. K.; Boyce, M.; Trengove, R. D. *Nephrology* **2012**, *17*, 104–110.

73. Romick-Rosendale, L. E.; Brunner, H. I.; Bennett, M. R.; Mina, R.; Nelson, S.; Petri, M.; Kiani, A.; Devarajan, P.; Kennedy, M. A. *Arthritis Res. Ther.* **2011**, *13*, R199.

74. Jia, L.; Wang, C.; Zhao, S.; Lu, X.; Xu, G. *J. Chromatogr. B Analyt. Technol. Biomed. Life Sci.* **2007**, *860*, 134–140.

75. Akira, K.; Masu, S.; Imachi, M.; Mitome, H.; Hashimoto, M.; Hashimoto, T. *J. Pharm. Biomed. Anal.* **2008**, *46*, 550–556.

76. de la Cuesta, F.; Alvarez-Llamas, G.; Gil-Dones, F.; Martin-Rojas, T.; Zubiri, I.; Pastor, C.; Barderas, M. G.; Vivanco, F. *Expert Rev. Proteomics* **2009**, *6* (4), 395–409.

77. Zhong, F.; Liu, X.; Zhou, Q.; Hao, X.; Lu, Y.; Guo, S.; Wang, W.; Lin, D.; Chen, N. *Nephrol. Dial. Transplant.* **2012**, *27*, 556–565.

78. Murphy, R. C.; Hankin, J. A.; Barkley, R. M. *J. Lipid Res.* **2009**, *50*, S317–S322.

79. Kaneko, Y.; Obata, Y.; Nishino, T.; Kakeya, H.; Miyazaki, Y.; Hayasaka, T.; Setou, M.; Furusu, A.; Kohno, S. *Exp. Mol. Pathol.* **2011**, *91* (2), 614–621.

80. Nilsson, A.; Forngren, B.; Bjurström, S.; Goodwin, R. J.; Basmaci, E.; Gustafsson, I.; Annas, A.; Hellgren, D.; Svanhagen, A.; Andrén, P. E.; Andrén, P. E.; Lindberg, J. *PLoS One* **2012**, *7* (10), e47353.

81. Grobe, N.; Elased, K. M.; Cool, D. R.; Morris, M. *Am. J. Physiol. Endocrinol. Metab.* **2012**, *302* (8), E1016–E1024.

82. Stenvinkel, P.; Carrero, J. J.; Axelsson, J.; Lindholm, B.; Heimbürger, O.; Massy, Z. *Clin. J. Am. Soc. Nephrol.* **2008**, *3*, 505–521.

83. Jun, M.; Lv, J.; Perkovic, V.; Jardine, M. J. *Ther. Adv. Chronic Dis.* **2011**, *2*, 265–278.

84. Sakuma, M.; Nakamura, M.; Tanaka, F.; Onoda, T.; Itai, K.; Tanno, K.; Ohsawa, M.; Sakata, K.; Yoshida, Y.; Kawamura, K.; Makita, S.; Okayama, A. *Circ. J.* **2010**, *74*, 792–797.

85. Seiler, S.; Reichart, B.; Roth, D.; Seibert, E.; Fliser, D.; Heine, G. H. *Nephrol. Dial. Transplant.* **2010**, *25*, 3983–3989.

86. Kanbay, M.; Nicoleta, M.; Selcoki, Y.; Ikizek, M.; Aydin, M.; Eryonucu, B.; Duranay, M.; Akcay, A.; Armutcu, F.; Covic, A. *Clin. J. Am. Soc. Nephrol.* **2010**, *5*, 1780–1786.

87. Mesquita, M.; Demulder, A.; Damry, N.; Mélot, C.; Wittersheim, E.; Willems, D.; Dratwa, M.; Bergmann, P. *Clin. Chem. Lab. Med.* **2009**, *47*, 339–346.

88. Soriano, S.; Gonzalez, L.; Martin-Malo, A.; Rodríguez, M.; Aljama, P. *Clin. Nephrol.* **2007**, *67*, 352–357.

89. Kanbay, M.; Yilmaz, M. I.; Sonmez, A.; Turgut, F.; Saglam, M.; Cakir, E.; Yenicesu, M.; Covic, A.; Jalal, D.; Johnson, R. J. *Am. J. Nephrol.* **2011**, *33*, 298–304.

90. Kanbay, M.; Ikizek, M.; Solak, Y.; Selcoki, Y.; Uysal, S.; Armutcu, F.; Eryonucu, B.; Covic, A.; Johnson, R. J. *Am. J. Nephrol.* **2011**, *33*, 325–331.

91. Guo, Q.; Carrero, J. J.; Yu, X.; Bárány, P.; Qureshi, A. R.; Eriksson, M.; Anderstam, B.; Chmielewski, M.; Heimbürger, O.; Stenvinkel, P.; Lindholm, B.; Axelsson, J. *Nephrol. Dial. Transplant.* **2009**, *24*, 3468–3473.

92. Goicoechea, M.; de Vinuesa, S. G.; Gómez-Campderá, F.; Aragoncillo, I.; Verdalles, U.; Mosse, A.; Luño, J. *Kidney Int. Suppl.* **2008**, *111*, S67–S70.

93. Lajer, M.; Tarnow, L.; Jorsal, A.; Teerlink, T.; Parving, H. H.; Rossing, P. *Diabetes Care* **2008**, *31*, 747–752.

94. Liabeuf, S.; Barreto, D. V.; Barreto, F. C.; Chasseraud, M.; Brazier, M.; Choukroun, G.; Kamel, S.; Massy, Z. A. *Nephrol. Dial. Transplant.* **2010**, *25*, 2596–2602.

95. Tonelli, M.; Wiebe, N.; Culleton, B.; House, A.; Rabbat, C.; Fok, M.; McAlister, F.; Garg, A. X. *J. Am. Soc. Nephrol.* **2006**, *17*, 2034–2047.

96. Fassett, R. G.; Venuthurupalli, S. K.; Gobe, G. C.; Coombes, J. S.; Cooper, M. A.; Hoy, W. E. *Kidney Int.* **2011**, *80*, 806–821.

97. Schiffer, E.; Liabeuf, S.; Lacroix, C.; Temmar, M.; Renard, C.; Monsarrat, B.; Choukroun, G.; Lemke, H.-D.; Vanholder, R.; Mischak, H.; Massy, Z. A. *J. Hypertens.* **2011**, *29*, 783–790.

98. Corrado, C.; Raimondo, S.; Chiesi, A.; Ciccia, F.; De Leo, G.; Alessandro, R. *Int. J. Mol. Sci.* **2013**, *14*, 5338–5366.

99. Del Boccio, P.; Raimondo, F.; Pieragostino, D.; Morosi, L.; Cozzi, G.; Sacchetta, P.; Magni, F.; Pitto, M.; Urbani, A. *Electrophoresis* **2012**, *33*, 689–696.

100. Beloribi, S.; Ristorcelli, E.; Breuzard, G.; Silvy, F.; Bertrand-Michel, J.; Beraud, E.; Verine, A.; Lombardo, D. *PLoS One* **2012**, *7* (10), e47480.

101. Esser, J.; Gehrmann, U.; LuizD'Alexandri, F.; Hidalgo-Estevez, A. M.; Wheelock, C. E.; Scheynius, A.; Gabrielsson, S.; Radmark, O. *J. Allergy Clin. Immunol.* **2010**, *126*, 1032–1040.

102. Conde-Vancells, J.; Gonzalez, E.; Lu, S. C.; Mato, J. M.; Falcon-Perez, J. M. *Expert Opin. Drug Metab. Toxicol.* **2010**, *6* (5), 543–554.

103. Nieuwland, R.; van der Post, J. A.; Lok, C. A.; Kenter, G.; Sturk, A. *Semin. Thromb. Hemost.* **2010**, *36* (8), 925–929.

104. Mause, S. F.; Weber, C. *Circ. Res.* **2010**, *107*, 1047–1057.

105. Mallat, Z.; Tedgui, A. *Circ. Res.* **2001**, *88* (10), 998–1003.

106. Falati, S.; Liu, Q.; Gross, P.; Merrill-Skoloff, G.; Chou, J.; Vandendries, E.; Celi, A.; Croce, K.; Furie, B. C.; Furie, B. *J. Exp. Med.* **2003**, *197* (11), 1585–1598.

107. Mayr, M.; Grainger, D.; Mayr, U.; Leroyer, A. S.; Leseche, G.; Sidibe, A.; Herbin, O.; Yin, X.; Gomes, A.; Madhu, B.; Griffiths, J. R.; Xu, Q.; Tedgui, A.; Boulanger, C. M. *Circ. Cardiovasc. Genet.* **2009**, *2* (4), 379–388.

Chapter 12

The Clinical Application of Proteomics and Metabolomics in Neonatal Medicine

Alan R. Spitzer and Donald H. Chace

The MEDNAX Center for Research, Education, and Quality, Pediatrix Medical Group of MEDNAX, Inc., Sunrise, Florida, USA

Chapter Outline

Comprehensive Analytical Chemistry, Vol. 64. http://dx.doi.org/10.1016/B978-0-444-62650-9.00012-9

1 INTRODUCTION

Genomics, proteomics, and metabolomics (the "omics") represent new approaches for estimating the genetic possibility of disease, for assessing the role of protein metabolism in health and disease, and for understanding the outcomes of metabolism in biological processes. One of the most potentially intriguing areas for application of these technologies occurs in the perinatal period, examining noninvasively the potential for disease and the manifestation of disease states in the fetus and the neonatal patient. Mass spectrometry (MS) is an important emerging analytical approach in the study of small molecules and metabolism. The use of MS, together with increasingly rapid sequencing of the genome and analysis of the metabolic foundation for disease, has resulted in the emergence of several novel fields of scientific investigation that essentially did not exist in medicine a little over a decade ago.

Proteomics refers to the study of proteins in health and disease states. Metabolomics, in its most broadly defined term, includes the study of both endogenous (amino acids, fatty acids, steroids, vitamins, etc.) and exogenous molecules (drugs, toxins, environmental pollutants, food additives, etc.) within biological systems. In the recent years, genomics and proteomics have received the lion's share of scientific interest and research funding for the past few decades. Metabolomics has been dominant primarily in the pharmaceutical space, with MS analysis being paramount in screening many hundreds of drugs and their metabolites for efficacy in treatment of disease processes. Metabolomics applied to the biomedical space is beginning to emerge, however, in part due to the success of new clinical applications that use MS and metabolic screening. Its importance as a part of an integrated evaluation of health and disease is being increasingly recognized. Figure 1 shows how DNA, RNA, and metabolites may interact in health and disease and be potentially modified by the environment. One measure of health is the monitoring of metabolites that result from gene expression and enzyme (protein) activity. Metabolic screening can determine in many instances whether a disease is present, whether treatment is effective, and if an alternative therapy may be indicated. In fact, medicine is progressively moving toward integrating the "omics" (genomics, proteomics, and metabolomics) to personalize care by providing the most effective treatment of disease in a particular clinical situation and not through a "one size fits all" approach. With increasing interest in "omic" medicine in adults, it is only natural that similar assessment and treatment of infants and children would soon follow.

One of the earliest and perhaps most rewarding introductions of metabolomics in pediatric medicine has occurred with the bioanalysis of endogenous substrates and altered metabolism that may be indicative or predictive of significant disease states. Newborn and metabolic screening of inherited disorders of metabolism has, in many respects, served as a cornerstone for innovative

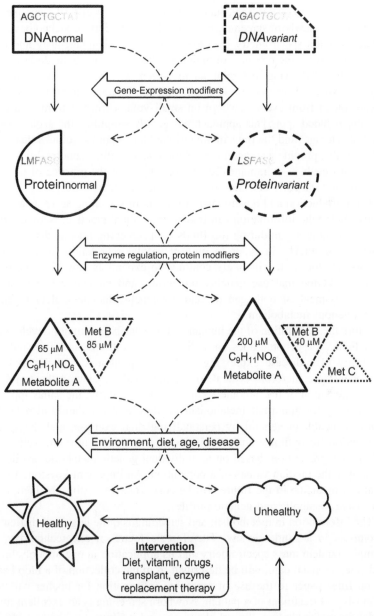

FIGURE 1 Interaction of genomics, proteomics, and metabolomics for healthy and disease states that are inherited and/or acquired.

methods in the practice of preventative medicine *(1–4)*. A critical component in the success of newborn screening has been the availability of technology, such as tandem mass spectrometry (MS/MS), which can simultaneously and cost-effectively test for the presence of multiple genetic and metabolic diseases with minimal blood drawing and minimal invasiveness, always critical issues for neonates. Much of the success in this regard can be traced to systems used to procure blood from every newborn on small volumes in the form of a dried filter paper blood spot. This approach has greatly simplified the acquisition of blood, its storage, and the transmission of the blood spot to facilities that could report results, provide accurate interpretation, and confirm abnormalities. Such innovations enabled a dramatically new paradigm for the practice of neonatal medicine and the early detection of serious disease states.

The introduction and development of newborn screening serve as a model of how metabolic information can transform medical practice. Neonatal metabolic screening began during the 1960s with screening for one disease, phenylketonuria (PKU), in order to prevent mental retardation, to more than 50 disorders today. It will likely continue to expand as new biomarkers for disease detection that use genetic, proteomic, and metabolic techniques to allow assessment of both genetic and nongenetic disorders of endogenous and exogenous metabolism.

While the elaboration of the human genome has been a major development in medicine, the genome does not indicate when a disease is active, only the possibility of a disease expressing itself. The proteomic and metabolomic components are therefore critical to evaluate and understand the timing of the manifestation of the genome's constituents. Similarly, the same approach to screening for abnormal metabolism in disease states should also lead to improved health in general by personalizing diet, exercise, and therapeutic interventions to optimize an individual's genetics, age, nutrition, well-being, and disease prevention, based on an individual genetic/metabolic profile and not simply the median values of a population. An important keystone to personalized medicine in the future will be cost-effective, easily obtained, comprehensive metabolic and genetic profiles.

The information in metabolism and screening is quite large and difficult to encompass in a single review. Therefore, a focus on one technology, for example, tandem mass spectrometry and its application in one area of clinical medicine, neonatal care, will be described from the collection of a blood sample on filter paper to the use of screening information for his/her part of a diagnosis and treatment of a patient. Newborn screening is an excellent initial model for understanding clinical chemistry and its impact at the bedside.

2   A BRIEF HISTORY OF NEWBORN SCREENING

The use of a biomarker for detecting metabolic disease began nearly 50 years ago when Robert Guthrie developed a bacterial inhibition assay to screen for

PKU *(5,6)*. This assay measured a single amino acid, phenylalanine, which is the metabolite that accumulates excessively as a result of reduced enzymatic activity of phenylalanine hydroxylase, which converts phenylalanine to tyrosine. During the subsequent 25 years, screening of the neonate expanded to include other diseases such as congenital hypothyroidism (CH), congenital adrenal hyperplasia (CAH), amino acid disorders such as maple syrup urine disease (MSUD), homocystinuria, tyrosinemias, biotinidase deficiencies, galactosemia, and sickle cell disease. For each disease situation, a single critical metabolite that would accumulate in abnormal concentrations was targeted, such as thyroxine in CH and 17-OH progesterone in CAH.

Improvements to disease screening occurred through more specific assays and improved efficiency from consolidation of multiple assays to the same testing platform by the late 1980s. None of these platforms for testing, however, have had such a dramatic and important impact as tandem mass spectrometry, which was developed in the 1990s specifically for newborn screening of multiple amino acid and fatty acid disorders in a single rapid analysis. This development resulted in a dramatic change in the approach to screening *(7)*. As such, it is interesting to note that the MS technology was not developed specifically for use in newborn screening. Much of MS technology was initially targeted to high-throughput metabolic studies in the pharmaceutical industry, a process that will be discussed later in this chapter with respect to its potential role in neonatal management. The technology was also important in the discovery of certain metabolic diseases, such as disorders of fatty acid metabolism, as well as other diagnostics. It was soon apparent, however, that MS/MS could also be adapted toward high-throughput multianalyte newborn screening coupled with a strong foundation in other areas of clinical chemistry *(1,8–10)*. Figure 2 illustrates how technology has improved from single-metabolite assays to multiple analytes for the detection of PKU.

2.1   The Role of Clinical MS

The foundation of clinical MS and its role in the diagnosis of inherited disorders of metabolism is chromatography. The earliest work in this field can be traced to the 1950s with the analysis of blood gases *(11,12)*. From thin-layer chromatography, the analysis of fatty acids such as methyl ester derivatives utilized gas chromatography (GC), which ultimately expanded with improved derivatization techniques and sample preparation strategies to include other biomarkers and metabolites such as phospholipids, amino acids, simple sugars, and nucleic acids *(13–18)*. The detection of metabolites eluting from GC was subsequently dramatically improved by coupling to a mass spectrometer (GC–MS) *(19,20)*. GC–MS is considered two distinct separation methods, one physical and the other mass, and in a sense is its own confirmatory technique. It is still widely used today as the major diagnostic tool for the screening of metabolic disorders.

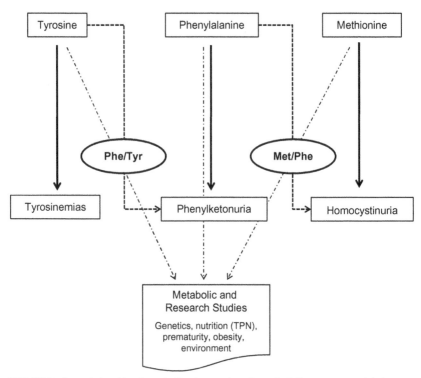

FIGURE 2 Interrelationship of metabolites in genetic and acquired disease states and their use as biomarkers for amino acid metabolism. Ratios of two metabolite concentrations (molar ratios) improve specificity for detection and characterization of disease.

The clinical assessment and characterization of a variety of metabolic diseases became feasible because of advances in GC–MS. GC–MS represents an ideal diagnostic method for assessing potentially affected individuals, due to the broad range of metabolites that can be detected and the great selectivity of evaluation provided by this technique. However, it is a less than ideal approach in newborn screening since it requires urine collection and is not capable of high-throughput population analysis. Newborn screening requires the evaluation of large numbers of normal, asymptomatic patients within a very short time period so that treatment can be initiated as early as possible to have the greatest efficacy. The subsequent development of a comparably selective technique, such as MS/MS, for the performance of metabolic studies in the late 1980s initiated its application to newborn screening for amino acid and fatty acid metabolic disorders during the early 1990s. MS separates protonated biomarkers such as alpha amino acids in the first mass analyzer, breaks them into characteristic fragment ions in the collision chamber, and detects the mass of these fragments in the second mass analyzer. The early application of MS/MS to population screening marked a new era of MS in clinical chemistry, according to Bruns, who first described the impact of the early screening studies *(21)*.

MS/MS was capable of detecting a wide range of metabolites such as fatty acids and organic acids such as acylcarnitines, in addition to amino acids, without the need for chromatography. This development enabled the screening of dozens of metabolites in approximately 2 min and immediately led to an expansion of neonatal metabolic screening. The early use of diagnostic acylcarnitines with MS/MS was adapted for newborn screening, while the analysis of key amino acids, such as phenylalanine for PKU, led to a similar expansion in newborn screening *(22,23)*. What started as the screening for several thousand specimens in pilot studies in North Carolina and a private laboratory in Western Pennsylvania evolved to permit the screening of nearly every infant in the United States and tens of millions of infants worldwide for dozens of metabolic disorders *(2,4)*. It is also interesting that with respect to acylcarnitines, the same method is used for both screening and diagnostic tests, the only difference being whether plasma (diagnostic) or a dried blood spot (screening) is being analyzed.

Today, many of the older methods used historically in clinical chemistry are being replaced by methods that utilize MS. In 2013, some areas of interest in clinical chemistry include steroid and vitamin D analyses that require a high selectivity and relative sensitivity. For example, the improved selectivity is essential in the analysis of testosterone in infants due to low concentrations. Immunoassays are not sufficiently selective to provide an accurate measurement of the true concentrations of these metabolites. There are many more assays being used or that are now in development in clinical chemistry, but a discussion of this growing use of metabolic screening is too broad and too evolutionary to be included here. Therefore, this chapter focuses on those proteins and metabolites (e.g., amino acids and acylcarnitines) that are part of a classic newborn screening panel.

2.2   The Dried Blood Spot

Among the most important aspects of any clinical chemistry method are the characteristic of the biological sample, its method of collection and stability, and the integrity at the point of sample preparation and analysis. With regard to newborn screening, one of its most interesting aspects and perhaps the area of greatest impact in clinical-based technology has been the unique manner in which blood is collected for the performance of this test. Blood is obtained using a cotton-based filter paper that is manufactured to absorb a reproducible quantity of blood that is applied to it. This filter paper collection, historically called "the Guthrie card or PKU card," was developed for use in a bacterial inhibition assay for PKU developed by Robert Guthrie *(24)*. Disks punched from the dried blood spots were placed on agars comprised of bacteria that would grow only in the presence of phenylalanine. Other punches from the card were developed for subsequent BIA tests such as that for MSUD and homocystinuria *(25,26)*. Screening tests developed for other assays such as CH and CAH were based on extraction of metabolites from the dried blood spot, since that was the manner in which blood was available to the screening laboratory. Currently, all newborn screening tests for >50 diseases with blood from four to six "zones," which are delineated by a dashed circle imprinted on the card. Figure 3A provides a photograph of one type of filter paper collection card with four circles filled with dried blood.

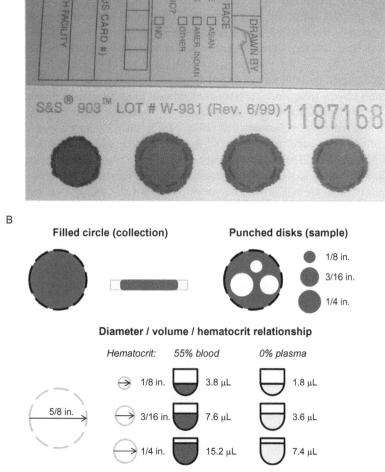

FIGURE 3 (A) Photograph of dried blood spots on filter paper (white portion) attached to a patient information form (partial photo, gold section (gray section in print version)). (B) Illustration of a dried blood spot from top and cross-sectional views (left panel) and paper punch sample sizes typically used in newborn screening (NBS) laboratories (right panel). The size of the collected blood spot, its hematocrit, and punch size determine the blood volume of the specimen analyzed (lower panel).

Figure 3B illustrates the relationship between blood volume and the area of a dried blood spot. Approximately, 50–75 µL of blood is usually collected on each large circle as marked by dashes, although in its practical application, it can vary substantially with very large or very small blood applications. From each dried blood spot, a disk is punched that is most often either a 1/8th or 3/16th inch punch. A quarter inch punch has also been used in some applications. These disks all fit in microtiter plate wells and therefore are highly satisfactory for use in newborn screening on large populations of infants. From the four to six circles of blood, most newborn screening labs will punch between 8 and 12 disks of various diameters leaving enough blood on the filter paper card for a few repeat analyses. In general, there is one punch per disease screened. With MS/MS analysis however, there is one punch for multiple diseases, since MS/MS analysis is a test of multiple metabolites (>50) from a single test. As screening expands, it is critical that the test becomes consolidated and at least duplexed, if not multiplexed, such that more than one disease screen is obtained from a single punch.

The dried blood spot is experiencing its own renaissance of sorts, as there is now major pharmaceutical interest in this mode of sample collection due to its advantages rather than limitations. Some of the advantages long recognized by the newborn screening community are the easy preservation of sample delivery to the lab and the space savings and increased stability for many analytes during storage of specimens. It is also important to recognize that the total amount of blood applied to a filter paper card is typically less than a half milliliter (500 µL). This modest volume is a critical consideration in the screening of very low-birth-weight infants or any research study where the total amount of blood that can be collected is very small. For example, a 600 g premature infant has a total blood volume of approximately 60 mL, or two ounces! This amount compares to the 2–3 mL of blood required in a typical collection using a syringe. Finally, the dried blood spot is a safer specimen to handle with regard to infectious disease and is one reason why it is possible to send a filter paper card through regular mail without the complexity associated with the transport and preservation of vials of blood.

As newborn screening by MS/MS is discussed in detail, one of the primary lessons learned has been that the method of sample collection is critical in directing the methods used in analysis and interpretation of results. In the case of dried blood spots, it was apparent that the blood volume indeed varies substantially from sample to sample and has a significant analytical impact on quantification and interpretation. The quantity of blood contained with each disk, as observed in Figure 3A and B, will vary by hematocrit, for example. Higher hematocrit increases the viscosity of blood and it will therefore spread less when applied to the filter paper. The net result is a high blood volume per area of filter paper. Lower hematocrit specimens, which spread more readily, have a larger area and a lower blood volume per area. Plasma, which has no red cells, spreads to nearly twice the area of blood and consequently has approximately half the volume of plasma per unit of area on the filter paper, as shown in Figure 3B.

Another factor is the size of the blood volume application. Typically, 25–50 µL of blood is obtained from a heel stick. However, some labs utilize capillary tubes and make many "dots" on the card. These smaller diameter blood spots (which have diameters of approximately 1/4 to 1/3 inch) have lower blood volumes than a filled circle because the blood does not appropriately saturate the card. In fact, a blood card is always punched from the back to ensure that the most completely saturated area is used in the punch. Finally, the site at which the disk is punched may have an effect on the quantity and quality of blood if it is oversaturated (double-spotted), has serum rings, or has been contaminated in some manner. The quality of the puncher may also have an effect on blood volume. Typically, a 1/8th or 3/16 inch punch is obtained using an automated puncher or a standard paper puncher. A paper puncher is typically used in low-volume laboratories, as clean cuts may not be possible with worn, older punchers as the diameter cannot be certified. The impact on quantification in terms of variance in blood volume is sufficiently large to mask other errors in the analysis. It is for this reason that the techniques using filter paper blood spots are considered screening techniques. The type of high precision that would be necessary for definitive diagnosis is reduced, even if highly accurate MS/MS analysis is used.

One part of newborn screening that cannot be controlled is the manner by which the specimen is collected, as described in the preceding text. It is important that any newborn or clinical screening, or research study, carefully controls how specimens are collected. The training of phlebotomy teams demands consistency in the manner of collection, as described in the previous paragraphs (Section 2.2). Newborn screening programs typically require a heel stick, but for various reasons, many specimens are collected by applying blood collected from venipuncture or from capillary tubes at a time when blood is being drawn for other reasons. This inconsistency is often mandated by consideration for reducing painful sticks to the tiny infant. An important consideration, however, is that collection of specimens in tubes might be contaminated with anticoagulants such as EDTA, or amino acids, from total parenteral nutrition (TPN) solutions previously administered in the same intravenous line. For example, some studies in a clinical research trial may take a sample of 1–2 mL of blood collected intravenously. The spots are then applied to the filter paper card by a pipette. It is recommended that the blood spots be applied in volumes from 25 to 75 µL. One final note for consideration is that not all specimens are dried properly. It is recommended that 3 h or more be utilized after collection, depending on humidity in the collection facility. Wet spots have residual enzyme activity and may adversely affect assays that measure enzyme activity or cause metabolites to degrade due to oxidation or residual enzyme action. Clinicians evaluating results from a dried blood spot should be aware that a 20% variance in quantification is typical. This variance can be reduced, however, by using stable isotope internal standards and calculation of molar or concentration ratios.

2.3 MS: Selectivity

The key issues in utilizing a mass spectrometer are selectivity and quantification. High selectivity is the single most important advantage of mass detection over other detectors and methodology (e.g., immunoassays), especially for small molecules. Immunoassays or other systems that require an antigen–antibody complex to be generated often have problems with selectivity especially at low concentrations of small molecules, as occurs with the analysis of metabolites in the circulation. That is not to say that MS is perfectly selective. Due to the sheer numbers of small metabolites, there are many compounds that have the same mass or elemental content or both. Therefore, they cannot be readily separated by using a mass spectrometer alone. Tandem MS, however, can separate compounds of the same mass provided that there are unique fragments that the tandem MS can analyze. However, in some cases, no MS or MS/MS technique can separate two compounds of identical elemental content and similar structure such as Leu and Ile without the use of chromatographic separation or special derivatization techniques that are structurally and isomerically selective. However, although it does occasionally have imperfect selectivity, it is not necessary to completely ignore the results obtained from MS/MS. In MSUD, for example, both leucine and isoleucine, isomeric compounds, are elevated, and the two metabolites are additive. Furthermore, in MSUD, another metabolite of significance is alloisoleucine, which is also additive but cannot be separated by mass from Leu and Ile. Nevertheless, the assay, based on millions of specimens analyzed, is effective at "screening" for MSUD. A diagnostic test that utilizes chromatography (GC or LC) is required to confirm MSUD and quantify the individual metabolites correctly.

There are many other "isomers" in MS/MS analysis, such as several acylcarnitines that share the same mass, such as many of the hydroxyacylcarnitines and dicarboxylic acid acylcarnitines. In addition, interfering products may, at times, share a common fragment ion or neutral loss and as a result will be detected in a profile. No method, however, is perfectly selective and a physician reviewing results from a newborn screen, whether analyzed using an immunoassay or MS, should be aware of the fact that in some cases, similar mass compounds may lead to potential sources of error.

2.4 MS: Isotope Dilution

Stable-isotope-labeled compounds are the most important standard reagents used in most MS-based assays that require accurate quantification. Measurement of metabolites using stable isotopes is commonly referred to as isotope dilution mass spectrometry (IDMS). The name is derived from the addition of a known amount of an enriched "labeled" form of the same compound that is targeted in a measurement. The ratio of the quantity of unlabeled to labeled compound can be measured, and the concentration then can be determined. For the measurement of the concentration of phenylalanine, for example, a known amount of a stable isotope standard ($^{13}C_6$-phenylalanine, an isotope enriched with six ^{13}C atoms and a subsequent mass shift of 6 Da) is added in known quantities to blood or plasma. The actual concentration of Phe present in blood or plasma is determined in part by the ratio of the ion intensity of the Phe to $^{13}C_6$-Phe. IDMS is considered the "gold standard" for accurate and precise quantification, because both labeled and unlabeled compounds differ by only enrichment or mass and not by chemical structure. However, when dried blood spots are used, as is common with newborn screening, the internal standard is typically added with the extraction solvent and mixed only with the unlabeled standard after extraction of the metabolite from the blood spot. Isotope dilution principles apply after the extraction only, and since extraction efficiency can vary and be less than 100%, the ideal conditions of isotope dilution do not apply. Extraction efficiency can be determined and quantification can be quite accurate; therefore, the term pseudoisotope dilution MS (*pseudo*IDMS) has been used for dried blood spots to indicate a less than ideal variant of IDMS.

The quantification of metabolites in dried blood spots primarily ensures that the quality of the isotopes standards is excellent in terms of chemical and isotopic purity. When using MS/MS, it is essential that the fragments produced by the collision cell and the product ions detected ensure that both labeled and unlabeled metabolites are identical. Most importantly, the choice of the isotope label and the structural positions must be such that they are stable and do not exchange with other isotopes during sample preparation. Finally, it is imperative that the mass shift is sufficiently high (at least 3 Da) for small molecules less than 1000 Da and that the label occurs at a mass free from other compound interference. Figure 4 illustrates the concepts of quantification using stable isotope with Phe measurement in a dried blood spot as an example.

2.5 Interpretation of Results

Interpretation of complex metabolic profiles and communication of lab results to clinicians has been one of the greatest challenges in dried blood spot analysis by MS/MS for acylcarnitines and amino acids. Following its introduction, MS/MS defined numerous diseases of fatty acid and organic acid metabolism that were uncommonly encountered in a typical practice and not well understood by most pediatricians. Furthermore, even in disorders such as PKU that were well known, the improvements made by MS/MS screening for PKU compared with older technology such as BIA or fluorometry were not well

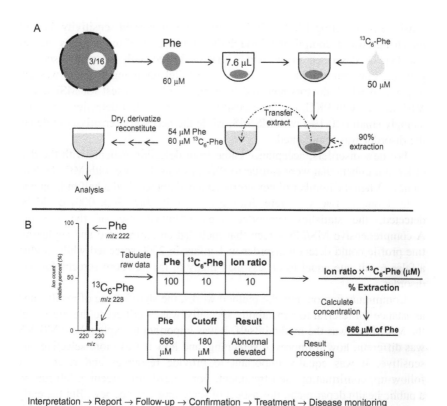

FIGURE 4   (A) The use of stable isotope internal standards for quantification of Phe extracted from a dried blood spot (DBS) and its role in sample preparation for MS/MS analysis. (B) MS/MS analysis of Phe showing quantification of Phe and its stable isotope internal standard ($^{13}C_6$-Phe). Ion counts (signal intensity) from each compound are calculated as an ion ratio. Concentrations are subsequently calculated based on the known amount of internal standard added and the extraction efficiency (recovery) of Phe from the DBS.

understood or appreciated. Lastly, screening was being significantly expanded without the appropriate base of knowledge or available therapies to ensure proper treatment prior to the appearance of symptoms. As a result, many pediatricians and public health policy officials were somewhat wary of MS/MS applications in newborn screening, and it has taken a number of years to universally expand the number of disorders screened as mandated by the states. Fortunately, MS/MS has gained acceptance by demonstrating positive results in screening at a much lower cost of analysis per sample than what was expected.

One example of the increasing acceptance of MS/MS was seen with comparison analysis of specific abnormalities. The classic example occurs with the detection of PKU using two methods, fluorometry and MS/MS. MS/MS was shown to be much more selective than fluorometry and was as clinically sensitive in detecting PKU with the same if not improved sensitivity. Part of the change was accomplished through the examination of molar ratios such as Phe/Tyr, as described in one study that evaluated early hospital discharge patients. Furthermore, in Pennsylvania, a parallel screening program was put in place in which doctors received both results of a BIA test for MSUD and MS/MS. As with PKU, the false-positive rate for MSUD detection was vanishingly small (<0.1%) with MS/MS, while the necessary sensitivity to detect the disorder was maintained.

For new disorders, acceptance came from detecting patients with the disorder in numbers that were similar to PKU, as was the case with MCAD deficiency. When the number of newborns screened approached a half million and at least one or two very rare disorders (1 in 100,000–300, 000) could be detected, the statistics supported the continued use of the method. A comprehensive MS/MS screen that included an amino acid and acylcarnitine profile could detect a disorder in 1 infant in 2500 screened while having a false result rate equal to or less than many of the newborn screens for one disorder.

Communication on how the method works, the significance of results, and assistance for referral to a metabolic specialist was as critical to the success of the MS/MS as was the method itself. It was important to explain how MS/MS was different, how it improved screening, and how it was more selective and sensitive. It was equally important to provide resources and referral for follow-up, confirmation, and treatment. For a newborn screening lab run by a public health department, this aspect of screening was not new. But the early screening and proof of method were done by a private laboratory and ultimately adapted by public health newborn screening. There were several less obvious lessons learned in this process of developing and validating newborn screening by MS/MS.

2.6 MS/MS Validation

Clinical validation of new methods is a very important issue in neonatal screening, and many concepts are still being appreciated with regard to MS/MS applications in this area of medicine. The critical standard for a newborn screening validation is perhaps the most essential outcome: accurate results. Only following the analysis of hundreds of thousands or perhaps millions of samples to assess the true positive and false result rates, can a method's accuracy be determined? Yet, there is no ability to attain these numbers without actually implementing a specific methodology. Clinical trials or pilot screening studies can provide some important data with respect to the likelihood of success, but only with implementation in multiple labs, thereby allowing screening of large and diverse populations with follow-up of presumptive and potential missed cases, can a method's true value be determined? Nevertheless, method validation for MS/MS did occur and has been evolving for newborn screening for the past several decades.

It is important to first note that many classical clinical assays have traditionally measured one metabolite to detect one disease. Consequently, many of the rules of method validation were designed around this premise. MS/MS, as originally designed, detected two classes of compounds, amino acids and acylcarnitines, in four to five different MS methods (known as scan modes such as neutral loss, precursor ion, or selected reaction monitoring), for approximately 500 distinct masses, more than 70 known compounds, and 20–30 stable isotope internal standards. How then did one approach such a complicated validation to gain acceptance as a reliable, useful method? The answer is quite simple – start simply and compare to what was already established.

First, it was important to define the need for development of MS/MS in newborn screening in the first place and to have a method in which a newborn screening could be modeled. The analysis of acylcarnitines in plasma using MS/MS was developed in a diagnostic and research setting. MS/MS was helpful in characterizing and detecting a range of new diseases of fatty acid and organic acid oxidation, and the patient studies and research helped investigators and physicians to detect, characterize, and treat these newly described diseases. The studies of affected infants and children and the effectiveness of treatment supported the concept of screening for disorders such as MCAD deficiency. Early surveys using DNA suggested that the frequency of the disorder was similar to that of PKU.

The next step that was required was the development of an MS/MS analysis of samples from newborns for MCAD deficiency. However, a plasma analysis could not be used, since this was not the preferred or existing format for biological specimens. It was therefore necessary to develop a blood spot technology directly for this purpose. In any method development, it is also helpful to compare a new technique to an older one. The challenge in this particular instance, however, was that there was no previous screening methodology for acylcarnitines. MS has been used, however, to analyze amino acids for decades through various GC–MS techniques and in newer studies using LC–MS approaches. Third, through a comparison to one of the most important and original assays in newborn screening, the detection of PKU, a MS/MS blood spot analysis could theoretically be established. For these reasons, the first method in MS/MS analysis of dried blood spots examined Phe. Many of the details of sensitivity, selectivity, extraction, and derivatization were optimized for this metabolite. During the validation, it was discovered that more than one amino acid could be detected in the same blood spot. Detecting multiple amino acids in a single analysis was critical to success, since it then enabled the analysis of closely related metabolites such as Phe and Tyr. Accurate measurement of Phe and other metabolites such as Tyr, so that a Phe/Tyr ratio could be calculated, was critical to demonstrate that

MS/MS not only could compare to existing newborn screening methods but also could potentially improve them. It was therefore easy to add acylcarnitine analysis to a screening panel since the extraction and derivatization schemes could be unified as butylesters (acylcarnitines were analyzed as methylesters in the diagnostic application).

The next step in validation was to test known positive specimens that had been archived. This approach was important in validating a variety of disorders. In today's environment, however, these specimens, critical to method validation, have become increasingly more difficult to obtain due to privacy concerns and regulations. For example, it was important to obtain blood spots analyzed previously by other methods and compare results in a blinded fashion in order to validate the newer methodology. A fundamental paper for demonstrating the improvement of PKU detection was carried out using this approach.

In terms of traditional quantitative validation, the most efficient, albeit slowest, approach was to validate by disease and metabolites. It was also helpful to focus on key metabolites in disease and not every marker potentially detectable from a quantitative perspective. Other metabolites that did not have a standard could be considered supportive qualitative markers. This approach could then be incorporated into a method for interpretation, a process that continues to refine itself based on the screening experience for rare disorders, method improvement, and secondary or confirmatory test.

An illustration of potential approaches to interpretation of multianalyte technologies such as MS/MS is shown in Figure 5. The key to interpretation is to communicate to the clinician the interpretation guidelines and to develop these guidelines in collaboration with the clinical specialist. Follow-up coordinators or genetic counselors can provide the results of screening effectiveness and outcomes for these patients. Ultimately, the success of a method in terms of the number of patients screened with the technique relates to the ability of multiple labs to reproduce the method, assure quality in testing, provide harmonization of data, and demonstrate the benefits from ordering the test. In this regard, MS/MS for newborn screening has been quite successful with the large numbers of newborns screened in both the United States and abroad.

3 CLINICAL CORRELATES: WHAT THE PHYSICIAN NEEDS TO KNOW AT THE BEDSIDE

What information does the physician need to know from a newborn screening report? First, the primary question is why is the physician assessing the newborn screening report in the first place? A normal newborn screening report on most infants typically gets filled in the medical record with little additional evaluation being required. Only abnormal results are reviewed and followed up. In the case of a presumptive positive result, however, the physician is required to submit a repeat specimen and/or confirmatory test to assist in a

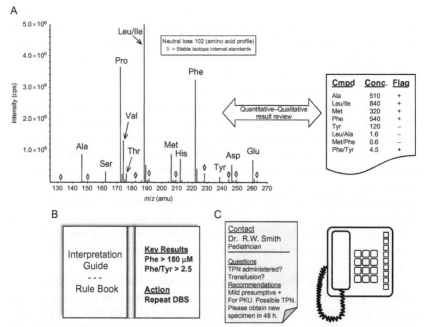

FIGURE 5 NBS analysis and follow-up illustration show how abnormal results are flagged automatically if settings are above certain concentrations (cutoffs and alert levels) and are reviewed using an interpretation system; repeat specimen and follow-up protocols are utilized, which results in a follow-up with a metabolic specialist for further testing, confirmation, and diagnosis.

diagnosis of a specific metabolic disorder and to coordinate treatment or follow-up with a metabolic specialist, hematologist, endocrinologist, or other specialist, depending upon what is found. The issues with respect to rare genetic disorders, however, require knowledge about which a typical pediatrician made limited background understanding. Nonetheless, the pediatrician of record is often the individual who is often required to communicate results to the family of the potentially affected infant. Screening has therefore often required careful coordination of follow-up either by genetic counselors from both private and public health screening laboratories. An entire network of genetic counselors, public health specialists, metabolic specialists, nutritionists, and the pediatricians of record are commonly required to assure that the newborn with a genetic disease is diagnosed and treated properly.

The issue is that the pediatrician at the bedside needs to understand more about rare disorders that he/she may never have encountered or have encountered only rarely in their entire clinical experience. The laboratory with the newest technology is required to explain the meaning of lab results, their reliability, and what changed in the methods and the results from older techniques. One of the biggest problems in newborn screening has been the false-positive results that are often seen and, if sufficiently high, a "cry wolf" syndrome often prevails. Metabolic disease requires timely and early treatment. Newborn screening is just one part of the potential world of clinical and metabolic screening. As medicine becomes more integrated and information more abundant and potentially complex, new systems will be required to assist the clinician/pediatrician in the understanding of the various and complex diseases that emerge through screening.

The essential lesson learned is that the laboratory cannot simply run a test and report "x" μM of metabolite "y" that is "z" % above normal. Information presented in this way presents no pathway for treatment, no sense of urgency, and no information to guide the clinician as to how best to proceed. For example, a newborn screening result from a premature infant may show an elevation of Phe of 240 μM and a Phe/Tyr ratio of 3.0. This result is above normal for both Phe and Phe/Tyr. Is this sample from a patient with PKU? Without further assessment or understanding, the parents of the child might be told of the potential for PKU, urgent follow-up, greater expense, and higher stress. In truth, the screening lab may have additional information that may change the way the physician responds.

Consider also the situation that occurs if other results showed milder elevation of Leu and Met as well. Also consider that the child was very premature and was receiving intravenous nutrition with a formula that had very little tyrosine in it. A more comprehensive report to a clinician might include and suggest a generalized elevation of amino acids in addition to Phe and an abnormal ratio that may reflect the milder higher levels of amino acids including Phe and a lower tyrosine due to the composition of TPN. It may therefore be the case that these results are an artifact of TPN administration and the infant's prematurity. Further, the parents might realize that newborn screening will be used during treatment to check for any signs indicating that the child has a metabolic disease rather than immature metabolism due to age or a contamination due to TPN. A laboratory should develop these guidelines and information for clinicians in any new method and work with specialist to communicate this properly.

For new assays that go beyond newborn screening and that are not mandated by public health departments, it is critical for the clinician to know what a test means and how it improves one's ability to care for his/her patients.

4 BEYOND NEWBORN SCREENING: ADDITIONAL CLINICAL APPLICATIONS AND THE POTENTIAL FOR THE FUTURE

The analysis of a dried blood spot for newborn screening represents a snapshot of an infant's metabolic status at a specific moment in time, namely, when the blood spot is being obtained. Metabolism, however, is a changing entity, often affected by a variety of circumstances encountered by preterm and term infants during the course of their early life. Because so much information and so many metabolites can be referenced from a single blood spot, one of the most promising aspects of this technique is the ability to follow metabolic change in even low-birth-weight neonates over a period of time. While the time period being assessed may only reflect normal development in most cases, in some instances, infants affected by a variety of circumstances, including acute disease states, may demonstrate metabolic changes that can be readily followed by repeated blood spots over time. Thus, the blood spot and the use of MS/MS technology may allow novel diagnostic and management techniques that previously could not be considered because of the higher volumes of blood that one would need to draw from a tiny neonatal patient. This exciting potential should therefore open new avenues for the pediatrician and neonatologist in the care of the neonatal intensive care unit (NICU) patient that may enhance outcomes in ways that previously would not be possible.

As previously noted, newborn screening using MS/MS is primarily an analytical method that measures acylcarnitines and amino acids in dried blood spots. One of the most important insights learned from the experience of screening was that a metabolic profile in an infant provided far more information than just the simple detection of a series of inherited disorders of intermediary metabolism. The same advantages of MS/MS analysis of amino acids or acylcarnitines from dried blood spots in a newborn profile, namely, clinical selectivity, sensitivity, and detection of multiple metabolites with a single analysis, simple sample collection, and delivery, could be applied to other fields of medicine. Multiple uses of the same assay but with different interpretation parameters help reduce laboratory costs and increase availability. Figure 6 illustrates how metabolites can be useful as biomarkers for many scientific and medical investigations, disease detections, and health assessments.

One of the initial examples of the application of this technology outside neonatal screening is seen in the postmortem metabolic evaluation of metabolic diseases from children with unknown or suspected cause of sudden death (27–29). A dried blood spot can be obtained at autopsy for this purpose. The acylcarnitine profile can be examined to determine if a metabolic disease contributed to the cause of death. In prior publications, as many as 1% of infants who die unexpectedly have been shown to have a metabolic disease that might have been detected in newborn screening. It is likely, however, that this rate has decreased substantially during the past several years because of expanded newborn screening throughout the United States and the world. The analysis of acylcarnitine and amino acid profiles in this postmortem situation is identical to that for blood spots in newborn screening. Data interpretation in this situation, however, is quite different. Metabolic profiles of postmortem blood spots are more complex and require experts who have analyzed samples from the population. In addition, this analysis has been used in the medicolegal investigation of infant death in cases such as suspected shaken baby syndrome. Its success occurred in part from the low cost of analysis at labs that already were performing newborn and metabolic screening.

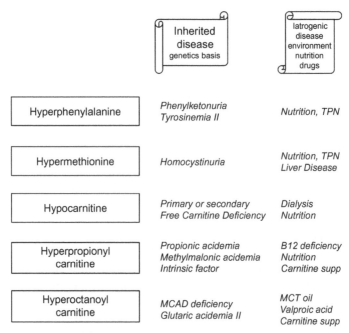

	Inherited disease genetics basis	Iatrogenic disease environment nutrition drugs
Hyperphenylalanine	Phenylketonuria Tyrosinemia II	Nutrition, TPN
Hypermethionine	Homocystinuria	Nutrition, TPN Liver Disease
Hypocarnitine	Primary or secondary Free Carnitine Deficiency	Dialysis Nutrition
Hyperpropionyl carnitine	Propionic acidemia Methylmalonic acidemia Intrinsic factor	B12 deficiency Nutrition Carnitine supp
Hyperoctanoyl carnitine	MCAD deficiency Glutaric acidemia II	MCT oil Valproic acid Carnitine supp

FIGURE 6 The concentration of metabolites as measured in a multiplex method such as MS/MS of amino acids and acylcarnitines can detect metabolic aberrations that are attributed to disorders with basis in genomics (inherited disease) or are due to environmental or developmental cause (nongenetic).

Abnormal amino acid or acylcarnitine profiles are often quite accurate but may be determined to be as a false-positive for a genetic disease if elevated levels did not result in the diagnosis of an inborn error of metabolism. That perspective might occur in a newborn screening laboratory whose mandate is primarily the detection of inherited disease. Many abnormal amino acid and acylcarnitine profiles, however, are actually true results and not false-positives *per se*. Recently, there has been additional research that has examined the abnormal profiles from premature and very low-birth-weight infants *(30–32)*. It has been well known that this population within newborn screening has the greatest frequency of false-positive results. What is the cause of the elevated amino acids or acylcarnitines in these profiles? Are these elevations significant and do they have adverse health effects, as with PKU and other metabolic diseases, or are they simply the result of a different nongenetic cause?

4.1 Nutritional Effects on Neonatal Metabolism

In a recent study of premature infants, it was found that the metabolic profiles of these patients are not dramatically different from that of other newborns if the blood sample was collected before intravenous nutrition was provided

(31). Additional recent work indicates a postnatal metabolism immediately after birth in both term and preterm infants that is very similar. The ability, however, of the early preterm infant to handle higher metabolic loads from enhanced nutrition has been called into question *(33)*. Premature (and very low-birth-weight) infants require intravenous amino acids, fatty acids, glucose, and other important electrolytes for survival and growth. The component that is the most variable and the subject of the greatest debate is the quantity of amino acids provided to these infants. All preterm infants have a period of time following birth during which significant weight is lost, as much as 10–15% of birth weight. Some clinicians believe that this situation, very different from what occurs *in utero*, can be altered through the provision of additional protein/amino acids. More proteins (amino acids) should theoretically promote growth, but this concept relies on the notion that a preterm infant has the metabolic capability to process the additional amino acids, which may be a fallacy. The inability to handle certain amino acids may produce levels that can be associated with some degree of neurotoxicity in infants with marginal metabolic pathways. In one controlled clinical trial, it was shown that high levels of amino acids provided in parenteral nutrition formulas produced high levels of amino acids for those compounds present in large quantities but lower levels of amino acids for others. That study also examined whether liver disease was more prominent in higher levels of amino acids provided. The results indicated that there did appear to be limitations in extremely low-birth-weight infants to utilize the additional amino acids provided.

An additional role for MS/MS technology that we have recently examined in our laboratory has been the identification of constituent components of nutritional products provided to neonates *(34)*. In examining the composition of commonly used liquids used for feeding neonates, one can determine that significant differences exist between premature infant mother's milk, mature mother's milk, cow's milk, and the premature formulas that are often employed in the NICU (Table 1). Theoretically, if one could establish through metabolomics the infant's ability to metabolize various amino acids and concurrently assess a series of nutritional products for the neonate, a far more appropriate and tailored approach to neonatal nutrition could be devised than what is presently available. In the current situation, the model used is more of a "one size fits all" approach to feeding, in which only intake volume changes rather than varying the core components (especially amino acids and lipids) of nutrition based upon the infant's metabolic capacity. Within this conceptual framework, it should then be possible to enhance amino acid administration based upon the measured needs/metabolic capacity of the infant in order to minimize weight loss following birth. It should also suggest the development of a greater series of gestational age-related products that are more ideally formulated for any infant's changing metabolic status.

The terminology of a false result should perhaps also be replaced with something more specific, such as presumptive positive or elevated amino

TABLE 1 Protein Concentrations (g/L) in Milk and Formula Specimens Measured Using Two Different Spectrophotometric Kit Methods and Manufacturer-Listed Concentrations (MFG List) (34)

Specimen	Pierce™ BCA	Bradford	MFG list (g/L)
Whole milk[a]	30.8	27.8	33.0
Standard formula 1[b]	22.6	9.4	14.0
Standard formula 2[b]	20.7	9.6	14.2
Standard formula 3[b]	22.6	2.9	14.9
Premature infant formula 1	25.1	11.3	20.3
Premature infant formula 2	27.4	11.4	21.0
Premature infant formula 3	31.7	14.1	24.2
Premature infant formula 4	24.8	9.7	14.0
Premature infant formula 5	29.3	13.2	20.1
Premature infant formula 6	35.9	14.0	26.8
Breast milk[c]	12.5	7.0	14

[a]$n=6$, [b]$n=3$, [c]$n=2$, where n is the number of milk or formula specimens that have different lots or production dates.

acids in a premature infant provided with intravenous (IV) nutrition. Newer studies suggest that some false-positives may actually be due to the contamination of blood with IV nutrition fluids and should be identified as an inadequate specimen rather than a false-positive (35). From the physician's perspective, the communication of derived results and their interpretation is essential. No doubt, specialized clinical laboratories must play a more thoughtful role in communicating the precise meaning of "abnormal" in the manner that was previously required by newborn screening and metabolic disease. In many of these cases, as noted before, the results are not abnormal, but rather reflect the clinical circumstances in which the infant is being treated.

Acylcarnitines or amino acids may also be important in disease monitoring and treatment or as markers for new therapies, toxicities, etc. In one application using dried plasma spots, carnitine and acylcarnitines may be useful in detecting possible carnitine deficiency as a result of kidney dialysis for patients with end-stage renal disease (36,37). A deficiency should result in carnitine supplementation in those patients that cannot replenish their levels fast enough. In fact, this is one of the first pharmaceutical-related applications of screening. The measurement of certain amino acids such as Phe and Tyr and their ratio is also routinely performed to monitor the effectiveness of dietary intervention in patients with PKU.

Several studies have recently examined the use of MS/MS biomarkers for infection in the maternal–fetal dyad and found that chorioamnionitis can be identified using this technology. While no neonatal studies to date have utilized similar investigations, it appears likely that this approach will emerge in the not-too-distant future as an important adjunct to NICU care. Ultimately, the combination of MS/MS for identification of the child at risk for septicemia, coupled with the identification of the DNA of bacteria, may fully replace our current conventions of drawing blood cultures for the diagnosis of sepsis. This approach should also have the benefit of a rapid turnaround time, much faster than waiting the 24–48 h commonly seen in culturing for neonatal infection.

4.2 Future Applications: Integrated Clinical Proteomic/Metabolomic Screening

As previously noted, MS technology was first applied in the broad area of metabolic screening for neonatal diseases. The value of this approach has been seen from the fact that it is still used extensively for this purpose around the world. This application, however, may ultimately be viewed as merely the tip of the iceberg, given the potential that analysis of proteins and metabolites has in refining our understanding of diseases and active disease states during clinical care.

While genomics has often been the primary focus of interest in novel approaches to diagnosis during the past decade in medicine, and the elegant elaboration of the human genome has been one of the great advances in medical science, the genome has serious limitations with respect to its ability to diagnose active disease. The identification of the presence of a certain gene, for example, only reveals that an individual has inherited that gene and that there is a potential for a disease to manifest itself at some future date. The gene, however, may never express itself clinically (especially true of the heterozygous state), and there is no indication to what extent it may express or whether the expression will result in clinical pathophysiology that produces a medical abnormality. It also does not indicate how an individual patient might respond to therapeutic interventions or metabolize certain drugs.

Beyond neonatal screening, however, proteomic and metabolomic assessment of the neonate has not been used to any significant degree to date. Recently, though, this situation has begun to change as the merits of evaluation in acute illnesses have become increasingly apparent. As a physiologically challenged patient, the clinical circumstances of the extremely low-birth-weight (ELBW) infant are constantly changing, yet he/she possesses limited volumes of available blood to measure the various disease processes that constantly affect these infants both before and after birth. Given these circumstances, proteomic and metabolomic assessment by tandem mass spectroscopy would appear to be an ideal methodology for the evaluation of the fetal and neonatal patient, and new uses of MS/MS are beginning to appear.

The use of mass spectrometric detection of abnormal metabolites or protein biomarkers in the blood is highly likely to emerge as a standard technique for the evaluation of an active disease or a pathophysiological process during the next decade. In addition, since protein synthesis and metabolism are the primary means through which pathophysiological processes express themselves in the body, the diagnostic assessment of protein biomarkers could conceivably anticipate the appearance of an abnormal clinical condition, prior to the actual clinical appearance of that disease. In other words, the detection of trace amounts of abnormal biomarkers may occur before a disease is apparent to the patient or the physician. Consequently, medicine may soon evolve to a new paradigm through proteomics/metabolomics. In this situation, one would no longer wait for a disease to appear before making a diagnosis. Instead, the concept of "screening" for possible diseases would be a routine part of care. In this clinical scenario, the clinician would pursue disease detection by examining low-level abnormalities of proteins or other biomarkers and intervening before the pathophysiology emerges clinically or the patient changes. Clearly, much work needs to be done prior to the establishment of this approach to medicine, but the concept is not as far-fetched as it may initially seem. As increasing work is introduced that yields specific disease biomarkers, the ability of using a filter paper to screen for thousands of these biomarkers in small samples becomes very attractive. In addition, every physician knows that it is far better to prevent disease from occurring as opposed to treating a disease that has become active. The value of this approach, therefore, is a goal that appears to be worth pursuing. It seems probable in the future that the overall understanding of disease will require both an evaluation of the genetic potential for disease and the development of a methodology for detecting appropriate biomarkers in the blood prior to or during the early stages of a disease so that intervention can dramatically limit morbidity.

5 INITIAL EMERGING AVENUES FOR FETAL AND NEONATAL ASSESSMENT WITH PROTEOMICS AND METABOLOMICS

5.1 Fetal Evaluation

Until relatively recent times, the fetus has been a black box for the physician, and any attempt at intervention was likely to result in a rapid demise. With the introduction, however, by Liley of amniotic fluid evaluation of elevated ΔOD levels as a marker for hyperbilirubinemia secondary to Rh disease, and the initiation of intrauterine transfusion for severe hemolytic disease, the era of fetal evaluation and intervention began *(38)*. In the succeeding years, the fetus has been the subject of increasing evaluation not only for the potential of disease but also for interventions as dramatic as exteriorization of the fetus outside the uterus for fetal surgery for a variety of clinical conditions *(39)*.

Consequently, the fetus has emerged as a true patient, whose well-being should be regarded as vital with respect to not only survival but also active disease states and morbidity. As a result, both the obstetrician and the neonatologist have become increasingly interested in technologies that can assess the status of the fetus.

Fetal assessment, however, is not without risk. In Harrison's efforts at fetal surgical intervention, premature delivery has been an ongoing concern, and it is well documented that even as commonplace, an intervention as amniocentesis is associated with a procedural risk of about 0.25–0.5% for fetal demise or miscarriage *(40)*. Furthermore, while amniotic fluid is relatively abundant in most pregnancies, fetal blood and tissue are notably sparse, so that access to direct fetal information still remains a technical challenge. The complexity of this situation is helped, however, by the fact that fetal cells, free fetal DNA, and protein do get into both the amniotic fluid and the maternal circulation, so that maternal blood or amniotic fluid assessment of the fetus can theoretically occur and several diagnostic tools have already been marketed for noninvasive fetal testing *(41)*. In this environment, therefore, the potential for proteomic/metabolomic assessment offers great promise and is now beginning to emerge *(42,43)*.

Initial studies have suggested the value of MS/MS technology in monitoring pregnancy and neonatal care. Gravett *et al.* described the identification of intra-amniotic infection (IAI) through proteomic profiling of novel biomarkers in both rhesus monkeys and women with preterm labor (PTL) *(44)*. Utilizing surface-enhanced laser desorption (SELDI)-time-of-flight (TOF)–MS, they were able to determine unique protein expression profiles in both women and rhesus monkeys with IAI (Figure 7). In particular, calgranulin B appeared to be overexpressed in both amniotic fluid and blood and some other novel immunoregulators. These investigators speculated that the ability to identify these biomarkers may ultimately lead to early detection of IAI and lead to interventions that could result in improved pregnancy outcome. Given the central role of IAI in preterm delivery, the ability to assess and treat such patients very early in the course of infection might be one of the primary methods in which preterm deliveries could be reduced. This work was recently confirmed by Ruetschi *et al.*, who found overexpression of 17 proteins in amniotic fluid, more commonly in women with PTL than those with rupture of membranes *(45)*. Furthermore, the promise of early detection of congenital abnormalities through similar techniques appears to be on the immediate horizon. The potential changes in the approach to pregnancy that these studies portend cannot be overemphasized.

Gravett *et al.* recently published further work that targets some of the areas most amenable to early detection and intervention during pregnancy, labor, and delivery, as a step toward alleviating the disease burden of these illnesses globally *(46)*. Early metabolomic/proteomic evaluation could theoretically lead to a marked reduction in adverse pregnancy outcome.

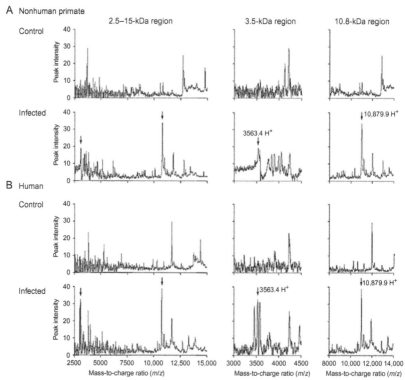

FIGURE 7 Group B streptococcus infection-induced differential protein expression in nonhuman primate (A) and human (B) amniotic fluid samples by surface-enhanced laser desorption/ionization (SELDI-TOF MS) using normal-phase protein chip arrays. Spectrum from 2.5 to 15 kDa collected at 235 nm laser intensity. Detailed spectra show increased expression of the 3.5 and 10.8 kDa peaks between control and infected. Arrows indicate the unique peaks represented by polypeptides overexpressed in infection.

Nagalla *et al.* had also expanded the potential for proteomic analysis in a series of publications, using multidimensional protein identification technology (MudPIT). In a study of novel protein biomarkers for PTL in maternal cervical and vaginal fluid, 28 proteins in cervical–vaginal fluid were determined to be present in differential concentrations in mothers with PTL compared with those with normal pregnancies. Calgranulins, annexins, S-100 calcium-binding protein A7, and epidermal fatty acid-binding protein were abundant in women with PTL, as were serum proteins alpha-1-antitrypisn, alpha-1-acid glycoprotein, haptoglobin, serotransferrin, and vitamin D-binding protein *(47)*. This new insight into prematurity is very exciting and portends an era in which mothers might be routinely screened for potential markers of preterm delivery on a regular basis. Should these biomarkers be detected, an aggressive approach to imminent PTL could be initiated and the pregnancy might be sustained for a longer duration of time, assuming that the indicators for PTL did not indicate infection.

A second study from this same group examined maternal serum in Down syndrome *(48)*. In this investigation, first- and second-trimester maternal trisomy 21 serum specimens were paired with gestational-matched controls. MudPIT LC/LC–MS/MS profiling was used, along with MALDI-TOF MS peptide profiling. Discrimination between trisomy 21 and nonaffected mothers was achieved with a 96% recognition capability and revealed that the majority of unique biomarkers were serum glycoproteins. Additional recent work has confirmed this capability of detecting aneuploidies in the maternal circulation *(49)*. Given that the current nonmetabolomic prenatal assessment for trisomy 21 achieves only an 80–85% recognition rate, proteomic evaluation would represent an important enhancement to early diagnosis of Down syndrome. These findings need to be confirmed in further prospective trials, which are currently under way, but clearly indicate what is on the immediate horizon for fetal identification of both abnormal processes and genetic diseases. The ability to detect such abnormalities without direct fetal intervention represents a major advance in medical care.

In the immediate future, many other common problems of pregnancy and the fetus are likely to be reexamined by proteomics and metabolomics, ultimately identifying the best predictive biomarkers for disease and adverse neonatal outcome. As we have begun to see, pregnancy-induced hypertension (PIH), growth restriction *in utero*, diabetic pregnancy, and placental insufficiency are but a few of the situations in which early proteomic assessment is likely to play a prominent role in the next several years (Figure 8).

5.2 Neonatal Evaluation

Until recently, the neonate has been a "black box," an eminently valued patient, yet one that is difficult to study and treat effectively. Neonatal research, in particular, is often plagued by ethical problems of invasiveness of protocols, volume of blood to be drawn, inability of the patient to speak on his/her own behalf, etc. Even simple studies that try to address neonatal outcome often pose problems in these areas, requiring volumes of blood that subsequently need to be replaced by transfusion. Because the perceived benefit to both individual patients and to society as a whole is deemed to be highly valuable, such downsides to neonatal research are often accepted as inevitable. The introduction, however, of MS analysis, utilizing low volumes of blood for multiple biomarkers, appears to be a very attractive alternative with great potential. A number of critical areas have already begun to be addressed and include such diverse entities as hypoxic–ischemic brain injury, the infant of a diabetic mother, neonatal renal function, neonatal nutrition, neonatal sepsis, and pharmaceutical development.

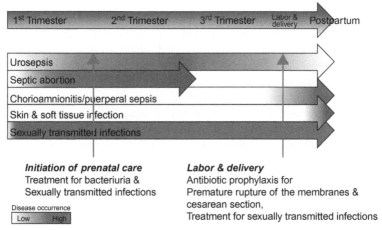

FIGURE 8   Areas of potentially bundled metabolomic and proteomic assessments/interventions in perinatal medicine. *From Gravett* et al. (46).

No aspect of neonatal outcome is more critical than brain function as it affects individuals for the remainder of their lives. Once a family has asked the question, "Will my baby survive?" the next question is always "How will he/she do?" This question almost inevitably refers to the neurodevelopmental outcome of the infant. Again, until recently, the physician could only make an educated clinical guess and try to direct care to achieve the best outcome. Studies such as head ultrasounds or magnetic resonance imaging (MRI) performed in an attempt to grasp the potential quality of life for an infant provided only "snapshots" of an infant's status and were far removed from the kind of dynamic evaluation that could truly gauge the neonate's NICU progress or needs. While studies such as ultrasound created dramatic enhancements in our understanding of the neonate's condition, they still reveal only a single moment in time and not much about the changing state of the infant. Infant nutrition and its effects on development, in particular, have also been a very difficult issue to assess, and it is probably a tribute to the neonate that our lack of understanding of nutrition had not obstructed outcome improvements during the past several decades. Nevertheless, it is apparent that many neonates are malnourished in the NICU, and there is postnatal growth restriction for many infants that is as problematic as *in utero* growth restriction *(50)*.

5.3   Nutritional Considerations

Optimal neonatal nutrition, however, is an ongoing puzzle and a major concern. While there is little question that breast milk provides optimal neonatal nutrition for the term infant, the nutritional needs of the premature infant are not as obvious. While there are significant advantages of breast milk with respect to immune protection for the premature neonate, it is well established that breast milk provides insufficient quantities of protein, sodium, calcium, iron, vitamin D, and other nutrients to sustain growth *(51)*. Neonatologists therefore routinely use human milk fortifier, which supplements breast milk, in an effort to enhance growth in the premature infant. Fortification of human milk, however, changes many of its unique characteristics, and there are also differences in the available fortifiers that the practicing neonatologist needs to be aware of while providing nutrition to the low-birth-weight infant, in particular. Furthermore, as seen in Table 2, in infants for whom breast milk is not available, substantial differences exist in the amino acid composition of available formulas *(34)*. Acylcarnitines also differ widely. It is therefore very advantageous to know what formulas best feed an infant's nutritional requirements at any time in order to ensure optimal growth. Because the manufacturing process itself alters proteins, it is unclear how bioavailable nutrients in formulas are in specific clinical circumstances. Again, knowing the composition of a specific formula, as well as the clinical ability of the infant to tolerate that product, may provide for enhanced feeding and subsequent growth. What is evident, however, is that the current array of products for feeding infants, especially low-birth-weight babies, while far more complete than ever, still has limitations that must be considered. Formula analysis provides some valuable insight into what an infant is receiving so that optimal supplementation can be provided.

In some instances, as well, the differences between formula and human milk are striking, even when some of the reasons for the composition of human milk may not be entirely understood. For example, most, though not all, human milk samples will have very high levels of glutamate, rarely seen in any commercially available formula. While glutamate has been associated with improved brain and gastrointestinal maturation, it is also apparent that not all human milk specimens are saturated with glutamate, raising the question of its exact nutritional role in neonatal growth and development. Clearly, significant additional work must be done to fully appreciate both the metabolic needs of the ELBW infant and the ability of various products.

Not all premature infants, however, can be enterally fed. Due to the ELBW neonate (<1000 g at birth), it may be days to weeks before enteral feedings can be initiated. In addition, the risk of necrotizing enterocolitis, and the common interruptions of enteral feeding for a variety of reasons, requires that intravenous nutrition be provided. Here, the situation becomes even murkier with respect to the provision of protein and fats (glucose is easier to assess, since one can simply follow blood glucose levels to ensure that glucose is being maintained at an appropriate level). Because of the known growth problems in premature infants, there has been a trend during the past decade to push protein to 4 g/kcal/day in an effort to maintain known

TABLE 2 Selected Free Amino Acid Concentrations (μM) in Milk Formula and Specimens Measured By Isotope Dilution MS/MS

	Alanine	Glutamine	Leucine/ Isoleucine	Methionine	Phenylalanine	Tyrosine	Aspartic acid	Glutamic acid	Arginine
Whole milk[a]	156	5	22	10	2	5	18	283	35
Standard formula 1[b]	53	8	5	2	1	1	6	70	9
Standard formula 2[b]	161	19	20	3	14	11	10	85	21
Standard formula 3[b]	99	16	75	4	21	22	25	27	745
Premature infant formula1	57	7	6	6	2	3	17	70	11
Premature infant formula 2	38	3	6	10	1	2	11	66	11
Premature infant formula 3	59	4	8	6	0	2	20	94	14
Premature infant formula 4	68	22	7	3	1	2	9	49	19
Premature infant formula 5	61	4	12	8	3	4	13	83	20
Premature infant formula 6	64	5	11	6	2	3	15	80	26
Breast milk 1	1206	27	289	12	46	89	74	1275	74
Breast milk 2	656	26	47	8	9	9	164	1568	13

[a]n = 6.
[b]n = 3.

intrauterine protein accretion rates. It is unclear, however, that this approach is either safe or appropriate, since amino acid levels have not been commonly measured in preterm infants on TPN, and the available solutions for protein administration in a 500 g neonate are essentially the same ones that are used for adults. Needless to say, metabolic differences are quite substantial between these two patient populations, and the supplementation requirements are also likely to be different.

Recently, in an effort to delineate the effects of parenteral protein administration, Clark *et al.* used MS/MS analysis in a multicenter randomized trial to determine the amino acids levels that were achieved in premature infants receiving TPN *(31)*. In addition, they tried to verify if increased rates of protein administration during the first month of life could achieve higher growth rates, as has been suggested previously *(52)*. The results of this study were surprising: infants who received 33% more protein did *not* grow at an accelerated rate during the first month of life. In addition, certain amino acids and amino acid acylcarnitines (e.g., isovalerylcarnitine) were elevated to levels that were of concern, while other amino acid levels (e.g., alanine) appeared to be inadequate. Essentially, some of the amino acid metabolic pathways in the premature infant seemed to be saturated, while others appeared to be undersupplied. The literature is at significant variance here, in that earlier and subsequent trials have shown that both positive effects upon growth and no effects when protein/amino acid supplementation have been increased *(53)*. Recently, our research work has indicated that infants may not tolerate increased protein/amino acid levels above a certain range *(34)*.

Intravenous therapy with elevated levels of protein may therefore result in levels of certain amino acids, such as phenylalanine, isovalerylcarnitine, or other metabolic byproducts, that could produce temporary, but important, toxicities. Such toxicities, especially if unrecognized and prolonged, may be one of the reasons why some infants, even in the absence of typical types of neurological injury such as intraventricular hemorrhage (IVH) or periventricular leukomalacia (PVL), may still manifest poorer than expected long-term developmental outcomes. In composite, these data suggest that the currently available TPN solutions might not be ideal for the very low-birth-weight (VLBW) neonate. Of greater interest, however, is the concept that it might actually be possible to increase protein administration in a way to mimic intrauterine growth, assuming that a more optimal solution or a set of intravenous protein/amino acid solutions were made available. The study also indicated the value of the filter paper blood spot to measure a large variety of compounds in the VLBW neonate with little blood loss to the infant and represented a significant potential technology that could enhance neonatal care in the near future. The ability to assess an infant's nutritional status and needs and to provide a more tailored nutritional product to meet those needs appears to be a concept whose time has arrived.

5.4   Other Avenues for Neonatal Critical Care Monitoring with Tandem Mass Spectroscopy

The potential for the use of neonatal MS/MS evaluation is exceptional, and this technology is likely to emerge as a valuable adjunct to care in the next decade. Several areas of neonatal intensive care would seem to benefit the most from proteomic assessment of the newborn infant. These include (but are not limited to) the following:

(1) The diagnosis and treatment of neonatal septicemia
(2) Assessment of the neonate with hypoxic–ischemic or other brain injury
(3) Inflammatory markers in the pathogenesis of chronic lung and liver disease
(4) Tailored drug therapy

5.5   Neonatal Septicemia

The diagnosis of neonatal infection remains one of the most elusive problems in neonatal intensive care. While much of the focus in the NICU concerns the premature infant, the most common overall admitting diagnosis to the NICU is the term newborn infant with suspected septicemia. Typically, such infants have been born following a somewhat difficult delivery in which the mother has had a fever, raising the suspicion of infection in the infant. In most instances, maternal fever is simply due to dehydration, prolonged labor, or spinal anesthesia, but often, the cause is unclear and the concern about the infant is substantial. Besides the blood culture, which may take 2 days or longer to confirm the diagnosis, other commonly used neonatal diagnostic tests do not add much in making the diagnosis. The CBC in a healthy baby may often show an elevated white blood cell count (up to 30,000/mm^3), and in cases of true septicemia in the neonatal period, the WBC count will actually decrease as the marrow becomes depressed. C-reactive protein results may also be unclear, especially in the mid-range. Furthermore, neonates rarely run a fever themselves, often becoming hypothermic during infection. As a result, the clinician often makes an educated guess about the likelihood of infection while awaiting a culture result. Antibiotics are used during this period of time, often unnecessarily, because of the fear that the child may have a life-threatening infection, a grave issue in an already immunocompromised patient (which all neonates are). The frequent and prolonged use of antibiotics, however, tends to breed increasingly resistant microorganisms, which may be devastating to future generations of neonates admitted to that NICU. Therefore, in the same manner that the proteome of maternal infection has been examined, the neonatal proteome during infection also needs assessment. To date, few studies have been done to confirm the merits of proteomic assessment, but some recent work has suggested that neonatal infection should also yield to proteomic evaluation (54). Part of the dilemma to date, however, has been the difficulty in identifying the most satisfactory biomarkers that could readily be analyzed by metabolomics testing. It does appear likely, however, that certain inflammatory mediators should be present in serum prior to the appearance of signs of full-blown septicemia, and else, there would not be an associated clinical syndrome. Further work in this area over the next decade should allow the identification and detection of appropriate biomarkers for this problem.

5.6   Hypoxic–Ischemic Brain Injury, Perinatal Asphyxia, and Chronic Lung Disease: The Potential for Proteomic Evaluation

At the present time, the primary methods for the evaluation of brain injury in the neonatal period involve some aspect of brain imaging. Cranial ultrasound *(55)*, computerized tomography, and MRI of various types (e.g., magnetic resonance angiography *(56)*, magnetic resonance spectroscopy *(57)*, and diffusion-weighted imaging *(58)*) are the tools used to define both the extent of injury during the neonatal period and the likelihood of residual damage following NICU discharge. More recently, other interesting techniques, such as cerebral function monitoring (CFM), have been used to examine both brain development in the neonate and the likelihood of neurological injury *(59,60)*. For many years, IVH was thought to have the greatest long-term consequence to the VLBW baby *(61)*. More recently, however, white matter disease (WMD) and PVL, thought to be alternative and related expressions of chronic hypoxic–ischemic injury in the neonate, have attracted attention as the leading events that are likely to produce more permanent neurological disability in the form of cerebral palsy (CP) and developmental delay *(62)*. Recent work indicates that biomarker detection may be possible in these early disease states to assist in guiding therapy *(63)*.

In term infants, acute hypoxic–ischemic injury and birth trauma appear to be more significant events in the evolution of brain injury *(64)*. It has been widely recognized, however, that although many neonates who ultimately develop CP sustain their injury during birth or in the NICU, most CP patients do not have evidence of acute neonatal events capable of producing permanent injury *(65–67)*. In term neonates, the large majority of infants with CP have no clearly discernible asphyxial event to which the injury can be attributed.

In the cases in which CP does not appear to be the result of some aspect of neonatal or perinatal care, the obvious question is when and how does it occur? It has become increasingly apparent that some brain injury is probably initiated before the neonatologist ever sees the patient. In examining those factors that result in PTL, Romero *et al.* began to notice that certain prostaglandins appeared to be increased in mothers delivering prematurely *(68,69)*. They speculated that PTL may be the result of an inflammatory process, with the most likely etiology of inflammation being intrauterine infection. This infection, most notably, may not necessarily be clinically apparent, and many women may have subclinical chorioamnionitis, which goes entirely unrecognized until premature labor ensues and the placenta is retrospectively examined. More importantly, they also recognized that if one examined the amniotic fluid of mothers of infants who develop CP, there were significant elevations of cytokines, which were not increased in neonates who did not develop CP *(70,71)*. While this was comforting to neonatologists who often agonized about what might have occurred to their patients that resulted in white matter injury, even though no clear-cut events in the NICU may have been observed, it made the diagnosis of maternal infection and inflammation much more important. These initial observations have been confirmed in numerous studies since that time *(72)*. Furthermore, it was also recognized that bronchopulmonary dysplasia, an inflammatory pulmonary process of the ventilated premature neonate, most probably had a similar onset prior to birth, which was subsequently aggravated by a variety of commonplace NICU events *(73)*. Consequently, the diagnosis of maternal inflammatory disease now occupies a central role in neonatal–perinatal research, and the specific timing of the onset of infection has become increasingly important. Lastly, the possibility of early diagnosis of intrauterine infection as a more treatable entity, as noted earlier, has emerged as a critical focus of perinatal investigation.

During the same time period in which the etiological factors of white matter injury in the VLBW infant were being examined in depth, new attention began to be focused on the term neonate with hypoxic–ischemic encephalopathy. Few problems have been as troubling to both the perinatologist and the neonatologist as birth asphyxia/hypoxic–ischemic injury. Because so many of these infants ultimately either die or have permanent neurological injury and because so many of these cases result in malpractice litigation, physicians have been desperate to understand the origins of this entity. Even the diagnosis has been an ongoing concern, and the American College of Obstetrics and Gynecology and the American Academy of Pediatrics have labored repeatedly in an attempt to better categorize this entity and the factors that result in an asphyxiated infant. But the primary motivation that drove the increased interest in making a rapid diagnosis of perinatal hypoxic–ischemic injury was the recognition that brain and body cooling appeared to limit the manifestations of the disease process *(74,75)*. It appeared, however, that brain and body hypothermia needed to be initiated prior to 6 h of life in order to discern any beneficial effect. Later application seemed to have little, if any, value to the asphyxiated infant.

Several trials have confirmed the benefit of hypothermia following perinatal hypoxic–ischemic injury *(76,77)*. It has been noted, however, that brain or body cooling has not produced improved outcomes that are as dramatic as those seen in animal studies, when the brain injury can be precisely timed. It is evident then that the most significant variable over which little control could be exerted was between the time of the onset of intrauterine hypoxemia and the time of birth. As with the VLBW baby and the development of WMD, one of the principal problems encountered in these trials was the inability to accurately determine the period during which the infant was initially becoming asphyxiated *in utero*. It seems likely that if a technique could evolve that would provide such information, the use of cooling could be more specifically introduced in selected patients, with a far more favorable outcome. Conceivably, one could trace the course of an injury by the increase and decrease in cerebral injury biomarkers, allowing for a better effort at timing an event and understanding the etiology. Proteomics or metabolomics would appear to offer a valuable methodology in this respect, and some work has recently suggested that certain biomarkers may be detectable in these clinical situations *(78)*.

6   TAILORED DRUG THERAPY FOR THE NEONATE

Neonatal patients have often been termed "therapeutic orphans," reflecting the fact that very few drugs have been either developed or tested in neonatal care. As a result, many medications that are currently used in the nursery have never fully been evaluated in neonates in order to discern the differences between adult and neonatal metabolic rates for a variety of commonly used medications. In addition, neonatal intensive care patients themselves vary enormously in size, maturity, and metabolic capability. A 24- to 25-week gestational age infant is clearly going to be a very different patient from the term 40-week gestational age infant in terms of maturity of organ systems involved in drug metabolism. Changes in liver and kidney function in particular change dramatically during the terminal stages of gestational development, with significant impact upon the rate upon drug metabolism and excretion. Because of these factors, many drugs used in neonatal intensive care have doses extrapolated from studies performed in adults, yet the dose per kilogram that is recommended may work well for one neonate, but poorly for another. As a result, the neonatologist is often left mystified with respect to why a diuretic dose has minimal effects in one patient, yet may be excessive for another. Having a better way of measuring drug response, therefore, has long been one of the "Holy Grails" of neonatal care.

With these issues in mind, it seems highly likely that many of the techniques previously described in this chapter might be applied to evaluate effects of drugs on the neonate in order to individually tailor dosing to the infant's needs. Unfortunately, little work along these lines has been carried out to date. The initial phases of such work, however, are beginning to appear. Takai *et al.* in Japan utilized imaging mass spectroscopy to quantify drug distribution in multiple organs *(79)*. In this study, they demonstrated in a mouse model that a dopamine D2 receptor drug could be followed by MALDI-MS to quantitate the amount of drug in various tissues simultaneously. In another mouse model, the various chemicals in cigarette smoke and their effect upon the liver proteome were identified by Canales *et al*. This group of investigators showed that exposure to cigarette smoke both pre-and postnatally resulted in decreased liver metabolic proteins, decreased gluconeogenic activity, and altered lipid metabolism, resulting in lower-birth-weight offspring *(80)*.

The ability to follow drug disposition, alteration of the proteome, and impact upon outcome appears to be a potentially valuable step forward in a more progressive approach to medication use in the NICU in the not-too-distant future. While much more work needs to be carried out to make this feasible, the first steps have been taken and appear to be promising.

7 SUMMARY

The introduction of proteomics into the NICU appears to be a coming event that will have significant implications for medical practice. It is highly likely that this approach will have substantial benefits for the neonate with many positive effects upon neonatal outcomes.

REFERENCES

1. Chace, D. In: *Contemporary Practice in Clinical Chemistry;* Clarke, W.; Dufour, R. D., Eds.; AACC Press: Washington, DC, 2006.
2. Chace, D. H.; Kalas, T. A.; Naylor, E. W. *Clin. Chem.* **2003**, *49*, 1797–1817.
3. Chace, D. H. *Clin. Chem.* **2003**, *49*, 1227–1228, author reply 1228–1229.
4. Chace, D. H. *J. Mass Spectrom.* **2009**, *44*, 163–170.
5. Guthrie, R. *Triangle* **1969**, *9*, 104–109.
6. Guthrie, R. *Screening* **1992**, *1*, 5–15.
7. Levy, H. L. *Clin. Chem.* **1998**, *44*, 2401–2402.
8. Millington, D.; Kodo, N.; Terada, N.; Roe, D.; Chace, D. *Int. J. Mass Spectrom. Ion Process.* **1991**, *111*, 211–228.
9. Chace, D. H.; DiPerna, J. C.; Naylor, E. W. *Acta Paediatr. Suppl.* **1999**, *88*, 45–47.
10. Chace, D. H. *Chem. Rev.* **2001**, *101*, 445–477.
11. James, A. T.; Martin, A. J. *Biochem. J.* **1951**, *48*, vii.
12. Fowler, K. T.; Hugh-Jones, P. *Br. Med. J.* **1957**, *1*, 1205–1211.
13. Brooks, C. J.; Horning, E. C.; Young, J. S. *Lipids* **1968**, *3*, 391–402.
14. Vandenheuvel, W. J.; Smith, J. L.; Putter, I.; Cohen, J. S. *J. Chromatogr.* **1970**, *50*, 405–412.
15. Mrochek, J. E.; Rainey, W. T., Jr. *Anal. Biochem.* **1974**, *57*, 173–189.
16. Mawhinney, T. P.; Robinett, R. S.; Atalay, A.; Madson, M. A. *J. Chromatogr.* **1986**, *358*, 231–242.
17. Sweetman, L.; Hoffmann, G.; Aramaki, S. *Enzyme* **1987**, *38*, 124–131.
18. Lehotay, D. C. *Biomed. Chromatogr.* **1991**, *5*, 113–121.
19. Clayton, P. T. *J. Inherit. Metab. Dis.* **2001**, *24*, 139–150.
20. Gerhards, P. *GC/MS in Clinical Chemistry*. Wiley-VCH: Weinheim, Federal Republic of Germany, 1999.
21. Bruns, D. E.; Lo, Y. M. D.; Wittwer, C. *Molecular Testing in Laboratory Medicine: Selections from Clinical Chemistry, 1998–2001, with Annotations and Updates*. AACC Press: Washington, DC, 2002.

22. Millington, D. S.; Kodo, N.; Norwood, D. L.; Roe, C. R. *J. Inherit. Metab. Dis.* **1990**, *13*, 321–324.
23. Chace, D. H.; Millington, D. S.; Terada, N.; Kahler, S. G.; Roe, C. R.; Hofman, L. F. *Clin. Chem.* **1993**, *39*, 66–71.
24. Guthrie, R.; Susi, A. *Pediatrics* **1963**, *32*, 338–343.
25. Naylor, E. W.; Guthrie, R. *Pediatrics* **1978**, *61*, 262–266.
26. Murphey, W. H.; Patchen, L.; Guthrie, R. *Biochem. Genet.* **1972**, *6*, 51–59.
27. Chace, D. H.; DiPerna, J. C.; Mitchell, B. L.; Sgroi, B.; Hofman, L. F.; Naylor, E. W. *Clin. Chem.* **2001**, *47*, 1166–1182.
28. Wilcox, R. L.; Nelson, C. C.; Stenzel, P.; Steiner, R. D. *J. Pediatr.* **2002**, *141*, 833–836.
29. Dott, M.; Chace, D.; Fierro, M.; Kalas, T. A.; Hannon, W. H.; Williams, J.; et al. *Am. J. Med. Genet. A* **2006**, *140*, 837–842.
30. Steinbach, M.; Clark, R. H.; Kelleher, A. S.; Flores, C.; White, R.; Chace, D. H.; et al. *J. Perinatol.* **2008**, *28*, 129–135.
31. Clark, R. H.; Chace, D. H.; Spitzer, A. R. *Pediatrics* **2007**, *120*, 1286–1296.
32. Kelleher, A. S.; Clark, R. H.; Steinbach, M.; Chace, D. H.; Spitzer, A. R. *J. Perinatol.* **2008**, *28*, 270–274.
33. Clark, R. H.; Kelleher, A. S.; Chace, D. H.; Spitzer, A. R., manuscript in review, Unpublished data.
34. Chace, D. H.; De Jesús, V. R.; Haynes, C. A.; Lanza, K. L.; Kelleher, A.; Clark, R.; Spitzer, A. R., Submitted for Publication, September 2013, Unpublished data.
35. Chace, D. H.; DeJesus, V. R.; Lim, T. H.; et al. *Clin. Chim. Acta* **2010**, *411*, 1806–1816.
36. Reuter, S. E.; Evans, A. M.; Chace, D. H.; Fornasini, G. *Ann. Clin. Biochem.* **2008**, *45*, 585–592.
37. Reuter, S. E.; Evans, A. M.; Faull, R. J.; Chace, D. H.; Fornasini, G. *Ann. Clin. Biochem.* **2005**, *42*, 387–393.
38. Liley, A. W. *Am. J. Obstet. Gynecol.* **1961**, *82*, 1359–1370.
39. Harrison, M. R.; Bressack, M. A.; Churg, A. M.; de Lorimier, A. A. *Surgery* **1980**, *88* (1), 174–182.
40. Trouffer, P. M. L.; Wapner, R. J.; Johnson, A. In: *Intensive Care of the Fetus and Neonate;* Spitzer, A., Ed, 2nd ed.; Elsevier, Inc: Philadelphia, PA, 2005; p 64.
41. Hyodo, M.; Samura, O.; Fujito, N.; et al. *Prenat. Diagn.* **2007**, *27* (8), 717–721.
42. Van Hove, J. L.; Chace, D. H.; Kahler, S. G.; Millington, D. S. *J. Inherit. Metab. Dis.* **1993**, *16*, 361–367.
43. Shigematsu, Y.; Hata, I.; Nakai, A.; et al. *Pediatr. Res.* **1996**, *39*, 680–684.
44. Gravett, M. G.; Novy, M. J.; Rosenfeld, R. G.; et al. *JAMA* **2004**, *292* (4), 462–469.
45. Ruetschi, U.; Rosen, A.; Karlsson, G.; et al. *J. Proteome Res.* **2005**, *4* (6), 2236–2242.
46. Gravett, C. A.; Gravett, M. G.; Martin, E. T.; et al. *PLoS Med.* **2012**, *9* (10), e1001324. http://dx.doi.org/10.1371/journal.pmed.1001324, Epub October 9, 2012.
47. Pereira, L.; Reddy, A. P.; Jacob, T.; et al. *J. Proteome Res.* **2007**, *6*, 1269–1276.
48. Nagalla, S. R.; Canick, J. A.; Jacob, T.; et al. *J. Proteome Res.* **2007**, *6*, 1245–1257.
49. Narasimhan, K.; Lin, S. L.; Tong, T.; et al. *Prenat. Diagn.* **2013**, *33* (3), 223–231. http://dx.doi.org/10.1002/pd.4047. Epub 2013 Feb 1.
50. Clark, R. H.; Wagner, C. L.; Merritt, R. J.; et al. *J. Perinatol.* **2003**, *23*, 337–344.
51. Pereira, G. R.; Chan, S. W. In: *Intensive Care of the Fetus and Neonate;* Spitzer, A., Ed, 2nd ed.; Elsevier, Inc: Philadelphia, PA, 2005; p 991.
52. Premji, S. S.; Fenton, T. R.; Sauve, R. S. *Cochrane Database Syst. Rev.* **2006**, *1*, CD003959.
53. Theile, A. R.; Radmacher, P. G.; Anschutz, T. W.; et al. *J. Perinatol.* **2012**, *32* (2), 117–122. http://dx.doi.org/10.1038/jp.2011.67. Epub 2011 May 26.

54. Buhimschi, C. S.; Bhandari, V.; Dulay, A. T.; et al. *PLoS One* **2011**, *6*, e26111.
55. Burdjalov, V.; Srinivasan, P.; Baumgart, S.; Spitzer, A. R. *J. Perinatol.* **2002**, *22* (6), 478–483.
56. Husain, A. M.; Smergel, E.; Legido, A.; et al. *Pediatr. Neurol.* **2000**, *23* (4), 307–311.
57. Vigneron, D. B. *Neuroimaging Clin. N. Am.* **2006**, *16* (1), 75–85.
58. Counsell, S. J.; Shen, Y.; Boardman, J. P.; et al. *Pediatrics* **2006**, *117* (2), 376–386.
59. Burdjalov, V. F.; Baumgart, S.; Spitzer, A. R. *Pediatrics* **2003**, *112*, 855–861.
60. West, C. R.; Groves, A. M.; Williams, C. E.; et al. *Pediatr. Res.* **2006**, *59* (4), 610–615.
61. Papile, L. A.; Burstein, J.; Burstein, R.; et al. *J. Pediatr.* **1978**, *92*, 529–534.
62. Volpe, J. J. *Pediatrics* **2003**, *112* (pt. 1), 176–180.
63. Douglas-Escobar, M.; Weiss, M. D. *Front. Neurol.* **2013**, *3*, 185. http://dx.doi.org/10.3389/fneur.2012.00185.
64. Berger, R.; Garnier, Y. *J. Perinat. Med.* **2000**, *28* (4), 261–285.
65. Tran, U.; Gray, P. H.; O'Callaghan, M. J. *Early Hum. Dev.* **2005**, *81* (6), 555–561.
66. Laptook, A. R.; O'Shea, T. M.; Shankaran, S.; Bhaskar, B. NICHD Neonatal Network. *Pediatrics* **2005**, *115* (3), 673–680.
67. Wood, N. S.; Costeloe, K.; Gibson, A. T.; et al. EPICure Study Group. *Arch. Dis. Child. Fetal Neonatal Ed.* **2005**, *90* (2), F134–F140.
68. Romero, R.; Parvizi, S. T.; Oyarzun, E.; et al. *J. Reprod. Med.* **1990**, *35* (3), 235–238.
69. Romero, R.; Avila, C.; Santhanam, U.; Sehgal, P. B. *J. Clin. Invest.* **1990**, *85* (5), 1392–1400.
70. Yoon, B. H.; Romero, R.; Yang, S. H.; et al. *Am. J. Obstet. Gynecol.* **1996**, *174* (5), 1433–1440.
71. Yoon, B. H.; Jun, J. K.; Romero, R.; et al. *Am. J. Obstet. Gynecol.* **1997**, *177* (1), 19–26.
72. Ellison, V. J.; Mocatta, T. J.; Winterbourn, C. C.; et al. *Pediatr. Res.* **2005**, *57* (2), 282–286.
73. Dammann, O.; Leviton, A.; Bartels, D. B.; Dammann, C. E. *Biol. Neonate* **2004**, *85* (4), 305–313.
74. Bona, E.; Hagberg, H.; Loberg, E. M.; et al. *Pediatr. Res.* **1998**, *43* (6), 738–745.
75. Gunn, A. J.; Gluckman, P. D.; Gunn, T. R. *Pediatrics* **1998**, *102* (4 Pt 1), 885–892.
76. Gluckman, P. D.; Wyatt, J. S.; Azzopardi, D.; Ballard, R.; Edwards, A. D.; Ferriero, D. M.; Polin, R. A.; Robertson, C. M.; Thoresen, M.; Whitelaw, A.; Gunn, A. J. *Lancet* **2005**, *365* (9460), 663–670.
77. Shankaran, S.; Laptook, A. R.; Ehrenkranz, R. A.; et al. for the National Institute of Child Health and Human Development Neonatal Research Network. *N. Engl. J. Med.* **2005**, *353* (15), 1574–1584.
78. Massaro, A. N.; Jeromin, A.; Kadom, N.; et al. *Pediatr. Crit. Care Med.* **2013**, *14* (3), 310–317. http://dx.doi.org/10.1097/PCC.0b013e3182720642.
79. Takai, N.; Tanaka, Y.; Inazawa, K.; Saji, H. *Rapid Commun. Mass Spectrom.* **2012**, *26*, 1549–1556.
80. Canales, L.; Chen, J.; Kelty, E.; et al. *Toxicology* **2012**, *300*, 1–11.

Part IV

Other Omics Strategies, Data Treatment, Integration and Systems Biology

Profiling of Genetically Modified Organisms Using Omics Technologies

Alberto Valdés, Carolina Simó, Clara Ibáñez and Virginia García-Cañas

Laboratory of Foodomics, Institute of Food Science Research (CIAL), CSIC. Nicolás Cabrera 9, Madrid, Spain

1 INTRODUCTION

For centuries, the production of foods with the desired quality has been a major goal in agriculture. To that aim, classical plant breeding has been applied to improve plant varieties with different techniques, such as plant crossing and selection, cell tissue culture, and mutagenesis based on irradiation, among others. On the other side, genetic engineering (or recombinant DNA technology) allows to transfer selected individual gene sequence from one organism into another, where the acceptor can be from the same species or not. Recombinant technology represents one of the most technological advances in the past decades in modern biotechnology, and the organism derived from this technology is termed genetically modified organism (GMO). In this sense, genetically modified (GM) foods are foods derived from organisms whose genetic material (DNA) has been modified in a way that does not occur naturally, for example, through the introduction of a gene from a different organism (1). The modifications incorporated through the

Comprehensive Analytical Chemistry, Vol. 64. http://dx.doi.org/10.1016/B978-0-444-62650-9.00013-0

recombinant technology generally represent some advantages in terms of agronomic productivity and industrial processing of the GM crops over their nonmodified counterparts.

The first GM crop was commercialized in 1996, and since then, 170 million hectares have been approved by 2012. At this time, 30 countries have approved GMO crops (2). Soybean, maize, cotton, and rapeseed are the most represented crops in terms of cultivated area, and by itself, soybean accounts for more than 50% of the GM crop production (mostly used for high-protein animal feed) (3). Besides these major crops, other minor GM crops that can be also found in the market include canola, potatoes, eggplant, carrots, etc. Regarding the GM traits in GMOs, the most frequent are herbicide resistance, insect resistance, and resistance to viral pathogens (4). Other important traits are resistance to virus, resistance to severe environmental conditions, or enhanced nutritional properties. Some GM products that are in the pipeline of commercialization in a near future include plants enriched in β-carotene (5), vitamin E (6), or omega-3 fatty acids (7), which are considered as second-generation GMOs.

2 DEBATED SAFETY ISSUES ON GMOs

Development, release into the environment, and commercialization of GMO have been greatly debated since the first GMO was sown, more than three decades ago (8). The main questionable aspects regarding GMOs have been centered on four areas, namely, environmental concerns (9,10), potential harm to human health (11,12), ethical concerns interferences with nature and individual choice (13), and patent issues (14,15).

Due to the abovementioned controversial safety aspects of GMOs, the European Union and other countries have established strict regulations including risk assessment, labeling, traceability, and marketing of GMOs. One commonly established concept for the evaluation of safety assessment of GM foods is the substantial equivalence (16). This approach is based on the assumption that traditional crop-plant varieties currently in the market that have been consumed for decades have gained a history of safe use (17). Consequently, they can be used as comparators for the safety assessment of new GMO crop varieties derived from established plant lines. However, as these regulations have not been established for all the countries, there is an "asynchronous approval" of these GM crops. In the same manner, GMO labeling and traceability differ between countries with different national legislation. For example, by the Regulation 1829/2003 of the European Union, it is mandatory to label as GMO containing when any food contain more than 0.9% of an authorized GMO, and the threshold is established at 0.5% when the GMO is nonauthorized. Meanwhile, in Australia and Japan, the threshold for labeling has been established at 1% and 5%, respectively.

Although recombinant DNA technology is considered highly accurate for genetic modification, one of the main controversial issues associated with

GMO safety are the possible unintended effects, which might occur during GMO development. The unintended effects can be defined as effects that go beyond the primary expected effects of the genetic modification, and represent statistically significant differences in a phenotype compared with an appropriate phenotype control *(18)*. Unintended effects can be originated by rearrangements, insertion, or deletions during the genetic transformations or during the tissue culture stages of GMO development *(19,20)*. Alterations linked to secondary and pleiotropic effects of gene expression are some examples of unintended changes, and they could be somehow explained considering the function of a transgene, the site of integration in the genome, or based on our current knowledge of plant metabolism *(21,22)*. In some cases, unintended effects will be observed if they result in a distinct phenotype, including compositional alterations. Thus, the comprehensive characterization of the plant at the molecular level would therefore facilitate detection and description of the potential unintended effects in GMOs *(23)*.

3   OMICS PROFILING IN GMO ANALYSIS

As it has pointed out above, any GMO or its derived products have to pass through an approval system before entering on the market *(24)*. In consequence, there is a need for analytical tools that facilitate comprehensive compositional studies of GMOs in order to effectively investigate substantial equivalence and the potential adverse effects on the human health, including the existence or not of unintended effects. Compositional equivalence between GM crops and conventional non-GM comparators is considered to provide an equal or increased assurance of the safety of foods derived from GM plants. To this regard, the selection of the comparator is a crucial aspect, and many questions about the need for comparing varieties grown in different areas and seasons, the key components to be analyzed, among others have been raised *(25)*. Following the recommendations from the Organization of Economic Cooperation and Development, compositional equivalence between GM and non-GM lines can be achieved by targeted analysis focused on macronutrients and micronutrients, antinutrients, and natural toxins for each crop variety *(18,26)*. With the application of this strategy, 95% of the crop composition is covered *(27)*, and some studies have proved that unintended effects could be identified *(5,28)*. However, it has been pointed out that this approach is biased and some unintended effects derived from genetic transformation may remain undetected *(29)*. To solve this problem, a comprehensive study of GMO composition would help in the recognition of unintended effects that could not be detected using targeted analysis. For this purpose, the development and use of profiling technologies such as genomics, transcriptomics, proteomics, and metabolomics have been recommended by the European Food Safety Agency (EFSA) *(24)*. Moreover, application of profiling analysis has been suggested by a panel of experts on risk assessment and management in those comparison studies where the most scientifically isogenic and conventional comparator would not grow, or not grow as well, under the relevant stress condition *(30)*.

The combination of transcriptomics, proteomics, and metabolomics technologies could provide a great coverage of genes, proteins, and metabolites. In this context, Foodomics, recently defined as "a new discipline that studies the food and nutrition domains through the application of advanced omics technologies in order to improve consumers' well-being and confidence" *(31,32)* can provide valuable information that could be essential for GMO detection, traceability, and characterization (Figure 1). Also, Foodomics offers unprecedented opportunities to study the molecular mechanism leading to a particular phenotype or the mechanism operating in important cellular processes, such as the response to different stresses *(33,34)*.

In spite of the recommendations and the great opportunities offered by omics technologies, there are some criticisms about the usefulness of molecular profiling in GMO risk assessment *(35)*. The main argument against their use is based on the problems of their routine use among laboratories, due to the lack of standardized and validated procedures. Another significant issue against profiling relies on the limited predictive capacity of the profiles for safety evaluation. Although molecular profiling can effectively measure relative differences in compounds between two varieties with high sensitivity, the biological relevance of such differences cannot be determined without previous knowledge of the natural variability of the crop composition *(36)*. Even knowing the natural variability of a compound, it is difficult to decode the biological meaning of the detected differences in terms of food safety risk *(35)*. Moreover, when comparing a GM crop to its non-GM isogenic variety,

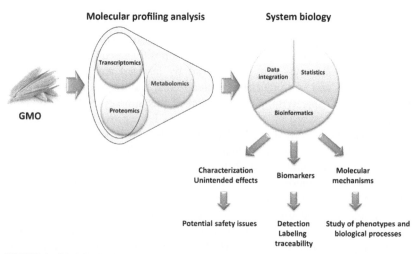

FIGURE 1   Ideal foodomics platform to analyze genetically modified organisms (GMOs).

it is important to grow both varieties under identical conditions to avoid the influence of other variability factors such as location, climate, season, and farming practices. Furthermore, all these factors along with the natural variation of the chemical composition of a crop have to be taken into account to make this overview as complete as possible *(37)*. Considering all the advantages and drawbacks of molecular profiling for GMO risk assessment, their use has been generally accepted by the scientific community, as shown by the number of works reported on profiling approaches for comparative profiling analysis and/or the investigation of unintended effects in GMOs *(33,38)*. In the published omics profiling studies on GMOs, it is interesting to note that some differences can be linked to genetic transformation. However, these studies demonstrate that differences between conventional varieties are in general more pronounced than the divergences observed between GM and non-GM crops. In addition, variations have been found when the same variety is grown in different environmental conditions. Representative examples of omics techniques, namely, transcriptomics, proteomics, and metabolomics, will be discussed in the following sections.

3.1   Transcriptomics

In the last years, gene expression profiling techniques, such as DNA microarray, have matured and experienced a great development in terms of high-throughput sensitivity and automatism. Linked to this aspect, the extensive optimization and standardization of gene expression microarray have put this technique at the forefront of transcriptomic techniques. Gene expression microarray technique is based on the hybridization of specific nucleic acids, and this feature can be used to measure the relative quantities of specific mRNA sequences in two or more conditions for thousands of genes simultaneously. Microarray technology is helpful for the identification of differences in comparative transcriptomic analysis, and for instance, it has been applied by Van Dijk *et al.* as a holistic approach to discover changes present in the natural variation of specific genes in different conditions *(39)*. In addition, transcriptomic analysis has also demonstrated to be a valuable profiling method to assess possible unintended effects of genetic transformation in wheat, maize, soybean, potato, and rice crops (Table 1).

Gene expression microarray technology has been applied by different research groups in comparative transcriptional studies between the transcriptional profiles in GM wheat and its untransformed counterpart. For example, Gregersen *et al.* analyzed the gene expression profile in developing seeds of wild-type wheat and wheat transformed for endosperm-specific expression of an *Aspergillus fumigatus* phytase *(40)*. In their study, authors concluded that the expression of the codon-modified *A. fumigatus* phytase gene in the wheat seed had no significant effects on the overall gene expression patterns in the developing seed. Later, Baudo *et al.* analyzed transgenic and

TABLE 1 Transcriptomics Profiling of GMOs Using Microarray Technique

GM Crop	Phenotype	Tissue	Donor Specie	Genetic Modification	Reference
Wheat	Endosperm-specific phytase expression	Seed	Aspergillus fumigatus	PhyA	(40)
	Nutritionally enhanced	Seed and leaf	Triticum aestivum	Glu-A1, Glu-D1	(41)
Maize	Insect resistance	Seed	Bacillus thuringiensis	Cry1Ab	(42)
	Insect resistance	Leaf	B. thuringiensis	Cry1Ab	(43)
Soybean	Herbicide tolerance	Leaf	Agrobacterium tumefaciens	CP4 EPSPS	(44)
	Herbicide tolerance	Leaf	A. tumefaciens	CP4 EPSPS	(45)
Potato	Starch metabolism	Tuber	Solanum tuberosum	Sus4	(46)
Rice	Free tryptophan accumulation	Seed	Oryza sativa	OASA1D	(47)
	Control stress-inducible genes	Seed	Hordeum vulgare	BCBF1	(48)
	Alanine aminotransferase overexpression	Root and shoot	H. vulgare	AlaAT	(49)
	Fungal resistance	Leaf	Aspergillus giganteus	afp	(50)
	Insect resistance	Leaf	B. thuringiensis	Cry1Ab	(51)

conventionally bred wheat lines expressing additional genes encoding high molecular weight subunits of glutenin, suggesting that the presence of the transgenes did not significantly alter gene expression *(41)*.

Transcriptomic profiles of GM maize have been also obtained using microarray technology. In a series of articles, Coll *et al.* reported the comparative study of different MON810 varieties with their near-isogenic counterpart *(42,43)*. In the first study, after *in vitro* culturing of the maize plantlets under highly controlled experimental conditions (to avoid changes due to environmental factors), high-density Affymetrix microarray technology was used to analyze the gene expression *(42)*. Of the 13,339 genes represented in the microarray, 307 and 25 differentially expressed genes (DEGs) were found when comparing two MON810 varieties and their near-isogenic counterparts. However, 693 and 832 DEGs were found when comparing between non-GM varieties and between GM varieties, respectively. These results suggest that the genetic background of each variety has a great influence when assessing the substantial equivalence of GM crops. In a later study, a similar methodology was applied to assess the effect of different field environments and cultural conditions over MON810 maize varieties and their counterparts *(43)*. Microarray data revealed a deregulation of 0.07–0.2% of the maize transcriptome when growing the plants under low-nitrogen and control conditions. The expression of 13 and 23 genes was altered between the transgenic and nontransgenic in control and low-level nitrogen conditions, respectively. However, a higher number of genes (31) were deregulated when comparing the transgenic line in low-nitrogen and control conditions, suggesting that the environmental conditions have higher influence than the genetic modification. As a common technique applied when handling microarray expression data because of its high sensitivity, real-time quantitative polymerase chain reaction (RT-qPCR) was used to confirm the gene expression. A total of 37 amplifying systems were designed and successfully applied, getting a 71.1% degree of coincidence between microarray and RT-qPCR data. The transcriptional profiles of these sequences were subjected to principal component analysis (PCA), and the results indicate that the natural variation of gene expression between the varieties and conditions is larger than the variation due to the genetic modification.

A comparative transcriptomic study on glyphosate-resistant soybean and its near-isogenic line was carried out by Zhu *et al. (44)*. Both lines were treated with glyphosate, and gene expression profiles were obtained at 1, 4, and 24 h posttreatment using a cDNA microarray representing 27,513 genes. After the treatment, 170 genes were rapidly altered in glyphosate-sensitive soybean, while transcript changes in glyphosate-resistant soybeans were minor or negligible due to the empirical false discovery rate. In addition, 2 genes out of 27,513 were found altered when comparing transcriptional profiles between cotyledons from resistant and sensitive lines, so the authors concluded that there were not unexpected consequences at the transcriptome level associated with the genetic modification. In a separate report, Cheng *et al.* investigated the genotype of five soybean cultivars, two transgenic and three conventional, grown in the same conditions *(45)*. In their study, authors used Affymetrix Soybean GeneChip to analyze 25 soybean samples, and the resultant gene expression profiles were subjected to PCA and unsupervised hierarchical clustering. Both multivariate analyses demonstrated that the GM cultivars did not cluster into a separate group from traditional cultivars. These results suggest that transgene insertion had negligible effects on global gene expression. Pairwise comparison between the GM cultivars and the non-GM counterparts showed that the number of DEGs identified was lower than those obtained in the comparison between non-GM varieties. Furthermore, only the gene cysteine protease inhibitor and dihydroflavonol-4-reductase were downregulated in both transgenic cultivars compared with their non-GM counterpart. However, it could not be concluded if these changes proceed from the natural variation of the parent genotype, an effect of the transgenic product or an effect of the insertion event.

In addition to these studies, Baroja-Fernández *et al.* have focused their studies on the characterization of GM potatoes *(46)*. Agilent microarray slides containing 46,345 genes were used to obtain the transcriptomic profiles from potatoes with modified sucrose synthase (SuSy) gene. In SuSy-antisensed tubers, the expression of 357 genes was found to be dysregulated compared with their non-GM counterpart; however, 118 genes were deregulated in SuSy-overexpressing tubers. In spite of these results, the SuSy-overexpressing tubers exhibited a substantial increase in starch, UDPglucose, and ADPglucose content when compared with controls. Nevertheless, there were no changes in the expression of genes that encode for enzymes directly involved in starch and sucrose metabolism.

Rice plants with genetic modifications have been studied several times by various groups. For instance, Batista *et al.* studied the extent of transcriptome modification through transgenesis and mutation breeding *(48)*. Gene expression of two stable transgenic plants, two γ-irradiated stable mutants, and the corresponding unmodified original genotypes were analyzed by Affymetrix GeneChip Rice Genome Array, which covers 51,279 genes. Hierarchical clustering of gene expression profiles showed that modified genotypes always grouped with the respective unmodified controls. Although the authors concluded that the use of either mutagenesis or transgenesis may cause alterations in the expression of untargeted genes, the alterations were more extensive in mutagenized than in transgenic plants for all the cases studied. Affymetrix GeneChip Rice Genome Array was also applied by Beatty *et al.* to investigate the transcriptional profiles of roots and shoots of GM rice overexpressing alanine aminotransferase (AlaAT) *(49)*. A higher number of DEGs were found in roots *(52)* than in shoots *(36)* in the transgenic line, corresponding to 0.11% and 0.07%, respectively, of the rice genome. Although authors could not find genes directly related with aminotransferase activity, nitrogen transport, and assimilation, higher levels of some amino acids, total nitrogen content, and grain yield were found in transgenic lines compared with control plants. Later, Montero *et al.* analyzed the transcriptome profile of GM rice expressing the AFP antifungal protein and its non-GM counterpart using the same microarray platform *(50)*. Of the 51,279 genes represented in the microarray chip, 0.4% DEGs with at least a twofold increase or decrease were found in the GM variety over the near-isogenic variety. The expression of 34 gene sequences was confirmed by RT-qPCR with 82% of agreement between the RT-qPCR and the microarray data. The analysis of the expression of the confirmed genes suggested that 35% and 15% of the detected differences could be attributed to procedure used to obtain GM plants and the transformation event, respectively. Thus, only around a 50% of the transcriptional unintended effects could be associated to the transgene itself. More recently, an insect-resistant rice variety has been investigated by Liu *et al.* to detect potential unintended effects (susceptibility to rice brown spot mimic lesion and sheath blight disease) of the insect-resistant transgenic KDM rice *(51)*. Using the Affymetrix GeneChip Rice Genome Array, 680 DEGs were found when comparing gene expression profiling of the GM and its non-GM counterpart. To know the pathways and biological functions more altered after the genetic modification, DEGs were subjected to functional enrichment analysis using the bioinformatics tool Plant MetGenMAP. Among the 17 significantly changed pathways, 8 were directly implicated in plant stress and defense responses, and the other 9 were directly associated with plant amino acid metabolism. Amino acid profiling using isobaric tags for relative and absolute quantification (iTRAQ) and liquid chromatography (LC) coupled to mass spectrometry (MS) (iTRAQ®-LC-MS/MS) technique was performed to confirm the transcriptomic results. These analyses showed changes in 10 amino acids and in γ-amino-*n*-butyric acid, a typical stress response amino acid in plants.

3.2   Proteomics

The study of proteins is especially interesting in food safety because they may act as toxins, antinutrients, or allergens *(53)*. Proteomics, a high-throughput technology able to quantify hundreds of proteins simultaneously, has become very important in comparative studies of GM plants and their nonmodified counterparts *(54)*. Two conceptually different strategies can be followed in comparative proteomics: the shotgun and the bottom-up approaches.

In the shotgun proteomics approach, protein digestion is performed without any prefractionation or separation of the proteome. The resulting peptides from the protein hydrolysis are generally separated by LC or capillary electrophoresis (CE) followed by MS analysis, providing rapid and automatic identification of proteins in the sample. Although this strategy has already been demonstrated to be a suitable strategy for protein profiling, it has been barely applied in the field of GMOs. Simó *et al.* investigated the unintended proteomics effects in herbicide-resistant GM soybean by the application of CE coupled to a time-of-flight (TOF) mass analyzer and electrospray ionization (ESI) *(55)*. Optimization of several parameters affecting the CE-ESI-TOF MS separation and detection was carried out in this work in the first stages of the method development. Once the conditions were reached, 151 peptides were automatically detected for each soybean line, but no statistically differences between the samples were found. Luo *et al.* quantified differences in protein profiles between GM rice and its near-isogenic line, combining the shotgun approach with the iTRAQ technique *(52)*. Four different digested samples were treated with four independent isobaric reagents, designed to react with all primary amines of protein hydrolysates. Rice-endosperm-treated samples were subsequently pooled and analyzed by tandem MS. The analyses revealed significant differences in 103 proteins out of the 1883 identified between GM and wild-type rice.

Unlike the shotgun approach, bottom-up proteomics approach has been widely applied to investigate the substantial equivalence and potential unintended effects in GMOs. This strategy involves the application of two-dimensional gel electrophoresis (2-DGE) followed by image analysis. Excision of the proteins from the gel spots and hydrolysis with trypsin is mandatory prior to the identification of the differentially expressed proteins (DEPs) by MS. 2-DGE analysis has some advantages and drawbacks: on the one hand, it provides the highest protein resolution capacity with a low cost of instrumentation; on the other hand, highly hydrophobic proteins with extreme isoelectric points or high molecular weight are difficult to analyze using this methodology. In addition, the gel-to-gel variation is the one of the major sources of error. For the protein identification, matrix-assisted laser desorption/ionization (MALDI) coupled to a TOF mass spectrometer, or different variants of LC-MS are used. 2-DGE separation technique has been applied for comparing the proteome between the GM and the non-GM counterpart of different crops, such as, maize, soybean, potato, and tomato, among others.

Albo *et al.* compared the proteomic profiles obtained from insect-resistant GM MON810 maize and its non-GM counterpart using 2-DGE/MALDI-TOF/TOF MS *(56)*. Some unintended effects were found among the GM and the non-GM maize, such as the glucose and ribitol dehydrogenase spots uniquely found in the GM maize, or the endochitinase A spot only found in non-GM maize. Also, some spots were overexpressed (triosephosphate isomerase 1 and globulin-1 S) and other spots were downregulated (cytosolic 3-phosphoglycerate kinase and aldose reductase) in the GM maize with respect to the non-GM counterpart. In a different study, Zolla *et al.* used 2-DGE/LC coupled to an ion trap (IT) mass spectrometer for the analysis of two subsequent generations of the MON810 maize variety and their counterparts in different environmental conditions *(57)*. The comparison between the non-GM and GM maize grown in different environmental conditions revealed 100 DEPs. On the other side, only the expression level of 43 proteins was altered in transgenic seeds with respect to their controls when controlled growing conditions were used. With these results, the authors concluded that environment plays the main influence on proteomic profiles of transgenic seeds. In a later work, 2-DGE/LC-ESI-IT MS was also used by Coll *et al.* to analyze the proteome profiles of two different MON810 maize varieties *(58)*. A small number ($\leq 1.2\%$ analyzed proteins) of quantitative differential spots were observed between the GM and non-GM varieties. In addition, all differences were variety-specific and thus could not directly be attributed to the MON810 modification.

Comparative protein profiling analysis has also been carried out in potatoes using different varieties, environmental conditions, and GM lines *(59)*. 2-DGE/LC-ESI-IT MS analysis showed statistically significant differences in 1077 of 1111 protein spots between different varieties and landraces, respectively. Comparing GM lines and their non-GM counterparts, only 9 proteins out of 730 exhibited significant differences. Multivariate PCA indicated clear separation between several genotypes, but GM potatoes could not be separated from their non-GM counterparts.

Brandão *et al.* also applied the bottom-up approach to compare seed–protein profiles of herbicide-tolerant soybean and its near-isogenic line *(60)*. Some 2-DGE parameters such as the loaded mass of the proteins, the pH separation range, and manual/automatic image editing were optimized prior to the evaluation (Figure 2). Of the 10 proteins with at least 90% variation found to be differentially expressed between the GM soybean and its counterpart, 8 proteins were successfully identified by MALDI coupled to a quadrupole (Q)-TOF MS.

2-DGE has been sometimes substituted by differential in gel electrophoresis (DIGE) technique to prevent gel-to-gel irreproducibility. In DIGE, different samples are labeled with ultrahigh-sensitive fluorescent dyes, typically Cy5 and Cy3, and then loaded in the same gel. After separation, gel images

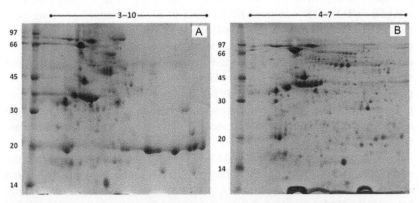

FIGURE 2 Two-dimensional gels of soybean seeds in the optimized conditions of the applied mass of protein at pH 3–10 range (A) and pH 4–7 range (B). *Reproduced from Ref. (60).*

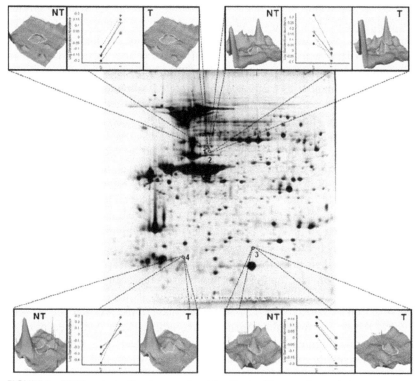

FIGURE 3 DIGE analysis of soybean seed proteins. Spots with expression variation among samples of transgenic (T) and nontransgenic (NT) soybean seeds. Downregulated proteins (2,4) and overexpressed proteins (1 and 3) are shown. *Reproduced from Ref.* (61).

obtained are processed with specific software for comparative analysis using two different detection channels, allowing the simultaneous detection of protein spots labeled with the two fluorescent dyes. DIGE in combination with MALDI-Q/TOF MS was used by Barbosa *et al.* to compare profiling proteomes of GM soybean and its non-GM control (Figure 3) *(61)*. The proteomic data obtained could be correlated with results from enzymatic determination of catalase, superoxide dismutase, glutathione reductase, and ascorbate peroxidase, suggesting higher oxidative stress in the transgenic soybean seeds. DIGE and MALDI-TOF/TOF MS techniques have also been combined to study the effect of transgenes, conventional genetic breeding, and natural genetic variation on the proteome of two insect-resistant transgenic rice *(62)*. Multivariate PCA of protein profiles could not differentiate between the GM varieties and their non-GM counterparts, but it could separate the non-GM varieties. The highest differences were found between the *indica* and *japonica* cultivars, followed by the three *indica* varieties, and finally between the GM rice and their counterparts. Univariate analysis of variance of the 6 rice proteomes allowed the detection of 443 DEPs and the identification of 234 proteins by the use of MALDI-TOF/TOF MS technology. Most of the identified proteins were related with metabolism, protein folding and modification, and defense response.

CE and LC techniques can be good substitutes of 2-DGE for protein separation. These techniques can be coupled directly to a mass spectrometer, require lower amount of starting material, provide full automation, high-throughput capabilities, and have better reproducibility than 2-DGE in terms of qualitative and quantitative analysis. Based on this concept, CE-ESI-MS with different mass spectrometers (TOF and IT) were compared for protein profiling of insect-resistant transgenic maize *(63)*. The performance of both analyzers showed similar sensitivity and repeatability when analyzing intact zein-protein fraction extracted from three different maize varieties and their near-isogenic lines. Although the CE-ESI-TOF MS provided more identified proteins, the differences between the GM and no-GM maize were not statistically significant. In another report, García-López *et al.* developed a LC-ESI-IT MS method to compare albumin, globulin, prolamin, and glutelin protein fractions isolated from several insect-resistant maize varieties and their nontransgenic lines *(64)*. Some differences could be found between maize from diverse regions, but not between the GM and non-GM lines.

3.3   Metabolomics

Metabolomics involves identification and quantification of a high number of metabolites that are substrates, intermediates, and end products of cellular activities. Metabolite profiling has been used to characterize the biological variation of the metabolic composition in commercial maize hybrids by the environment and/or genotype *(65)*. Also, this technology could be used to investigate the effectiveness of the genetic engineering procedure, as it is frequently used to obtain optimal production of plant metabolites, which may directly benefit human health and plant growth *(66)*. In addition, metabolomics has the potential to play an important role in GMO analysis, allowing the detection of intended or unintended effects, which may occur due to the genetic transformation *(26)*.

A typical metabolite profiling analysis involves the following steps: (i) metabolite extraction that often has to be adapted on a case-by-case basis depending on the type of sample and analytical platform chosen; (ii) sample preparation which may include partial purification and derivatization steps; (iii) instrumental analysis of sample; (iv) detection and quantification of metabolite signals to generate a data matrix that summarizes the detected signals and their intensity data; and (v) statistical analysis of metabolite profiles *(67)*.

Owing to the extraordinary diversity of the chemical structure and physicochemical properties of metabolites, there is no a single analytical platform or methodology capable to detect, quantify, and identify all metabolites in the same analysis. Two major analytical platforms are currently used for metabolomics analysis: nuclear magnetic resonance (NMR) and MS. These techniques, stand alone or combined with separation techniques (LC-NMR, gas chromatography (GC)-MS, LC-MS, and CE-MS), are complementary and frequently used in parallel in metabolomics research *(68)*.

3.3.1 Nuclear Magnetic Resonance

NMR has been applied for metabolite profiling in different crops, such as maize, potato, wheat, and lettuce. For instance, Manetti *et al.* studied the different accumulation of metabolites in hydroalcoholic extracts of GM MON810 maize seeds and their non-GM control. Authors of this work found that the levels of choline, asparagine, histidine, and trigonelline were lower in GM maize than their controls *(69)*. Hydroalcoholic extracts of GM MON810 maize seeds were also analyzed by Piccioni *et al.* with 1D and 2D NMR techniques *(70)*. In this case, 40 water-soluble metabolites were identified, and ethanol, lactic acid, citric acid, lysine, arginine, raffinose, trehalose, R-galactose, and adenine were identified for the first time in the ^{1}H NMR spectrum in maize seeds. Also, PCA carried out with the metabolite profile enabled the discrimination of the transgenic seeds from the nontransgenic ones.

In another report, Defernez *et al.* applied a similar approach to study different lines of GM and control potato samples *(71)*. Metabolite profiles were firstly subjected to a multivariate analysis for an initial exploration, followed by univariate analysis to confirm which compounds were mainly responsible for the differences found. Proline, trigonelline, and other phenolic compounds were statistically different between the GM and control potato samples. However, the most obvious differences were seen between the non-GM varieties studied. Transgenic potatoes expressing human beta amyloid, curdlan synthase, or glycogen synthase were also analyzed by Kim *et al.* using ^{1}H NMR *(72)*. The data obtained were submitted to PCA, and no differences were obtained between transgenic and control lines. In the case of GM pea plant, ^{1}H NMR profiles, analyzed by PCA and other multivariate tools, failed on providing an acceptable classification of the GM pea plant, the null segregant control without the transgene, and the parental line *(73)*. Similar results were found by Sobolev *et al.* in transgenic lettuce with enhanced growth properties *(74)*. The comparison of 24 hydrosoluble metabolites detected by NMR in 180 samples could not differentiate between the GM and the non-GM counterparts.

3.3.2 MS-Based Technologies

Compared to NMR, MS-based procedures have been more used in metabolite profiling of GMOs. The main advantage of MS is its higher sensitivity, and when coupled to GC, LC, or CE, higher resolution and sensitivity could be achieved for low-abundance metabolites *(75)*.

GC-MS combines high separation efficiency and reproducibility due to the stable ionization achieved by electron impact (EI), and is one of the most popular profiling techniques to study the metabolome of GMOs. The first report using this technique was carried out by Roessner *et al.* that characterized the metabolite profile of a transgenic potato tuber variety with modified sugar or starch metabolism *(76)*. After polar extraction followed by a methoximation and silylation, 77 out of 150 compounds detected were identified by comparison of the obtained spectra with commercially available spectra from MS libraries. The identified compounds provided valuable information of the altered metabolic pathways and the unexpected changes in the GM potatoes. When using the same methodology for the analysis of transgenic potatoes with altered sucrose catabolism, the same group reported increased levels of amino acids *(77)*. PCA was also applied in other studies of the same group, allowing the discrimination of the GM crop lines from the respective non-GM line *(78)*. Multivariate analysis has been also applied for the statistical analysis of GC-MS data obtained from a transgenic potato designed to contain high levels of inulin-type fructans *(79)*. Flow injection analysis (FIA)-ESI MS was used to analyze 600 potato extracts, and 2000 tuber samples were analyzed by GC-TOF MS. Among the 242 metabolites detected, the chemometrics analysis could not differentiate between the GM and the non-GM potatoes.

The derivatization and analysis of GC-MS method developed by Roessner *et al. (76)* was also applied by other authors to investigate the metabolomics profiling of a tryptophan (Trp)-enriched GM soybean *(80)*. It was reported that out of the 37 total organic acids, sugars, alcohols, and phenolic compounds identified, fructose, myo-inositol, and shikimic were found in higher concentration in GM leaves. Also, the metabolomic analyses of embryogenic cultures exhibited higher levels of malonic acid and urea and lower levels of β-hydroxybenzoic acid and galactose in GM soybean.

Other GC-MS methods have been developed to investigate transgenic rice. For instance, Zhou *et al.* applied a GC-flame ionization detection (FID) and GC-MS to study the unintended effects of different insect-resistant transgenic rice *(81)*. The resulted data were analyzed by partial least squares-discriminant analysis (PLS-DA) and PCA, suggesting that both the environment and the gene manipulation had remarkable impacts on the contents of different compounds. In addition, the levels of sucrose, mannitol, and glutamic were increased in GM rice. In another study, carotenoid-biofortified GM rice and five conventional rice cultivars were analyzed by GC-TOF MS *(82)*. PCA carried out with the 52 identified metabolites could separate the pigmented and nonpigmented rice samples (Figure 4). However, transgenic rice could not be distinguished from the nontransgenic counterpart, suggesting that natural variation between varieties is higher than the differences between GM and non-GM isogenic lines.

Owing to its relevancy in metabolite profiling analysis, several groups have investigated the suitability of different metabolite extraction procedures for GMO analysis. Selective extraction techniques, such as supercritical fluids

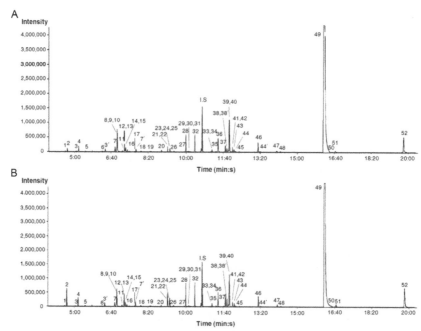

FIGURE 4 Selected ion chromatograms of metabolites extracted from nontransgenic rice (cv. NDB) (A) and transgenic rice (PAC) (B) as methoxime/trimethylsilyl derivatives separated on a 30 m × 0.25-mm ID fused-silica capillary column coated with 0.25-lm CP-SIL 8 CB low bleed. Peak identification: 1, pyruvic acid; 2, lactic acid; 3, valine; 4, alanine; 5, oxalic acid; 6, glycolic acid; 3′ valine; 7, serine; 8, ethanolamine; 9, glycerol; 10, leucine; 11, isoleucine; 12, proline; 13, nicotinic acid; 14, glycine; 15, succinic acid; 16, glyceric acid; 17, fumaric acid; 7′ serine; 18, threonine; 19, b-alanine; 20, malic acid; 21, salicylic acid; 22, aspartic acid; 23, methionine; 24, pyroglutamic acid; 25, 4-aminobutyric acid; 26, threonic acid; 27, arginine; 28, glutamic acid; 29, phenylalanine; 30, p-hydroxybenzoic acid; 31, xylose; 32, asparagine; 33, vanillic acid; 34, glutamine; 35, shikimic acid; 36, citric acid; 37, quinic acid; 38, fructose; 38′, fructose; 39, galactose; 40, glucose; 41, syringic acid; 42, mannose; 43, mannitol; 44, ferulic acid; 45, p-coumaric acid; 46, inositol; 44′ ferulic acid; 47, tryptophan; 48, sinapic acid; 49, sucrose; 50, cellobiose; 51, trehalose; 52, raffinose; IS, internal standard (ribitol). *Reproduced from Ref. (82).*

or accelerated solvents, have been applied to investigate unintended effects in GM soybean *(83)* and maize *(84)*. These techniques enabled the selectively extraction of amino acids and fatty acids, and the combination with GC-MS enabled their quantification. More recently, Frank *et al.* designed a complex extraction scheme to obtain four fractions containing major lipids, minor lipids, sugars, sugar alcohols and acids, and amino acids and amines *(85)*. The extraction method was applied to insect-resistant and herbicide-tolerant GM maize and their nonmodified counterparts, grown at distinct locations and in different seasons. The fractions obtained were independently analyzed using GC-EI-Q MS, and PCA of the data indicated that environmental influences had more impact on the maize metabolite profiles than the genetic modification.

LC-MS has been also used for metabolite profiling of GM crops providing advantages such as versatility, wide dynamic range, and reproducible quantitative analysis. LC-MS is able to separate and to analyze complex samples. LC-MS is frequently used to profile polar/nonvolatile, large and thermolabile compounds, demonstrating good performance on profiling secondary metabolites and complex lipids. In addition, LC-MS can resolve and quantify multiple components in crude biological extracts typically down to the nanomolar or picomolar range.

Some interesting examples of LC-MS application in this field are the studies on transgenic rice with different modifications: endosperm flavonoid production *(86)*, enhanced starch synthesis *(87)*, increased tryptophan production *(88)*, or insect resistance *(89)*. In the latter work, a LC-ESI-Q/TOF MS method was developed to compare acetonitrile/water, acetone/water, and methanol/water in terms of metabolite extraction *(89)*. The PLS-DA enabled the classification of the GM and non-GM samples by using 15 metabolite levels. However, when including samples grown in different sites and dates, it was shown that environmental factors played a greater role than gene modification for most of the metabolites. Also, LC-Q/IT MS coupling with different interfaces (ESI or atmospheric-pressure chemical ionization) has been used to characterize wheat overexpressing genes that confer increased fungal resistance *(23)*. Similarly, LC-ESI-Q MS has been used to investigate GM tomato overexpressing a grapevine gene that encodes the enzyme stilbene synthase *(90)*.

CE coupled to MS has been also successfully applied for metabolite profiling of GMOs. It is considered complementary to LC-MS and GC-MS, as ionic and polar thermolabile compounds can be analyzed. Although high efficiency, speed, and resolution can be achieved, only moderate sensitivity is reached due to the minimum amount of sample injected. Rice overexpressing dihydroflavonol-4-reductase, which enhances H_2O_2 tolerance, submergence, and infection by *Magnaporthe grisea*, was explored by CE-ESI-Q MS *(91)*. Identification of chemical compounds was performed by comparison of their *m/z* values and migration times with standard metabolites. *cis*-Aconitate, isocitrate, and 2-oxoglutarate were higher in GM leaves, whereas fructose-1,6-bisphosphate and glyceraldehyde-3-phosphate were lower in GM roots. CE-ESI-TOF MS has been used for metabolite profiling of GM maize *(92)* and soybean *(93,94)*. Levandi *et al.* carried out a multivariate statistical analysis of the maize metabolic profiles, finding statistically differences between the GM and non-GM lines *(92)*. In their study, 2 metabolites (L-carnitine and stachydrine) out of the 27 tentatively identified were found statistically significant. A method developed by Garcia-Villalba *et al.* enabled the tentative identification of 45 metabolites (including isoflavones, amino acids, and carboxylic acids) in herbicide-resistant transgenic soybean *(93)*. Interestingly, 4-hydroxy-l-threonine seemed to disappear in the transgenic soybean compared to its parental nontransgenic line. In a separate report, herbicide-resistant transgenic soybean was characterized by Giuffrida *et al.* using a novel chiral CE-ESI-TOF MS method *(94)*. In that research, the obtained D/L-amino acid profiles were very similar for conventional and GM soybean.

Fourier transform ion-cyclotron (FT-ICR-MS) provides the highest mass resolution and accuracy, and enables the determination of the elemental compositions of metabolites, which facilitates annotation procedures for unknown compounds *(95)*. Direct infusion analysis of plant extract without a previous separation and/or derivatization can be achieved; however, its use is very restricted due to the equipment cost, the difficulties in hardware handling, and the extremely large amount of data generated. Takahashi *et al.* applied this technique to elucidate the effects of the overexpression of the YK1 gene in stress-tolerant GM rice *(96)*. More than 850 metabolites could be determined, and the metabolomics fingerprint in callus, leaf, and panicle was significantly different from one another.

3.3.3   Multi-platform Strategies

The combination of more than one analytical platform for metabolomics profiling generally provides complementary results, which enables the comprehensive analysis of GMO metabolome. As an example of this, León *et al.* combined FT-ICR-MS with CE-TOF MS for the metabolic profiling of six varieties of maize, three transgenic insect-resistant lines, and their corresponding near-isogenic lines *(97)*. The spectral data obtained in both positive and negative ESI modes with FT-ICR-MS were uploaded into MassTRIX server in order to identify maize-specific metabolites annotated in the KEGG database *(98)*. Interestingly, electrophoretic mobilities and *m/z* values provided by CE-TOF MS were very helpful in the identification of those compounds that could not be unequivocally identified by FT-ICR-MS, such as isomeric compounds. LC-MS and GC-MS have been used for the comparative analysis of grapevine varieties with enhanced response to abiotic stress and their nonmodified counterparts *(99)*. Differences in hydroxycinnamic acid, quercetin-3-glucoside, quercetol-3-glucuronide, and in the degree of polymerization in proanthocyanidins were found when comparing the profiles of phenolic compounds carried out by LC-ESI-IT MS. However, volatile secondary metabolites that belong to the classes of monoterpenes, C12-norisoprenoids, and shikimates were profiled by GC-EI-IT MS, and no differences were found between GM and non-GM lines. Two transgenic ringspot virus-resistant papaya varieties and their nonmodified counterparts have also been analyzed by LC-MS and GC-MS *(100)*. LC-ESI-Q MS enabled the detection of organic acids, carotenoids, and alkaloids, whereas GC-EI-IT MS was applied for the detection of volatile organic compounds and sugar/polyals. GM and non-GM lines could not be differentiated with the multivariate analysis of the both data platforms; however, it could separate papaya

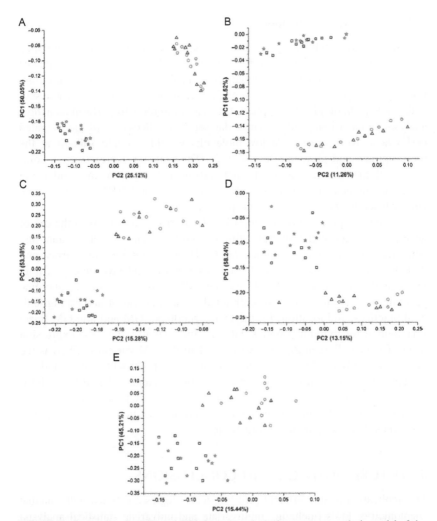

FIGURE 5 Score plots for PC1 and PC2 from the principal component analysis model of the total dataset of *cv*. MZH colored in four different ways: (A) volatile organic compounds; (B) sugars; (C) organic acids; (D) carotenoids; (E) alkaloids. (□) Transgenic in May, (*) nontransgenic in May, (△) transgenic in September, (○) nontransgenic in September. *Reproduced from Ref*. (100). (For the color version of this figure, the reader is referred to the online version of this chapter.)

samples from different harvesting times (Figure 5). Kusano *et al.* applied a broader approach to compare two GM tomato varieties overexpressing miraculin glycoprotein *(101)*. The multivariate analysis using orthogonal PLS-DA of LC-ESI-Q/TOF MS, GC-EI-TOF MS, and CE-ESI-Q/TOF MS data provided lower differences between the transgenic lines and the controls than differences observed among ripening stages and traditional cultivars.

3.4 Cross-Omics Studies

The information obtained by the different omics techniques can be combined to generate a broader view of GMO composition. This approach may enhance the opportunities to identify potential unintended effect and also the interrelationships between the different levels of information. In the studies where several omics technologies are applied, the general trend is to perform statistical analysis on each independent omics dataset. Thus, Scossa *et al.* investigated GM wheat overexpressing a low-molecular-weight glutenin subunit at the transcriptome and proteome levels *(102)*. The glutenin overexpression coincided with the downregulation of other classes of storage proteins, evidencing the complementary potential of cDNA microarrays and 2-DGE to assess the concordance between the RNA and protein levels.

A more complete study accomplished by Barros *et al.* involved the use of gene, protein, and metabolite profiling of two transgenic insect-resistant and herbicide-tolerant maize varieties *(103)*. Gene expression microarray and 2-DGE analyses were performed for transcriptome and proteome profiling, respectively, whereas ^{1}H NMR and GC-MS were used for metabolome profiling. Univariate analysis of individual variables (year of harvest, agricultural practices, and location) for the factor genotype (GM and non-GM lines) was applied for individual omics datasets. Interestingly, the gene expression level of maize allergen Zeam14 was lower in the GM varieties, whereas the glucose and fructose were increased in the insect-resistant maize, and the γ-tocopherol and inositol were decreased in the herbicide-tolerant line. Multivariate analysis indicated that growing seasons as well as locations had a stronger overall influence on the three levels of information of the three maize genotypes than the genetic modification.

4 FUTURE OUTLOOK AND CONCLUSIONS

The application of the different profiling technologies combined with suitable chemometrics tools (including multivariate and univariate statistical analysis) provides valuable information about the possible effects of genetic modification at transcriptomics, proteomics, and metabolomics levels. However, the lack of common standardized experimental protocols is still limiting its use in any stage of the safety assessment of GMOs. Nonetheless, implementation of these techniques in control laboratories will diminish the high costs, and the unification and validation of analytical platforms and protocols will enable the comparison of experiments performed in laboratories around the world. In the near future, the development and application of novel methodologies, such as next-generation sequencing in gene-profiling, advances of protein coverage in protein profiling, or the application of multidimensional techniques such as $GC \times GC$ or $LC \times LC$ in metabolite profiling, could make the analysis of GMOs more reliable and, in turn, more challenging. Besides, there is a demand for the availability of advanced bioinformatics tools capable of the integrating and interpreting the huge amount of data provided by these high-throughput technologies.

ACKNOWLEDGMENTS

This work was supported by AGL2011-29857-C03-01 project (Ministerio de Economía y Competitividad, Spain) and CSD2007-00063 FUN-C-FOOD (Programa CONSOLIDER, Ministerio de Educación y Ciencia, Spain). A. V. thanks the Ministerio de Economía y Competitividad for his FPI predoctoral fellowship.

REFERENCES

1. World Health Organization (WHO). http:/www.who.int/foodsafety/en/.
2. James, C. *Global Status of Commercialized Biotech/GM Crops: 2012. ISAAA Brief No. 44.* ISAAA: Ithaca, NY, 2012.
3. Jacobsen, S.; Sorensen, M.; Pedersen, S. M.; Weiner, J. *Agron. Sustain. Dev.* **2013**, *33*, 651–662.
4. Hails, R. S. *Trends Ecol. Evol.* **2000**, *15*, 14–18.
5. Ye, X.; Al-Babili, S.; Kloti, A.; Zhang, J.; Lucca, P.; Beyer, P.; Potrykus, I. *Science* **2000**, *287*, 303–305.
6. Cahoon, E. B.; Hall, S. E.; Ripp, K. G.; Ganzke, T. S.; Hitz, W. D.; Coughland, S. J. *Nat. Biotechnol.* **2003**, *21*, 1082–1087.
7. Kinney, A. J. *Curr. Opin. Biotechnol.* **2006**, *17*, 130–138.
8. Berg, P.; Baltimore, D.; Brenner, S.; Roblin, R. O.; Singer, M. F. *Science* **1975**, *188*, 991–994.
9. Thomson, J. *Trends Food Sci. Technol.* **2003**, *14*, 210–228.
10. Wolfenbarger, L. L.; Phifer, P. R. *Science* **2000**, *290*, 2088–2093.
11. Craig, W.; Tepfer, M.; Degrassi, G.; Ripandelli, D. *Euphytica* **2008**, *164*, 853–880.
12. Domingo, J. L. *Crit. Rev. Food Sci.* **2007**, *47*, 721–733.
13. Frewer, L.; Lassen, J.; Kettlitz, B.; Scholderer, J.; Beekman, V.; Berdal, K. G. *Food Chem. Toxicol.* **2004**, *42*, 1181–1193.
14. Herring, R. J. *Nat. Rev. Genet.* **2008**, *9*, 458–463.
15. Vergragt, P. J.; Brown, H. S. *Technol. Forecast. Soc.* **2008**, *75*, 783–798.
16. Organization of Economic Cooperation and Development (OECD). *Safety Evaluation of Foods Derived by Modern Biotechnology: Concept and Principles;* OECD: Paris, France, 1993.
17. Kok, E. J.; Kuiper, H. A. *Trends Biotechnol.* **2003**, *21*, 439–444.
18. Cellini, F.; Chesson, A.; Colquhoun, I.; Constable, A.; Davies, H. V.; Engel, K. H.; Gatehouse, A. M. R.; Kärenlampi, S.; Kok, E. J.; Leguay, J. J.; Lehesranta, S.; Noteborn, H. P.; Pedersen, J.; Smith, M. *Food Chem. Toxicol.* **2004**, *42*, 1089–1125.
19. Latham, J. R.; Wilson, A. K.; Steinbrecher, R. A. *J. Biomed. Biogeosci.* **2006**, *25376*, 1–7.
20. Rosati, A.; Bogani, P. G.; Santarlasci, A.; Buiatti, M. *Plant Mol. Biol.* **2008**, *67*, 271–281.
21. Ali, S.; Zafar, Y.; Xianyin, Z.; Ali, G. M.; Jumin, T. *Afr. J. Biotechnol.* **2008**, *7*, 4667–4676.
22. Kuiper, H. A.; Kleter, G. A. *Trends Food Sci. Technol.* **2003**, *14*, 277–293.
23. Ioset, J. R.; Urbaniak, B.; Ndjoko-Ioset, K.; Wirth, J.; Martin, F.; Gruissem, W.; Hostettmann, K.; Sautter, C. *Plant Mol. Biol.* **2007**, *65*, 645–654.

24. European Food Safety Agency. *Guidance Document of the Scientific Panel on Genetically Modified Organisms for the Risk Assessment of Genetically Modified Plants and Derived Food and Feed;* EFSA Communications Department: Parma, Italy, 2006.

25. Kok, E. J.; Keijer, J.; Kleter, G. A.; Kuiper, H. A. *Regul. Toxicol. Pharmacol.* **2008**, *50*, 439–444.

26. Shepherd, L. V. T.; McNicol, J. W.; Razzo, R.; Taylor, M. A.; Davies, H. V. *Transgenic Res.* **2006**, *15*, 409–425.

27. Chassy, B. M. *Regul. Toxicol. Pharmacol.* **2010**, *58*, S62–S70.

28. Hashimoto, W.; Momma, K.; Katsube, T.; Ohkawa, Y.; Ishige, T.; Kito, M.; Utsumi, S.; Murata, K. *J. Sci. Food Agric.* **1999**, *79*, 1607–1612.

29. Millstone, E.; Brunner, E.; Mayer, S. *Nature* **1999**, *401*, 525–526.

30. Ad Hoc Technical Expert Group (AHTEG), 2010. https://www.cbd.int/doc/meetings/bs/bsrarm-02/official/bsrarm-02-05-en.pdf, United Nations Environment Programme Convention for Biodiversity.

31. Cifuentes, A. *J. Chromatogr. A* **2009**, *1216*, 7109.

32. Herrero, M.; García-Cañas, V.; Simo, C.; Cifuentes, A. *Electrophoresis* **2010**, *31*, 205–228.

33. García-Cañas, V.; Simó, C.; León, C.; Ibáñez, E.; Cifuentes, A. *Mass Spectrom. Rev.* **2011**, *30*, 396–416.

34. García-Cañas, V.; Simó, C.; Herrero, M.; Ibáñez, E.; Cifuentes, A. *Anal. Chem.* **2012**, *84*, 10150–10159.

35. Doerrer, N.; Ladics, G.; McClain, S.; Herouet-Guicheney, C.; Poulsen, L. K.; Privalle, L.; Stagg, H. *Regul. Toxicol. Pharmacol.* **2010**, *58*, S2–S7.

36. Ricroch, A. E.; Berge, J. B.; Kuntz, M. *Plant Physiol.* **2011**, *155*, 1752–1761.

37. Davies, H. *Food Control* **2010**, *21*, 1601–1610.

38. Heinemann, J. A.; Kurenbach, B.; Quist, D. *Environ. Int.* **2011**, *37*, 1285–1293.

39. van Dijk, J. P.; Cankar, K.; Scheffer, S. J.; Beenen, H. G.; Shepherd, L. V.; Stewart, D.; Davies, H. V.; Wilkockson, S. J.; Leifert, C.; Gruden, K.; Kok, E. J. *J. Agric. Food Chem.* **2009**, *57*, 1612–1623.

40. Gregersen, P. L.; Brinch-Pedersen, H.; Holm, P. B. *Transgenic Res.* **2005**, *14*, 887–905.

41. Baudo, M. M.; Lyons, R.; Powers, S.; Pastori, G. M.; Edwards, K. J.; Holdsworth, M. J.; Shewry, P. R. *Plant Biotechnol. J.* **2006**, *4*, 369–380.

42. Coll, A.; Nadal, A.; Palaudelmas, M.; Messeguer, J.; Mele, E.; Puigdomenech, P.; Pla, M. *Plant Mol. Biol.* **2008**, *68*, 105–117.

43. Coll, A.; Nadal, A.; Collado, R.; Capellades, G.; Kubista, M.; Messeguer, J.; Pla, M. *Plant Mol. Biol.* **2010**, *73*, 349–362.

44. Zhu, J.; Patzoldt, W. L.; Shealy, R. T.; Vodkin, L. O.; Clough, S. J.; Tranel, P. J. *J. Agric. Food Chem.* **2008**, *56*, 6355–6363.

45. Cheng, K. C.; Beaulieu, J.; Iquira, E.; Belzile, F. J.; Fortin, M. G.; Stromvik, M. V. *J. Agric. Food Chem.* **2008**, *56*, 3057–3067.

46. Baroja-Fernández, E.; Muñoz, F. J.; Montero, M.; Etxeberria, E.; Sesma, M. T.; Ovecka, M.; Bahaji, A.; Ezquer, I.; Li, J.; Prat, S.; Pozueta-Romero, J. *Plant Cell Physiol.* **2009**, *50*, 1651–1662.

47. Dubouzet, J. G.; Ishihara, A.; Matsuda, F.; Miyagawa, H.; Iwata, H.; Wakasa, K. *J. Exp. Bot.* **2007**, *58*, 3309–3321.

48. Batista, R.; Saibo, N.; Lourenc, T.; Oliveira, M. M. *Proc. Natl. Acad. Sci. U.S.A.* **2008**, *105*, 3640–3645.

49. Beatty, P. H.; Shrawat, A. K.; Carroll, R. T.; Zhu, T.; Good, A. G. *Plant Biotechnol. J.* **2009**, *7*, 562–576.

50. Montero, M.; Coll, A.; Nadal, A.; Messeguer, J.; Pla, M. *Plant Biotechnol. J.* **2011**, *9*, 693–702.
51. Liu, Z.; Li, Y.; Zhao, J.; Chen, X.; Jian, G.; Peng, Y.; Qi, F. *Int. J. Biol. Sci.* **2012**, *8*, 953–963.
52. Luo, J.; Ning, T.; Sun, Y.; Zhu, J.; Zhu, Y.; Lin, Q.; Yang, D. *J. Proteome Res.* **2009**, *8*, 829–837.
53. D'Alessandro, A.; Zolla, L. *J. Proteome Res.* **2012**, *11*, 26–36.
54. Gong, C. Y.; Wang, T. *Front. Plant Sci.* **2013**, *4*, Article 41.
55. Simó, C.; Domínguez-Vega, E.; Marina, M. L.; García, M. C.; Dinelli, G.; Cifuentes, A. *Electrophoresis* **2010**, *31*, 1175–1183.
56. Albo, A. G.; Mila, S.; Digilio, G.; Motto, M.; Aime, S.; Corpillo, D. *Maydica* **2007**, *52*, 443–455.
57. Zolla, L.; Rinalducci, S.; Antonioli, P.; Righetti, P. G. *J. Proteome Res.* **2008**, *7*, 1850–1861.
58. Coll, A.; Nadal, A.; Rossignol, M.; Puigdomenech, P.; Pla, M. *Transgenic Res.* **2011**, *20*, 939–949.
59. Lehesranta, S. J.; Davies, H. V.; Shepherd, L. V.; Nunan, N.; McNicol, J. W.; Auriola, S.; Koistinen, K. M.; Suomalainen, S.; Kokko, H. I.; Kärenlampi, S. O. *Plant Physiol.* **2005**, *138*, 1690–1699.
60. Brandão, A. R.; Barbosa, H. S.; Arruda, M. A. Z. *J. Proteomics* **2010**, *73*, 1433–1440.
61. Barbosa, H. S.; Arruda, S. C. C.; Azevedo, R. A.; Arruda, M. A. Z. *Anal. Bioanal. Chem.* **2012**, *402*, 299–314.
62. Gong, C. Y.; Li, Q.; Yu, H. T.; Wang, Z.; Wang, T. *J. Proteome Res.* **2012**, *11*, 3019–3029.
63. Erny, G. L.; León, C.; Marina, M. L.; Cifuentes, A. *J. Sep. Sci.* **2008**, *31*, 1810–1818.
64. García-López, M. C.; García-Cañas, V.; Marina, M. L. *J. Chromatogr. A* **2009**, *1216*, 7222–7228.
65. Asiago, V. M.; Hazebroek, J.; Harp, T.; Zhong, C. *J. Agric. Food Chem.* **2012**, *60*, 11498–11508.
66. Okazaki, Y.; Saito, K. *Plant Biotechnol. Rep.* **2012**, *6*, 1–15.
67. Saito, K.; Matsuda, F. *Annu. Rev. Plant Biol.* **2010**, *61*, 463–489.
68. Shulaev, V. *Brief. Bioinform.* **2006**, *7*, 128–139.
69. Manetti, C.; Bianchetti, C.; Bizzarri, M.; Casciani, L.; Castro, C.; D'Ascenzo, G.; Delfini, M.; Di Cocco, M. E.; Lagana, A.; Miccheli, A.; Motto, M.; Conti, F. *Phytochemistry* **2004**, *65*, 3187–3198.
70. Piccioni, F.; Capitani, D.; Zolla, L.; Mannina, L. *J. Agric. Food Chem.* **2009**, *57*, 6041–6049.
71. Defernez, M.; Gunning, Y. M.; Parr, A. J.; Shepherd, L. V. T.; Davies, H. V.; Colquhoun, I. J. *J. Agric. Food Chem.* **2004**, *52*, 6075–6085.
72. Kim, H. S.; Kim, S. W.; Park, Y. S.; Kwon, S. Y.; Liu, J. R.; Joung, H.; Jeon, J. H. *Biotechnol. Bioprocess Eng.* **2009**, *14*, 738–747.
73. Charlton, A.; Allnutt, T.; Holmes, S.; Chisholm, J.; Bean, S.; Ellis, N.; Mullineaux, P.; Oehlschlager, S. *Plant Biotechnol. J.* **2004**, *2*, 27–35.
74. Sobolev, A. P.; Testone, G.; Santoro, F.; Nicolodi, C.; Iannelli, M. A.; Amato, M. E.; Ianniello, A.; Brosio, E.; Giannino, D.; Mannina, L. *J. Agric. Food Chem.* **2010**, *58*, 6928–6936.
75. Issaq, H. J.; Van, Q. N.; Waybright, T. J.; Muschik, G. M.; Veenstra, T. D. *J. Sep. Sci.* **2009**, *32*, 2183–2199.
76. Roessner, U.; Wagner, C.; Kopka, J.; Trethewney, R. N.; Willmitzer, L. *Plant J.* **2000**, *23*, 131–142.
77. Roessner, U.; Willmitzer, L.; Fernie, A. R. *Plant Physiol.* **2001**, *127*, 749–764.

78. Roessner, U.; Luedemann, A.; Brust, D.; Fiehn, O.; Linke, T.; Willmitzer, L.; Fernie, A. R. *Plant Cell* **2001**, *13*, 11–29.
79. Catchpole, G. S.; Beckmann, M.; Enot, D. P.; Mondhe, M.; Zywicki, B.; Taylor, J.; Hardy, N.; Smith, A.; King, R. D.; Kell, D. B.; Fiehn, O.; Draper, J. *Proc. Natl. Acad. Sci. U.S.A.* **2005**, *102*, 14458–14462.
80. Inaba, Y.; Brotherton, J. E.; Ulanov, A.; Widholm, J. M. *Plant Cell Rep.* **2007**, *26*, 1763–1771.
81. Zhou, J.; Ma, C.; Xu, H.; Yuan, K.; Lu, X.; Zhu, Z.; Wu, Y.; Xu, G. *J. Chromatogr. B* **2009**, *877*, 725–732.
82. Kim, J. K.; Park, S.; Lee, S. M.; Lim, S.; Kim, H. J.; Oh, S.; Yeo, Y.; Cho, H. S.; Ha, S. *Plant Biotechnol. Rep.* **2013**, *7*, 121–128.
83. Bernal, J. L.; Nozal, M. J.; Toribio, L.; Diego, C.; Mayo, R.; Maestre, R. *J. Chromatogr. A* **2008**, *1192*, 266–272.
84. Jimenez, J. J.; Bernal, J. L.; Nozal, M. J.; Toribio, L.; Bernal, J. *J. Chromatogr. A* **2009**, *1216*, 7288–7295.
85. Frank, T.; Röhlig, R. M.; Davies, H. V.; Barros, E.; Engel, K. *J. Agric. Food Chem.* **2012**, *60*, 3005–3012.
86. Shin, Y. M.; Park, H. J.; Yim, S. D.; Baek, N. I.; Lee, C. H.; An, G.; Woo, Y. M. *Plant Biol.* **2006**, *4*, 303–315.
87. Nagai, Y. S.; Sakulsinghroj, C.; Edwards, G. E.; Satoh, H.; Greene, T. W.; Blakeslee, B.; Okita, T. W. *Plant Cell Physiol.* **2009**, *50*, 635–643.
88. Matsuda, F.; Ishihara, A.; Takanashi, K.; Morino, K.; Miyazawa, H.; Wakasa, K.; Miyagawa, H. *Plant Biotechnol.* **2010**, *27*, 17–27.
89. Chang, Y.; Zhao, C.; Zhu, Z.; Wu, Z.; Zhou, J.; Zhao, Y.; Lu, X.; Xu, G. *Plant Mol. Biol.* **2012**, *78*, 477–487.
90. Nicoletti, I.; De Rossi, A.; Giovinazzo, G.; Corradini, D. *J. Agric. Food Chem.* **2007**, *55*, 3304–3311.
91. Takahashi, H.; Hayashi, M.; Goto, F.; Sato, S.; Soga, T.; Nishioka, T.; Tomita, M.; Kawai-Yamada, M.; Uchimiya, H. *Ann. Bot.* **2006**, *98*, 819–825.
92. Levandi, T.; Leon, C.; Kaljurand, M.; García-Cañas, V.; Cifuentes, A. *Anal. Chem.* **2008**, *80*, 6329–6335.
93. Garcia-Villalba, R.; León, C.; Dinelli, G.; Segura-Carretero, A.; Fernandez-Gutierrez, A.; García-Cañas, V.; Cifuentes, A. *J. Chromatogr. A* **2008**, *1195*, 164–173.
94. Giuffrida, A.; León, C.; García-Cañas, V.; Cucinotta, V.; Cifuentes, A. *Electrophoresis* **2009**, *30*, 1734–1742.
95. Xu, Y.; Heilier, J. F.; Madalinski, G.; Genin, E.; Ezan, E.; Tabet, J. C.; Junot, C. *Anal. Chem.* **2010**, *82*, 5490–5501.
96. Takahashi, H.; Hotta, Y.; Hayashi, M.; Kawai-Yamada, M.; Komatsu, S.; Uchimiya, H. *Plant Biotechnol.* **2005**, *22*, 47–50.
97. León, C.; Rodriguez-Meizoso, I.; Lucio, M.; García-Cañas, V.; Ibañez, E.; Schmitt-Kopplin, P.; Cifuentes, A. *J. Chromatogr. A* **2009**, *1216*, 7314–7323.
98. Suhre, K.; Schmitt-Kopplin, P. *Nucleic Acids Res.* **2008**, *36*, W481–W484.
99. Tesniere, C.; Torregrosa, L.; Pradal, M.; Souquet, J. M.; Gilles, C.; Dos Santos, K.; Chatelet, P.; Gunata, Z. *J. Exp. Bot.* **2006**, *57*, 91–99.
100. Jiao, Z.; Deng, J. C.; Li, G. K.; Zhang, Z. M.; Cai, Z. W. *J. Food Compos. Anal.* **2010**, *23*, 640–647.

101. Kusano, M.; Redestig, H.; Hirai, T.; Oikawa, A.; Matsuda, F.; Fukushima, A.; Arita, M.; Watanabe, S.; Yano, M.; Hiwasa-Tanase, K.; Ezura, H.; Saito, K. *PLoS One* **2011**, *16*, 16989.
102. Scossa, F.; Laudencia-Chingcuanco, D.; Anderson, O. D.; Vensel, W. H.; Lafiandra, D.; D'Ovidio, R.; Masci, S. *Proteomics* **2008**, *8*, 2948–2966.
103. Barros, E.; Lezar, S.; Anttonen, M. J.; van Dijk, J. P.; Rohlig, R. M.; Kok, E. J.; Engel, K. *Plant Biotechnol. J.* **2010**, *8*, 436–451.

Chapter 14

MS-Based Lipidomics

Päivi Pöhö, Matej Oresic and Tuulia Hyötyläinen

Steno Reseach Center, Niels Steensens Vej 2, 2820 Gentofte, Denmark

1 INTRODUCTION

Lipids are a vital class of essential metabolites which have many key biological functions. They are structural components of cell membranes, energy storage sources, and intermediates in signaling pathways. Lipids are both functionally and structurally a very diverse group of compounds and the number of molecular lipid species found in biological systems has been estimated to be at the order of hundreds of thousands. Overall, lipids have an extremely large number of species and different kind of structures, and over 30,000 lipid structures are already in the LIPID MAPS database *(1)*. Quehenberger *et al.* have defined the human plasma lipidome, containing quantitative levels of over 500 different lipid molecular species from the six major lipid categories *(2)*. The most widely used classification divides lipids into eight main categories, namely fatty acyls (FAs), glycerolipids (GLs), glycerophospholipids (GPs), sphingolipids (SLs), sterol lipids, prenol lipids, saccharolipids, and polyketides. Each category contains distinct classes and subclasses of lipids as presented in Table 1.

The simplest class of lipids are FAs. This group includes various types of fatty acids, eicosanoids, fatty alcohols, fatty aldehydes, fatty esters, fatty amides, fatty nitriles, fatty ethers, and hydrocarbons. Many FAs, especially the eicosanoids derived from *n*-6 and *n*-3 polyunsaturated fatty acids, have

Comprehensive Analytical Chemistry, Vol. 64. http://dx.doi.org/10.1016/B978-0-444-62650-9.00014-2

TABLE 1 Classification of Lipids

Category	Abbreviation	Subcategory	Example
Fatty acyls	FA	Fatty acids and conjugates Octadecanoids Eicosanoids Docosanoids Fatty alcohols Fatty aldehydes Fatty esters	 **Fatty acid: palmitic acid**
Glycerolipids	GL	Monoacylglycerols Diacylglycerols Triacylglycerols	 **Triglyceride: TG(18:0/18:1/18:3)**
Glycerophospholipids	GP	Phosphatidic acids Phosphatidylcholines Phosphatidylserines Phosphatidylglycerols Phosphatidylethanolamines Phosphatidylinositols Phosphatidylinositides Cardiolipins	 **1-palmitoyl-2-oleoylphosphatidylcholine**
Sphingolipids	SP	Sphingoid bases Ceramides Phosphosphingolipids Phosphonosphingolipids Neutral glycosphingolipids Acidic glycosphingolipids	 **C16 ceramide**

Sterol lipids	ST	Sterols Steroids Secosteroids Bile acids and derivatives	 cholesterol
Prenol lipids	PR	Isoprenoids Quinones and hydroquinones Polyphenols	 2E,6E-farnesol
Saccharolipids	SL	Acylaminosugars Acylaminosugar glycans Acyltrehaloses Acyltrehalose glycans	 2,3-Di-O-acyl-trehalose
Polyketides	PK	Macrolide polyketides Aromatic polyketides Nonribosomal Peptide/polyketide hybrids	 alfatoxin B1

distinct biological activities. FAs are also the main lipid building blocks of more complex lipids, such as GLs including monoacylglycerides (MGs), diacylglycerides (DGs), and triacylglycerides (TGs). These neutral lipids have a glycerol backbone with fatty acid chains attached to the glycerol group. GPs, also referred to as phospholipids (PLs), are key components of the lipid bilayer of cells and are also involved in metabolism and signaling. The most common GPs found in biological membranes are phosphatidylcholines (PCs), phosphatidylethanolamines (PEs), and phosphatidylserines (PSs). SLs are a complex family of lipids which all have a sphingoid base backbone. The sterol lipids contain a common steroid nucleus of a fused four-ring structure with a hydrocarbon side chain and an alcohol group. The main sterol lipid in animal fat is cholesterol which also has an important role in the lipid membrane. The prenol lipid category consists of subclasses like isoprenoids, polyprenols, quinones, and hydroquinones. All prenols are synthesized for the five-carbon precursors: isopentenyl diphosphate and dimethylallyl diphosphate. The saccharolipids are group of compounds in which fatty acids are linked directly to a sugar backbone and they have structures that are compatible with membrane bilayers. The polyketides are synthesized by polymerization of acetyl and propionyl subunits by classic enzymes or iterative and multimodular enzymes.

Lipid analyses include both global profiling of the main lipid classes and targeted methods for specific lipid classes which are usually not covered by the global methods due to their low concentrations (e.g., eicosanoids, steroids), instability, or other physicochemical features. Different analytical methods are typically needed for these two approaches.

Advanced analytical techniques, particularly mass spectrometry (MS), often combined with liquid chromatography (LC) or gas chromatography (GC), are requisite for lipid analysis and they have played the crucial role in the emergence as well as the progresses of lipidomics. MS is the principal choice for the lipid analysis, particularly using electrospray ionization (ESI) and sometimes also atmospheric pressure chemical ionization or laser-based MS methods for surface analysis. The MS-based techniques are the best choice for lipidomics due to their superior sensitivity and molecular specificity, and because they provide the ability to resolve the extensive compositional and structural diversity of lipids in biological systems.

For global nontargeted screening, the high-resolution accurate MS systems capable of tandem mass measurements are typically applied. Particularly, Orbitrap MS and (quadrupole) time-of-flight MS ((Q)-TOF MS) systems are extensively used in lipidomics. Recently, ion mobility has been introduced into the commercial Q-TOF MS systems (3). Ion mobility separation as an integral part of the detection system offers an additional separation dimension, which greatly increases detectable and identifiable components in complex biological samples as well as enables the separation of isobaric molecules without increasing the analysis time. Triple quadrupole MS instruments

(QqQ MS) are, on the other hand, a good option for sensitive, quantitative targeted analysis using multiple reaction monitoring (MRM), and they can also be used for class-specific detection through precursor ion and neutral loss scanning *(4–6)*.

2   SAMPLE PREPARATION

For most MS-based analyses, sample preparation is required prior the analysis. Typically, extraction of lipids from the sample matrix is required.

The choice of the extraction solvent is dependent on two functional features of lipids that control their solubility in organic solvents: the hydrophobic hydrocarbon chains of the fatty acid or other aliphatic moieties and any polar functional groups, such as phosphate or sugar residues, which are markedly hydrophilic. Triacylglycerols (TGs) or cholesterol esters are neutral lipids that do not have any polar groups and they are readily soluble in nonpolar solvents (e.g., hexane, toluene, or cyclohexane) and also in moderately polar solvents (e.g., diethyl ether or chloroform), while they have relatively low solubility in polar solvents such as methanol. More polar lipids, such as PLs and glycosphingolipids, are only slightly soluble in hydrocarbon-based solvents, but they have good solubility in more polar solvents like methanol, ethanol, or chloroform. Also the pH does play an important role in the extraction of specific lipid subclasses, either due to degradation at specific pH conditions or due to their acid–base characteristics. For example, acidic lipids, such as PSs and phosphatidic acids, are generally more efficiently extracted at acidic conditions. In addition, careful optimization of the extraction conditions is necessary for the more polar lipids such as gangliosides or polyphosphoinositides.

Several different solvents have been used for lipid extraction such as alcohols (methanol, ethanol, isopropanol, *n*-butanol), acetone, acetonitrile, ethers (diethyl ether, isopropyl ether, dioxane, tetrahydrofuran), halocarbons (chloroform, dichloromethane), hydrocarbons (hexane, benzene, cyclohexane, isooctane), and their mixtures. Best results have been obtained by using mixtures of solvents, the most common mixture being chloroform and methanol in the ratio of 2:1 (v/v), as suggested by Folch *(7)* and which has been modified slightly by Bligh and Dyer *(8)*. While the Folch extraction is widely used, the use of chloroform makes its use in the modern robotic sample preparation challenging. Thus, use of methyl-*tert*-butyl ether (MTBE)/methanol/water has been suggested, and it has been shown to give the same or slightly better recoveries as the established Bligh and Dyer method *(9)*. It should be noted that the ratio of the extraction solvent to the amount of sample is also important. The ratio should be 25% w/v tissue-to-solvent to allow satisfactory recovery of the lipids. However, it is not possible to obtain quantitative recovery of all lipid subclasses in one single extraction method. The differences in the recoveries can be at least partially corrected by the use of class-specific standards. Of course, the class-specific standards do not account the different extraction behavior within the lipid subclass due to differences in the fatty acid chain lengths and the degree of saturation; however, these differences are typically relatively minor.

Other extraction methods used in the lipid extraction include supercritical fluid extraction (SFE) and pressurized liquid extraction (PLE). With SFE, good extraction yields have been obtained for nonpolar lipids including ester-ified fatty acids, acylglycerols, and unsaponifiable matter. However, complex polar lipids are only sparingly soluble in supercritical carbon dioxide alone and polar modifiers, such as methanol, ethanol, or even water is required to improve the extraction of polar lipids (10). SFE has been used for the extrac-tion of lipids especially from various food matrices, such as different nuts, edible oils, and seeds (11). The recoveries of lipids in SFE were on the same levels than with conventional solvent extraction methods (12,13); no signifi-cant differences between the fatty acids extracted were observed. PLE has also been used in lipid extraction, although only in very few applications (14). The elevated temperatures used in PLE can cause alteration of the lipid composition.

3   LIPID ANALYSIS

Global lipid profiling methods typically cover molecular lipid species from the major classes of lipids such as cholesteryl esters, ceramides, mono-(MG), di- (DG), and TGs, and membrane PLs, for example, sphingomyelins (SMs), PCs, phosphatidylethanolamines (PEs), PSs, and lysophospholipids. In targeted lipid analysis, specific lipid classes that are poorly covered by the global profiling methods are usually analyzed. These lipids include ster-oids, sterols, bile acids, fatty acids, signaling lipids such as eicosanoids, and ceramides, as well as polar lipids and inositol lipids.

In lipidomics, MS can be used either by direct infusion, that is, by the so-called shotgun MS, or in combination with chromatographic separation, typically LC and sometimes also with GC. Both approaches have their own advantages and limitations. Most targeted lipid analyses are performed with liquid chromatography coupled to mass spectrometry (LC-MS), while the use of gas chromatography–mass spectrometry (GC-MS) is utilized only for the analysis of fatty acids and some steroids. In addition, surface analysis by MS has been applied in lipid analysis of intact tissues.

In lipid analysis, the most common ionization technique is ESI. ESI can be applied in both positive and negative ion mode. While the positive ion mode is more universal, the negative ion mode is better to some polar lipids, such as glycolipids and cardiolipins as shown in Table 2.

The global lipidomics analyses by MS can be further divided to "untargeted," "focused," and "targeted" lipidomics. The untargeted lipidomics method aims to detect all lipids (or a certain type of lipid) contained in an

TABLE 2 Major Lipid Classes Detected with ESI-MS *(4)*

Lipid Class	Positive Mode	Negative Mode
Lysophospholipids	+	+
Monoacylglycerols	+	−
Phospholipids	+	+
Sphingomyelins	+	+
Ceramides	+	+
Diacylglycerols	+	−
Cholesterol esters	+	−
Triacylglycerols	+	−

extracted lipid sample without preliminary information of the molecular ions or their fragmentations. The strategy in the untargeted lipidomics is to subject all detected peaks to further data processing. High-resolution mass spectrometer like FTMS and Orbitrap provides exceptional resolution and accuracy for use in untargeted lipidomics. In addition, QIT and Q-TOF can be utilized for this type of global lipid profiling. The focused lipidomics strategy is used for detecting lipids in some groups while utilizing specific fragments (product and precursor ion scan) or neutral loss (neutral loss scan) caused by a specific feature of the partial structures of the molecules. The methods applying focused lipidomics strategy are used for comprehensive analysis of lipids with structural similarities. The focused lipidomics is typically done with an instrument utilizing a quadrupole as the first mass analyzer, a collision cell for the fragmentation, and an ion trap (IT) or TOF as the second mass analyzer. Table 3 lists common fragment ions or neutral losses used in focused lipidomics. The targeted lipidomics is applied to predetermined molecules by using information of parent ions and their specific fragment ions, by using selected reaction monitoring (SRM) or MRM modes.

3.1 Shotgun Lipidomics

In shotgun lipidomics, the sample extract is directly infused to MS. The current trend is to use the highly multidimensional MS approach, including multiplex lipid extractions, different tandem MS scan modes, and various fragments and chemical modifications *(4,15)*. Detailed tandem mass spectrometric mapping with multiple fragmentation strategies is possible during the direct infusion. The major advantage of shotgun lipidomics is that mass spectra of an individual lipid can be measured at constant concentration during the direct infusion of the extract. Selected examples of lipidomics by shotgun analyses are presented in Table 4.

TABLE 3 Common Fragment Ions or Neutral Losses of Different Lipid Groups in ESI-MS/MS

Lipid Class	Major Precursor Ion	MS/MS Mode	Product Ion or Neutral Loss (m/z)	ESI Polarity
Phosphatidylcholine (PC)	$[M+H]^+$	PIS	184	+
Phosphatidylethanolamine (PE)	$[M+H]^+$	NLS	141	+
	$[M+H]^-$	PIS	196	−
Phosphatidic acid (PA)	$[M+H]^-$	PIS	153	−
	$[M+NH_4]^+$	NLS	115	+
Phosphatidylglycerols (PGs)	$[M+H]^-$	PIS	153	−
	$[M+NH_4]^+$	NLS	189	+
Phosphatidylinositols (PIs)	$[M+H]^-$	PIS	241	−
	$[M+NH_4]^+$	NLS	277	+
Phosphatidylserines (PSs)	$[M+H]^-$	NLS	87	−
	$[M+H]^+$	NLS	185	+
Phosphatidylinositol monophosphates (PIPs)	$[M+H]^-$	PIS	321	−
Phosphatidylinositol bisphosphate (PIP_2)	$[M+H]^-$	PIS	401	−
Phosphatidylinositol trisphosphates (PIPs)	$[M+H]^-$	PIS	481	−
Diacylglycerols (DGs)	$[M+NH_4]^+$	NLS	FA + NH3	+
Triacylglycerols (TGs)	$[M+NH_4]^+$	NLS	FA + NH3	+
Cholesteryl esters (CEs)	$[M+NH_4]^+$	NLS	FA + NH3	+
	$[M+NH_4]^+$	PIS	369	+
Sphingomyelin (SM)	$[M+H]^+$	PIS	184	+
Ceramides with d18:1 long-chain base	$[M+H]^+$	PIS	264	+
Sulfatides	$[M+H]^-$	PIS	97	−
Monogalactosyl diacylglycerols	$[M+NH_4]^+$	NLS	179	+
Digalactosyl diacylglycerols	$[M+NH_4]^+$	NLS	341	+

A neutral loss (FA+NH_3) means that mass of neutral loss is sum of fatty acid chain of the lipid and ammonia (15,16).

TABLE 4 Selected Examples of Shot Gun Lipidomics

Mass Spectrometer	Ionization Mode	MS Scans	Details	Reference
QqQ	ESI negative and positive ion mode	Full-scan MS, PIS, NLS	NanoMate	*(17)*
QqQ	ESI negative and positive ion mode	Full-scan MS, PIS, NLS		*(18)*
Q-TOF, LTQ-Orbitrap	ESI negative and positive ion mode	Full-scan MS, PIS, MRM	NanoMate	*(19)*
QqQ	ESI negative and positive ion mode	PIS, NLS	NanoMate	*(20)*
QqQ	ESI negative and positive ion mode	Full-scan MS, PIS, NLS, product ion scan	–	*(16)*
Q-TOF, QqQ	ESI positive ion mode	DDA	NanoMate	*(21)*
Q-TOF	ESI negative ion mode	Product ion scan	–	*(22)*
QqQ/LIT	ESI negative and positive ion mode	PIS, NLS	–	*(23)*
LTQ-Orbitrap, Q-TOF	ESI negative and positive ion mode	DDA	NanoMate	*(24)*
QqQ, Q-TOF, IT	ESI negative and positive ion mode	PIS, NLS	–	*(25)*
LTQ-Orbitrap	ESI positive ion mode	Full-scan MS	NanoMate	*(26)*
LTQ-Orbitrap	ESI negative and positive ion mode	DDA, PIS	NanoMate, HCD for fragmentation	*(27)*
LTQ-Orbitrap, QqQ	ESI negative ion mode	Full-scan MS, product ion scan	NanoMate	*(28)*
IM-TOF	MALDI positive ion mode	Full-scan MS	Ion mobility separation	*(29)*

In shotgun lipidomics, two different approaches are used. Top-down shot-gun lipidomics strategy is good for high-throughput screening, and it is based on full scan and accurately determined masses of lipids with modern high-resolution mass spectrometer. Top-down lipidomics recognizes lipids from major classes, and peak in spectrum is a sum of lipids with exact same mass-to-charge ratio. With top-down strategy, individual lipid species cannot necessarily be identified. Bottom-up lipidomics strategy is complementary and aims to accurate quantification of individual molecular species and relies on detection of characteristic fragment ions by MS/MS.

ESI is widely used in shotgun lipidomics since the source can be used for lipid separation, which simplifies the spectra and interpretation. Lipids can be classified into different groups according to their electrical properties, and those properties regulate their behavior in ESI-MS analyses. The first cate-gory is anionic lipids that contain one or more net negative charge at neutral pH. The second category is weak anionic acids that carry a net negative charge only at alkaline pH. This group tends to form negative ions in negative ion mode under neutral pH, but usually their ionization efficiency is poor at least when compared to the first category. The third category contains polar lipids which are electrically neutral in alkaline and neutral pH conditions. The fourth category contains some specialized lipids which have varying elec-trical properties. To avoid suppression, total lipid amount should not exceed 50 µM, while the concentration of major lipid species should be maintained below 1 µM.

However, even with the recent advances in MS instrumentation, the shot-gun lipidomics suffers from a limited in-spectrum dynamic range and ion sup-pression. Also, it is not possible to differentiate the various lipid isomers with MS alone. The main limitation is, however, the reliable determination of those lipids that are present at low concentrations. In quantitative analysis, it is not possible to include isotope-labeled standards for each lipid, and thus, class-specific standards are used in the lipid analysis with shotgun lipidomics. For polar lipids, class-specific standards have been shown to be sufficient in quan-titation because ionization of polar lipids is largely dependent on the class-specific head group (16). However, there are also contradictory results about the influence of FA chain length and unsaturation on the ionization efficiency, showing that these factors do have an effect on the quantitative results (30,31). It should also be noted that accurate quantification of lipid species by shotgun lipidomics can be potentially hindered by interferences arising from isotopes, isobars, and isomers of the lipids. In addition, it is important to keep the total lipid concentration below the concentration that favors aggre-gate formation, that is, to have sufficiently dilute samples. As this may be challenging if the lipid extract contains nonpolar lipids such as triglycerides in high concentrations, lipid fractionation before the analysis is often required to avoid decrease in the sensitivity of the analysis. Lipid molecules with short and/or polyunsaturated acyl chains may show higher apparent response factors than those in the same class containing long and/or saturated acyl chains at a concentration that lipid aggregates form *(32)*. Thus, lipid aggregation may substantially affect ionization efficiency in a species-dependent fashion. Another limitation of shotgun lipidomics are isobaric overlaps of the M+2 isotope with the monoisotopic peak of the compound with one double bond less.

3.2   LC-MS

The novel approaches in LC-MS-based lipidomics, such as ultra-high performance LC (UHPLC), combined with high-resolution MS methodologies allow fast and reliable analysis of various classes of lipids *(33)*. The modern instruments, with fast duty cycles, allow acquisition of high and low collision energy data simultaneously to provide both data on precursor ions and fragmentation data for all detectable molecular ions *(34)*.

In lipid analyses with LC-MS, mass scanning methods have included MS full-scan acquisition with accurate mass, acquiring precursor and fragment exact mass data simultaneously with MS^e scan, precursor ion scan and MRM. The most popular fragmentation method in MS/MS experiments has been collision-induced dissociation, but also higher energy collision dissociation has been utilized.

In nontargeted screening, quantitative data can be obtained in a similar manner as in the shotgun approach, that is, by using sufficient number of class-specific internal standards in combination with external calibration *(17,35)*. One of the main limitations with the LC-MS-based methods is the same as in shotgun MS, that is, the ion suppression. However, the ion suppression has a substantially smaller effect in LC-MS than in the shotgun approach because the chromatographic separation reduces sample complexity, thereby alleviating matrix interferences in the ionization process. However, the composition of the mobile phase does have an effect on the ionization efficiency, and thus, the matrix effects may vary during the analysis. This type of systematic variation can be minimized by using a proper set of internal standards. Memory effects, as well as loss of specific lipids, may also cause problems in the LC-MS analysis if the solvent composition is not optimally selected, that is, the solvent strength is not sufficient to elute all the lipids from the column.

Selected examples of analytical methods used for the determination of global profiling of lipids are listed in Table 5. Extraction is usually based on simple liquid extraction, using modified Folch or Blight and Dyer extraction *(4,5)*. For more acidic lipids, such as PSs and phosphatidic acids, adjustment of the pH in the aqueous phase is required. The analysis is most typically performed with LC-MS in RPLC mode, with the UHPLC methods gradually replacing the conventional HPLC methods. HRMS systems, such

TABLE 5 Typical Applications on Global Profiling of Lipids by LC-MS

Sample Type	Sample Preparation	Analysis	Detection	LOD/LOQ	Analysis Time (min)	Reference
Serum	Modified Blight and Dyer extraction	Micro HPLC (C_{18})	MS; ESI hybrid quadrupole time of flight	–/–	58	(36)
Rat liver mitochondria	Modified Blight and Dyer extraction, sonication	HPLC (C_{18})	MS; ESI Orbitrap	–/–	30	(37)
Plasma	Modified Blight and Dyer extraction	UHPLC (C_{18})	MS; ESI hybrid quadrupole time of flight	–/–	15	(38)
Rat plasma	Extraction/protein precipitation with acetonitrile	UHPLC (C_8)	MS; ESI hybrid quadrupole time of flight	–/–	12	(17)
Plasma, cells, tissues	Modified Folch extraction, sonication, homogenization	UPHLC (C_{18})	MS; ESI hybrid quadrupole time of flight	–/–	18	(39)
Lipid droplets	Extraction to methanol and MTBE, fractionation with HILIC-HPLC	HPLC (C_{18})	MS; ESI hybrid linear ion trap Fourier transform ion cyclotron resonance	–/–	35	(40)
Rat peritoneal surface layer	Modified Folch extraction, sonication, homogenization	2D HPLC (1D; silica, 2D; C_8)	MS; ESI hybrid quadrupole time of flight	–/–	150	(41)
Egg, soya, and porcine	Folch extraction, filtration	2D HPLC (1D; HILIC/silica, 2D; C_{18})	MS; ESI ion trap	–/–	260	(42)
Plasma	Extraction with chloroform/methanol	HPLC (C_{18})	MS; ESI hybrid quadrupole time of flight	–/–	95	(43)
Cells, rat liver	Modified Blight and Dyer extraction, fractionation with DEAE cellulose	HPLC (C_{18})	MS; ESI hybrid linear ion trap Orbitrap	2 fmol/µl (PA and PS)	45	(44)

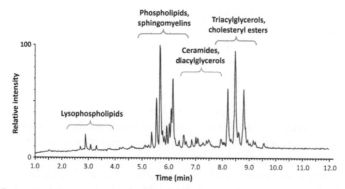

FIGURE 1 Total ion chromatogram from UHPLC–TOFMS lipidomics analysis of serum showing chromatographic separation of major lipid classes using ESI+ mode (for more details, see ref. 23). (For color version of this figure, the reader is referred to the online version of this chapter.)

as TOFMS, QTOFMS, and Orbitrap, are needed in the global profiling. UHPLC–HRMS analysis typically allows detection of several hundreds of lipids in a single run, analysis time varying from 10 to 20 min. In lipidomics analyses, the solvent strength is an important parameter and the organic solvent should have relatively strong solvent strength in order to obtain good peak shapes for late-eluting TGs and also to avoid any significant carry-over due to complex biological samples. Often, acetonitrile/isopropanol mixtures are used as the organic solvent, and also elevated temperatures can be utilized for the enhancement of the elution of the late-eluting lipids.

An example of UPLC–QTOFMS analysis of lipids in plasma in positive ionization mode is shown in Figure 1. In the UHPLC separation, PLs are separated based on their fatty acid composition and degree of desaturation. Thus, lysophospholipids elute early, with lysoPCs containing one fatty acid attached to the glycerol backbone eluting before diacylphospholipids which have two fatty acids. PLs and SMs elute in the same region while ceramides which do not contain PC head group as their sphingomyelin counterparts elute after PCs and SMs. Also, DGs elute close to PCs and SMs. Cholesterol esters and TGs elute last.

3.3   Surface Analysis of Lipids

Several new methodologies have been developed in recent years for the surface analysis of lipids. These methods allow direct detection of lipids from surfaces such as tissue sections and intact cells without prior extraction and thus enable the determination of spatial distribution of the lipids. There are three basic requirements for the surface analysis. The analytes of interest must be desorbed from the surface by the interaction of a sampling probe such as spray, laser, or plasma. The desorbed analytes must first be ionized and then analyzed in gas phase in the MS analyzer. Most surface analysis methods have a soft ionization, producing intact ions as dominant ionic products, thus making the interpretation and identification more straightforward. Traditional surface analysis methods by MS include matrix-assisted laser desorption/ ionization (MALDI) *(18)* and secondary ion mass spectrometry (SIMS) *(19)*. More recently, desorption electrospray ionization *(20)* and direct analysis in real time *(21)* have been developed. These methods, unlike MALDI and SIMS, allow direct surface analysis under ambient conditions, thus eliminating challenges related to analyzing the samples under vacuum. The different surface analysis techniques have been recently reviewed in detail *(22)*.

An example of surface analysis is shown in Figure 2 *(23)*. In this study, MALDI mass spectrometry imaging (MALDI MSI) using ultra-high-resolution Fourier transform ion cyclotron resonance mass spectrometry (FTMS) was utilized to determine the spatial distribution of GPs in porcine lens. The spatial and concentration-dependent distributions of 20 glycerophosphocholines, 4 glycerophosphates, and 6 glycerophosphoethanolamines were determined, and based on the planar molecular images of the lipids, the organization of fiber cell membranes within the ocular lens could be characterized.

4   IDENTIFICATION

The identification of the lipids is a very challenging task. The lack of comprehensive mass spectral libraries often limits the identification of compounds in LC-MS and shotgun methods. Some spectral libraries are available, such as the Human Metabolome Database (http:/www.hmdb.ca), the METLIN Metabolite Database (http:/metlin.scripps.edu) *(24)*, and the MassBank (http:/www.massbank.jp) *(25)*. However, construction of universal spectral databases for API-MS is challenging due to the poor reproducibility and high interinstrument variability of fragmentation patterns.

In the interpretation of the mass spectra from both shotgun MS and LC-MS studies, the first step is to ensure that the signal of interest really corresponds to a monoisotopic ion and not to a natural isotopologue ion or an adduct ion. Elemental compositions can be deduced from accurate mass measurements and used for further database queries. To aid the identification, a number of chemical databases containing huge amount of chemical structures to aid in structural elucidation after fragmentation of the metabolite are available (e.g., ChemSpider, http:/www.chemspider.com, http:/msbi.ipb-halle.de/ MetFrag) *(26)*. Also, several *in silico* software packages for the prediction of *in silico* spectra for compounds are available *(27,28)*.

In LC-MS analysis, identification of lipids requires the use of accurate mass measurements for the determination of elemental composition, combined with MS^n experiments to get information of the molecular fragments. In addition, in LC-MS, one metabolite may produce multiple peaks due to the presence of isotopes, adducts, and neutral loss fragments. Therefore, ion annotation is crucial in order to recognize a group of ions likely to originate from the same

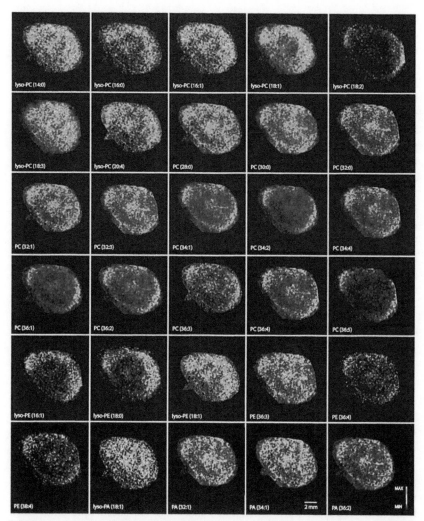

FIGURE 2   Spatial distribution of PCs, PEs, and Pas in porcine lens visualized in false colors and measured with MALDI-MS (from ref. 23). (For color version of this figure, the reader is referred to the online version of this chapter.)

compound. The isotopic abundance pattern can serve as a powerful additional constraint for removing wrong elemental composition candidates. The actual ratios of the stable isotopes differ slightly for each element within narrow range. Therefore, each monoisotopic molecular ion is always accompanied by additional isotope ions, and the abundance of the isotope ions is dependent on the actual elemental composition. Thus, this can be used as a powerful filter in calculating unique elemental compositions from mass spectral data.

In the identification of lipids, MS^n is usually needed, using both ESI+ and ESI−. In ESI+, only limited information can be obtained, mainly from the polar head group of the lipid while the ESI− mode gives more detailed identification of the lipid structure and of the fatty acid side chains *(29,45,46)*. For example, in negative-ion mode PCs are detected mainly as formate adducts, [M+COO]− which will fragment to yield abundant [M−15]− ion. Fragmentation of [M−15]− ion (MS^3) reveals information on fatty acid composition of the PC lipids. The most abundant negative ions in MS^3 spectra correspond to carboxylate anion from *sn*-2 and *sn*-1 position. Additionally, MS^3 spectra typically show a complementary ion which comprise of FA (from *sn*-1 position) and glycerophosphocholine head group, [M−15-R2COOH]−. MS^4 analysis of [M−15-R2COOH]− ions reveals critical information regarding the two etherlipid subclasses, plasmalogens and *O*-alkyl ethers. While the plasmalogen PC species, for example, produce a class-specific ion at *m/z* 224, the *O*-alkyl ethers yield instead a phosphatidic acid anion, [M−15-R2COOH-89]− (due to neutral loss of dimethylamino ethylene). Therefore, all three subclasses of important membrane lipids, plasmalogens and *O*-alkyl ethers as well as diacyl PCs, can readily be distinguished using ESI− and MS^n when the components are chromatographically separated. In the similar manner, also the PE ether lipid subclasses can be characterized based on unique fragments in MS^3 spectra *(47)*.

5   DATA PROCESSING FOR MS-BASED LIPIDOMICS

Data handling in lipidomics is an important stage since MS and LC-MS datasets are complex and data processing has a significant effect on the quality of a dataset. The data processing contains raw data processing with signal processing methods and combining data between measurements. The data processing steps can include conversion of raw data files, spectral filtering, or smoothing, peak detection, peak deconvolution, isotope correction, alignment, identification, and normalization.

Several software tools have been applied for lipidomics. Popular software packages have been developed including LipidXplorer *(48)* for handling mainly shotgun lipidomics data as well as MZmine *(36)*, XCMS *(37)*, and Lipid Data Analyzer *(38)* which can handle LC-MS-based lipidomics data. Several other tools and algorithms have been developed specifically for handling lipidomics data *(17,39,40)*. Several MS vendors also offer specialized software packages for this purpose including SIEVE *(41)* from Thermo Scientific, Analyst *(42)* from AB Sciex, and MassLynx *(43)* from Waters as well as open source software like LIMSA and MzMine2 *(38)*. Also built-in programs have been used for data processing. Software like Marker-View *(44)*, MarkerLynx *(49)*, and SIMCA-P *(50)* have been applied to multivariate analysis of lipidomics datasets.

6   CONCLUSIONS

The MS techniques have been developing fast during the last decade, and the novel MS instruments allow highly sensitive and accurate measurements of lipids. The current developments in the HRMS instruments have already made the quantitative analyses possible with relatively large dynamic range. However, several challenges still exist in lipidomics.

Analysis of lipids in large-scale studies, particularly in nontargeted screening, is challenging, even if only semi-quantitative analysis is applied. Careful optimization of the analytical workflow and strict quality control are essential, and so is the standardization of the sample collection. The methodological variation can be assessed by analyzing suitable set of standards, blank samples, as well as technical replicates of pooled samples. In practice, both shotgun MS and UHPLC-MS can be successfully used for large-scale analysis, taking into account the limitations of each methodology. In shotgun approach, it is challenging to determine isomeric compounds or lipids that are present in low concentration. In addition, the identification of unknown lipids is difficult with shotgun methods. In UHPLC-MS analysis, the methodology has to be optimized carefully in order to avoid memory and matrix effects, library of spectra needs to be constructed linked to specific lipid peaks, and follow-up tandem MS analyses may be needed in order to confirm the identities of specific lipid peaks. Regardless of the analytical approach chosen, the proper setup and optimization of the data processing workflow also play a crucial role.

The exact structural and spatial characterization of the lipids is still very limited, particularly in routine determination of lipids. As an example, routine determination of sn1/sn2 acyl positions in PLs is not possible, nor is the determination of positions of double bonds or acyl chain branching. Such information would be important to understand the data in the biochemical context (e.g., affinity for specific lipid enzymes). While promising efforts are under way, it may still take time before the emerging tools for lipid structural elucidation are introduced in routine lipidomic analyses. Big advances have also been made over the past years in the developments of methods for surface analysis of lipids. These approaches will be crucial for in-depth elucidation of the spatial complexity of cellular and subcellular lipidomes.

REFERENCES

1. Sud, M.; Fahy, E.; Cotter, D.; Dennis, E. A.; Subramaniam, S. *J. Chem. Educ.* **2012**, *89*, 291–292.
2. Quehenberger, O.; Armando, A. M.; Brown, A. H.; Milne, S. B.; Myers, D. S.; Merrill, A. H.; Bandyopadhyay, S.; Jones, K. N.; Kelly, S.; Shaner, R. L.; Sullards, C. M.; Wang, E.; Murphy, R. C.; Barkley, R. M.; Leiker, T. J.; Raetz, C. R. H.; Guan, Z.; Laird, G. M.; Six, D. A.; Russell, D. W.; McDonald, J. G.; Subramaniam, S.; Fahy, E.; Dennis, E. A. *J. Lipid Res.* **2010**, *51*, 3299–3305.
3. Zhong, Y.; Hyung, S. J.; Ruotolo, B. T. *Analyst* **2011**, *136*, 3534–3541.
4. Han, X.; Yang, K.; Gross, R. W. *Mass Spectrom. Rev.* **2012**, *31*, 134–178.
5. Xiao, J. F.; Zhou, B.; Ressom, H. W. *Trends Anal. Chem.* **2012**, *32*, 1–14.
6. Balgoma, D.; Checa, A.; Sar, D. G.; Snowden, S.; Wheelock, C. E. *Mol. Nutr. Food Res.* **2013**, *57*, 1359–1377.

7. Folch, J.; Lees, M.; Sloane Stanley, G. H. *J. Biol. Chem.* **1957**, *226*, 497–509.

8. Bligh, E. G.; Dyer, W. J. *Can. J. Biochem. Physiol.* **1959**, *37*, 911–917.

9. Matyash, V.; Liebisch, G.; Kurzchalia, T. V.; Shevchenko, A.; Schwudke, D. *J. Lipid Res.* **2008**, *49*, 1137–1146.

10. Cocero, M. J.; Calvo, L. *J. Am. Oil Chem. Soc.* **1996**, *73*, 1573–1578.

11. Brunner, G. *J. Food Eng.* **2005**, *67*, 21–33.

12. Taylor, S. L.; King, J. W.; List, G. R. *J. Am. Oil Chem. Soc.* **1993**, *70*, 437–439.

13. Hopia, A. I.; Ollilainen, V.-M. *J. Liq. Chromatogr.* **1993**, *16*, 2469–2482.

14. Pieber, S.; Schober, S.; Mittelbach, M. *Biomass Bioenergy* **2012**, *47*, 474–482.

15. Schwudke, D.; Liebisch, G.; Herzog, R.; Schmitz, G.; Shevchenko, A. *Methods Enzymol.* **2007**, *433*, 175–191.

16. Han, X.; Gross, R. W. *Proc. Natl. Acad. Sci. U.S.A.* **1994**, *91*, 10635–10639.

17. Myers, D. S.; Ivanova, P. T.; Milne, S. B.; Brown, H. A. *Biochim. Biophys. Acta* **2011**, *1811*, 748–757.

18. Karas, M.; Hillenkamp, F. *Anal. Chem.* **1988**, *60*, 2299–2301.

19. Werner, H. W. *Surf. Sci.* **1975**, *47*, 301–323.

20. Zoltan, T.; Justin, M. W.; Bodgan, G.; Cooks, R. G. *Science* **2004**, *306*, 471–473.

21. Cody, R. B.; Laramee, J. A.; Durst, H. D. *Anal. Chem.* **2005**, *77*, 2297–2302.

22. Ellis, S. R.; Brown, S. H.; Het Panhuis, M.; Blanksby, S. J.; Mitchell, T. W. *Prog. Lipid Res.* **2013**, *52*, 329–353.

23. Pól, J.; Vidová, V.; Hyötyläinen, T.; Volný, M.; Novák, P.; Strohalm, M.; Kostiainen, R.; Havlíček, V.; Wiedmer, S. K.; Holopainen, J. M. *PLoS One* **2011**, *6*, e19441.

24. Smith, C. A.; O´Maille, G.; Want, E. J.; Qin, C.; Trauger, S. A.; Brandon, T. R.; Custodio, D. E.; Abaqyan, R.; Siuzdak, G. *Ther. Drug Monit.* **2005**, *27*, 747–751.

25. Horai, H.; Arita, M.; Kanay, S.; Nihei, Y.; Ikeda, T.; Suwa, K.; Ohima, Y.; Tanaka, K.; Tanaka, S.; Aoshima, K.; Osa, Y.; Kakazu, Y.; Kusano, M.; Tohge, T.; Matsuda, F.; Sawada, Y.; Yokota Hirai, M.; Nakanishi, H.; Ikdea, K.; Akimoto, N.; Maoka, T.; Takahashi, T.; Ara, T.; Sakurai, N.; Suzuki, H.; Shibata, D.; Neumann, S.; Ida, T.; Tanaka, K.; Funatsu, K.; Matsuura, F.; Soga, T.; Taguchi, R.; Saito, K.; Nishioka, T. *J. Mass Spectrom.* **2010**, *45*, 703–714.

26. Wolf, S.; Schmidt, S.; Muller-Hannemann, M.; Neumann, S. *BMC Bioinf.* **2010**, *11*, 148.

27. Neumann, S.; Böcker, S. *Anal. Bioanal. Chem.* **2010**, *398*, 2779–2788.

28. Schymanski, E. L.; Meringer, M.; Brack, W. *Anal. Chem.* **2009**, *81*, 3608–3617.

29. Khaselev, N.; Murphy, R. C. *J. Am. Soc. Mass Spectrom.* **2000**, *11*, 283–291.

30. Koivusalo, M.; Haimi, P.; Heikinheimo, L.; Kostiainen, R.; Somerharju, P. *J. Lipid Res.* **2001**, *42*, 663–672.

31. Brugger, B.; Erben, G.; Sandhoff, R.; Wieland, F. T.; Lehmann, W. D. *Proc. Natl. Acad. Sci. U.S.A.* **1997**, *94*, 2339–2344.

32. Yang, K.; Han, X. *Metabolites* **2011**, *1*, 21–40.

33. Sandra, K.; Sandra, P. *Curr. Opin. Chem. Biol.* **2013**, *7*, 847–853.

34. Wrona, M.; Mauriala, T.; Bateman, K. P.; Mortishire-Smith, R. J.; O'Connor, D.; Schuhmann, K.; Herzog, R.; Schwudke, D.; Metelmann-Strupat, W.; Bornstein, S. R.; Shevchenko, A. *Anal. Chem.* **2011**, *83*, 5480–5487.

35. Nygren, H.; Seppänen-Laakso, T.; Castillo, S.; Hyötyläinen, T.; Oresic, M. *Methods Mol. Biol.* **2011**, *708*, 247–257.

36. Pluskal, T.; Castillo, S.; Villar-Briones, A.; Oresic, M. *BMC Bioinformatics* **2010**, *11*, 395.

37. Nordstrom, A.; O'Maille, G.; Qin, C.; Siuzdak, G. *Anal. Chem.* **2006**, *78*, 3289–3295.

38. Hartler, J.; Trotzmuller, M.; Chitraju, C.; Spener, F.; Kofeler, H. C.; Thallinger, G. G. *Bioinformatics* **2011**, *27*, 572–577.

39. Jung, H. R.; Sylvanne, T.; Koistinen, K. M.; Tarasov, K.; Kauhanen, D.; Ekroos, K. *Biochim. Biophys. Acta* **2011**, *1811*, 925–934.

40. Song, H.; Ladenson, J.; Turk, J. *J. Chromatogr. B* **2009**, *877*, 2847–2854.

41. Bird, S. S.; Marur, V. R.; Sniatynski, M. J.; Greenberg, H. K.; Kristal, B. S. *Anal. Chem.* **2011**, *83*, 940–949.

42. Bartz, R.; Li, W.-H.; Venables, B.; Zehmer, J. K.; Roth, M. R.; Welti, R.; Anderson, R. G. W.; Liu, P.; Chapman, K. D. *J. Lipid Res.* **2007**, *48*, 837–847.

43. Esch, S. W.; Tamura, P.; Sparks, A. A.; Roth, M. R.; Devaiah, S. P.; Heinz, E.; Wang, X.; Williams, T. D.; Welti, R. *J. Lipid Res.* **2007**, *48*, 235–241.

44. Schwudge, D.; Hannich, J. T.; Surendranath, V.; Grimard, V.; Moehring, T.; Burton, L.; Kurzchalia, T.; Shevchenco, A. *Anal. Chem.* **2007**, *79*, 4083–4093.

45. Hsu, F. F.; Turk, J. *J. Am. Soc. Mass Spectrom.* **2007**, *18*, 2065–2073.

46. Nygren, H.; Pöhö, P.; Seppänen-Laakso, T.; Lahtinen, U.; j Orešic, M.; Hyötyläinen, T. *LCGC Europe* **2013**, 142–148, March issue.

47. Zemski Berry, K. A.; Murphy, R. C. *J. Am. Soc. Mass Spectrom.* **2004**, *15*, 1499–1508.

48. Herzog, R.; Schwudke, D.; Schuhmann, K.; Sampaio, J. L.; Bornstein, S. R.; Schroeder, M.; Shevchenko, A. *Genome Biol.* **2011**, *12*, R8.

49. Rainville, P. D.; Stumpf, C. L.; Shockcor, J. P.; Plumb, R. S.; Nicholson, J. K. *J. Proteome Res.* **2007**, *6*, 552–558.

50. Wang, C.; Kong, H.; Guan, Y.; Yang, J.; Gu, J.; Yang, S.; Xu, G. *Anal. Chem.* **2005**, *77*, 4108–4116.

Chapter 15

Foodomics: Food Science and Nutrition in the Postgenomic Era

Elena Ibáñez and Alejandro Cifuentes

Laboratory of Foodomics, Institute of Food Science Research (CIAL), CSIC. Nicolás Cabrera 9, Madrid, Spain

Chapter Outline

1 INTRODUCTION TO FOODOMICS

Food safety is by far the first request still demanded by consumers. Today, food safety is becoming a more and more difficult challenge for food researchers due to the so-called Globalization and the worldwide movement of food and related raw materials, which are repeatedly demonstrated to be able to generate global contamination episodes. Moreover, many products contain multiple and processed ingredients, which are often shipped from different parts of the world, and share common storage spaces and production lines. As a result, ensuring the safety, quality, and traceability of food has never become more complicated and crucial than today, making necessary the development of new, more advanced, and powerful analytical strategies.

On the other hand, research in food science and nutrition is growing parallel to the consumers concern about what is in their food, the safety of the food they eat, and the possible health's benefits or risks of the food they consume. Thus, food is now considered not only a source of energy but also an affordable way to prevent future diseases. However, in order to scientifically demonstrate and understand the healthy effect of food and food ingredients, analytical strategies have to face important difficulties derived, among others, from food complexity, the huge natural variability, the large number of different nutrients and bioactive food compounds, their very different concentrations, their bioavailability and transformation in the human tract, the numerous targets with different affinities and specificities that might exist in the human body, etc. The number of opportunities (e.g., new methodologies, new generated knowledge, new products, etc.) derived from this trend are impressive and include, for example, the possibility to account for food products tailored to promote the health and well-being of groups of population identified on the basis of their individual genomes. Interaction of modern food science and nutrition with disciplines such as pharmacology, medicine, or biotechnology provides impressive new challenges and opportunities.

Besides, the new European regulations in the European Union (e.g., Regulation EC 258/97 or EN 29000 and subsequent issues), the Nutrition Labeling and Education Act in the United States and the Montreal Protocol have had a major impact on food laboratories since preserving sustainability as well as food quality is also a way to improve consumer's well-being and confidence. Consequently, more powerful, cleaner, and cheaper analytical procedures are now required by food chemists, regulatory agencies, and quality control laboratories. These demands have increased the need for more sophisticated instrumentation and more appropriate methods able to offer better qualitative and quantitative results while increasing the sensitivity, precision, specificity, and/or speed of analysis.

As a result, researchers in food science and nutrition are moving from classical methodologies to more advanced strategies, and usually borrow methods well established in medical, pharmacological, and/or biotechnology research. As a result, advanced analytical methodologies, "omics" approaches and bioinformatics—frequently together with *in vitro*, *in vivo*, and/or clinical assays—are applied to investigate topics in food science and nutrition that were considered unapproachable few years ago.

This trend in modern food science and nutrition has generated the appearance of new terms such as nutrigenomics, nutrigenetics, nutriepigenetics, nutritional genomics, transgenics, functional foods, nutraceuticals, genetically modified (GM) foods, microbiomics, toxicogenomics, nutritranscriptomics, nutriproteomics, nutrimetabolomics, systems biology, etc. This novelty has also brought about some problems related to the poor definition of part of this terminology or their low acceptance, probably due to the difficulty to work in a developing field in which several emerging strategies are frequently put together.

Although the term Foodomics is being used in different Web pages and scientific meetings since 2007 (see, e.g., Refs. *(1,2)*), Foodomics was for the first time defined in a SCI journal in 2009 as "a discipline that studies the food and nutrition domains through the application of advanced omics technologies to improve consumer's well-being, health, and confidence" *(3–6)*. Thus, Foodomics is not only a more universal framework that comprises in a simple and straightforward concept all of the aforementioned emerging terms (e.g., nutrigenomics, nutrigenetics, nutriepigenetics, microbiomics, toxicogenomics, nutritranscriptomics, nutriproteomics, nutrimetabolomics, etc.), but more importantly, Foodomics is a global approach that makes possible to investigate all the possible connections among food (including composition, quality, and safety), diet, and the individual (including food impact on health/illness).

A representation of the areas covered by Foodomics and the analytical tools (omics) employed can be seen in Figure 1. As mentioned, analytical strategies used in Foodomics have to face important difficulties derived, among others, from food complexity, the huge natural variability, the large number of different nutrients and bioactive food compounds, their very different concentrations, and the numerous targets with different affinities and specificities that they might have. In this context, transcriptomics, proteomics, and metabolomics represent powerful analytical platforms developed for the analysis of gene expression (transcripts), proteins, and metabolites, respectively. Foodomics can help to investigate and solve crucial topics in food science and nutrition in a short- and long-term perspective *(8)*, for instance: (a) to understand the gene-based differences among individuals in response to a specific dietary pattern following nutrigenetic approaches; (b) to understand the biochemical, molecular, and cellular mechanisms that underlies the beneficial or adverse effects of certain bioactive food components following nutrigenomic approaches; (c) to determine the effect of bioactive food constituents on crucial molecular pathways; (d) to know the identity of the genes involved in the previous stage to the onset of the disease, and therefore, to discover possible molecular biomarkers; (e) to establish the global role and functions of gut microbiome, a topic that is expected to open an impressive field of research; (f) to carry out the investigation on unintended effects in GM crops; (g) to understand the stress adaptation responses of food-borne pathogens to ensure food hygiene, processing, and preservation; (h) to investigate the use of food microorganisms as delivery systems including the impact of gene inactivation and deletion systems; (i) the comprehensive assessment of food safety, quality, and traceability ideally as a whole; (j) to understand the molecular basis of biological processes with agronomic interest and economic relevance, such as the interaction between crops and its pathogens, as well as

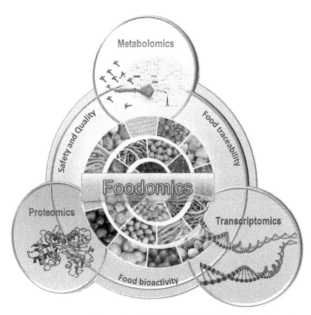

FIGURE 1 Scheme of the new discipline of Foodomics including the main tools employed and the main fields of application. *Reproduced from Ref. (7).*

physicochemical changes that take place during fruit ripening; (k) to fully understand postharvest phenomena through a global approach that links genetic and environmental responses and identifies the underlying biological networks. In this regard, it is expected that the new omics technologies combined with systems biology, as proposed by Foodomics, can lead postharvest research into a new era. The interest in Foodomics also coincides with a clear shift in medicine and biosciences toward prevention of future diseases through adequate food intakes, and the development of the so-called functional foods that will be discussed below.

2   OMICS TOOLS USED IN FOODOMICS

Foodomics involves the use of multiple tools to deal with its different applications. Thus, the use of omics tools (as e.g., transcriptomics, proteomics, or metabolomics) is a must in this new discipline. Although a detailed description on these tools is out of the scope of this chapter, some fundamental concepts on different omics techniques are provided below.

2.1   Transcriptomics

In the expression process of genomic information, several steps may be regulated by nutrients and other bioactive compounds in food. Consequently, the analysis of changes in messenger RNA (mRNA) expression by nutrients and bioactive food constituents is often the first step to study the flow of molecular information from the genome to the proteome and metabolome and one of the main goals in Foodomics research *(9,10)*. The global analysis of gene expression can therefore offer impressive opportunities in Foodomics (e.g., for the identification of the effect of bioactive food constituents on homeostatic regulation and how this regulation is potentially altered in the development of certain chronic diseases). Two conceptually different analytical approaches have emerged to allow quantitative and comprehensive analysis of changes in mRNA expression levels of hundreds or thousands of genes. One approach is based on microarray technology *(9,11,12)* and the other group of techniques is based on DNA sequencing *(13)*.

2.1.1 Gene Expression Microarray Technology

DNA microarrays are collections of oligonucleotides or probes, representing thousands of genes, attached to a substrate, usually a glass slide, at predefined locations within a grid pattern. This technique is based on specific nucleic acid hybridization, and it can be used to measure the relative quantities of specific mRNAs in two or more samples for thousands of genes simultaneously. Regardless the platform used for the analysis, the typical experimental procedure is based on the same analytical steps: RNA is extracted from a source of interest (tissue, cells, or other materials), labeled with a detectable marker (typically, fluorescent dye) and allowed to hybridize to the microarrays with individual DNA sequences hybridizing to their complementary gene-specific probes on the microarray. Once hybridization is complete, samples are washed and imaged using a confocal laser scanner. Theoretically, the fluorescent signal of derivatized-nucleic acids bound to any probe is a function of their concentration. The relative fluorescence (FL) intensity for each gene is extracted and transformed to a numeric value *(14)*.

Gene expression microarrays are powerful, but limited detection range and variability arising throughout the measurement process can obscure the biological signals of interest. In order to fully exploit the possibilities of DNA microarrays, a careful study of the experimental design is required. Data acquisition and preprocessing are important posttechnical steps in microarray experiments. The latter involves data normalization, which consists of adjusting the individual hybridization intensities in order to remove variation derived from unequal quantities of starting RNA, differences in labeling or detection efficiencies between the fluorescent dyes used, and systematic biases in the measured expression levels. The most common approaches adopted for normalization are total intensity normalization, Robust Multiarray Average, Locally Weighted Scatterplot Smoothing, and the use of housekeeping genes *(15)*.

The analysis of the vast amount of microarray data for extracting biologically meaningful information is perhaps the most challenging and daunting task. The fundamental goal of microarray expression profiling is to identify genes that are differentially expressed in the condition of interest. Data analysis process can be divided into three main parts: identification of significantly regulated genes, identification of global patterns of gene expression, and determination of the biological meaning of both, individual genes and group genes. First, filtering criteria, including fold change and statistical significance determined by comparison statistics is necessary in order to identify candidate genes that are differentially expressed. After filtering data, the differentially expressed genes are then classified into discrete groups or clusters, based on expression pattern. This step allows the identification of groups of coregulated genes. To achieve this, sophisticated bioinformatics tools such as unsupervised clustering, principal component analysis, and self-organizing maps are the most widely used (16).

2.1.2 Sequencing-Based Technologies

Sequencing-based techniques consist on counting tags of DNA fragments to provide "digital" representation of gene expression levels using sequencing. The development of "next-generation" sequencing methods, also referred to as either "massive parallel" or "ultra-deep" sequencing, is changing the way in which gene expression is studied (17). These novel technologies have had an enormous impact on research in a short time period, and it is likely to increase further in the future. Next-generation sequencing (NGS) technologies apply distinct concepts and procedures aimed to increase sequencing throughput in a cost-effective and rapid manner. Last advances in emulsion PCR and noncloning-based methods for DNA amplification, together with improved imaging instruments have facilitated the development of sequencers capable of reading up to tens of millions of bases per run in a massively parallel fashion. The read lengths typically achievable by these technology range from 30 to 300 bp, depending on the sequencing platform used. These short reads will represent a challenge for either mapping them to reference genome or *de novo* assembling without genomic reference (18,19).

2.2 Proteomics

The proteome is the set of expressed proteins at a given time under defined conditions; it is dynamic and varies according to the cell type and functional state. One of the main differences when working with proteins is that there is not an amplification methodology for proteins comparable to PCR. Physical and chemical diversity of proteins are also higher than nucleic acids. They differ among individuals, cell types, and within the same cell depending on cell activity and state. In addition, there are hundreds of different types of post-translational modifications (PTMs), which evidently will influence chemical properties and functions of proteins. PTMs are key to the control and modulation of many processes inside the cell. The selected analytical strategy for the detection of PTMs will depend on the type of modification: acetylation, phosphorylation, ubiquitination, sumoylation, glycosylation, etc. Besides, the huge dynamic concentration range of proteins in biological samples (a 10^{10} dynamic range has been estimated in serum for protein concentration) causes many detection difficulties because many proteins are below the sensitivity threshold of the most advanced instruments used in proteomics. For this reason, fractionation and subsequent concentration of the proteome is often needed. Fractionation of the proteome can be done by differential centrifugation into different populations of organelles and also attending solubility properties into different cellular compartments. The chemical tagging strategies, involving the modification of functional groups of amino acid residues, including PTM in proteins (and peptides), is the methodology of choice when a specific population of the proteome is under study. Another strategy for proteome fractionation is the sequential or simultaneous immunoaffinity depletion of the most-abundant proteins; the main drawback regarding this approach is that codepletion of a certain fraction of the proteome with the abundant proteins can occur. The use of a library of combinatorial ligands, acting by reducing the signal of high-abundance proteins while increasing the level of the low-abundance ones to bring their signal within the detection limit of the present-day analytical instruments has been also reported *(10)*.

The use and development of high-resolving separation techniques as well as highly accurate mass spectrometers is nowadays essential to solve the proteome complexity. Currently, more than a single electrophoretic or chromatographic step is used to separate the thousands of proteins found in a biological sample. This separation step is followed by analysis of the isolated proteins (or peptides) by mass spectrometry (MS) via the so-called "soft ionization" techniques, such as electrospray ionization (ESI) and matrix-assisted laser desorption/ionization (MALDI) combined with the everyday more powerful mass spectrometers. Two fundamental analytical strategies can be employed: the bottom-up and the top-down approach.

2.2.1 Bottom-Up Approach

The bottom-up approach is the most widely used in proteomics since it can apply conventional or modern methodologies. In a conventional approach, large-scale analyses of proteomes is accomplished by the combination of two-dimensional gel electrophoresis (2DE) followed by MS analysis (see Figure 2, for an example). 2DE is the methodology that currently provides the highest protein species resolution capacity with low-instrumentation cost. However, 2DE is laborious, time-consuming, and presents low sensitivity, depending strongly on staining and visualization techniques. 2DE has also some limitations to separate highly hydrophobic biomolecules or proteins with extreme isoelectric point or molecular weight values. Moreover, one of

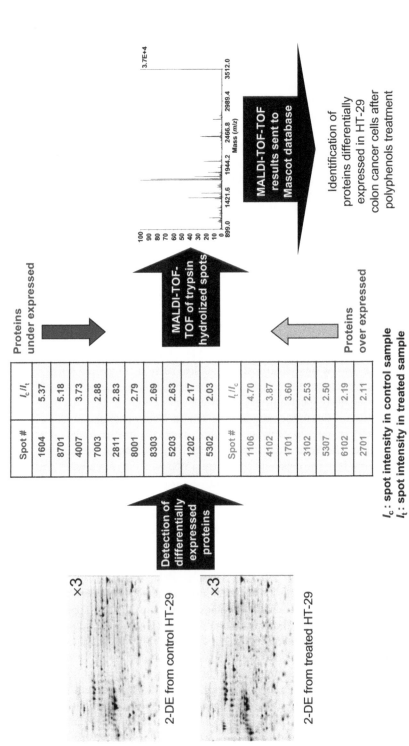

I_c : spot intensity in control sample
I_t : spot intensity in treated sample

FIGURE 2 Scheme of a *bottom-up* proteomics analysis of HT-29 colon cancer cells treated with dietary polyphenols. Minimum three biological replicates were used for each experiment (treated vs. control, i.e., minimum six 2-DE gels were used for each experiment). Differentially expressed proteins were determined by image analysis of the 2-DE gels determining the ratio between the spots intensity in control and treated samples (I_c/I_t, where I_c is the spot intensity in the control sample and I_t in the treated sample). Differentially expressed proteins were hydrolyzed with trypsin and then the peptidic fragments analyzed by MALDI-TOF/TOF. Protein identification was done using the Mascot database. *Reproduced from Ref.* (20).

the major sources of error in 2DE is gel-to-gel variation. In this sense, the introduction of difference gel electrophoresis (DIGE), in which gel variation is eliminated by loading different samples in the same gel, has brought about an important improvement. DIGE methodology is based on the use of novel ultra-high sensitive fluorescent dyes (typically, Cy3, Cy5, and Cy2) to label up to three different protein samples that will be separated in the same 2DE run. After image analysis of the 2DE gels, the protein spots of interest are subsequently submitted to an in-gel digestion step with a protease enzyme. MS and databases of protein sequences are then used in different ways for protein identification, following: (i) the peptide mass fingerprint approach using MS data, in which the molecular masses of the peptides from the protein digest are compared with those simulated of already sequenced proteins or (ii) by tandem MS in a sequence tag search in which a few peptidic sequences obtained from MS/MS analysis (using collisional activation or some other energy deposition process) can be used to search in protein databases. The main limitation of the bottom-up approach is that the information obtained is related to a fraction of the protein, loosing information about PTM.

Shotgun proteomics is an advanced bottom-up approach in which a complex protein mixture is digested with endoproteinases of known specificity, providing a comprehensive, rapid, and automatic identification of complex protein mixtures. In spite of the higher sample complexity due to the huge number of peptidic species obtained from the hydrolyzed protein mixture, peptides are, in general, better separated and analyzed by MS than proteins. To carry out shotgun proteomics, most efforts have been focused on the development of on-line combination of various chromatographic and/or electrokinetic separation methods, as well as in multidimensional protein identification technology, coupled with MS or tandem MS. The vast majority is based on the combination of ion exchange chromatography and reversed-phase chromatography, although other combinations have been reported (e.g., capillary electrophoresis (CE) or LC based on monolithic columns or HILIC, hydrophilic interaction). ESI has been the interface of choice to ionize peptides eluted from LC or CE columns and to analyze them by MS: from the lower resolving power ion trap instruments able to generate MSn spectra to the more advanced hybrid instrumentation (such as quadrupole-time of flight (Q-TOF), TOF-TOF, ion trap-TOF, Q-ion trap, TOF-TOF-TOF, Orbitrap, ion trap-FTMS, etc.) *(21)*.

2.2.2 Top-Down Approach

A growing number of researchers are focusing on the use of top-down proteomics, a relatively new approach compared to bottom-up, in which structure of proteins is studied through measurement of their intact mass followed by direct ion dissociation in the gas phase. The main advantages over the bottom-up approach are that higher sequence coverage is obtained, it permits the study of PTM, and it makes possible to discern between biomolecules with a high degree of sequence identity.

Today, Fourier transform ion cyclotron resonance (FTICR)-MS offers the highest mass resolution, resolving power, accuracy, and sensitivity among present MS technologies (5). Nevertheless, its high purchase and maintenance costs, mostly derived from the expensive superconducting magnet and liquid helium supply required, precludes its general use. Thus, a variety of instruments have been used for top-down proteomics, as for example, MALDI-TOF/TOF, ESI-Q/TOF, ESI-IT, and Orbitrap. In those cases, in which the molecular mass of proteins exceeds the analytical power of the mass spectrometer, they can be subjected to a limited proteolysis to produce large polypeptides that are then analyzed. A key to the top-down approach is the capacity to fragment intact proteins inside the mass spectrometer. Besides, top-down proteomics approaches are usually limited to simple protein mixtures since multiple charged proteins generate very complex spectra. In general, a protein isolated from a previous fractionation or purification step is directly infused to the high-resolution MS (21).

2.2.3 Other Proteomic Approaches

New proteomic approaches based on array technology are also being employed. Protein microarrays can be composed by recombinant protein molecules or antibodies immobilized in a high-density format on the surface of a substrate material. There are two major classes of protein micro- (or nano-) arrays: analytical and functional protein microarrays, the antibody-based microarray being the most common platform in proteomic studies. These miniaturized arrays can be fabricated with an almost infinite number of antibodies (in the particular case of antibody microarrays) carrying the desired specificities and being capable of simultaneously profiling numerous low-abundant protein analytes in complex proteomes, while consuming only a few microliters of sample. The microarray patterns generated can then be transformed into proteomic maps, or detailed molecular fingerprints, revealing the composition of the proteome. After importing data files from the microarray scanning sources and normalization and quantification of protein microarrays, a statistical analysis has to be carried out.

Surface-enhanced laser desorption ionization time-of-flight mass spectrometry (SELDI-TOF-MS), can also be used as a complementary, rapid, and high-throughput proteomic approach to identify differentially expressed peptides and proteins and, therefore, can help to save time and research effort in studies about the effect of bioactive food components on the proteome. This technology enables to speed up the process significantly with the capacity to run and analyze hundreds of samples in parallel per day. Different SELDI chip types have different surfaces, depending on the type of proteins to be analyzed, ranging from chromatographic chemistries, to surfaces with a specific affinity (e.g., antibodies, receptors, enzymes, and ligands) that bind one specific protein or group of proteins (21).

2.3   Metabolomics

Metabolome is usually defined as the full set of endogenous or exogenous low-molecular weight metabolic entities of approximately <1000 Da (metabolites) and the small pathway motifs that are present in a biological system (cell, tissue, organ, organism, or species). Besides, unlike nucleic acid or protein-based omics techniques, which intend to determine a single chemical class of compounds, metabolomics has to deal with very different compounds of very diverse chemical and physical properties. Moreover, the relative concentration of metabolites in the biological fluids can vary from millimolar concentration (or higher) to picomolar, making it easy to exceed the linear range of the analytical techniques employed.

Metabolites are, in general, the final downstream products of the genome and reflect most closely the operation of the biological system, its phenotype. Application of metabolomics in Foodomics can be considered even more challenging than in other areas of research. Humans feed on other organisms, each with its own metabolome, so the number of different metabolomes that shape our diet is significantly complex. Moreover, the intricate interactions between gut microflora and host metabolism as well as other extrinsic factors such as food habits, diet, and other lifestyle parameters can also give rise to metabolic variability. There are three basic approaches that can be used in metabolomics: target analysis, metabolic profiling, and metabolic fingerprinting. Target analysis aims the quantitative measurement of selected analytes, such as specific biomarkers or reaction products. Metabolic profiling is a nontarget strategy that focuses on the study of a group of related metabolites or a specific metabolic pathway. It is one of the basic approaches to phenotyping, because the study of metabolic profiles of a cell gives a more accurate description of a phenotype. Meanwhile, metabolic fingerprinting does not aim to identify all metabolites, but to compare patterns of metabolites that change in response to the cellular environment. One of the main challenges in metabolomics is to face the complexity of any metabolome, usually composed by a huge number of compounds of very diverse chemical and physical properties (sugars, amines, amino acids, peptides, organic acids, nucleic acid, or steroids). Sample preparation is especially important in metabolomics, because the procedure used for metabolite extraction has to be robust and highly reproducible. Sample preparation will depend on the sample type and the targeted metabolites of interest (fingerprinting or profiling approach). Moreover, no single analytical methodology or platform is applicable to detect, quantify, and identify all metabolites in a certain sample. Two analytical platforms are currently used for metabolomic analyses: MS and NMR-based systems. These techniques either stand alone or combined with separation techniques (typically, LC-NMR, GC-MS, LC-MS, and CE-MS) can produce complementary analytical information to attain more extensive metabolome coverage.

2.3.1 Mass Spectrometry Approaches for Metabolomics

Direct injection MS is sometimes used for metabolomics studies but it requires high or ultra-high resolution mass analyzers as TOF-MS, Q-TOF-MS, Orbitrap (all with a mass accuracy typically <10 ppm), or FTICR-MS which provides a mass accuracy <1 ppm and detection limits lower than attomole or femtomole levels. Hybrid analyzers, as for example Q-TOF, have the advantages of mass accuracy given by a TOF analyzer combined with the possibility to fragment the ions and thus, provide information about the structure of the detected metabolites. The most usual ionization techniques are the ones that use atmospheric pressure ionization, especially ESI interfaces, as it can work with a wide range of polarities. Some new sample ionization sources that have been recently used for metabolomics applications are, for example, Direct Analysis in Real Time, Proton Transfer Reaction, and Ion Mobility. Although direct injection techniques have been successfully used in metabolomics, mass accuracy and resolution provided by MS instruments are usually not enough to undoubtedly separate and identify all metabolites in a sample by direct infusion. To overcome these limitations, in metabolomics, high-resolution MS analyzers are usually used coupled to separation techniques as gas chromatography (GC), high-performance liquid chromatography (HPLC), or CE.

2.3.2 Nuclear Magnetic Resonance Approaches for Metabolomics

Nuclear magnetic resonance (NMR) is a high-reproducibility technique that has demonstrated its great potential in metabolomics studies (22). Coupling of NMR with separation techniques is less extended than in MS, but can also give interesting results in metabolomics. Offline coupling with solid-phase extraction or on-line coupling with HPLC are the most extended in metabolomic research. The lack of sensitivity of NMR is one of its main drawbacks for its general use in metabolomics since it usually cannot provide information on metabolites that are at low or very low concentration.

2.4 Other Omic Approaches

2.4.1 Mirnomics

Mirnomics refers to the global analysis of microRNAs (miRNAs) and sometimes it is considered as a subdiscipline within epigenetics. miRNAs are post-transcriptional regulators that bind to complementary sequences on target mRNAs transcripts resulting either in translational repression or degradation of the target with consequent gene silencing (23). It is now recognized that more than half of the cellular transcripts expressed are regulated by miRNAs

(24). High-throughput methodologies for detecting global miRNA alterations are now available (as e.g., high-throughput miRNAs arrays *(25)*) although it is more complex to elucidate the function of miRNA compared to mRNA, since a single miRNA may be responsible for regulating several target sequences and thus many gene transcripts *(26)*. Several studies examining the role of nutrients and bioactive components in modulating miRNAs have already been published *(27)*.

2.4.2   Epigenomics

Epigenomics studies the mechanisms of gene expression that can be maintained across cell divisions, and thus the life of the organism, without changing the DNA sequence. The epigenetic mechanisms are related to changes induced (e.g., by toxins or bioactive food ingredients) in gene expression via altered DNA methylation patterns, altered histone modifications, or noncoding RNAs, including small RNAs. In mammals, many dietary components, including folate, vitamin B6, vitamin B12, betaine, methionine, and choline, have been linked to changes in DNA methylation. Although it is too early to apply epigenetic alterations that are induced by dietary ingredients as biomarkers in public health and medicine, research in this area is expected to be boosted by the expanding use of next-generation DNA sequencing technologies. Applications include chromatin immunoprecipitation followed by DNA sequencing (ChIP-seq) to assess the genomic distribution of histone modifications, histone variants and nuclear proteins, and global DNA methylation analysis through the sequencing of bisulphite-converted genomic DNA. Combined with appropriate statistical and bioinformatic tools, these methods will permit the identification of all the loci that are epigenetically altered. A more detailed description on dietary factors and epigenetic modifications can be found in Ref. *(28)*.

2.4.3   Peptidomics

Food peptidomics deals with the identification and quantification of nutritionally relevant peptides usually called "bioactive peptides" *(29)*. These peptides can show different biological activities (e.g., antioxidative, antihypertensive, anti-inflammatory, immune boosting, etc.). Interestingly, more than half of the peptides described in the literature (and found in on-line databases as, e.g., BioPEP, PepBank, EROP, or APD) *(29)* are of a size smaller than six amino acids. Hence, they represent a challenge in terms of analysis when using current proteomics techniques. Therefore, development of peptidomics field in Foodomics through high-throughput large-scale assays for these molecules is mandatory to better conduct research in this field. The explanation to this small size of bioactive peptides (less than five to six amino acids) is that many industrial processes using enzymatic hydrolysis do liberate small peptides. Therefore many *in vitro* studies are performed on such ingredients and highlight activities related to such small peptides. But more importantly, whichever extent of hydrolysis the ingredient has, it will finally have to first transit through the digestive system in order to deliver its benefit. Considering the variety of enzymes available in the digestive systems (e.g., pepsin, trypsin, chymotrypsin, elastase, carboxy-peptidases, amino-peptidases, di- and tripeptidases), it is not surprising that small peptides are the largest category with known biological activity and therefore represent the most promising candidates for any food application targeting to deliver a health benefit. Initiatives to develop methods to identify such small peptides in complex food matrices have not started yet and although few publications have already described methods for few peptides *(29)*, an MRM atlas (as is the case in proteomics for key organisms) is not available yet.

2.4.4 Lipidomics

Lipids are a diverse group of compounds with multiple key biological functions (e.g., they can act as energy storage, participate in signaling pathways, constitute the cellular structural building blocks in both cell and organelle membranes, etc.). The diversity in lipid function is reflected by a huge variation in the structures of lipid molecules, which can comprise hundreds of thousands distinct lipid molecules *(30)*. Lipidomics is a subdiscipline of metabolomics, with the focus on the global study of molecular lipids (i.e., the complete lipid profile within a cell, tissue, or organism), including pathways and networks of cellular lipids in biological systems *(30)*. It covers not only the analysis of lipid species and their abundance but also their biological activities, subcellular localization, and tissue distribution. The main analytical techniques used in lipidomics include GC-MS, UPLC-MS, NMR, and shotgun MS. A more detailed description of lipidomics and its applications can be found elsewhere *(30)*.

2.4.5 Glycomics

The glycome is defined as the set of all glycans in an analyzed tissue or organism. Nearly all membrane and secreted proteins as well as numerous intracellular proteins are modified by covalent addition of complex oligosaccharides, also called glycans, that play a role in almost every biological process. Changes in the activity and/or localization of any of the enzymes involved in glycan biosynthesis will affect the final structure of a glycan. Therefore, in addition to being defined by an individual's genetic background, the glycome is shaped by environmentally induced changes in gene expression *(31)*. Until only a few years ago, glycan analysis was extremely laborious and complex, hampering large-scale studies of the glycome. However, significant progress has been made in the past few years and several methods for high-throughput glycan analysis now exist *(31)*. In principle, three main analytical techniques are being used for high-throughput analysis of glycans:

liquid chromatography, CE, and MS. More detailed description on glycomics and its applications can be found in Ref. *(31)*.

3   DATA ANALYSIS AND INTEGRATION: FOODOMICS AND SYSTEMS BIOLOGY

Due to the huge amount of data usually obtained from omics studies, it has been necessary to develop strategies to convert the complex raw data obtained into useful information. Thus, bioinformatics has become also a crucial tool in Foodomics, although nowadays it is also considered one of the main bottle-necks that are delaying the fast development of this approach. Over the last years, the use of biological knowledge accumulated in public databases by means of bioinformatics allows to systematically analyze large data lists in an attempt to assemble a summary of the most significant biological aspects. Also, statistical tools are usually applied, for example, for exploratory data analysis to determine correlations among samples (which can be caused by either a biological difference or a methodological bias), for discriminating the complete data list and reducing it to the most relevant ones for biomarkers discovery, etc.

Data treatment from a Foodomics experiment can be different if the data comes from a single omic platform or if integration of data from several omic platforms is required. In the case of considering only data from a transcrip-tomics experiment, the combined use of public databases and bioinformatics allows the systematic analysis of large gene lists in order to assemble a sum-mary of the most enriched and significant biological aspects. The principle behind enrichment analysis used in transcriptomics is that if a certain biological process is occurring in a given study, the co-functioning genes involved should have a higher (enriched) potential to be selected as a relevant group by high-throughput screening technologies. This approach increases the probability for researchers to identify the correct biological processes most pertinent to the biological mechanism under study. Thus, a variety of high-throughput enrichment tools (e.g., DAVID, Onto-Express, FatiGO, GOminer, EASE, ProfCom, etc.) have been developed since 2002 in order to assist microarray end user understanding the biological mechanisms behind the large set of regulated genes. These bioinformatics resources systematically map the list of interesting (differentially expressed) genes to the associated biological annotation terms and then statistically examine the enrichment of gene members for each of the terms by comparing them to a control (or refer-ence). Enrichment analysis would not be possible without appropriately structured databases such as Gene Ontology *(32,33)*. Gene Ontology provides a systematic and controlled language, or ontology, for the consistent descrip-tion of attributes of genes and gene products, in three key biological domains that are shared by all organisms: molecular function, biological process, and cellular component. A scheme of a procedure developed for data processing

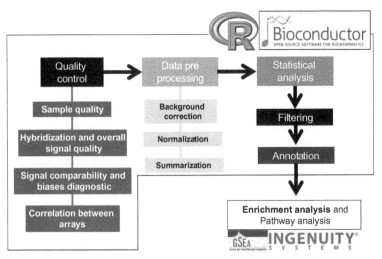

FIGURE 3 Scheme of a procedure developed for the processing of data from microarrays based on the open-source R programming language. Minimum three biological replicates were used for each experimental condition (i.e., treated vs. control colon cancer cells). A file with extension .cel (raw expression data) is obtained for each replicate from the microarray. The main packages for data processing used in R were affy, annotate, limma, genefilter, affyQCReport, affyPLM, and ArrayTools. *Reproduced from Ref.* (20).

from microarrays based on the open-source R programming language combined with Gene Set Enrichment Analysis is given in Figure 3. In this work (20), the use of Affymetrix microarray platform and stringent data analysis provided interesting insights into the antiproliferative activity of dietary polyphenols against colon cancer cells. Readers interested in data analysis for metabolomics or proteomics can find the information elsewhere (34).

The difficulty of data analysis in Foodomics increases enormously when data from different "omics" platforms needs to be integrated in order to understand the biological meaning of the results obtained from the investigated system (e.g., cell, tissue, organ) with a global perspective. This can ideally be done through a holistic strategy as proposed by systems biology. Systems biology is the analysis of the relationships among the elements in a system in response to genetic or environmental perturbations, with the goal of understanding the system or the emergent properties on the system (35). Thus, systems biology approaches may encompass molecules, cells, organs, individuals, or even ecosystems, and it is regarded as an integrative approach of all information at the different levels of genomic expression (mRNA, protein, metabolite). However, in Foodomics studies, biologic responses to a bioactive food component may be subtle and, therefore, careful attention should be given to the methodologies used to identify these responses. Unlike any reductionist approach that would take these techniques individually, systems biology exploits global data sets to derive useful information *(36,37)*. Each large data set contains sufficient noise to preclude the identification of multiple minor but relevant changes that could be unnoticed without adequate statistical tools since the researcher is focused on the changes that are really significant within the whole data set. Systems biology, however, by confining the information can provide a filter for "distracting" noise generated in each individual platform and minimize the data to be interpreted by focusing on only those endpoints common between the various experimental platforms. To achieve this, appropriate statistical models have to be used in order to filter through the large data sets and highlight only those important changes. The difficulty of this global approach is huge as demonstrated by the fact that, to the best of our knowledge, only two Foodomics papers had been published till 2013 *(20,38)* where data from transcriptomics, proteomics, and metabolomic platforms were generated to investigate the bioactivity of food ingredients at molecular level, and the huge amount of information from the three different expression levels (gen, protein, and metabolite variations) was integrated in a rational way.

The challenge in the combination of Foodomics and systems biology is not only at the technological level, where great improvements are being made and expected in the "omics" technologies but also in our limited knowledge on many biological processes that can have place at molecular level. Besides, as mentioned by Oresic *(37)*, systems biology in the context of food and nutrition research requires bridging across multiple levels and concepts, for example, cellular–organismal, host–microbial, and short-term versus long-term effects. Although it is now assumed that no single model can encompass the complexity of human metabolism in the context of nutrition, one possibility would be to consider a platform-based approach where multiple physiological levels relevant to diet are studied in the context of nutrition. According to Oresic *(37)*, such a platform would need to include at least the following four levels:

1. "Omics" platforms for the detailed characterization of food components.
2. Platform to characterize and model the production of food-derived molecules (e.g., metabolites) in the gut, which enter the enterohepatic circulation and thus affect the host physiology.
3. Platforms to characterize and model the effects of food-derived molecules on host physiology at the cellular, tissue, as well as organismal levels.
4. Sensitive platform to characterize the host physiology and health status.

Last, but no least, bioinformatics (including data processing, clustering, dynamics, or integration of the various "omics" levels) will have to progress for systems biology to demonstrate all its potential. In this regard, it is also interesting to mention that the traditional medical world has often noted that, although many of the omics tools and Foodomics approaches provide academically interesting research, they have not been translated to methods or approaches with medicinal impact and value because data integration when dealing with such complex systems is not straightforward. In the future, the combination of Foodomics and systems biology can provide crucial information on, for example, host–microbiome interactions, nutritional immunology, food microorganisms including pathogens resistance, postharvest, plant biotechnology, farm animal production, etc.

4   PRESENT APPLICATIONS OF FOODOMICS

4.1   Foodomics in Food Safety, Quality, and Traceability Studies

As mentioned above, food safety is by far the first request demanded by consumers. A vast legislation is focused on the substances that can be or cannot be found or added to a particular food, and if so, usually stating maximum limits of utilization. Typical examples of these kinds of components are antibiotics and pesticides used in farming and agriculture to increase the production rates and benefits and to assure the correct growth of animals and crops, respectively. As a consequence, accurate, precise, and robust analytical methods are needed in order to determine if the products found in the market comply with the legislation. Foodomics has brought about a great increase in sensitivity and selectivity, allowing frequently the detection of these potentially harmful components at even lower concentrations than the official requirements. Moreover, within this field, not only the detection of totally or partially forbidden compounds is sought but also the determination of potentially dangerous compounds (e.g., pollutants and toxins) or organisms that could be present in food, and which might pose a risk for consumers' health (39).

Transcriptomic, proteomic, and/or metabolomic approaches have already been shown to be extremely useful to assess food safety and quality at every stage of production to ensure food safety for human consumption. They are also valuable tools to distinguish between similar food products and to detect food frauds (adulteration, origin, authenticity, etc.), food-borne pathogens, toxic species, food allergens, etc.

In the context of food safety, several DNA microarray chips have been developed for the detection of food-borne pathogens (40,41) and toxigenic microorganisms (42). DNA microarrays for gene expression analysis have also demonstrated to be helpful on detecting mycotoxins in foods. This approach offers new possibilities to study the influence of environmental and technological parameters like pH, temperature, and water activity on the activation of mycotoxin biosynthesis (43). The effect of the consumption of hypoallergenic wheat flour on the expression of a wide-spectrum of genes is another example on the use of gene expression microarray technology in food safety. This technique can be applied in animal and cellular models demonstrating to be an efficient strategy for evaluating different aspects concerning food safety (44).

A lot of work is being carried out by the scientific community to achieve early detection of prion diseases in animals which have not only a known risk of transmission to humans but an economic impact on animal production. A particular case of the application of "omics" approaches to guarantee food safety is the biomarker discovery in body fluids and tissues regarding prion diseases as reviewed by Huzarewich *et al.* *(45)*.

GC-MS-based metabolomics has also demonstrated its potential to detect the presence of *Salmonella* and *E. coli* O157:H7 at levels of 7 ± 2 CFU/25 g of food *(46)*. A headspace solid-phase microextraction protocol was employed to sampling the generated compounds. Using chemometrics, it was possible to construct prediction models based on the profile of the metabolites detected in those samples. Although a single compound could not be found associated to the growth of a particular bacteria, the analysis of the whole profile allowed the correct classification of ground beef and chicken samples contaminated with the studied microorganisms.

The concept "food quality" includes multiple aspects of food. Because this term is not a completely objective parameter, the consideration of appropriate quality will strongly depend on the food, in particular, on the product itself, or even on the consumer. Among the aspects related to food quality, food composition, aroma, flavor, taste, or food properties can be pointed out. For this reason, assessing food quality is a complex task that may imply different types of analysis depending on the particular food or product. As in other fields of food science, as the complexity increases, Foodomics approaches are gaining attention *(39)*.

Thus, Foodomics can help to solve some of the new challenges that modern food safety, quality, and traceability have to face. These challenges encompass the multiple analyses of contaminants, allergens, the establishment of more-powerful analytical methodologies to guarantee food origin, traceability and quality, the discovery of biomarkers to detect unsafe products or the capability to detect food safety problems before they grow and affect more consumers, etc.

Regarding food allergens, as mentioned by Picariello *et al.* *(47)*, the need for reliable methods allowing accurate structural identification and dosimetry of the offending allergens in food has prompted researchers to develop new methods for their unambiguous characterizations. The integrated approaches based on proteomics, peptidomics, and metabolomics are merged into the new discipline of Foodomics and have proved to be essential at various levels in the study of food allergens, from the structural characterization of novel food allergens to the controversial feature of the resistance to digestion of allergenic proteins or to the efficiency of epitopes removal from a food intended to some particular patients. Readers interested in the different applications of -omic tools to food allergen analysis (including, e.g., the understanding of the structural features of epitopes, the elucidation of the mechanisms triggering individual response to allergens, the monitoring of the formation/fate of allergens along the technological processes leading from raw materials to finished foods, etc.) can find it elsewhere (47).

Transcriptomics, proteomics, and metabolomics represent powerful analytical platforms to acquire more detailed and complete information on food composition even beyond the traditional food component analysis. This comprehensive knowledge of biochemical composition of foods will provide a better understanding of metabolic networks allowing food research community for a better insight of the molecular basis of important food characteristics such as flavor, color, texture, aroma, added-value nutrition, as well as the discovery of novel bioactive compounds in foods. In this regard, the presence of specific volatile and aminothiols in wine is associated with quality, worth, price, and taste. A novel Foodomics assay of thiol-containing compounds, such as free aminothiols and related conjugates, was developed using ultra-performance liquid chromatography (UPLC) with FL and ESI time-of-flight mass spectrometric (TOF/MS) detections (48). FL-specific derivatization was applied along with multivariate statistical analysis. The screening assay consisted of monitoring the UPLC-TOF/MS peaks of unknown thiols, which decreased due to the derivatization as compared to the nonderivatized thiols. The principal component analysis of the UPLC-TOF/MS data could be well differentiated and categorized into two groups. The orthogonal signal correction partial least-squares-discriminant analysis, the so-called S-plot, showed that the quality differentiation was directly related to the decrease of native thiols and increase of derivatized thiols. With this strategy, the mass difference from the derivatization reagent ($+m/z$ 198) could be utilized for the identification of these thiols using the FL peaks retention time and metabolomics databases. The presence of L-glutathione in rice wine was for the first time reported on the basis of the available metabolomics databases and standard matching. The authors mentioned that this novel concept based on Foodomics could be applied in food analysis for the ready screening of specific functional compounds by exploiting the various derivatization modes available.

Metabolite profiling using GC-MS has been applied to analyze a broad spectrum of low-molecular weight rice constituents for the investigation of time-dependent metabolic changes in the course of rice germination, establishing the basis of the research of the potential advantageous nutritional properties of germinated rice (49). Also, by using GC-MS-based metabolomics, it is also possible to determine a particular metabolite profile specific of a vegetable variety. This technique has been employed in combination with chemometrics to characterize different potato (Solanum tuberosum) varieties and cultivars (50).

As mentioned above, Foodomics tools also offer enormous potential to study the molecular basis of biological processes with agronomic interest and economic relevance, such as the interaction between crops and its pathogens, as well as physicochemical changes that take place during fruit ripening.

In this line, successful examples of transcriptome analysis are based on recently developed microarray chips covering the genome of important crop species such as watermelon, citrus, melon, canola, etc *(51,52)*.

Proteomic and metabolic changes also occur during crops growing, food processing/preparation (fermentation, baking, boiling, etc.), and food conservation/storage (freezing, smoking, drying, etc.). These tools are, therefore, very useful for getting a deeper understanding of molecular details of foods and food-related matrices. Metabolomic analysis based on the combined use of UPLC with a high-resolution Q-TOF-MS was employed to monitor how water-soluble metabolites changed during a 2-month fermentation of soybean-based products *(53)*. The information obtained was relevant, demonstrating the significant influence of metabolites, including amino acids, urea cycle intermediates, nucleosides, and organic acids, on the overall nutritional and sensory quality of the resulting fermented products. Also, the usefulness of metabolomic studies has been demonstrated to improve quality control of beverage production. In this sense, NMR was successfully employed for the detection of adulteration of orange juices by the addition of lower cost grapefruit juices *(54)*, and for monitoring beer production, studying the effect on beer metabolic composition of site- and time-related variables during brewing process *(55)*. In a recent work, microbial spoilage of beef was studied by analyzing the release of volatile organic compounds by GC-MS *(56)*.

4.2 Foodomics of Transgenic Foods

Genetic engineering, or recombinant DNA technology, allows selected individual gene sequences to be transferred from an organism into another and also between nonrelated species. Genetic engineering has been used in agriculture and food industries in the last years in order to improve the performance of plant varieties (resistance to plagues, herbicides, and hydric or saline stresses), improve technological properties during storage and processing (firmness of fruits), or improve the sensorial and nutritional properties of food products (starch quality, content on vitamins or essential amino acids). The organisms derived from recombinant DNA technology are termed genetically modified organisms (GMOs). A transgenic food is a food that is derived from or contains GMOs.

The use of genetic engineering in the production of foods is constantly growing since the past years as well as the concern in part of the public opinion. This is due to the increasing impact of this technology in foodstuff production, by one side, and to the continued campaign against GMO crops leaded by ecologist organizations, by the other. Claims about advantages derived of GMO crops include those from biotechnology companies and most of the scientific community, stressing the benefits for the agriculture and the food industry and the lack of scientific evidence on any detrimental effects on human health. On the other side, ecologists groups are concerned about the impact of GM plants on human health and on the environment. In spite of the controversy of whether GMOs are beneficial or harmful for humans, animals, and environment, there is an increasing number of GMOs developed every year. Namely, in the last years, over 150 GMOs, representing 24 different crops, have been approved by regulatory agencies in different countries. In the near future, this number is expected to rise and a second generation of GMOs with nutritionally enhanced traits, such as, for instance, plants enriched in beta-carotene, vitamin E, or omega-3 fatty acids could likely obtain commercialization approval (57). In this context, most governments have approved regulations on the use, spreading, and marketing of GMOs, in order to regain the confidence of the consumers. Owing to the complexity that entails the compositional study of a biological system such as GMO, the study of substantial equivalence as well as the detection of any unintended effects needs to be approached.

In order to face the mentioned analytical challenges, two conceptually and methodologically different strategies have been proposed for the analysis of GMOs, that is, targeted analysis and profiling. In the last decade, the application of targeted analysis has been the prevailing strategy for quantitative and qualitative detection of GMOs in food samples, as well as for comparative safety assessment of a GM crop with its nonmodified counterpart. In the context of substantial equivalence, a group of 50–150 compounds are typically analyzed for each crop variety, following recommendations in the OECD consensus documents that include macro- and micronutrients, antinutrients, and natural toxins. Although it has been indicated that this strategy may cover more than 95% of the crop composition (58), its application to compare the composition of GMOs with their conventional counterparts has raised numerous concerns. More precisely, it has been pointed out that this strategy is biased, and presents many limitations, such as the possible occurrence of unknown toxicants and antinutrients, particularly in food-plant species with no history of (safe) use (59). Moreover, although a few studies have identified unintended effects with targeted approaches (60), this strategy might restrict the possibilities to detect other unpredictable effects that could result directly or indirectly from the genetic modification.

The mentioned limitations of targeted analysis have encouraged the development and application of new and more powerful analytical approaches to face the complexity of this problem and to improve the chances to detect unintended effects. To this regard, European Food Safety Agency (EFSA) has recommended the development and use of profiling technologies such as the ones prosed by Foodomics, with the potential to improve the breadth of comparative analyses (61). More recently, a panel of experts on risk assessment and management has recommended profiling especially in cases where the most scientifically valid isogenic and conventional comparator would not grow, or not grow as well, under the relevant stress condition (62).

It is well known that there is no single technique currently available to acquire significant amounts of data in a single experimental analysis to detect all compounds found in GMOs or any other organism. In consequence, multiple analytical techniques have to be combined to improve analytical coverage of proteins and metabolites, in good agreement with the EFSA recommendation on using advanced analytical techniques including -omics techniques to provide a broad profile of these GM foods including the monitoring of the composition, traceability, and quality of these GM foods *(61,63–65)*. The usefulness of Foodomics for achieving a complete characterization of GMO following the corresponding European regulations on GMO labeling and traceability in food and feed has already been mentioned *(66)*. A scheme of this approach can be seen in Figure 4. The development of global Foodomics strategies to investigate GMOs will provide extraordinary opportunities to increase our understanding about transgenic foods, including the investigation on unintended effects in GM crops, or the development of the so-called second-generation GM foods. Besides, Foodomics has to deal with the particular difficulties commonly found in food analysis, such as the huge dynamic concentration range of food components as well as the heterogeneity of food matrices, natural variability, and the analytical interferences typically found in these complex matrices. A number of reports demonstrating the suitability and applicability of different profiling approaches for comparative analysis of GMOs suggest good acceptance of these fast-evolving techniques by the scientific community *(57,65,67)*.

Gene microarray technology has demonstrated to have impressive multiplexing capabilities for GMO analysis and many examples can be found in the literature *(68)*. With this technology, detailed information has also been

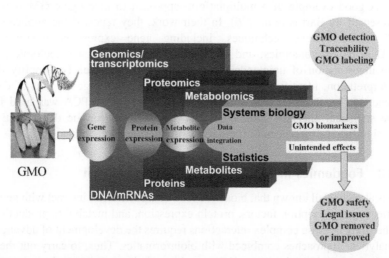

FIGURE 4   Ideal Foodomics platform to analyze genetically modified foods. GMO, genetically modified organism. *Reproduced from Ref. (66).*

obtained on nontargeted effects of transgenes in several plant species including potato, rice, wheat, and maize. In these cases, the genetic modification did not considerably alter overall gene expression that falls within the range of natural variation of landraces and varieties, supporting the possibility of producing transgenic plants that are substantially equivalent to nontransformed plants at transcriptomic level. Also, MS-based technologies have wide possibilities to evaluate GM crops based on their proteomic and metabolic profiles, as demonstrated through the large number of applications that use GC-MS, LC-MS, CE-MS, and/or MS as a stand-alone technique (57). An example of the great potential of these MS-based Foodomics approaches was presented by Kusano et al. (69). In their study, two GM tomato varieties overexpressing miraculin glycoprotein and a panel of traditional tomato cultivars were selected to prove a methodology based on three different metabolomics platforms. More specifically, data from GC-TOF-MS, LC-Q-TOF-MS, and CE-TOF-MS were summarized in single consensus data sets for further multivariate analysis. This combined data set included 175 unique identified metabolites and 1460 peaks with no or imprecise metabolite annotation. Next, PCA was performed using the physicochemical properties of 160 identified metabolites from the data set to evaluate its chemical diversity using a tomato metabolism database (LycoCyc) as reference. Using this method, the data set accounted for 85% of the chemical diversity. In addition to this, multivariate analysis based on orthogonal projections to latent structure's discriminant analysis was applied to study the differences between GM and control tomato lines. The changes between the GM and nonmodified lines were small compared to the changes observed between ripening stages and traditional cultivars.

A good example of a multiplatform approach to investigate GMOs was presented by Barros et al. (70). In their work, they reported the application of various omics techniques, including gene expression microarray, 2-DGE-based-proteomics, and NMR- and GC-MS-based metabolomics, to the investigation of unintended effects in two GM maize. To simplify the interpretation, identification, and presentations of the large amount of data generated by these technologies, chemometric tools like PCA was used for the statistical analysis, complemented with ANOVA for the determination of significant differences in transcripts, proteins, or metabolites.

4.3 Foodomics in Nutrition and Health Research

It is already well known that bioactive food compounds can interact with genes affecting transcription factors, protein expression, and metabolite production. The study of these complex interactions requires the development of advanced analytical approaches combined with bioinformatics. Thus, to carry out these studies transcriptomics, proteomics, and metabolomics approaches are now being employed together with an adequate integration of the information that they provide. One of the main challenges in this type of studies is to improve our limited understanding of the roles of nutritional compounds at molecular level (i.e., their interaction with genes and their subsequent effect on proteins and metabolites) for the rational design of strategies to manipulate cell functions through diet, which is expected to have an extraordinary impact on our health. The problem to solve is huge and it includes the study of the individual variations in gene sequences, particularly in single-nucleotide polymorphisms (SNPs), and their expected different answer to nutrients. Moreover, nutrients can be considered as signaling molecules that are recognized by specific cellular-sensing mechanisms. However, unlike pharmaceuticals, the simultaneous presence of a variety of nutrients with diverse chemical structures and concentrations and having numerous targets with different affinities and specificities increases enormously the complexity of the problem. Therefore, it is necessary to look at hundreds of test compounds simultaneously and observe the diverse temporal and spatial responses. As mentioned by Capozzi and Bordoni: "Thanks to the omics approach, researchers are now facing a new science which in theory can connect food components, foods, the diet, the individual, the health and the diseases. In practice, we are still far from this connection, which needs not only technologies but mainly a broad vision of the problem, since there are many actors playing the comedy. A broad vision means not only a broad expertise and the application of advanced technologies, but also the ability of looking at the problem with a different approach, a *foodomics approach*" *(71)*. The use of omics technologies in food science and nutrition research has given rise to the emergence of several specialties that are considered within the more global area covered by Foodomics. This is the case of nutritional genomics that includes nutrigenomics and nutrigenetics *(72)*. Nutrigenomics deals with the interactions between dietary components and the genome as well as the resulting changes in proteins and other metabolites, while the main goal of nutrigenetics is to understand the gene-based differences in response to a specific dietary pattern, a functional food, or a supplement for a specific health outcome.

4.3.1 Foodomics, Functional Foods, and Nutraceuticals

There is a clear trend in medicine and biosciences toward prevention of future diseases through adequate food intakes and the development of new functional foods and nutraceuticals. Although there is no officially accepted definition of functional foods, the definition proposed by Diplock *et al. (73)* is commonly used in the European Union and considers that a food is functional when it beneficially affects one or more target functions in the body beyond adequate nutrition in a way that is relevant to an improved state of health and well-being and/or reduction of risk of disease. Functional foods may include both natural (unmodified) foods and foods in which a component has been added, removed, or modified (including the bioavailability) by technological or biotechnological means (73). The term "nutraceutical" was originally defined by Stephen L. DeFelice, founder and chairman of the Foundation for Innovation in Medicine (Crawford, NJ, USA, 1989), as a combination of the words "nutrition" and "pharmaceutical": it refers to nutritional products that are generally sold in medicinal forms which have effects that are relevant to health. In contrast to pharmaceuticals, however, they are not synthetic substances or chemical compounds formulated for specific indications. According to the European Nutraceutical Association, nutraceuticals are "naturally derived bioactive compounds that are found in foods, dietary supplements and herbal products, and have health-promoting, disease-preventing, or medicinal properties" (74).

Foodomics can be an adequate strategy to investigate the complex issues related to prevention of future diseases and health promotion through the mentioned functional foods and nutraceuticals. It is now well known that health is heavily influenced by genetics. However, diet, lifestyle, and environment can have a crucial influence on the epigenome, gut microbiome and, by association, the transcriptome, proteome and, ultimately, the metabolome. When the combination of genetics and nutrition/lifestyle/environment is not properly balanced, poor health is a result.

Foodomics can be a major tool for detecting small changes induced by functional food ingredient(s) at different expression levels. A representation of an ideal Foodomics strategy to investigate the effect of functional food ingredient(s) on a given system (cell, tissue, organ, or organism) is shown in Figure 5. Following this Foodomics strategy, results on the effect of food ingredient(s) at genomic/transcriptomic/proteomic and/or metabolomic level are obtained, making possible new investigations on food bioactivity and its effect on human health at molecular level. It has been mentioned that it is probably too early to conclude on the value of many substances for health and the same can apply to other health relationships that are still under study. In this regard, it is interesting to remark that several of the health benefits assigned to many dietary constituents are still under controversy as can be deduced from the large number of applications rejected by different regulatory institutions on health claims of new foods and ingredients (75,76). Foodomics approaches can help to overcome these important limitations related to the controversial demonstration about the health claims on different foods, nutraceuticals, and food ingredients. Thus, as discussed in a recent article (77), the announcement by the Panel on Dietetic Products, Nutrition and Allergies (NDA) at the EFSA (75) indicates that no evidence has been provided to establish that a food ingredient having antioxidant activity/content and/or antioxidant properties can have a beneficial physiological effect on human health. This decision will not only have important economic impact on an antioxidants market that is calculated to move billions of Euros per year but also will have a negative influence on the consumers' confidence about the possibility to promote health through foods. Therefore, more sophisticated

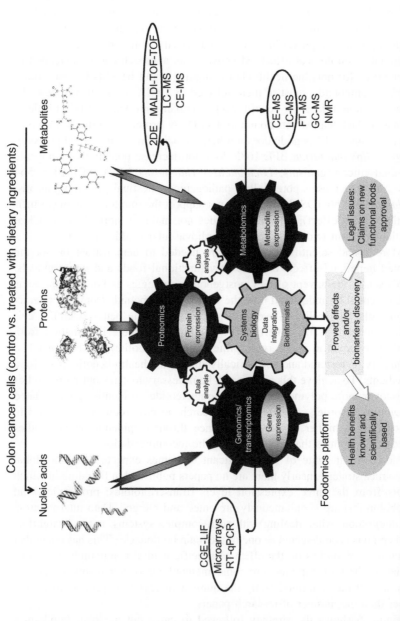

FIGURE 5 Scheme of an ideal Foodomics strategy to investigate the health benefits from dietary constituents, including methodologies and expected outcomes. Technologies used in this work are marked with a circle. *Reproduced from Ref.* (20).

approaches than the ones used so far will be required to demonstrate with strong scientific evidences the positive health effect derived from the intake of determined foods and food ingredients. In this regard, Foodomics applied to study the *in vitro* and *in vivo* mechanisms of these compounds (including, if required, Foodomics of biological samples from randomized clinical trials) are the right way to prove that antioxidant activity is indeed associated with a beneficial physiological effect. Moreover, this approach can be extended to better prove (or not) the health claims linking health benefits to many other different compounds, most of them rejected by EFSA so far. Just to describe a few: melatonin does not benefit sleep; xanthan gum does not boost satiety; green and black tea extracts do not protect DNA, proteins and lipids from oxidative damage; C12-peptide does not help to maintain normal blood pressure; *Lactobacillus plantarum* BFE 1685 does not decrease potentially pathogenic intestinal microorganisms; *Lactobacillus rhamnosus* LB21 NCIMB 40564 does not decrease potentially pathogenic intestinal microorganisms; *L. plantarum* 299v (DSM 9843) does not support the immune system (insufficiently defined effect); linoleic acid does not maintain normal neurological function; vitamin D does not benefit cardiovascular health; etc *(75,78)*.

More sound scientific evidences are needed to demonstrate or not the claimed beneficial effects of these new functional foods and nutraceuticals. In this sense, the advent of new postgenomic strategies as Foodomics seems to be essential to understand how the bioactive compounds from diet interact at molecular and cellular level, as well as to provide better scientific evidences on their health benefits. The combination of the information from the three expression levels (gen, protein, and metabolite) can be crucial to adequately understand and scientifically sustain the health benefits from food ingredients. To achieve this goal, it will be necessary to carry out more studies to discover more polymorphisms of one nucleotide, to identify genes related to complex disorders, to extend the research on new food products, and to demonstrate a higher degree of evidence through epidemiological studies based in Foodomics that can lead to public recommendations.

Moreover, in spite of the significant outcomes expected from a global Foodomics strategy, nearly there are no papers published in literature in which results from the three expression levels (transcriptomics, proteomics, and metabolomics) are simultaneously presented and merged. Data interpretation and integration, when dealing with such complex systems, is not straightforward and has been detected as one of the main bottlenecks. This has made that the number of studies on the effect of specific natural compounds, nutrients, or diet on the transcriptome–proteome–metabolome of organisms, tissues, or cells, is still rather limited being the number of review papers on this topic higher than the number of research papers.

Figure 6 shows the strategy followed to carry out a global Foodomics study on the chemopreventive effect of dietary polyphenols against HT-29 colon cancer cells *(20)*. In that work, we exploited the benefits of analytical

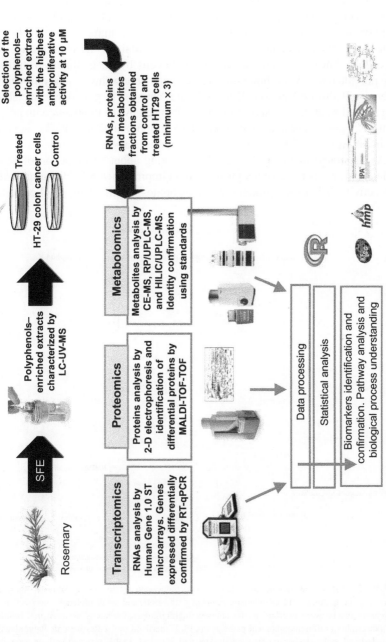

FIGURE 6 Global Foodomics strategy used to investigate the activity of rosemary polyphenols against colon cancer HT-29 cells proliferation at molecular level. *Reproduced from Ref. (20).*

multiplatforms to improve the description of the metabolome status in cultured HT-29 cells. In that study, CE-TOF-MS analyses were complemented with UPLC-TOF-MS using different chromatographic modes to carry out a broad metabolomic study on the antiproliferative effect of dietary polyphenols on HT-29 cells. Using this strategy, the intracellullar levels of 22 highly related metabolites, known to be essential for the maintenance of the cellular functions, were altered in HT-29 cells by the treatment with dietary polyphenols. After functional enrichment and pathway analysis using Ingenuity Pathway Analysis (IPA) software, changes in glutathione metabolism as well as in the levels of polyamines induced by the treatment with polyphenols were suggested that may have important implications in the proliferation of HT-29 cells as shown in Figures 7 and 8.

As illustrated in Figure 9, polyphenols bring about an induction of cell-cycle arrest, an increase of apoptosis, and an improvement of cellular antioxidant activity. The genes, proteins, and metabolites identified and implicated in

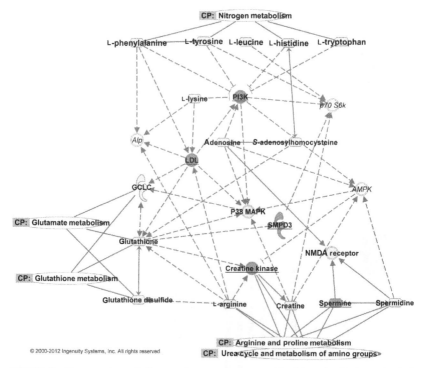

FIGURE 7 Top-scored metabolite-centric network generated by Ingenuity Pathway Analysis (IPA) describing amino acid metabolism, molecular transport, and small-molecule biochemistry. Microarray results were overlaid in the network highlighting associations between metabolites and transcripts in different canonical pathways (CP). Underlined molecules were downregulated; not underlined molecules were upregulated; Alp, p70 S6k, and AMPK (marked with an asterisk) were not altered. *Reproduced from Ref. (20).*

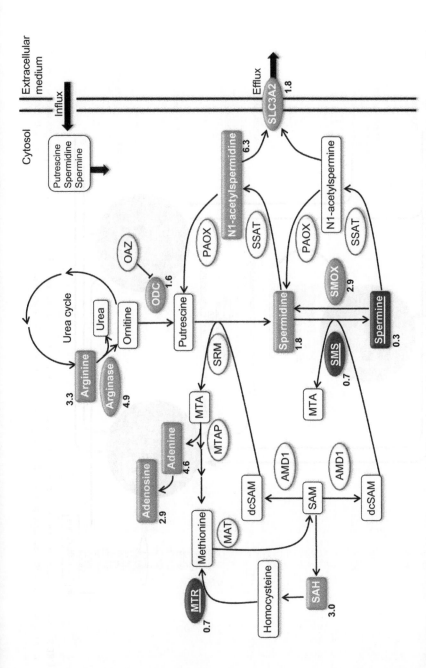

FIGURE 8 Polyamine pathway mapping transcriptomics and metabolomics results showing upregulation of acetylated polyamine efflux, as well as down- and upregulation of spermine biosynthesis and catabolism, respectively. Oval shape nodes represent the expression of genes and rectangular nodes represent metabolites. Downregulated components are underlined on a black background; upregulated are shown on a gray background. Numbers indicate gene/metabolite expression (fold change, natural scale) of the significant altered molecules. *Reproduced from Ref.* (20).

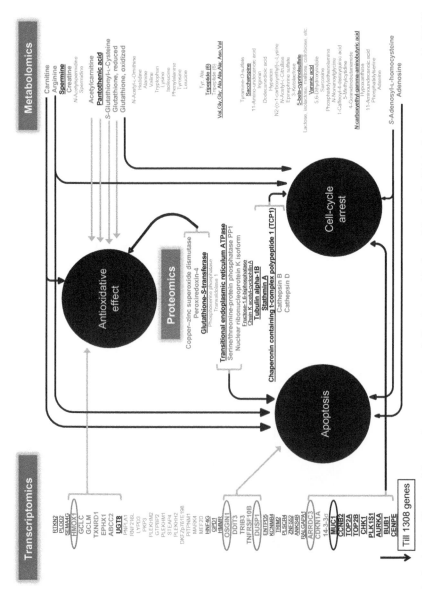

FIGURE 9 Foodomics identification of the proteins, genes, and metabolites involved in three of the principal biological processes altered in HT-29 colon cancer cells after the treatment with rosemary polyphenols. Underlined, downregulated; No underlined, upregulated. *Reproduced from Ref. (20).*

these three processes are shown in Figure 9 (detected up- and downregulated compounds are shown in Figure 9, downregulated compounds are underlined). Induction of apoptosis is especially relevant in colon cancer, since the renewal of the colon epithelium via apoptosis is the way used by our organism to eliminate deteriorated cells that can mutate to carcinogenic. Thus, significant variations observed for the metabolites arginine, *S*-adenosylhomocysteine, adenosine, and carnitine, together with changes induced in the proteins transitional endoplasmic reticulum ATPase, serine/threonine-protein phosphatase PP1, and nuclear ribonucleoprotein K isoform, and the variations observed in the genes *OSGIN1*, *DDIT3*, *TRIB3*, *TNFRSF10B*, and *DUSP1* can explain the antiproliferative effect of the rosemary polyphenols against HT-29 colon cancer cells due to stress-induced apoptosis. In the same way, cell-cycle arrest is also implicated in the antiproliferative effect from rosemary polyphenols, as can be deduced from changes in *S*-adenosylhomocysteine, oxidized glutathione, arginine, tubulin alpha-1B, stathmin A, chaperonin containing t-complex polypeptide 1, cathepsin B, cathepsin D, *ARRDC3*, *CDKN1A*, *14-3-3σ*, *MUC1*, *CCNB2*, *TOP2A*, *TOP2B*, *CHK1*, *PLK1S1*, *AURKA*, *BUB1*, and *CENPE* as detected by the three omics technologies used in this work and related to alteration of cell-cycle control by polyphenols. Additionally, significant variations observed for metabolites carnitine, acetylcarnitine, reduced glutathione, arginine, pantothenic acid, and *S*-glutathionylcysteine, together with changes in proteins copper–zinc superoxide dismutase, peroxiredoxin-4 and, glutathione-*S*-transferase, in conjunction with the variations detected in genes *HMOX1*, *GCLC*, *GCLM*, *TXNRD1*, *EPHX1*, *ABCC2*, and *UGT8*, can explain the chemopreventive effect from rosemary polyphenols via modulation of endogenous antioxidant enzymes and metabolites. Moreover, the relationship between each compound and the mentioned biological mechanism could be confirmed through the IPA information and/or literature for all the mentioned cases. In that report *(20)*, we remarked not only the impressive analytical power of the global Foodomics approach but also the limitations and challenges, mainly related to the lack of bioinformatic tools able to handle and integrate complex multidimensional data generated by the different omics platforms.

In another global Foodomics study on the antiproliferative effect of dietary polyphenols on human leukemia lines, Valdés *et al*. *(79)* combined whole-transcriptome microarray data with MS-based metabolomics data (obtained via CE-TOF-MS and UPLC-TOF-MS analysis). Functional enrichment and pathway analyses using IPA software as a previous step for a reliable interpretation of transcriptomic and metabolomic profiles were performed based on the results from each single platform. Rosemary polyphenols altered the expression of approximately 1% of the genes covered by the whole-transcriptome microarray in both leukemia cell lines. Overall, differences in the transcriptional induction of a number of genes encoding phase II detoxifying and antioxidant genes, as well as differences in the metabolic profiles observed in the two leukemia cell lines indicate that rosemary poly-phenols may exert a differential chemopreventive effect in leukemia cells with different phenotypes. IPA predictions on transcription factor analysis highlighted inhibition of *Myc* transcription factor function by rosemary poly-phenols, which may explain the observed antiproliferative effect of rosemary extract in the leukemia cells. Metabolomics analysis indicated that rosemary polyphenols affected differently the intracellular levels of some metabolites in two leukemia cell sublines. Integration of data obtained from transcrip-tomics and metabolomics platforms was attempted by overlaying datasets on canonical (defined) metabolic pathways using IPA software. This strategy enabled the identification of several differentially expressed genes in the met-abolic pathways modulated by rosemary polyphenols providing more evi-dences on the effect of these compounds. Using this Foodomics approach, stronger evidences on the chemopreventive and antioxidant response of leuke-mia cells to dietary polyphenols were provided.

4.3.2   Foodomics and the Human Gut Microbiome

The human gastrointestinal tract is home to an estimated 10^{14} bacteria, termed the gut microbiota, which surpasses by a factor of 10 the estimated 10^{13} human cells and represent by far the largest microbial community associated with the human body *(80)*. The intestinal bacteria in humans comprise 500–1000 bacterial species and complex metabolic relationships exist between bacterial communities and their host. Microbial profiles vary among individuals, with the average person harboring about 160 different species. The different compartments of the human gastrointestinal tract contain differ-ent bacterial species. Very little is known about these interbacterial relation-ships and how they relate to host health, but it is known that increasing diversity of the large bowel bacteria promotes metabolic homeostasis, stimu-lates development of the immune system, and the ability to resist invading pathogens. Likewise, the resident microbiota may also be an important factor in a number of diseases or conditions, including colon cancer, inflammatory bowel diseases, and obesity.

It is now well known that community composition and metabolic activity of the microbiota are influenced by a variety of factors including the diet and health status of the host. Intestinal bacteria live in an endosymbiotic relation-ship with a healthy host and regulate vital processes in the host related to metabolism (e.g., fermentation of digestion resistant carbohydrates leading to production of short-chain fatty acids), mucosal barrier function (e.g., inhi-bition of pathogen invasion and enhancement of epithelial barrier integrity), and immunity (e.g., priming mucosal immune system to maintain intestinal epithelium homeostasis). Nutrients interact directly with the host genome and indirectly through the collective genomes of the entire intestinal microbial community, commonly referred to as the metagenome, which therefore acts as an interface between food and host genome. Nutrients can be metabolized by intestinal microbes to produce metabolites that regulate target genes in the host metabolism. Changes in dietary intake of food components (e.g., fatty acids, carbohydrates, proteins and peptides, prebiotics, and probiotics) modulate gene expression in host intestine, as well as in liver, adipose tissue, and muscle and change the intestinal microbiota composition *(81)*.

The importance of the gut microbiota has pushed researchers to develop advanced analytical methods to investigate it. Thus, high-throughput NGS technologies have shown to enable comprehensive analyses of the effects of nutrients on mucosal immune responses (host genome) and microbial communities (microbiome) of the intestinal tract. Metagenomic methods have allowed the analysis of microbiota biological function and capability through the analysis of all bacterial genes present in microbiota indicating the most-abundant molecular functions (i.e., what genes are present and their functional potential). Metatranscriptomic analysis using NGS of mRNA (RNA-seq) has identified the genes that are most highly expressed in the community, while metabolomic analyses using HPLC-MS or NMR have provided a snapshot of metabolites resulting from host and microbiota interactions (i.e., the end products of metabolic interaction between host and microbe). Recently, a framework for gut metagenomic variation analysis was developed and applied it to 252 fecal metagenomes of 207 individuals from Europe and North America *(82)*. Using 7.4 billion reads aligned to 101 reference species, the authors detected 10.3 million SNPs, 107,991 short insertions/deletions, and 1051 structural variants. The average ratio of nonsynonymous to synonymous polymorphism rates of 0.11 was more variable between gut microbial species than across human hosts. Subjects sampled at varying time intervals exhibited individuality and temporal stability of SNP variation patterns, despite considerable composition changes of their gut microbiota. This indicates that individual-specific strains are not easily replaced and that an individual might have a unique metagenomic genotype, which may be exploitable for personalized diet or drug intake.

Millions of years of evolution have led to the acquisition of a complex intestinal microbiota that was selected for its capacity to maintain a symbiotic relationship with the host. This biota has formed through a complex set of environmental factors including dietary habits. Evidence suggests that this biota not only prevents pathogenic bacteria from accessing the epithelial barrier but also actively promotes the state of a healthy barrier through the action of their metabolism. Some bacteria such as *L. plantarum* appear to modulate the epithelial barrier through the action of secreted protein (*LGG* p40), whereas other such as *Clostridium* likely influences the barrier through production of metabolites (short-chain fatty acids). Much attention has been given to prebiotics (i.e., nondigestible nutrients), probiotics (i.e., beneficial microbes), and symbiotics (i.e., synergistic combination of pro- and prebiotics) to target changes in intestinal microbial composition, but variable results have been described, likely due to genetic heterogeneity of the study subjects. Despite the tremendous market value of the functional foods industry, claims of efficacy are often not supported by scientific evidence. In particular, it is now recognized that additional studies are required to support many of the health claims attributed to prebiotics and probiotics (83). Indeed, the EFSA issued negative opinions on the health claims of 416 products in 2010, including claims for the use of several probiotic products for maintaining digestive health and β-glucans for maintaining healthy blood glucose levels (84–86). In this regard, Foodomics studies can help to understand the biological mechanisms of action, and they can have a crucial role in providing information to enable informed decisions to be made on supplement use.

In view of the richness and diversity of the microbiota, it would be important to investigate in detail this biota and identify microorganisms with barrier protective function. Because of the interplay between diet and microbial composition, identification of nutritional components that contribute to barrier function should also be a forefront priority. Integration of microbial genomic, metabolomics, proteomics, and transcriptomic technology as proposed by Foodomics would be essential to carry this mission forward. Alterations in the host or microbiota transcriptome may not always result in changes in the proteome or metabolome in response to diet. Analysis of the metabolite profile in urine or blood may further elucidate the interactions between microbiota and host and how microbial activities are related to host phenotype and metabolism. Characterizing host or microbiota proteomes may also reveal changes in gene products that cannot be explained by alterations in transcription, such as those resulting from PTMs or alternate splicing of transcripts. Metagenomic analysis of the bacterial genes present in the large bowel may provide useful information regarding the selection pressures exerted in this ecosystem by determining which genes are enriched under certain situations. Understanding the intricate relationship between epithelial barrier, microbe, and diet would undeniably contribute key knowledge that could be harness for therapeutic purpose (87). The decreasing costs associated with high-throughput technologies and the advent of new single-molecule sequencing platforms provide great potential for extending existing knowledge. Despite the complex nature of food, host, and microbe interactions, an integrated Foodomics approach for studying these systems offers the potential for effective and personalized strategies for alleviating different illness in which gut microbiota plays an important role including colon cancer, inflammatory bowel diseases, and obesity. In summary, the microbiome is a key nutritional target today and might also become the foundation of future drug targeting and interventions.

4.4   Green Foodomics

As mentioned throughout the text, Foodomics can be understood as a global framework that gathers all the new challenges that the food science and nutrition domains will be facing in the current postgenomic era and providing new answers through the development and application of new strategies, mainly based on "omics" approaches for large-scale analysis. In this regard, one of the challenges that can impact future generations is sustainability, understood as a rational way of improving processes to maximize production while minimizing the environmental impact or, in the words of the Environmental Protection Agency (EPA), "sustainability creates and maintains the conditions under which humans and nature can exist in productive harmony, that permit fulfilling the social, economic and other requirements of present and future generations." Thus, Foodomics can be considered a green discipline in its goals since one of its main challenges is to preserve sustainability and green Foodomics tries to highlight Foodomics' goals with regard to green chemistry principles.

Topics related to environmental sustainability in Foodomics deal with the advancement and development of green analytical chemistry methods, with the design of novel foods and food ingredients but also with food production systems. In this world with an increasing population and therefore, with an increasing growing requirement for food, many aspects should be considered such as those related to intensive farming, pollution, desertification, climate change, energy requirements, etc.; green Foodomics would, ideally, gather all this information and help addressing the problems and the global solutions needed to handle the complexity of food supply in the world. Undoubtedly, this is a challenge that needs more focus and a special commitment from the different actors (researchers, politicians, industries, regulatory agencies, etc.) to become a reality in a near future.

Foodomics' environmental impact can be improved by greening the tools and protocols used applying basic concepts of Green Chemistry and Green Analytical Chemistry. In this sense, there are three main aspects that dominate the 12 principles that rule Green Chemistry: waste, hazard (health, environmental, and safety), and energy. Green chemistry has set itself the goal of making chemical technology more environmentally benign by "efficiently using (preferably renewable) raw materials, eliminating waste and avoiding the use of toxic and/or hazardous reagents and solvents in the manufacture and application of chemical products" (88). These basic rules should mark the track to follow in terms of processes involved in Foodomics. In this sense, the development of green processes involved in the production of, for example, functional food ingredients for their use in Foodomics' studies has been suggested (89) through the employment of environmentally clean techniques able to overcome limitations of classical extraction processes (such as those related to long extraction times, high volumes of toxic organic solvents, large amount of sample, low extraction yields, and low selectivity). These new processes deal with the application of assisted extraction techniques that use physical processes (such as high-frequency acoustic waves, microwaves, pressure, temperature, among others) to improve the efficiency and speed up the extraction process; moreover, these techniques allow to reduce the amount of wastes generated, help to decrease energy consumption, and offer the possibility of eliminating postextraction procedures while allowing the development of integrated or hyphenated processes in a single one; this area constitutes a very active field of research in the Foodomics discipline.

As for Green Analytical Chemistry's definition by L.H. Lawrence: "the use of analytical chemistry techniques and methodologies that reduce or eliminate solvents, reagents, preservatives and other chemicals that are hazardous to human health or the environment and that may also enable faster and more energy-efficient analysis without compromising performance criteria" *(90)*, it is clear that there are two key aspects that have to be considered: hazard reduction, as compared to traditional methods, and analytical performance maintenance or improvement. Therefore, the key aspects that should be considered when regarding the adverse environmental impact of analytical methods deal with reducing the amount and toxicity of solvents during sample pretreatment, minimizing solvents and reagents during the separation and measurement steps and developing alternative direct analytical methods that do not require solvents or reagents. Moreover, they should also consider developing methods able to consume fewer resources (such as energy), reduce costs in terms of waste generation and management, and reduce health risks. And, as mentioned, this has to be done while maintaining or improving the analytical performance of the method. Undoubtedly, this aspect has been suggested as the responsible for a limited translation of conventional analytical methods to greener ones. For a deeper knowledge on the essential guidelines for making analytical laboratories greener, readers are referred to Galuszka *et al. (91)*.

Different approaches have been suggested to improve greenness of Foodomics platforms involved in these types of studies *(89)*, mainly dealing with the direct analysis of samples (through the use of NMR- and MS-based methods), the improvement of sample preparation techniques by reducing the use of organic solvents and energy; the greenness of the separation techniques through the replacement of organic solvents, the minimization of solvents' use through miniaturization of the technique, or by employing new technologies in separation columns; and the generalized use of analytical microsystems, that is, integrated analytical systems at microscale to perform in one device all analytical steps (sample preparation, analytes separation and detection).

The development of green Foodomics runs parallel to the improvement and design of techniques able to assess the environmental impact of the different protocols/processes/operations involved. At present several techniques can be found in the literature to test, for instance, the impact of analytical chemistry methods (such as the Greeness profile, the HPLC-EAT, or the Analytical Eco-Scale) and the environmental impacts associated with a product or process, over its entire life cycle (such as Life Cycle Assessment). Nevertheless, techniques able to provide a more holistic view of the different aspects covered by Foodomics are needed to approach to the sustainability goal of this discipline.

Recently, Weckwerth *(92)* reviewed the concept and opportunities of green systems biology as a way to investigate plants genotype–phenotype relationship using and holistic approach, combining the information provided by genome sequencing, gene reconstruction, and annotation with the genome-scale molecular analysis using -omics technologies and computer-assisted analysis, modeling and interpretation of biological data. By using this approach, important advancements in biodiversity preservation, improvements of plant productivity, and diversity for world feeding and renewable energy resources toward sustainability are expected.

5 FUTURE TRENDS AND OPPORTUNITIES FOR FOODOMICS

Apart from the challenges to be faced in the near future in food science and nutrition already described above (e.g., the production of new functional foods with scientifically proved claims; the improvement of food safety, quality, and traceability; the development, production, and monitoring of new transgenic foods, including the so-called second-generation GM crops; environmental sustainability, etc.), there are other interesting trends and opportunities that can be anticipated for Foodomics that are described below.

5.1 Foodomics and Personalized Nutrition: How Far We Are?

One of the main long-term goals of Foodomics in the global frame of food, nutrition, and health is to provide recommendations for personalized diet that lead to a specific phenotype and subsequently to prevent the onset and progression of disease. Goals may include weight loss/gain (or optimization of body composition), physical performance, or reduction in disease risk. Numerous factors contribute to variation in nutritional requirements and responses to diet from each individual, including sex, stage of life cycle, disease, physical activity level, genetic (and epigenetic) background, gut microbial community, and environmental exposures. Although several of these are nowadays considered in the construction of personalized nutritional recommendations (e.g., sex, age, adiposity, and activity level), to date, the more complex factors such as genomics (and epigenomics), host–microbial community structure, and environmental exposures are often not included in the equation *(93)*. In this regard, although observational studies have provided some insights into associative interactions between genetic or phenotypic variation and diet and their impact on health, very few human experimental studies have addressed these relationships. Given the growth in our knowledge, there is the potential to develop personalized dietary recommendations. However, developing these recommendations assumes that an improved understanding of the phenotypic complexities of individuals and their responses to the complexities of their diets will lead to a sustainable, effective approach to promote health and prevent disease. To achieve this goal, dietary interventions that test prescribed diets in well-characterized study populations and that monitor system-wide responses (ideally using several omics platforms as proposed by Foodomics) are needed to make correlation–causation connections and to characterize phenotypes under controlled conditions. Advances in analytical techniques now make possible to collect large amounts of genetic, epigenetic, metabolomic, and gut microbiome data. One of the main challenges here is integrating such data in a systematic way, in order to provide a holistic view of biological systems. These data have the potential to transform approaches toward nutrition advising by allowing us to recognize and embrace the metabolic, physiologic, and genetic differences among individuals. The ultimate goal is to be able to integrate these multidimensional data so as to characterize the health status and disease risk of an individual and to provide personalized recommendations to maximize health (93). As more and more measurements become available, the complexity of the analysis, that is, the number of variables in statistical models, increases. This poses additional challenges for design of trials and requires the use of advanced statistical models appropriate for analysis of high-dimensional problems. Randomized, controlled, double-blind, clinical intervention trials are the ones which can provide the required scientific evidence (level I of evidence) and, to some extent, large cohort studies (level II of evidence) can also do so (94). It is no surprising that individuals exhibit different physiologies, even if challenged with analogous diets. Nutritional studies need to deal with a large interindividual variability, especially in human cohorts. The establishment of baselines by introducing control groups in nutritional studies is pivotal in evaluating nutritional outcomes. Lifestyle, age, gender, genetic and epigenetic background, food habits, gut microbiota, and body composition are some of the variables that characterize the discrepancies between individuals' response. The absorption of nutrients is influenced by food preparation and cooking, as well as food combinations. The digestion, transport, and temporal effects of food ingredients in the human body is complex, given its compartmentalization and organ organization, leading to differential biotransformation, partition, availability, and excretion. Another major challenge is related to the quantification of a nutritional effect, especially on health (in opposition to disease effects). A definition of health is thus difficult to establish, making an improvement in health status a difficult evaluation. Moreover, differences between subjects are often much larger than differences between nutritional interventions. Nutrition provokes changes of minimum magnitude, as a result of multiple minor genetic differences, low gene expression, and protein alterations, modulated by numerous ingredients, with an immensity of confounding factors (95).

While nutritional advice from public health authorities is getting more specific for subpopulations, it is still a long way from one of the claimed ultimate goals of personalized nutrition: to tailor dietary advice for the individual to optimally achieve a lifestyle goal based on diet. To date, commercialized whole-genome sequencing is available and coming down rapidly in price. However, the type of advice that is provided based on SNP profiling is questionable, given the current lack of understanding of the complexity of multigene-based differences (disease cannot be considered if one considers that a personalized nutrition approach would ultimately seek to mitigate disease risk, rather than treat disease) and the difficulty in replicating SNP associations across different populations and cohorts.

Introduction of genetic tests seems therefore premature, especially when considering the fact that any recommendations and decisions may be based on insufficient or even inadequate scientific data. Robust and validated data are difficult to obtain, because of the limited availability of biomarkers of health as well as the high cost of human Foodomics intervention studies *(96)*.

Nutritional studies need systems-wide approaches, as proposed by Foodomics, to deal with population heterogeneity (genome, epigenome, gut microbiome), diet heterogeneity and complexity (as multi-ingredient ensembles), and low amplitude effects on metabolism. It has been also proposed *(95)* that in contrast to whole-genome sequencing, metabolomics may appear to be an ideal tool to address issues in nutrition related to excesses or deficiencies in known nutrients or changes in key diet-related metabolic pathways such as central energy metabolism and gut microbiota metabolism. Metabolomics could be used both as a way of identifying biomarkers/biomarker patterns that could be used as diagnostic tests or as a diagnostic tool in its own right, measuring multiple metabolic pathways simultaneously. There remains much that is unknown or controversial, and there is much to be learned from the experience of pharmaceutical research in this field. However, fundamental differences between drugs and food also mean that novel approaches will be needed.

A growing trend toward systemic approaches in which different Foodomics approaches are combined and applied is foreseen. This will allow physiological changes to be assessed very robustly throughout the various molecular layers of mRNA, protein, and metabolite changes. Foodomics is still growing as a branch of the life sciences and is gaining significant recognition in the scientific community. In summary, personalized nutrition may be a great opportunity for Foodomics to demonstrate all its potential, but (1) it will be necessary to carry out more studies in terms of, for example, discovering more polymorphisms of one nucleotide, identifying genes related to complex disorders, and demonstrating a higher degree of evidence through epidemiological studies based on Foodomics that can lead to public recommendations; (2) it will be necessary to better analyze the interrelationship among genetic variants, epigenome, nutrients, and environmental factors including their effect on microbiota; (3) it will be mandatory to support a better nutritional education and to extend the research on new food products. This knowledge can only be generated using multidisciplinary approaches, considering international consortia and working on Foodomics based on extensive populations. International agreement on consistency of study designs, technologies, data analysis, and interpretation will help to ensure comparability of data across study populations. As a long-term result, a personalized diet depending on a particular phenotypic profile can become a reality.

5.2 The Role of Foodomics in Pangenomics

Today, thousands of tons of commercial starter cultures (i.e., *Lactococcus lactis, Streptococcus thermophilus*, and a variety of *Lactobacillus* species) are used each year to produce millions of tons of cheese, yogurt, and other dairy products. Important features of starters for cheese-making are rapid acidification, robustness toward bacteriophage, and the development of the desired flavor characteristic of each particular cheese variety. In yogurt production, the development of the desired texture is also an essential consideration. Probiotics are living microorganisms that are consumed specifically to obtain a health benefit. Most probiotics are from the genera *Lactobacillus* and *Bifidobacterium*. They can be consumed directly as part of a dietary supplement or as a component of a health-promoting food, most often a fermented dairy product. Benefits of consumption of probiotics include improved gastrointestinal tract health and improved immune function.

With the continuous reductions in the cost and time involved in DNA sequencing, a new approach to characterize bacteria (as the ones used as starter cultures and probiotics) is emerging. It is based on a comparison of complete genome sequences of a number of members of the same species, which has been called pangenomics *(97)*. The pangenome is the sum of all genes present within the species while the core genome is the set of genes shared by all the members of the species. Sequences present in the pangenome but absent in the core genome are considered variable genes or dispensable genes and in some cases represent genes acquired by horizontal gene transfer. As more genome sequences are determined, the size of the core genome is likely to decrease, while the pangenome will increase. Pangenomics opens an array of new opportunities for understanding and improving industrial starter cultures and probiotics. These include understanding the formation of texture and flavor in dairy products, understanding the functionality of probiotics (including healthy or pathogenic relationships with the host) as well as providing information that can be used for strain screening, strain improvement, safety assessments, and process improvements *(98)*. For instance, comparing genomes and pangenomes makes possible to thoughtfully select combinations of starter cultures to provide metabolic cascades that produce tailored flavor development in cheese and other fermented products. Low cost, rapid genome sequencing is a reality and has become a magnificent tool for strain characterization (see, e.g., the homepage http:/www.ebi.ac.uk/genomes/ from the European Bioinformatics Institute's, which contains the complete genome of nearly 100 strains of lactic acid bacteria and *Bifidobacterium* in a database of more than 2000 bacterial entries). Analysis of the pangenome is a relatively new concept for industrial starter cultures and probiotics with only a limited number of studies reported. Therefore, the opportunities for Foodomics in this field are really huge, considering the low number of works published and the interest of the above-mentioned topics (improving strain screening and dairy products quality based on genome selection, enhancing safety assessments and dairy process, understanding the functionality and relationships of probiotics with the host). Moreover, genome sequence comparisons will play an important role in connecting genotypes to phenotypes, especially when the phenotype is complex or poorly understood, while comparative genomics will inevitably be matched with comparative proteomics to gain even deeper insights into gene functions in probiotics *(98)*. These new approaches are expected to bring microbiology to a new era.

ACKNOWLEDGMENTS

This work was supported by AGL2011-29857-C03-01 (Ministerio de Ciencia e Innovación, Spain) and CSD2007-00063 FUN-C-FOOD (Programa CONSOLIDER, Ministerio de Educacion y Ciencia, Spain) projects.

REFERENCES

1. Slater, N.; Wilson, I. http:/www.ceb.cam.ac.uk/pages/foodomics.html (accessed on August 26, 2013).
2. Capozzi, F.; Placucci, G. *1st International conference in Foodomics, Cesena, Italy* 2009.
3. Cifuentes, A. *J. Chromatogr. A* **2009**, *1216*, 7109–7110.
4. Herrero, M.; García-Cañas, V.; Simó, C.; Cifuentes, A. *Electrophoresis* **2010**, *31*, 205–228.
5. Herrero, M.; Simó, C.; García-Cañas, V.; Ibáñez, E.; Cifuentes, A. *Mass Spec. Rev.* **2012**, *31*, 49–69.
6. García-Cañas, V.; Simó, C.; Herrero, M.; Ibáñez, E.; Cifuentes, A. *Anal. Chem.* **2012**, *84*, 10150–10159.
7. García-Cañas, V.; Simó, C.; Castro-Puyana, M.; Cifuentes, A. *Electrophoresis* **2014**, *35*, 147–169. http://dx.doi.org/10.1002/elps.201300315.
8. Cifuentes, A. In: *Foodomics;* Cifuentes, A., Ed.; John Wiley & Sons: New Jersey, 2013; pp 1–13.
9. Valdés, A.; García-Cañas, V.; Rocamora-Reverte, L.; Gómez-Martínez, A.; Ferragut, J. A.; Cifuentes, A. *Genes Nutr.* **2013**, *8*, 43–60.
10. García-Cañas, V.; Simó, C.; León, C.; Cifuentes, A. *J. Pharm. Biomed.* **2010**, *51*, 290–304.
11. Anderle, P.; Farmer, P.; Berger, A.; Roberts, M. A. *Nutrition* **2004**, *20*, 103–108.
12. Gohil, K.; Chakraborty, A. A. *Nutrition* **2004**, *20*, 50–55.
13. Morozova, O.; Marra, M. A. *Genomics* **2008**, *92*, 255–264.
14. Storhoff, J. J.; Marla, S. S.; Mirkin, C. A. In: *Microarray Technology and Its Applications;* Müller, U. R.; Nicolau, D. V., Eds.; Springer-Verlag: Berlin, Heidelgerg, 2005; pp 147–180.
15. Stafford, P.; Tak, Y. In: *Methods in Microarray Normalization;* Safford, P., Ed.; CRC Press: Boca Raton, FL, 2008; pp 151–203.
16. Santos, Dos In: *Bioinformatics. A Practical Approach;* Ye, S. Q., Ed.; Chapman & Hall/CRC: Boca Raton, FL, 2008; pp 131–187.
17. Mardis, E. R. *Annu. Rev. Genomics Hum. Genet.* **2008**, *9*, 387–402.
18. Margulies, M.; Egholm, M.; Altman, W. E.; Attiya, S.; Bader, J. S.; Bemden, L. A.; Berka, J.; Braverman, M. S.; Chen, Y. J.; Chen, Z.; Dewell, S. B.; Du, L.; Fierro, J. M.; Gomes, X. V.;

Godwin, B. C.; He, W.; Helgesen, S.; Ho, C. H.; Irzyk, G. P.; Jando, S. C.; Alenquer, M. L.; Jarvie, T. P.; Jirage, K. B.; Kim, J. B.; Knight, J. R.; Lanza, J. R.; Leamon, J. H.; Lefkowitz, S. M.; Lei, M.; Li, J.; Lohman, K. L.; Lu, H.; Makhijani, V. B.; McDade, K. E.; McKenna, M. P.; Myers, E. W.; Nickerson, E.; Nobile, J. R.; Plant, R.; Puc, B. P.; Ronan, M. T.; Roth, G. T.; Sarkis, G. J.; Simons, J. F.; Simpson, J. W.; Srinivasan, M.; Tartaro, K. R.; Tomasz, A.; Vogt, K. A.; Volkmer, G. A.; Wang, S. H.; Wang, Y.; Weiner, M. P.; Yu, P.; Begley, R. F.; Rothberg, J. M. *Nature* 2005, *437*, 376–380.

19. Schendure, J.; Porreca, G. J.; Reppas, N. B.; Lin, X.; McCutcheon, J. P.; Rosenbaum, A. M.; Wang, M. D.; Zhang, K.; Mitra, R. D.; Church, G. M. *Science* 2005, *309*, 1728–1732.

20. Ibáñez, C.; Valdés, A.; García-Cañas, V.; Simó, C.; Celebier, M.; Rocamora, L.; Gómez, A.; Herrero, M.; Castro, M.; Segura-Carretero, A.; Ibáñez, E.; Ferragut, J. A.; Cifuentes, A. *J. Chromatogr. A* 2012, *1248*, 139–153.

21. Mena, M.; Albar, J. P. In: *Foodomics;* Cifuentes, A., Ed.; John Wiley & Sons: New Jersey, 2013; pp 15–67.

22. Dumas, M. E.; Maibaum, E. C.; Teague, C.; Ueshima, H.; Zhou, B.; Lindon, J. C.; Nicholson, J. K.; Stamler, J.; Elliot, P.; Chan, Q.; Holmes, E. *Anal. Chem.* 2006, *78*, 2199–2208.

23. Bartel, D. P. *Cell* 2009, *136*, 215–233.

24. Friedman, R. C.; Farh, K. K.; Burge, C. B.; Bartel, D. P. *Genome Res.* 2009, *19*, 92–105.

25. Giovannetti, E.; van der Velde, A.; Funel, N.; Vasile, E.; Perrone, V.; Leon, L. G.; De Lio, N.; Avan, A.; Caponi, S.; Pollina, L. E.; Gallá, V.; Sudo, H.; Falcone, A.; Campani, D.; Boggi, U.; Peters, G. J. *PLoS One* 2012, *7*, e49145.

26. Varma, V.; Kaput, J. In: *Nutrigenomics and Nutrigenetics in Functional Foods and Personalized Nutrition;* Ferguson, L. R., Ed.; CRC Press: Boca Raton, 2014.

27. Mathers, J. C.; Strathdee, G.; Relton, C. L. *Adv. Genet.* 2010, *71*, 3–39.

28. Milagro, F. I.; Mansego, M. L.; De Miguel, C.; Martínez, J. A. *Mol. Asp. Med.* 2013, *34*, 781–812.

29. Lahrichi, S. L.; Affolter, M.; Zolezzi, I. S.; Panchaud, A. *J. Proteomics* 2013, *8* (8), 83–91.

30. Bondia-Pons, I.; Hyotylainen, T. Lipidomics. In: *Foodomics;* Cifuentes, A., Ed.; John Wiley & Sons: New Jersey, 2013; pp 351–379.

31. Zoldoš, V.; Horvat, T.; Lauc, G. *Curr. Opin. Chem. Biol.* 2013, *17*, 34–40.

32. The Gene Ontology Consortium. *Nucleic Acids Res.* 2008, *36*, D440–D444.

33. The Gene Ontology Consortium. *Nat. Gen.* 2000, *25*, 25–29.

34. Cifuentes, A., Ed.; *Foodomics*; John Wiley & Sons: New Jersey, 2013.

35. (a) Hood, L.; Perlmutter, R. M. *Nat. Biotechnol.* 2004, *22*, 1215–1216; (b) Rigoutsos, I. In: *Systems Biology;* Rigoutsos, I., Ed.; Oxford University Press: Cary, NC, USA, 2006.

36. Panagiotou, G.; Nielsen, J. *Annu. Rev. Nutr.* 2009, *29*, 329–339.

37. Oresic, M. In: *Foodomics;* Cifuentes, A., Ed.; John Wiley & Sons: New Jersey, 2013; pp 539–550.

38. Bakker, G. C. M.; van Erk, M. J.; Pellis, L.; Wopereis, S.; Rubingh, C. M.; Cnubben, N. H. P.; Kooistra, T.; van Ommen, B.; Hendriks, H. F. *J. Am. J. Clin. Nutr.* 2010, *91*, 1044–1051.

39. Castro-Puyana, M.; Mendiola, J. A.; Ibáñez, E.; Herrero, M. In: *Foodomics;* Cifuentes, A., Ed.; John Wiley & Sons: New Jersey, 2013; pp 453–470.

40. Kim, H. J.; Park, S. H.; Lee, T. H.; Hanm, B. H.; Kim, R.; Kim, H. Y. *Biosens. Bioelectron.* 2008, *24*, 238–246.

41. Wang, X. W.; Zhang, L.; Jin, L. Q. *Appl. Microbiol. Biotechnol.* 2007, *76*, 225–233.

42. Liu, Y.; Elsholz, B.; Enfors, S. O.; Gabig-Ciminska, M. *J. Microbiol. Methods* 2007, *70*, 55–64.

43. Schmidt-Heydt, M.; Geisen, R. *Int. J. Food Microbiol.* 2007, *117*, 131–140.

44. Narasaka, S.; Endo, Y.; Fu, Z. W. *Biosci. Biotechnol. Biochem.* **2006**, *70*, 1464–1470.
45. Huzarewich, R. L.; Siemens, C. G.; Booth, S. A. *J. Biomed. Biotechnol.* **2010**, *2010*, 613504.
46. Cevallos, J. M.; Danyluk, M. D.; Reyes-De-Corcuera, J. I. *J. Food Sci.* **2011**, *76*, M238–M246.
47. Picariello, G.; Mamone, G.; Addeo, F.; Nitride, C.; Ferranti, P. In: *Foodomics;* Cifuentes, A., Ed.; John Wiley & Sons: New Jersey, 2013; pp 69–100.
48. Inoue, K.; Nishimura, M.; Tsutsui, H.; Min, J. Z.; Todoroki, K.; Kauffmann, J. M.; Toyo'oka, T. *J. Agric. Food Chem.* **2013**, *61*, 1228–1234.
49. Shu, X. L.; Frank, T.; Shu, Q. Y.; Engel, K. H. *J. Agric. Food Chem.* **2008**, *6*, 11612–11620.
50. Dobson, G.; Shepherd, T.; Verrall, S. R.; Griffiths, W. D.; Ramsay, G.; McNicol, J. W.; Davies, H. V.; Stewart, D. *J. Agric. Food Chem.* **2010**, *58*, 1214–1223.
51. Wechter, W. P.; Levi, A.; Harris, K. R. *BMC Genomics* **2008**, *9*, 275–280.
52. Martinez-Godoy, M. A.; Mauri, N.; Juarez, J. *BMC Genomics* **2008**, *9*, 318–324.
53. Kang, H. J.; Yang, H. J.; Kim, M. J.; Han, E. S.; Kim, H. J.; Kwon, D. Y. *Food Chem.* **2011**, *127*, 1056–1064.
54. Cuny, M.; Vigneau, E.; Le Gall, G.; Colquhoun, I.; Lees, M.; Rutledge, D. N. *Anal. Bioanal. Chem.* **2008**, *390*, 419–427.
55. Almeida, C.; Duarte, I. F.; Barros, A.; Rodrigues, J.; Spraul, M.; Gil, A. A. *J. Agric. Food Chem.* **2006**, *54*, 700–706.
56. Ercolini, D.; Russo, F.; Nasi, A.; Ferranti, P.; Villani, F. *Appl. Environ. Microbiol.* **2009**, *75*, 1990–2001.
57. Valdés, A.; García-Cañas, V. In: *Foodomics;* Cifuentes, A., Ed.; John Wiley & Sons: New Jersey, 2013; pp 191–220.
58. Chassy, B. M. *Reg. Toxicol. Pharm.* **2010**, *58*, S62–S70.
59. Kuiper, H. A.; Kleter, G. A.; Hub, P. J.; Kok, E. J. *Plant J.* **2001**, *27*, 503–528.
60. Ye, X.; Al-Babili, S.; Kloti, A.; Zhang, J.; Lucca, P.; Beyer, P.; Potrykus, I. *Science* **2000**, *287*, 303–305.
61. EFSA. *Guidance Document of the Scientific Panel on Genetically Modified Organisms for the Risk Assessment of Genetically Modified Plants and Derived Food and Feed.* EFSA Communications Departmente: Parma, Italy, 2006.
62. Ad Hoc Technical Expert Group (AHTEG). *Guidance Document on Risk Assessment of Living Modified Organisms*; UnitedNations Environment Programme Convention for Biodiversity, Ljubljana, 2010. http://www.cbd.int/doc/meetings/bs/bsrarm-02/official/bsrarm-02-05-en.pdf.
63. García-Villalba, R.; León, C.; Dinelli, G.; Segura-Carretero, A.; Fernández-Gutiérrez, A.; García-Cañas, V.; Cifuentes, A. *J. Chromatogr. A* **2008**, *1195*, 164–173.
64. Levandi, T.; León, C.; Kaljurand, M.; García-Cañas, V.; Cifuentes, A. *Anal. Chem.* **2008**, *80*, 6329–6335.
65. García-Cañas, V.; Simó, C.; León, C.; Ibáñez, E.; Cifuentes, A. *Mass Spec. Rev.* **2011**, *30*, 396–416.
66. García-Cañas, V.; Simó, C.; Herrero, M.; Ibáñez, E.; Cifuentes, A. In: *Omics Technologies: Tools for Food Science;* Benkeblia, N., Ed.; CRC Press: Boca Raton, 2012.
67. (a) Heinemann, J. A.; Kurenbach, B.; Quist, D. *Environ. Int.* **2011**, *37*, 1285–1293; (b) Ricoh, A. E. *New Biotechnol.* **2013**, *30*, 349–354.
68. Ricoh, A. E. *New Biotechnol.* **2013**, *30*, 349–354.
69. Kusano, M.; Redestig, H.; Hirai, T.; Oikawa, A.; Matsuda, F.; Fukushima, A.; Arita, M.; Watanabe, S.; Yano, M.; Hiwasa-Tanase, K.; Ezura, H.; Saito, K. *PLoS One* **2011**, *16*, 16989.
70. Barros, E.; Lezar, S.; Anttonen, M. J.; van Dijk, J. P.; Rohlig, R. M.; Kok, E. J.; Engel, K. H. *Plant Biotechnol. J.* **2010**, *8*, 436–451.

71. Capozzi, F.; Bordoni, A. *Genes Nutr.* **2013**, *8*, 1–4.

72. Ferguson, L. R., Ed.; *Nutrigenomics and Nutrigenetics in Functional Foods and Personalized Nutrition*; CRC Press: Boca Raton, 2014.

73. Diplock, A. T.; Aggett, P. J.; Ashwell, M.; Bornet, F.; Fern, E. B.; Roberfroid, M. B. *Br. J. Nutr.* **1999**, *81*, S1–S27.

74. Mondello, L. *Anal. Bioanal. Chem.* **2013**, *405*, 4589–4590.

75. EFSA. *Opinions of the NDA Panel Published on 2009 and 2010*; http:/www.efsa.europa.eu/cs/Satellite; 2010, (accessed on August 26, 2013).

76. Gilsenan, M. B. *Trends Food Sci. Technol.* **2011**, *22*, 536–542.

77. Daniells, S. *EFSA's Antioxidant Rejections Could be Blessing in Disguise*; http:/www.foodqualitynews.com/Legislation/EFSA-s-antioxidant-rejections-could-be-blessing-in-disguise; 2010, (accessed on August 26th, 2013).

78. Starling, S. *EFSA Mass Rejects Probiotics and Antioxidants as Article 13.1 Batch Two Published*; http:/www.beveragedaily.com/Product-Categories/Ingredients-and-additives/EFSA-mass-rejects-probiotics-and-antioxidants-as-article-13.1-batch-two-published; 2010, (accessed on August 26, 2013).

79. Valdés, A.; Simó, C.; Ibáñez, C.; Rocamora-Reverte, L.; Ferragut, J. A.; García-Cañas, V.; Cifuentes, A. *Electrophoresis* **2012**, *33*, 2314–2327.

80. Young, W.; Knoch, B.; Roy, N. C. In: *Nutrigenomics and Nutrigenetics in Functional Foods and Personalized Nutrition;* Ferguson, L. R., Ed.; CRC Press: Boca Raton, 2014.

81. Delzenne, N. M.; Neyrinck, A. M.; Cani, P. D. *Microb. Cell Fact.* **2011**, *10* (Suppl. 1), S10.

82. Schloissnig, S.; Arumugam, M.; Sunagawa, S.; Mitreva, M.; Tap, J.; Zhu, A.; Waller, A.; Mende, D. R.; Kultima, J. R.; Martin, J.; Kota, K.; Sunyaev, S. R.; Weinstock, G. M.; Bork, P. *Nature* **2013**, *493*, 45–50.

83. Sanders, M. E.; Tompkins, T.; Heimbach, J. T.; Kolida, S. *Eur. J. Nutr.* **2005**, *44*, 303–310.

84. EFSA Panel on Dietetic Products, N. a. A. N. *EFSA J.* **2010**, *8*, 1487.

85. EFSA Panel on Dietetic Products, N. a. A. N. *EFSA J.* **2010**, *8*, 1482.

86. EFSA Panel on Dietetic Products, N. a. A. N. *EFSA J.* **2010**, *8*, 1488.

87. Guzman, J. R.; Conlin, V. S.; Jobin, C. *BioMed Res. Int.* **2013**, *425146*, 2013.

88. Sheldon, R. A. *Chemistry* **2000**, *3*, 541–551.

89. Mendiola, J. A.; Castro-Puyana, M.; Herrero, M.; Ibáñez, E. In: *Foodomics;* Cifuentes, A., Ed.; John Wiley & Sons: New Jersey, 2013; pp 471–506.

90. Sandra, P.; Vanhoenacker, G.; David, F.; Sandra, K.; Pereira, A. *LCGC-Eur.* **2010**, *23*, 242–259.

91. Gałuszka, A.; Migaszewski, Z.; Namieśnik, J. *TrAC Trends Anal. Chem.* **2013**, *50*, 78–84.

92. Weckwerth, W. *J. Proteomics* **2011**, *75*, 284–305.

93. Lampe, J. W.; Navarro, S. L.; Hullar, M. A. J.; Shojaie, A. *Proc. Nutr. Soc.* **2013**, *72*, 207–218.

94. Konstantinidou, V.; Covas, M. I.; Sola, R.; Fito, M. *Mol. Nutr. Food Res.* **2013**, *57*, 772–783.

95. Scherer, M.; Ross, A.; Moco, S.; Collino, S.; Martin, F. P.; Godin, J. P.; Kastenmayer, P.; Montoliu, I.; Rezzi, S. In: *Foodomics;* Cifuentes, A., Ed.; John Wiley & Sons: New Jersey, 2013; pp 271–301.

96. Roos, B. *Proc. Nutr. Soc.* **2013**, *72*, 48–52.

97. Medini, D.; Donati, C.; Tettelin, H.; Masignani, V.; Rappuoli, R. *Curr. Opin. Genet. Dev.* **2005**, *15*, 589–594.

98. Garrigues, C.; Johansen, E.; Crittenden, R. *Curr. Opin. Biotechnol.* **2013**, *24*, 187–191.

Chapter 16

Omics Data Integration in Systems Biology: Methods and Applications

Ana Conesa and Rafael Hernández

Genomics of Gene Expression Laboratory, Centro de Investigación Príncipe Felipe, Valencia, Spain

1 INTRODUCTION

The advances of sequencing technologies of the 1990s made possible the sequencing of the first genomes and were also the germ for a new conception in biological research characterized by the comprehensive measurement of the nucleic acid molecular components that make up the living systems. The subsequent development of high-throughput technologies targeting specific molecular types gave rise to a new range of research disciplines named genomics, epigenomics, transcriptomics, proteomics, and metabolomics that are collectively referred to as "omics." Thanks to these technological advances, it is now practical and affordable to sequence entire genomes to search for variants associated to diseases; to measure the activities of genes and proteins in response to an environmental challenge; to map epigenetic modifications during development; to scan for protein–protein or nucleic acid–protein interactions on a genome-wide scale; and to measure sugars, lipids, amino acids,

Comprehensive Analytical Chemistry, Vol. 64. http://dx.doi.org/10.1016/B978-0-444-62650-9.00016-6

and proteins in intra- and extracellular compartments. Each of these technologies provides a specific insight into the biological system that has been used to understand the relationship between cellular elements and their phenotypic manifestations at the organismal level.

A logical consequence of the implementation of omics research is the consideration that these complementing genome-wide measurements could be combined to provide an even more complete picture of life at the molecular level. Such an approach would hold great potential since the simultaneous measurements of different types of features on the same set of samples would facilitate a joint analysis of the data that would bring us closer to the understanding of cellular physiology, which is the ultimate goal of Systems Biology (SB). However, the transition from omics or integrated omics to SB is not straightforward.

SB arose as the discipline that investigates the behavior and relationships of all the elements of a particular living system while it is active (1). A fundamental idea in SB is that perturbations (whether genetic, chemical, or biological) reveal the functioning of the system and that by monitoring gene, protein, or metabolic changes, we obtain the information pieces needed to formulate mathematical models that describe the structure and dynamics of the biological entity (1). Basic concepts in SB are comprehensiveness (incorporation of "all" molecular elements of the system), interpretability (SB has a clear goal in understanding biology), and predictive power (mathematical modeling is considered a fundamental aspect of SB). However, very few studies in this field can be found that are able to achieve simultaneously these three goals. This is because the intrinsic difficulty of deriving accurate and descriptive mathematical models in systems composed of thousands if not millions of components (2). Hence, different flavors of SB have developed as scientists focused on addressing particular aspects of the problems at hand.

The high dimensionality of omics becomes a burden when building predictive models of biology, and hence most of the mathematical SB models restrict themselves to analyzing systems with a reduced number of variables (e.g., specific signaling or metabolic pathways) and where the model topology is imposed beforehand. This paradigm also called "reverse engineering" has been solved by different mathematical approaches ranging from ordinary differential equations (3–5), Bayesian inference (6), Petri Nets (4), Linear and Nonlinear Programming (7), and Boolean networks (8). Flux balance analysis (FBA), which is the genome-wide analysis of metabolic regulation (9), achieves a higher level of modeling complexity. FBA relies on simple stoichiometry rather than on difficult-to-measure enzyme kinetics to analyze the behavior of metabolomics networks and also rests on previous knowledge about the network topology (10). FBA fits the experimental data under a set of physicochemical and thermodynamic constraints, to generate a flux distribution that is optimized toward a particular "objective," normally cell growth (11). FBA, despite being a genome-wide approach, has been mostly applied to the modeling of simple bacterial metabolic systems (12,13) and only marginally to modeling in higher eukaryotes *(14)*. Finally, many methodologies have been proposed to infer gene regulatory networks (GRNs) at the genome-wide scale using gene expression measurements. GRNs aim to discover coexpression, causal or interactive relationships between genes and join all these interactions in global regulatory network. GRNs are typically descriptive, rather than predictive, require large datasets, and are often difficult to interpret due to the magnitude of the networks. Examples of methods for large-scale GRN reconstruction are the Probabilistic Module Networks *(15)*, Weighted Gene Correlation Network Analysis *(16)*, ARACNE *(17)*, and GeneNet *(18)*. Finally, gene enrichment and pathway analysis represent the most "pure" interpretative SB *(19,20)*. Functional enrichment methods are applied on genome-wide datasets and are based on the rationale that genes with similar behavior might be functionally related. Functional enrichment is useful to understand which cellular functions are activated or deactivated under a certain perturbation, and although in some instances, it might incorporate network information *(21,22)*, they are normally also not predictive.

Within these different scenarios of SB methodological resources, integrative omics analysis emerges as a very attractive approach to address the different goals of SB. Indeed, various studies have been published during the last two decades that incorporate measurements from different omics platforms on the same experiments such as gene expression and metabolomics *(23,24)*, proteomics and transcriptomics *(25,26)*, genome variation analysis and gene expression *(27–29)*, or methylation and expression arrays *(30)*. The increasing availability of public and commercial providers that offer omics platforms at decreasing costs and the recent boost of high-throughput sequencing-based functional genomics assays are generating a new wave of omics studies where multiple analytical platforms are combined. The two basic aims of these multi-omics studies are on one hand to increase discovery power in the identification of biomarkers and on the other hand to better understand the function of the biological system by using a wider diversity of descriptors.

Not surprisingly, the traditional analysis paradigms of SB equally apply to the simultaneous analysis of multiple omics datasets, and we can identify integration modalities that differently match the SB properties of comprehensiveness, prediction power, and interpretability. Moreover, additional challenges and opportunities arise when different data types are available. Comprehensiveness is the defining characteristic of integration databases that gather different omics types in one centralized repository to establish interconnections among them and to provide a unified interface for information retrieval. Integrative statistical methods that incorporate different omics data types in their modelling strategy aim to unravel relationships between diverse molecular entities to create comprehensive predictive models of the biological system under study. Finally, many if not most of the studies incorporating different omics data types address the integration challenge at the level visualization using tools that recognize multiple types of molecular entities and display them in the form of interaction networks or biological pathways. This approach to integration puts the focus on the interpretation of the systems structure.

In this chapter, we cover these three aspects of integrative omics in relation with SB, namely, the existing integrated omics databases resources, the mathematical methods used for the integrative analysis of different omics data types, and the tools available for the integrated visualization of omics data. We do not aim to be exhaustive in any of these three areas, but rather illustrate the characteristics, potentials, and limitations of the current efforts in this field. Integrative approaches will be illustrated with application examples covering diverse biological domains, including biomedical, agricultural, microbial, and fundamental research. Finally, we introduce an ongoing EU project on data integration that is likely to generate a new array of methodologies and resources for an omics-based SB.

2   INTEGRATED DATABASE RESOURCES

Much effort has been invested in the creation of organism-specific or field-specific integrative databases that collect different types of omics data to make them available in an integrated fashion to support SB research. Virtually, every large international genomics consortia have created and maintain species-specific databases that act as a reference repository of the genomic information and resources of their targeted organisms. Some of these resources such as the UCSC Human Genome Browser (31), Ensembl (32), contain mostly genomic, annotation, gene expression, and variant information, while others such as Wormbase (33), Saccharomyces Genome Database (34), and Tair (35) also include extensive collections of biological accessions. On the other hand, different data warehousing frameworks have been developed to facilitate the easy querying of omics databases for existing and novel organisms, such as BioMart (36), PathwayTools (37), BioXRT (38), and Inter-Mine (39). These frameworks can be installed locally, populated with user-specific data and used for the integration and retrieval of omics information. Their performances have been recently evaluated by Triplet and Butler (40), who used a set of 14 queries to assess the capability of each system to accept complex interrogation and provide accurate answers. The authors concluded that different options represent different trade-offs between ease of deployment and versatility for querying the data and that, in general, these resources do not support basic statistical analysis of the data contained (40).

Next to these major organisms and general purpose databases, the availability of public genomics repositories of different types has triggered efforts to create specialized resources that collect omics data from different sources on a specific biology domain and offer them in an integrated fashion. Although there are not yet many of such databases that succeed in including both multiple omics data types and extensive phenotypic information, the number and richness of such resources are deemed to proliferate as the availability and diversity of omics data continue expanding. An added valued of these resources is to enhance the unified access to domain-specific omics data with integrative analysis and data mining on the collected data to provide novel and valuable information at the systems level. In the next paragraphs, we discuss a number of examples of this type of databases and highlight their contribution to SB research.

The YeastNet database integrates functional data for 5818 yeast genes, including gene expression, protein–protein interactions, domain co-occurrence, gene neighbor, genetic interaction, phylogenetic profiles, and tertiary structure of proteins *(41)*. The data created 362,512 functional links among the yeast genome and were used to infer a measure of "functional association" between each pair of genes. The YeastNet database proved to be useful for novel SB applications, including the network-based reconstruction of the functional annotation of yeast genes *(42)*, the prediction of the phenotypic consequences of individual genome variations in yeast strains *(43)*, and the prediction of epistasis *(44)*. Recently, we used the YeastNet functional association measure to validate genome-wide models of pathway interactions during the yeast cell cycle *(45)* and showed that connected biological pathways show functional association that could not be inferred by one single source of functional data *(45)*.

The systems genetics resource (SGR) contains genotypes, clinical, and gene expression data from human and mouse studies, and it is focused in the study of complex disease traits such as obesity, diabetes, and atherosclerosis *(46)*. The data mining and querying options implemented in the SGR Web site allow for the study of genomic variants controlling metabolic traits and for the identification of novel candidate genes functionally and phenotypically related to those involved in the database target diseases. Another domain-specific SB database is the LiverAtlas *(47)* that integrates human hepatic data. LiverAtlas covers data on liver-related genomics, transcriptomics, proteomics, metabolomics, and hepatic diseases, as well as curated literature information. Similar to SGR, LiverAtlas has proved to be a useful resource for the identification of candidate genes of hepatic diseases and for improving differential diagnosis of hepatocellular carcinoma based in SB models *(47)*.

Finally, the plant functional genomics community has been particularly active in the development of resources to integrate multiple omics types (several species *(48)*, trichomes *(49)*, crops *(50)*, grape *(51)*, soybean *(52)*, and tomato *(53)*). Optimas-DW *(48)* and GabiPD *(50)* store comprehensive data on transcriptomics, proteomics, metabolomics, and other high-throughput measurements used to investigate plant physiology. Both databases include pathway representation tools that allow to visualize omics data in the context of plant-specific pathways. Particularly, GabiPD developed the concept "GreenCard" as a gene-centric record that displays all available omics data on a given plant gene. "GreenCards" enable users to query GabiPD by different gene IDs or keywords and extend the search across species, providing an interesting opportunity for plant comparative integrative omics research.

3   INTEGRATIVE OMICS ANALYSIS

By integrative analysis of omics data, we understand the algorithmic and statistical approaches that pursue the combination of different omics data in one analysis. We can identify two major general strategies *(24,54)*:

(a) One is to determine first separately the association of the features in each data type to the conditions under study and then combine in a second stage the results of the different analyses in one interpretative model. Most of the published omics integration reports follow the first route.
(b) The other is to combine all data initially in one mathematical model to infer relationships between different omics variables that jointly better explain the conditions analyzed.

Furthermore, when talking about omics data integration, two different cases can be considered, depending on the goal of the integrative effort. The first form of integrative analysis is the combination of different datasets of the same type of measurement to improve the power of the statistical analysis and leverage from the results previously reported by the community. For example, data from different microarray platforms or separate studies can be combined in one joint analysis for greater power. In this case, most or many variables (e.g., genes) will be shared across datasets although they might not be measured in the same way (e.g., different probe sets). This type of integrative analysis is typically called *meta-analysis*, and a key analytical task here is to set disparate datasets on the same baseline and scale such that combining the data becomes possible.

Another situation is when different data types (i.e., gene expression and metabolomics) are to be combined in one model. Apart from normalization issues, the analysis has to deal with different variable sets that might have different properties, that is, span different dynamic ranges, follow different data distributions, or bear different levels of noise. Moreover, the integrative analysis may need to consider the biological relationship between variables, which may not always be known, or would need to pursue to precisely infer those. This is of the typical case for *omics data integration in SB* where we seek to model more distinct elements of the system. This section discusses methodological approaches to both meta-analysis and integrative SB. We distinguish three major themes that differ in the integration goal and type of statistical approach commonly followed: inference of GRNs from multiple omics data, where typically some kind of pairwise correlation metric underlies the definition of the network; integration of transcriptomics, metabolomics, and proteomics data, frequently analyzed by dimension reduction multivariate methods; and integration of gene expression and genomics data using machine learning approaches.

3.1 Meta-Analysis

Meta-analysis is the combined analysis of multiple datasets—typically collected from public databases—of one type or biological scope and is one of the most extended forms of integrative omics. Meta-analysis benefits from the utilization of a large number of observations to enhance statistical and discovery power. Correlation meta-analysis across thousands of gene expression microarray datasets can be used to identify distinct coexpression modules associated with breast cancer subtypes *(55)* or to dissect the association between gene coregulation and function *(56)*. Other statistical strategies can be used to mine large datasets that reveal associations other than coexpression. For example, Bassel *et al.* *(57)* applied a rule-based machine learning approach to infer functional networks in Arabidopsis. The approach, termed "coprediction," is based on the collective ability of groups of genes co-occurring within rules to accurately predict the developmental outcome of a biological system, irrespective of their expression patterns *(57)*. Nguyen *(58)* introduced a novel metric called the BayesGen that analyzes the probability of coexpression of a pair of genes using prior knowledge on the characteristics of the dataset and taking into account nonlinearity and different noise levels across datasets. Moreover, as a large number of datasets of other types, beyond gene expression, become available, meta-analysis is being extended to additional omics technologies. For example, a self-organizing map (SOM) is an unsupervised machine learning technique that has been used in two recent publications to analyze a large number of ChIP-seq (and DNase-seq) datasets *(59,60)*. SOM creates maps that consist of thousands of units (or "neurons") arranged in a two-dimensional grid where the top and the bottom as well as the left and right edges of the map are connected (effectively giving rise to a toroid) to avoid boundary effects. Each unit of the map has an associated vector that is initialized randomly, and it is associated with a subset of the data through the training. Each neuron of the map contains a number of transcription factor sites that can be mined by functional enrichment properties. These maps typically reveal very specific colocalization patterns between transcription factor partners.

Another approach to meta-analysis was introduced in 2005 by Hwang and coworkers *(61)* in one of the pioneering works of integration for SB. The authors present an integration strategy for multiplatform genomics studies based on a weighting scheme for dataset-specific *p*-values that result from dataset-optimized statistical tests. The method is a general integration framework that only requires that molecular entities can be identified across datasets, and has the typical characteristics of meta-analysis. The authors applied this method to the integration of 18 different datasets related to galactose utilization in yeast, that comprised gene expression, protein abundance, protein–protein interaction, and DNA–protein interaction data, and they were able to recapitulate known and novel features of this particular system *(62)*. In this work, the subnetwork representing the targeted system—glucose utilization—was selected using previous interactome information and integrative analysis was restricted to the components of this network. This illustrates how statistical analysis and database information can work together to focus the data integration challenge and to improve interpretability of the results. Still, this strategy suffers from some of the limitations of univariate statistics when applied to high-dimensional data: strong statistical significance corrections must be applied to account for the multiple testing situation and the correlation structure between biomolecules—a fundamental property of the system—is largely ignored.

3.2   Multivariate Analysis

Multivariate statistics are better suited to handle large, noisy, and correlated datasets, and they have also been applied to address the omics data integration problem (Figure 1). Multivariate dimension reduction methods analyze the covariance structure of the dataset to identify latent variables that reveal the relationships between the original features. A very simply approach was proposed by Smilde and coworkers *(63)* who introduced the RV coefficient, a multivariate measurement of the correlation between two data matrices. This parameter indicates when two omics datasets contain similar or complementary information and can be used as a distance of the similarity of the datasets. However, it does not return a true integrative model of the data. Co-inertial-analysis *(64,65)*, canonical correlation analysis *(66)*, and sparse partial least squares (PLS) *(67)* are other established dimension reduction techniques that have been successfully applied to integrate gene expression platforms or combine gene expression data with proteomics or genomics measurements. Similar to Hwang's approach for reducing the analysis space using previous interaction data, we have combined protein functional information stored in public databases with multivariate projection techniques to improve statistical power and interpretative capacity of the omics integration models *(24)*. We identified major functional expression signatures involved in liver detoxification processes in rats using principal component analysis (PCA) applied to the gene expression submatrix formed by all genes belonging to the same pathway. These signatures were then used as independent variables to define a PLS model on metabolic measurements obtained in the same biological samples. In this approach, biological information helps to extract relevant features of the data, reduce noise, and at the same time produce a direct pathway-oriented interpretation of the multivariate transcriptomics/metabolomics integration model.

A very interesting approach was proposed by Bylesjö and coworkers who used O2PLS for the integration of gene expression and metabolomics data in

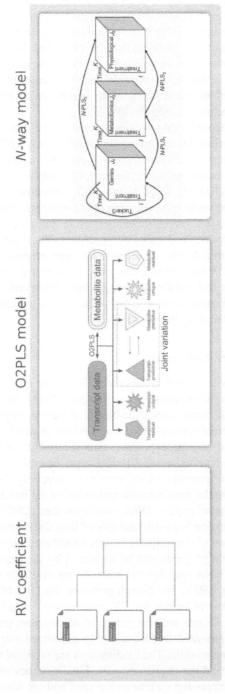

FIGURE 1 Multivariate approaches for omics data integration. (A) The RV coefficient is a correlation measure between datasets that can be used as distance metric. (B) The O2PLS method dissects gene expression and metabolomics datasets for shared and data type-specific variation. (C) The N-way approach accommodates experimental factors in a multidimensional block. Tucker3 is used to study intradataset covariation and NPLS analyzes between-block covariation. *Panel (B): Reproduced from Conesa et al. (24). Panel (C): Reproduced from Bylesjö et al. (23). (For color version of this figure, the reader is referred to the online version of this chapter.)*

plant integrative genomics *(23)*. The O2PLS methodology dissects the covariation structure of two omics datasets in three parts, namely, shared covariation and dataset-specific covariation. With this methodology, systematic variation that overlaps across analytical platforms can be separated from platform-specific systematic variation. By analyzing the loads of the gene expression and metabolite features in the latent variables of the shared and platform-specific models, the method can discern which biomolecules interact with one another and which behave independently *(23)*.

Finally, we have proposed using *N*-way approaches to data integration when multifactorial experimental designs are involved. *N*-way or Block analysis accommodates a given omics dataset in an *N*-dimensional structure with *N* being the number of factors of the experiment. We applied Tucker3 and NPLS, an *N*-way generalization of PCA and PLS, respectively, to model and integrate the same rat toxicogenomics example mentioned above. In this study, different treatments and bromobenzene exposure times were applied to the rats before transcriptomics and metabolic profiling. Thus, the study follows a double three-factorial design (time, dose, and gene expression/metabolite concentration). The *N*-way model not only showed the relationships between gene expression and metabolite changes but also revealed a curved relationship between exposure time and high-dose levels indicative of the accumulative effects of the drug treatment *(24)*.

3.3 Gene Regulatory Networks

One of the prototypic points of overlap between SB and the omics lies in the inference GRNs from high-throughput measurements. Many methods have been proposed that rank from correlation-based, Bayesian, regression, or mutual information, or combinations thereof and which primarily base their analysis in the utilization of gene expression data solely (for an interesting review, see Ref. *(68)*). However, using different types of omics data as part of GRN inference has the potential of defining more precisely the genomic elements that contribute to the gene network and reducing the number of false positives. The combination of protein–DNA interaction data (ChIP-on-chip or Chip-seq) and gene expression is a clear example of this. ChIP data reveal the sites of physical interaction of transcription factors with their target genes, and the expression of these target genes can be measured by the transcriptome-profiling assays such as RNA-seq. Several methods are available to join the two data sources to infer GRNs. Bar-Joseph proposed the GRAM algorithm that starts with an exhaustive search for TF combinations based on DNA-binding data *(69)*. These TF sets indicate groups of downstream genes whose expression patterns are analyzed to find a highly correlated subset, which serves as seed for the gene module. The binding data are revisited with a less stringent criterion to expand the regulatory core. Finally, genes are incorporated that have similar expression pattern as the seed and are also bound by the set of regulators. The GRAM method based the definition of gene modules on correlation among a sufficient number of samples. However, this measure has limitations when the coexpression module is only active in a subset on conditions. This problem is specifically addressed by biclustering, a data mining methodology that is able to find subsets of genes and samples that are highly correlated. Reiss *et al. (70)* proposed an adaptation of biclustering to GRN inference that incorporates *cis*-regulatory motif information. The promoter regions of the biclustered genes are screened for the occurrence of conserved patterns which are then used to search for neighboring gene candidates to be incorporated into the block *(70)*. Finally, Marbach *(71)* adds an additional layer of information integration such as functional data on chromatin status. In this approach, TF regulatory networks are generated from four different data sources: conserved motifs, ChIP data, gene expression, and histone modifications. In each network, each TF-target gene association is given with a weight obtained derived from the analysis of the dynamic or binding data. Weights are then combined either by addition or by logistic regression to generate a consensus network that captures the different types of regulatory evidence.

3.4   Integrating Genome Variation Data

The analysis of genome variation to identify genetic markers of traits and diseases is one of the most popular but also underpowered areas of genomics, as millions of variants populate individual genomes and finding causative variants requires large or very specific experimental designs. Integration of additional omics types or the application of SB methods will improve power in these analyses.

The typical example of integration of variant and gene expression data is eQTL (expression quantitative trait loci) analysis *(27)*. eQTLs are genomic loci that affect the expression of genes and are identified by the simultaneous analysis of gene expression and genome variation (i.e., single nucleotide polymorphisms, SNPs) in a population sample, followed by testing the linkage of changes in expression and genetic polymorphisms. The significance of the association of the variant to the trait can be improved by a multiple step procedure where pairwise association of the SNP, expression changes, phenotype tested, and significant calls are superimposed to reduce the number of false positives and provide a gene expression-related interpretation to the polymorphism–phenotype relationship *(72)*.

Another efficient way to empower SNP analysis is to use pathway information as scaffold to simultaneously test multiple polymorphisms. The underlying idea is that nucleotide variants may impact different components (genes) of the same pathway all affecting pathway functionality, which is what ultimately relates to the trait of interest. Individual variations may not have sufficient penetrance to reach significance that could be achieved by globally testing variation in all pathway components. This concept, developed initially for gene expression analysis, has been also applied to genome variant analysis *(73)* either by first selecting candidate polymorphisms and then testing for pathway significance *(73)* or by ranking variants for its association to the trait of interest and applying a gene-set approach *(74)*.

To conclude, we discuss a couple of approaches to the integration of genome variation and gene expression data and that fall within the realm of what is known as machine learning methods.

Random Forest (RF) is a supervised machine learning method that creates a set of classification trees obtained by the random selection of a group of variables from the variable space and a bootstrap procedure that recurrently selects a fraction of the sample space to fit the model. At each step of the tree building, the most discriminative variables are chosen until the tips of the tree contain pure classes. One advantage of RFs is that they naturally accommodate different data types as well as both quantitative and qualitative variables and can provide simplified explicative models of the relationship between molecular features and their relation with the phenotype. Moreover, interactions between variables can be inferred if a variable systematically makes the split of another variable significantly more or less likely than expected *(75)*. RFs have been used to integrate SNP and cytokine proteomics data to predict adverse reaction to smallpox vaccination *(25)*. This approach obtained a reduced molecular signature, composed of three proteins and one SNP that would predict outcome in 75% of the cases. Also, RFs have been used to create predictive models composed of gene expression and metabolite data that explain phenotypic traits in tomato *(76)*. RF selected variables included unknown metabolites and enabled the construction of meaningful networks that related known and novel components with established metabolite pathways.

Graphical models (e.g., Bayesian networks) have been used to describe complex interaction patterns among nucleotide variants, genes, and other phenotypes. The basic principle is to construct a small subnetwork that connects a target gene with a group of interacting genes and a set of *cis*-acting variants for the interacting genes that lie within a defined chromosomal radius. The eQTL association testing is conditional to the developed local subnetwork. This methodology allows the creation of a midsize network around a reduced number of target genes. Li *et al.* used this approach to identify a network of 65 target genes and 1330 modulator genes located within 72 QTL intervals involved in the modulation of gene expression in the mouse brain *(28)*. Similarly, Chu *et al.* were able to build complex gene–SNP networks around two candidate asthma genes (Interferon Receptor 12B2 and Interleukin 1B) using a similar strategy *(29)*.

4 TOOLS FOR INTEGRATIVE VISUALIZATION OF OMICS DATA

Integrated visualization of different omics data types is probably the most powerful tool for the interpretation of SB results. While mathematical models can reveal significant associations between system components or can predict the behavior of the system, the graphical display of omics data can lead to better insights of global functional properties. Many software tools are today available that accept diverse omics measurements and generate join visual representations of the data. Here, we discuss several of them for their ability to combine visualization with analysis and for the type of data they work with Figure 2.

A first major distinction is whether the tool is devoted to display genome or network information. Genomic information is often visualized using Web-based genome browsers (GBs), which typically display genome features continuously along chromosome coordinates. Many different layers of information can be displayed on the same locus using tracks that can often be customized. For example, one may include gene, transcript, and nucleotide variant annotations, but also dynamic features such as ChIP-seq or RNA-seq read coverage and calls from a particular biological sample. The UCSC Genome Browser *(31)*, which was originally developed to visualize the human genome, now holds hundreds of human and mouse ENCODE datasets as tracks that can be displayed along with additional user-supplied datasets. Gbrowse *(79)* is a portable GB that forms the basis of several other public GB sites such as worm-base, flybase, TAIR, and gramene and is a popular choice for custom installations and for new genomes. However, most of these GBs require being connected online to access them. Integrative genomics viewer (IGV) *(80)* is a typical representative another major set of GBs that work and store data on the local user computer and allow offline viewing and analysis. IGV can interact with online tracks as well as custom user tracks. Another interesting genome viewer is Genome Maps *(81)*, which implements a hybrid approach using highly efficient HTML5 data transfer technologies that allow fast upload of large volumes of high-throughput sequencing data and is particularly well suited for the analysis of large data collections, such as cancer or population studies with the data cached locally and visualized in real time on the client side.

A very different visualization approach is necessary to represent interactions between molecular features such as genes, proteins, and metabolites. These data are best analyzed in the form of graphs where nodes correspond to features and edges represent the interaction between them. Tools such as Cytoscape *(82)* provide a general framework for the representation of biological networks where virtually any feature can be represented, linked to existing databases, and visually enhanced with different graphical elements. Additionally, the Cytoscape framework includes many external plug-ins that allow further analysis on topological properties of the network, such as centrality or modularity, or functional characteristics, such as the enrichment of network genes in Gene Ontology categories. Other tools map the data provided by the user onto an interaction template, such as metabolic or signaling pathways that guide visualization and facilitate interpretation. KaPPA-View *(83)* and MapMan *(84)* are plant-specific tools that display metabolite

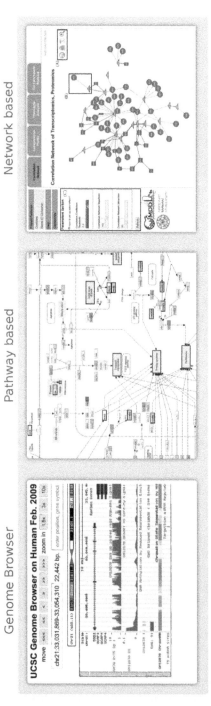

FIGURE 2 Overview of the three types of tools for integrated visualization of omics data. (A) Genome browsers display user tracks to display measurements and annotation along genome coordinates (UCSC GB view (31)). (B) Pathway-based methods used predefined pathways to map multi-omics data and show them jointly (Paintomics picture (77)). (C) Network-based methods build data-specific networks integrating relationships between different types of features (3Omics graph (78)). (For color version of this figure, the reader is referred to the online version of this chapter.)

and transcript levels on predefined pathway blocks. The MassTRIX software *(85)* translates NMR spectra into metabolic compounds and maps them onto KEGG pathways together with genome and transcriptome information. Pro-MeTra *(86)* accepts precomputed and custom-made pathway maps in scalable vector graphics (SVG) format and is able to display dynamics data although application is restricted microbial genomes and offers direct access to different omics experimental databases. Paintomics *(77)* represents multiple gene expression and metabolomics datasets onto KEGG pathways and performs joined pathway analysis considering both types of data. As these solutions use a known scaffold to analyze omics data, they are limited in their network inference and data mining functionalities, which is typically restricted to some kind of functional enrichment analysis. Finally, another set of tools incorporate extensive analysis options and focus on the discovery of integrated networks. For example, 3Omics *(78)* accepts proteomics, metabolics, and transcriptomics data on a sufficient number of samples to compute pairwise correlations and is able to create correlation networks that contain elements of the three molecular types. VANTED v2 *(87)* is a comprehensive omics integration framework that supports different types of topological, functional, and statistical analysis on the supported data, and has also the possibility of running simulation tasks on a predefined network to study behavioral aspects of the network. SteinerNet *(88)* accepts proteomics and transcriptomics data and maps these to a database of protein–protein interaction networks and transcription factor–target gene interactions, to create a network that maximally connects the submitted data incorporating additional database elements when needed.

5   CONCLUSIONS AND FUTURE PROSPECTS

The significant development of integration methodologies over the last two decades has brought high-throughput omics technologies into the realm of SB. Many methods and resources are now available that join two or more omics data types and combine genomics, transcriptomics, proteomics, metabolomics, and epigenomics into integrated models that improve the prediction or discovery power over single data type analyses. However, there is still a noticeable gap in the integration of genome-based and pathway-related data. Statistical methods integrate genomics features with gene expression either to infer GRNs and investigate genotype-phenotype relations or to join metabolism, proteome, and gene expression data to understand the dynamics of signaling and metabolic pathways. Visualization tools have been developed separately for these two domains. This implies that the connection between chromatin and epigenetic states on one side and cellular physiology in the other has been poorly explored. The current challenge of SB is to extend the integration challenge to many layers of cellular organization measurable using available omics and NGS technologies. The FP7 European STATegra (www.stategra.eu) project has this aim. This project will capture up to seven different omics measurements, expanding the molecular types discussed in this chapter, on the same set of samples in a replicated time course experimental design and will develop statistical and visualization tools that convey them into one integrated model. The ultimate goal is to create the data and analytical tool for the next generation omics SB.

ACKNOWLEDGMENTS

This work has been funded by the FP7 Health project STATegra (Grant number 306000) and the BIO2012-40244 grant of the Spanish Ministry of Economy and Competitiveness.

REFERENCES

1. Ideker, T.; Galitski, T.; Hood, L. *Annu. Rev. Genomics Hum. Genet.* **2001**, *2*, 343–372.
2. Erguler, K.; Stumpf, M. P. *Mol. Biosyst.* **2011**, *7*, 1593–1602.
3. Tegner, J.; Yeung, M. K.; Hasty, J.; Collins, J. J. *Proc. Natl. Acad. Sci. U. S. A.* **2003**, *100*, 5944–5949.
4. Peng, S. C.; Wong, D. S.; Tung, K. C.; Chen, Y. Y.; Chao, C. C.; Peng, C. H.; Chuang, Y. J.; Tang, C. Y. *BMC Bioinformatics* **2010**, *11*, 308.
5. Gomez-Cabrero, D.; Compte, A.; Tegner, J. *Interface Focus* **2011**, *1*, 438–449.
6. Lecca, P. *Drug Discov. Today* **2013**, *18*(11–12), 560–566.
7. Zhan, C.; Yeung, L. F. *BMC Syst. Biol.* **2011**, *5*, 14.
8. Wittmann, D. M.; Krumsiek, J.; Saez-Rodriguez, J.; Lauffenburger, D. A.; Klamt, S.; Theis, F. J. *BMC Syst. Biol.* **2009**, *3*, 98.
9. Raman, K.; Chandra, N. *Brief. Bioinform.* **2009**, *10*, 435–449.
10. Conesa, A.; Mortazavi, A. *BMC Syst. Biol.* **2014**, *8*(Suppl 2), S1.
11. Gianchandani, E. P.; Chavali, A. K.; Papin, J. A. *Wiley Interdiscip. Rev. Syst. Biol. Med.* **2010**, *2*, 372–382.
12. Edwards, J. S.; Covert, M.; Palsson, B. *Environ. Microbiol.* **2002**, *4*, 133–140.
13. Toya, Y.; Shimizu, H. *Biotechnol. Adv.* **2013**, *31*, 818–826.
14. Jamshidi, N.; Miller, F. J.; Mandel, J.; Evans, T.; Kuo, M. D. *BMC Syst. Biol.* **2011**, *20*, 200.
15. Segal, E.; Shapira, M.; Regev, A.; Pe'er, D.; Botstein, D.; Koller, D.; Friedman, N. *Nat. Genet.* **2003**, *34*, 166–176.
16. Langfelder, P.; Horvath, S. *BMC Bioinformatics* **2008**, *9*, 559.
17. Margolin, A. A.; Nemenman, I.; Basso, K.; Wiggins, C.; Stolovitzky, G. *BMC Bioinformatics* **2006**, *7* (Suppl. 1), S7.
18. Schafer, J.; Strimmer, K. *Bioinformatics* **2005**, *21*, 754–764.
19. Al-Shahrour, F.; Díaz-Uriarte, R.; Dopazo, J. *Bioinformatics* **2004**, *20*, 578–580.
20. Subramanian, A.; Tamayo, P.; Mootha, V. K.; Mukherjee, S.; Ebert, B. L.; Gillette, M. A.; Paulovich, A.; Pomeroy, S. L.; Golub, T. R.; Lander, E. S.; Mesirov, J. P. *Proc. Natl. Acad. Sci. U. S. A.* **2005**, *102*, 15545–15550.
21. Tarca, A. L.; Draghici, S.; Khatri, P.; Hassan, S. S.; Mittal, P.; Kim, J. S.; Kim, C. J.; Kusanovic, J. P.; Romero, R. *Bioinformatics* **2009**, *25*, 75–82.
22. Glaab, E.; Baudot, A.; Krasnogor, N.; Schneider, R.; Valencia, A. *Bioinformatics* **2012**, *28*, i451–i457.

23. Bylesjö, M.; Eriksson, D.; Kusano, M.; Moritz, T.; Trygg, J. *Plant J.* **2007**, *52*, 1181–1191.

24. Conesa, A.; Prats-Montalbán, J. M.; Nueda, M. J.; Tarazona, S.; Ferrer, A. *Chemometr. Intell. Syst. Lab.* **2008**, *104*, 101–111.

25. Reif, D. M.; Motsinger-Reif, A. A.; McKinney, B. A.; Rock, M. T.; Crowe, J. E., Jr.; Moore, H. H. *Genes Immun.* **2009**, *10*, 112–119.

26. Huang, S. S.; Fraenkel, E. *Sci. Signal.* **2009**, *2*, 40.

27. Jansen, R. C.; Nap, J. P. *Trends Genet.* **2001**, *17*, 388–391.

28. Li, H.; Lu, L.; Manly, K. F.; Chesler, E. J.; Bao, L.; Wang, J.; Zhou, M.; Williams, R. W.; Cui, Y. *Hum. Mol. Genet.* **2005**, *14*, 1119–1125.

29. Chu, J. H.; Weiss, S. T.; Carey, V. J.; Raby, B. A. *BMC Syst. Biol.* **2009**, *3*, 55.

30. Alashwal, H.; Dosunmu, R.; Zawia, N. H. *Neurotoxicology* **2012**, *33*, 1450–1453.

31. Kent, W. J.; Sugnet, C. W.; Furey, T. S.; Roskin, K. M.; Pringle, T. H.; Zahler, A. M.; Haussler, D. *Genome Res.* **2002**, *12*, 996–1006.

32. Flicek, P.; Amode, M. R.; Barrell, D.; Beal, K.; Billis, K.; Brent, S.; Carvalho-Silva, D.; Clapham, P.; Coates, G.; Fitzgerald, S.; Gil, L.; Girón, C. G.; Gordon, L.; Hourlier, T.; Hunt, S.; Johnson, N.; Juettemann, T.; Kähäri, A. K.; Keenan, S.; Kulesha, E.; Martin, F. J.; Maurel, T.; McLaren, W. M.; Murphy, D. N.; Nag, R.; Overduin, B.; Pignatelli, M.; Pritchard, B.; Pritchard, E.; Riat, H. S.; Ruffier, M.; Sheppard, D.; Taylor, K.; Thormann, A.; Trevanion, S. J.; Vullo, A.; Wilder, S. P.; Wilson, M.; Zadissa, A.; Aken, B. L.; Birney, E.; Cunningham, F.; Harrow, J.; Herrero, J.; Hubbard, T. J.; Kinsella, R.; Muffato, M.; Parker, A.; Spudich, G.; Yates, A.; Zerbino, D. R.; Searle, S. M. *Nucleic Acids Res.* **2013**, *42*(D1), D749–D755.

33. Harris, T. W.; Baran, J.; Bieri, T.; Cabunoc, A.; Chan, J.; Chen, W. J.; Davis, P.; Done, J.; Grove, C.; Howe, K.; Kishore, R.; Lee, R.; Li, Y.; Muller, H. M.; Nakamura, C.; Ozersky, P.; Paulini, M.; Raciti, D.; Schindelman, G.; Tuli, M. A.; Auken, K. V.; Wang, D.; Wang, X.; Williams, G.; Wong, J. D.; Yook, K.; Schedl, T.; Hodgkin, J.; Berriman, M.; Kersey, P.; Spieth, J.; Stein, L.; Sternberg, P. W. *Nucleic Acids Res.* **2013**, *42*(D1), D789–D793.

34. Costanzo, M. C.; Engel, S. R.; Wong, E. D.; Lloyd, P.; Karra, K.; Chan, E. T.; Weng, S.; Paskov, K. M.; Roe, G. R.; Binkley, G.; Hitz, B. C.; Cherry, J. M. *Nucleic Acids Res.* **2013**, *42*(D1), D717–D725.

35. Lamesch, P.; Berardini, T. Z.; Li, D.; Swarbreck, D.; Wilks, C.; Sasidharan, R.; Muller, R.; Dreher, K.; Alexander, D. L.; Garcia-Hernandez, M.; Karthikeyan, A. S.; Lee, C. H.; Nelson, W. D.; Ploetz, L.; Singh, S.; Wensel, A.; Huala, E. *Nucleic Acids Res.* **2012**, *40*, D1202–D1210.

36. Zhang, J.; Haider, S.; Baran, J.; Cros, A.; Guberman, J. M.; Hsu, J.; Liang, Y.; Yao, L.; Kasprzyk, A. *Database* **2011**, *2011*, bar 038. http://dx.doi.org/10.1093/database/bar038.

37. Karp, P. D.; Paley, S. M.; Krummenacker, M.; Latendresse, M.; Dale, J. M.; Lee, T. J.; Kaipa, P.; Gilham, F.; Spaulding, A.; Popescu, L.; Altman, T.; Paulsen, I.; Keseler, I. M.; Caspi, R. *Brief. Bioinform.* **2010**, *11*, 40–79.

38. Zhang, J.; Duggan, G. E.; Khaja, R. In: *3rd Canadian Working Conference on Computational Biology* 2004.

39. Smith, R. N.; Aleksic, J.; Butano, D.; Carr, A.; Contrino, S.; Hu, F.; Lyne, M.; Lyne, R.; Kalderimis, A.; Rutherford, K.; Stepan, R.; Sullivan, J.; Wakeling, M.; Watkins, X.; Micklem, G. *Bioinformatics* **2012**, *28*, 3163–3165.

40. Triplet, T.; Butler, G. *Brief. Bioinform.* **2013**, http://dx.doi.org/10.1093/bib/bbt031.

41. Kim, H.; Shin, J.; Kim, E.; Kim, H.; Hwang, S.; Shim, J. E.; Lee, I. *Nucleic Acids Res.* **2013**, *42*(D1), D731–D736. http://dx.doi.org/10.1093/nar/gkt981.

42. Dutkowski, J.; Kramer, M.; Surma, M. A.; Balakrishnan, R.; Cherry, J. M.; Krogan, N. J.; Ideker, T. *Nat. Biotechnol.* **2013**, *31*, 8–45.

43. Jelier, R.; Semple, J. I.; Garcia-Verdugo, R.; Lehner, B. *Nat. Genet.* **2011**, *43*, 1270–1274.

44. Lee, I.; Lehner, B.; Vavouri, T.; Shin, J.; Fraser, A. G.; Marcotte, E. M. *Genome Res.* **2010**, *20*, 1143–1153.

45. Ponzoni, I.; Tarazona, S.; Götz, S.; Montaner, D.; Dussaut, J. S.; Dopazo, J.; Conesa, A. *BMC Syst. Biol.* **2014**, *8*(Suppl 2), S7.

46. van Nas, A.; Pan, C.; Ingram-Drake, L. A.; Ghazalpour, A.; Drake, T. A.; Sobel, E. M.; Papp, J. C.; Lusis, A. J. *Front. Genet.* **2013**, *20*, 84.

47. Zhang, Y.; Yang, C.; Wang, S.; Chen, T.; Li, M.; Wang, X.; Li, D.; Wang, K.; Ma, J.; Wu, S.; Zhang, X.; Zhu, Y.; Wu, J.; He, F. *Liver Int.* **2013**, *33*, 1239–1248.

48. Colmsee, C.; Mascher, M.; Czauderna, M.; Hartmann, A.; Schlüter, U.; Zellerhoff, N.; Schmitz, J.; Bräutigam, A.; Pick, T. R.; Alter, P.; Gahrtz, M.; Witt, S.; Fernie, A. R.; Börnke, F.; Fahnenstich, H.; Bucher, M.; Dresselhaus, T.; Weber, A. P.; Schreiber, F.; Scholz, U.; Sonnewald, U. *BMC Plant Biol.* **2012**, *12*, 245.

49. Dai, X.; Wang, G.; Yang, D. S.; Tang, Y.; Broun, P.; Marks, M. D.; Sumner, L. W.; Dixon, R. A.; Zhao, P. X. *Plant Physiol.* **2010**, *152*, 44–54.

50. Riaño-Pachón, D. M.; Nagel, A.; Neigenfind, J.; Wagner, R.; Basekow, R.; Weber, E.; Mueller-Roeber, B.; Diehl, S.; Kersten, B. *Nucleic Acids Res.* **2009**, *37*, D954–D959.

51. Grimplet, J.; Cramer, G. R.; Dickerson, J. A.; Mathiason, K.; Van Hemert, J.; Fennell, A. Y. *PLoS One* **2009**, *21*, e8365.

52. Sakata, K.; Ohyanagi, H.; Nobori, H.; Nakamura, T.; Hashiguchi, A.; Nanjo, Y.; Mikami, Y.; Yunokawa, H.; Komatsu, S. *J. Proteome Res.* **2009**, *8*, 3539–3548.

53. Chiusano, M. L.; D'Agostino, N.; Traini, A.; Licciardello, C.; Raimondo, E.; Aversano, M.; Frusciante, L.; Monti, L. *BMC Bioinformatics* **2008**, *26* (Suppl. 2), S7.

54. Holzinger, E. R.; Ritchie, M. D. *Pharmacogenomics* **2012**, *13*, 213–222.

55. Wirapati, P.; Sotiriou, C.; Kunkel, S.; Farmer, P.; Pradervand, S.; Haibe-Kains, B.; Desmedt, C.; Ignatiadis, M.; Sengstag, T.; Schütz, F.; Goldstein, D. R.; Piccart, M.; Delorenzi, M. *Breast Cancer Res.* **2008**, *10*, R65.

56. Montaner, D.; Minguez, P.; Al-Shahrour, F.; Dopazo, J. *BMC Genomics* **2009**, *27*, 197.

57. Bassel, G. W.; Glaab, E.; Marquez, J.; Holdsworth, M. J.; Bacardit, J. *Plant Cell* **2011**, *23*, 3101–3116.

58. Nguyen, V. A.; Lió P, P. *BMC Genomics* **2009**, *10* (Suppl. 3), S14.

59. Mortazavi, A.; Pepke, S.; Jansen, C.; Marinov, G. K.; Ernt, J.; Kellis, M.; Hardison, R. C.; Myers, R. M.; Wold, B. *J. Genome Res.* **2013**, *23*, 2136–2148.

60. Xie, D.; Boyle, A. P.; Wu, L.; Zhai, J.; Kawli, T.; Snyder, M. *Cell* **2013**, *155*, 713–724.

61. Hwang, D.; Rust, A. G.; Ramsey, S.; Smith, J. J.; Leslie, D. M.; Weston, A. D.; de Atauri, P.; Aitchison, J. D.; Hood, L.; Siegel, A. F.; Bolouri, H. A. *Proc. Natl. Acad. Sci. U. S. A.* **2005**, *102*, 17296–17301.

62. Hwang, D.; Smith, J. J.; Leslie, D. M.; Weston, A. D.; Rust, A. G.; Ramsey, S.; de Atauri, P.; Siegel, A. F.; Bolouri, H.; Aitchison, J. D.; Hood, L. *Proc. Natl. Acad. Sci. U. S. A.* **2005**, *102*, 17302–17307.

63. Smilde, A. K.; Kiers, H. A.; Bijlsma, S.; Rubingh, C. M.; van Erk, M. J. *Bioinformatics* **2009**, *25*, 401–405. http://dx.doi.org/10.1093/bioinformatics/btn634.

64. Culhane, A. C.; Perriere, G.; Higgins, D. G. *BMC Bioinformatics* **2003**, *4*, 59.

65. Fagan, A.; Culhane, A. C.; Higgins, D. G. *Proteomics* **2007**, *7*, 2162–2171.

66. Waaijenborg, S.; Verselewel de Witt Hamer, P. C.; Zwinderman, A. H. *Stat. Appl. Genet. Mol. Biol.* **2008**, *7*, 1.

67. Lê Cao, K.; Martin, P. G.; Robert-Granié, C.; Besse, P. *BMC Bioinformatics* **2009**, *10*, 34.
68. Marbach, D.; Costello, J. C.; Küffner, R.; Vega, N. M.; Prill, R. J.; Camacho, D. M.; Allison, K. R. *Nat. Methods* **2012**, *9*, 796–804.
69. Bar-Joseph, Z.; Gerber, G. K.; Lee, T. I.; Rinaldi, N. J.; Yoo, J. Y.; Robert, F.; Gordon, D. B.; Fraenkel, E.; Jaakkola, T. S.; Young, R. A.; Gifford, D. K. *Nat. Biotechnol.* **2003**, *21*, 1337–1342.
70. Reiss, D. J.; Baliga, N. S.; Bonneau, R. *BMC Bioinformatics* **2006**, *7*, 280.
71. Marbach, D.; Roy, S.; Ay, F.; Meyer, P. E.; Candeias, R.; Kahveci, T.; Bristow, C. A.; Kellis, M. *Genome Res.* **2012**, *22*, 1334–1349.
72. Choy, E.; Yelensky, R.; Bonakdar, S.; Plenge, R. M.; Saxena, R.; De Jager, P. L.; Shaw, S. Y.; Wolfish, C. S.; Slavik, J. M.; Cotsapas, C.; Rivas, M.; Dermitzakis, E. T.; Cahir-McFarland, E.; Kieff, E.; Hafler, D.; Daly, M. J.; Altshuler, D. *PLoS Genet.* **2008**, *4*, e1000287.
73. Wang, K.; Li, M.; Hakonarson, H. *Nat. Rev. Genet.* **2010**, *11*, 843–854.
74. Medina, I.; Montaner, D.; Bonifaci, N.; Pujana, M. A.; Carbonell, J.; Tarraga, J.; Al-Shahrour, F.; Dopazo, J. *Nucleic Acids Res.* **2009**, *37*, W340–W344.
75. Touw, W. G.; Bayjanov, J. R.; Overmars, L.; Backus, L.; Boekhorst, J.; Wels, M.; van Hijum, S. A. *Brief. Bioinform.* **2013**, *14*, 315–326.
76. Acharjee, A.; Kloosterman, B.; de Vos, R. C.; Werij, J. S.; Bachem, C. W.; Visser, R. G.; Maliepaard, C. *Anal. Chim. Acta* **2011**, *705*, 56–63.
77. García-Alcalde, F.; García-López, F.; Dopazo, J.; Conesa, A. *Bioinformatics* **2011**, *27*, 137–139.
78. Kuo, T. C.; Tian, T. F.; Tseng, Y. J. *BMC Syst. Biol.* **2013**, *7*, 64.
79. Donlin, M. J. *Curr. Protoc. Bioinformatics* **2009**, http://dx.doi.org/10.1002/0471250953.bi0909s28.
80. Thorvaldsdóttir, H.; Robinson, J. T.; Mesirov, J. P. *Brief. Bioinform.* **2013**, *14*, 178–192.
81. Medina, I.; Salavert, F.; Sanchez, R.; de Maria, A.; Alonso, R.; Escobar, P.; Bleda, M.; Dopazo, J. *Nucleic Acids Res.* **2013**, *41*, W41–W46.
82. Saito, R.; Smoot, M. E.; Ono, K.; Ruscheinski, J.; Wang, P. L.; Lotia, S.; Pico, A. R.; Bader, G. D.; Ideker, T. *Nat. Methods* **2012**, *9*, 1069–1076.
83. Tokimatsu, T.; Sakurai, N.; Suzuki, H.; Ohta, H.; Nishitani, K.; Koyama, T.; Umezawa, T.; Misawa, N.; Saito, K.; Shibata, D. *Plant Physiol.* **2005**, *138*, 1289–1300.
84. Thimm, O.; Bläsing, O.; Gibon, Y.; Nagel, A.; Meyer, S.; Krüger, P.; Selbig, J.; Müller, L. A.; Rhee, S. Y.; Stitt, M. *Plant J.* **2004**, *37*, 914–939.
85. Suhre, K.; Schmitt-Kopplin, P. *Nucleic Acids Res.* **2008**, *36*, W481–W484.
86. Neuweger, H.; Persicke, M.; Albaum, S. P.; Bekel, T.; Dondrup, M.; Hüser, A. T.; Winnebald, J.; Schneider, J.; Kalinowski, J.; Goesmann, A. *BMC Syst. Biol.* **2009**, *23*, 82.
87. Rohn, H.; Junker, A.; Hartmann, A.; Grafahrend-Belau, E.; Treutler, H.; Klapperstück, M.; Czauderna, T.; Klukas, C.; Schreiber, F. *BMC Syst. Biol.* **2012**, *6*, 139.
88. Tuncbag, N.; McCallum, S.; Huang, S. S.; Fraenkel, E. *Nucleic Acids Res.* **2012**, *40*, W505–W509.

Index

Note: Page numbers followed by "*f*" indicate figures, and "*t*" indicate tables.